THE CORRESPONDENCE OF ISAAC NEWTON

VOLUME V

1709–1713

Machinery of the Mint, as illustrated in the *Encyclopédie* (Planches, VIII, Paris 1771, art. 'Monnoyage', plates XV and XVI). The upper figure shows the great screw press (a fly-press in modern terms) which impressed the round blank; the lower figure shows the machine which impressed letters or indentations ('milling') on the edge of the impressed coin to discourage clippers.

THE CORRESPONDENCE OF
ISAAC NEWTON

VOLUME V
1709–1713

EDITED BY

A. RUPERT HALL

AND

LAURA TILLING

CAMBRIDGE
PUBLISHED FOR THE ROYAL SOCIETY
AT THE UNIVERSITY PRESS
1975

CAMBRIDGE UNIVERSITY PRESS
Cambridge, New York, Melbourne, Madrid, Cape Town, Singapore, São Paulo, Delhi

Cambridge University Press
The Edinburgh Building, Cambridge CB2 8RU, UK

Published in the United States of America by Cambridge University Press, New York

www.cambridge.org
Information on this title: www.cambridge.org/9780521087216

First published 1975
This digitally printed version 2008

A catalogue record for this publication is available from the British Library

Library of Congress Catalogue Card Number: 59–65134

ISBN 978-0-521-08721-6 hardback
ISBN 978-0-521-08593-9 paperback

CONTENTS

Preface . . . . *page* xiv

A note on the manuscripts used in this volume xvii

Short titles and abbreviations for published works and manuscripts . . . xviii

Introduction xxi

THE CORRESPONDENCE

763 NEWTON *to* [LAUDERDALE], [?July/August 1709] 1
764 THE MINT *to* GODOLPHIN, 10 August 1709 2
765 COTES *to* NEWTON, 18 August 1709 3
766 NEWTON *to* COTES, 11 October 1709 5
767 BENTLEY *to* NEWTON, 20 October 1709 7
768 FLAMSTEED *to* SHARP, 25 October 1709 8
769 LOWNDES *to* THE MINT, 9 November 1709 9
770 THE MINT *to* GODOLPHIN, 16 November 1709 11
771 TAYLOUR *to* THE MINT, 28 January 1709/10 12
772 THE MINT *to* GODOLPHIN, 31 January 1709/10 13
773 NEWTON *to* GODOLPHIN, 16 February 1709/10 14
773*a* EXTRACTS ENCLOSED WITH THE PREVIOUS LETTER . 15
774 —— *to* NEWTON, 28 March 1710 17
775 COTES *to* NEWTON, 15 April 1710 24
776 LOWNDES *to* THE MINT, 28 April 1710 28
777 COTES *to* NEWTON, 29 April 1710 29
778 NEWTON *to* COTES, 1 & 2 May 1710 31
779 COTES *to* NEWTON, 7 May 1710 33
780 NEWTON *to* COTES, 13 May 1710 35
781 COTES *to* NEWTON, 17 May 1710 37
782 THE MINT *to* GODOLPHIN, 17 May 1710 37

783 COTES *to* NEWTON, 20 May 1710 . . . . . . *page* 39
784 NEWTON *to* COTES, 30 May 1710 42
785 WOODWARD *to* NEWTON, 30 May 1710 44
786 COTES *to* NEWTON, 1 June 1710 45
787 NEWTON *to* COTES, 8 June 1710 47
788 COTES *to* NEWTON, 11 June 1710 48
789 NEWTON *to* COTES, 15 June 1710 50
790 NEWTON *to* [LAUDERDALE], 22 June 1710 51
791 LOWNDES *to* THE MINT, 28 June 1710 53
792 THE MINT *to* GODOLPHIN, 29 June 1710 53
793 COTES *to* NEWTON, 30 June 1710 54
794 NEWTON *to* COTES, 1 July 1710 55
795 GODOLPHIN *to* THE MINT, 4 July 1710 56
796 NEWTON *to* MONTGOMERIE, [July 1710] 57
797 LOWNDES *to* THE MINT, 25 July 1710 58
798 SEAFIELD *to* NEWTON, 2 August 1710 58
799 CHAMBERLAYNE *to* NEWTON, 21 August 1710 59
800 COTES *to* NEWTON, 4 September 1710 60
801 LOWNDES *to* THE MINT, 5 September 1710 60
802 NEWTON *to* SLOANE, [13 September 1710] 61
802*a* MINUTES OF A GRESHAM COMMITTEE MEETING, 13 December 1704 62
803 NEWTON *to* COTES, 13 September 1710 64
804 NEWTON AND ELLIS *to* THE TREASURY, 18 September 1710 65
805 COTES *to* NEWTON, 21 September 1710 65
806 NEWTON AND ELLIS *to* ST JOHN, [late September 1710] . 68
807 NEWTON *to* COTES, 30 September 1710 70
808 NEWTON AND ELLIS *to* THE TREASURY, 4 October 1710 . 71
809 COTES *to* NEWTON, 5 October 1710 73
810 COTES *to* NEWTON, 26 October 1710 74
811 NEWTON *to* COTES, 27 October 1710 75

812 An Account of the Late Proceedings at the Royal Society, 22 November 1710 . . . . . . *page* 76
813 The Treasury *to* The Mint, 29 November 1710 . . . 79
814 Royal Warrant *to* Newton, 12 December 1710 . . . 79
815 Newton *to* The Treasury, 13 December 1710 81
816 Newton *to* The Treasury, [*c.* 31 December 1710] . . . 82
816*a* Memorandum by Newton 84
817 Kerridge *to* Newton and Sloane, 16 January 1710/11 . 90
818 Flamsteed *to* Sharp, 23 January 1710/11 91
819 Newton *to* The Treasury, 23 January 1710/11 92
820 The Mint *to* The Treasury, 2 February 1710/11 . . . 93
821 Cotes *to* Jones, 15 February 1710/11 94
822 Leibniz *to* Sloane, 21 February 1710/11 96
823 Arbuthnot *to* Flamsteed, 14 March 1710/11 99
824 Flamsteed *to* Arbuthnot, 23/25 March 1710/11 100
825 Newton *to* Flamsteed, [*c.* 24 March 1710/11] 102
826 Newton *to* Cotes, 24 March 1710/11 103
827 Arbuthnot *to* Flamsteed, 26 March 1711 105
828 Flamsteed *to* Arbuthnot, 28 March 1711 106
829 Cotes *to* Newton, 31 March 1711 107
829*a* Draft of a Preface to the New Edition [? Autumn 1712] 112
830 Keill *to* Newton, 3 April 1711 115
830*a* Newton *to* [? Sloane], [? April 1711] 117
831 Arbuthnot *to* Flamsteed, 6 April 1711 118
832 Arbuthnot *to* Flamsteed, 16 April 1711 119
833 Flamsteed *to* Arbuthnot, 19 April 1711 120
834 Arbuthnot *to* Flamsteed, 21 April 1711 122
835 The Mint *to* The Treasury, 25 April 1711 123
836 Lowndes *to* The Mint, 8 May 1711 125
837 Royal Warrant *to* The Mint, 10 May 1711 126
838 The Mint *to* The Treasury, 14 May 1711 127

839 LOWNDES *to* THE MINT, 14 May 1711 . . . . . *page* 128
840 NEWTON *to* THE TREASURY, [15 May 1711] 129
841 FLAMSTEED *to* SHARP, 15 May 1711 129
842 NEWTON, SLOANE, AND MEAD *to* FLAMSTEED, 30 May 1711 131
843 SLOANE *to* LEIBNIZ, [May 1711] 132
843*a* KEILL *to* SLOANE *for* LEIBNIZ, [May 1711] 133
844 COTES *to* NEWTON, 4 June 1711 152
845 THE MINT *to* OXFORD, 6 June 1711 154
846 NEWTON *to* COTES, 7 June 1711 155
847 COTES *to* NEWTON, 9 June 1711 156
848 NEWTON *to* OXFORD, 10 June 1711 160
849 LOWNDES *to* NEWTON, 14 June 1711 162
850 NEWTON *to* OXFORD, [*c.* 15 June 1711] 162
851 NEWTON *to* COTES, 18 June 1711 164
852 NEWTON *to* OXFORD, [*c.* 20 June 1711] 165
853 HALLEY *to* FLAMSTEED, 23 June 1711 165
854 COTES *to* NEWTON, 23 June 1711 166
855 NEWTON *to* OXFORD, 6 July 1711 172
856 CROWNFIELD *to* NEWTON, 11 July 1711 173
857 COTES *to* NEWTON, 19 July 1711 174
858 MEMORANDUM BY NEWTON, [before 24 July 1711] . . . 175
859 DRAFTS BY NEWTON, [before 27 July 1711] 178
860 NEWTON *to* COTES, 28 July 1711 179
861 MEMORANDUM BY NEWTON, [*c.* 28 July 1711] 180
862 ROYAL WARRANT, 30 July 1711 182
863 COTES *to* NEWTON, 30 July 1711 183
864 DRAFT WARRANT BY NEWTON, [*c.* 31 July 1711] . . . 186
865 NEWTON *to* OXFORD, 7 August 1711 186
866 LOWNDES *to* THE MINT, 16 August 1711 187
867 NEWTON AND PEYTON *to* OXFORD, 20 August 1711 . . . 188
868 THE MINT *to* OXFORD, 21 August 1711 189

869 NEWTON *to* OXFORD, 28 August 1711 . . . . . *page* 190
870 NEWTON *to* OXFORD, 28 August 1711 191
871 COTES *to* NEWTON, 4 September 1711 193
872 FLAMSTEED *to* SHARP, 20 September 1711 194
873 COTES *to* JONES, 30 September 1711 194
874 T. HARLEY *to* NEWTON, 3 October 1711 195
875 [NEWTON *to* OXFORD], [5 October 1711] 196
876 T. HARLEY *to* NEWTON, 6 October 1711 196
877 MEMORANDUM BY NEWTON, [mid October 1711] . . . 197
878 NEWTON *to* GREENWOOD, 9 October 1711 199
878*a* NEWTON *to* MARTIN, [before 1711] 201
879 COTES *to* NEWTON, 25 October 1711 202
880 JONES *to* COTES, 25 October 1711 203
881 COTES *to* JONES, 11 November 1711 204
882 JONES *to* COTES, 15 November 1711 204
883 FINAL STATEMENTS OF ACCOUNT, 12 December 1711 . . 205
884 LEIBNIZ *to* SLOANE, 18 December 1711 207
885 FLAMSTEED *to* SHARP, 22 December 1711 209
885A LOWNDES *to* THE MINT, 7 January 1711/12 211
886 TAYLOUR *to* NEWTON, 11 January 1711/12 211
887 T. HARLEY *to* NEWTON, 15 January 1711/12 212
888 NEWTON *to* [? SLOANE], [? February 1711/12] 212
889 NEWTON *to* COTES, 2 February 1711/12 215
889*a* DRAFT SCHOLIUM TO PROPOSITION 4, Book III 216
890 COTES *to* NEWTON, 7 February 1711/12 220
891 NEWTON *to* COTES, 12 February 1711/12 222
892 NEWTON *to* OXFORD, 14 February 1711/12 224
893 COTES *to* NEWTON, 16 February 1711/12 225
894 NEWTON *to* COTES, 19 February 1711/12 230
895 OXFORD *to* NEWTON, 23 February 1711/12 231
896 COTES *to* NEWTON, 23 February 1711/12 232
897 T. HARLEY *to* NEWTON, 26 February 1711/12 238

897*a* LORDS JUSTICES AND PRIVY COUNCIL OF IRELAND *to* ORMONDE, 8 January 1711/12 . . . . . . *page* 238
898 NEWTON *to* COTES, 26 February 1711/12 240
899 COTES *to* NEWTON, 28 February 1711/12 242
900 NEWTON *to* OXFORD, 3 March 1711/12 245
901 COTES *to* NEWTON, 13 March 1711/12 246
902 SAUNDERSON *to* JONES, 16 March [1711/12] 247
903 NEWTON *to* COTES, 18 March 1711/12 248
904 LOWNDES *to* THE MINT, 18 March 1711/12 250
905 OXFORD *to* NEWTON, 21 March 1711/12 250
906 MARY PILKINGTON *to* NEWTON, 22 March 1711/12 . . . 251
906*a* NEWTON *to* TODD, [n.d.] 253
907 ANSTIS *to* NEWTON, 24 March 1711/12 254
908 NEWTON *to* COTES, 3 April 1712 255
909 NEWTON *to* COTES, 8 April 1712 263
909*a* DRAFT FOR PROPOSITION 37 264
910 COTES *to* NEWTON, 14 April 1712 269
911 COTES *to* NEWTON, 15 April 1712 271
912 NEWTON *to* COTES, 22 April 1712 273
912*a* THE ENCLOSED PAPER 274
913 COTES *to* NEWTON, 24 April 1712 275
914 COTES *to* NEWTON, [26 April 1712] 278
915 NEWTON *to* COTES, [29 April 1712] 281
916 COTES *to* NEWTON, 1 May 1712 282
917 COTES *to* NEWTON, 3 May 1712 284
917*a* FIRST DRAFT OF SCHOLIUM TO PROPOSITION 35 . . . 287
918 NEWTON *to* THE EDITOR OF THE *MEMOIRS OF LITERATURE*, [after 5 May 1712] 298
919 NEWTON *to* COTES, 10 May 1712 302
920 COTES *to* NEWTON, 13 May 1712 303
921 COTES *to* NEWTON, 25 May 1712 305
922 NEWTON *to* COTES, 27 May 1712 307

923 THE MINT *to* OXFORD, 16 June 1712 . . . . . *page* 307
924 TAYLOUR *to* NEWTON, 16 June 1712 309
925 NEWTON *to* OXFORD, 23 June 1712 309
926 ROYAL SOCIETY *to* FLAMSTEED, 3 July 1712 313
927 OXFORD *to* NEWTON, 10 July 1712 313
928 THE MINT *to* OXFORD, 16 July 1712 314
929 COTES *to* NEWTON, 20 July 1712 315
930 HERCULES SCOTT *to* [NEWTON], 26 July 1712 316
931 COTES *to* NEWTON, 10 August 1712 318
932 NEWTON *to* COTES, 12 August 1712 320
933 NEWTON *to* COTES, [14 August 1712] 323
934 CERTIFICATE BY NEWTON, 14 August 1712 323
935 COTES *to* NEWTON, 17 August 1712 324
936 TAYLOUR *to* THE COMPTROLLERS OF THE ACCOUNTS OF THE ARMY, 21 August 1712 326
937 NEWTON *to* COTES, 26 August 1712 327
937*a* THE ENCLOSED REVISED DRAFT 328
938 COTES *to* NEWTON, 28 August 1712 329
939 H. SCOTT *to* NEWTON, 28 August 1712 331
940 OXFORD *to* NEWTON AND HALLEY, 28 August 1712 . . . 332
940*a* NEWTON *to* CAWOOD, [after 28 August 1712] 333
941 NEWTON *to* COTES, 2 September 1712 333
942 COTES *to* NEWTON, 6 September 1712 335
943 NEWTON *to* COTES, 13 September 1712 337
944 COTES *to* NEWTON, 15 September 1712 338
945 NEWTON AND HALLEY *to* OXFORD, 18 September 1712 . . 340
946 NEWTON *to* COTES, 23 September 1712 341
947 NEWTON *to* [? MEDOWS AND BRUCE], [? September 1712] . 341
948 MEDOWS, BRUCE AND NEWTON *to* OXFORD, 7 October 1712 . 342
948*a* NEWTON'S ANNEXED MEMORANDUM, [early October 1712] . 343
949 [NEWTON *to* OXFORD], [? October 1712] 345
950 NEWTON *to* HENRY INGLE, 13 October 1712 346

51 Newton *to* Cotes, 14 October 1712 . . . . . *page* 347
951*a* Newton *to* Nikolaus Bernoulli, [? *c.* 1 October 1712] . 348
952 Newton *to* Cotes, 21 October 1712 350
953 Cotes *to* Newton, 23 October 1712 351
954 Newton *to* Oxford, 29 October 1712 352
955 Mary Pilkington *to* Newton, 30 October 1712 . . 353
956 Cotes *to* Newton, 1 November 1712 354
957 Royal Warrant *to* Newton, 20 November 1712 . . 355
958 Cotes *to* Newton, 23 November 1712 356
959 T. Harley *to* The Mint, 5 December 1712 . . . 357
960 The Mint *to* Oxford, December 1712 357
961 Newton *to* Cotes, 6 January 1712/13 361
961*a* De Vi Electrica 362
962 Cotes *to* Newton, 13 January 1712/13 369
963 Newton *to* The Chancellor of the Exchequer, 16 January 1712/13 370
964 Newton *to* Hercules Scott, [28 January 1712/13] . . 371
965 H. Scott *to* Fauquier, 29 January 1712/13 373
966 Newton *to* Oxford, 29 January 1712/13 374
967 Lowndes *to* The Mint, 31 January 1712/13 375
968 Lowndes *to* The Mint, 3 February 1712/13 375
969 Lowndes *to* The Mint, 4 February 1712/13 376
970 Bolingbroke *to* The Royal Society of London, 7 February 1712/13 376
971 Lowndes *to* The Mint, 12 February 1712/13 378
972 Treasury Reference *to* The Mint, 12 February 1712/13 378
973 Derham *to* Newton, 20 February 1712/13 379
974 Lowndes *to* The Mint, 26 February 1712/13 381
975 Newton at the Treasury, 27 February 1712/13 . . 381
976 Newton *to* [The Editors of the *Acta Eruditorum*], February 1712/13 383
977 Newton *to* Cotes, 2 March 1712/13 384

978 Newton and the Committee *to* Flamsteed, 5 March 1712/13 . . . . *page* 385
979 Newton and Bentley *to* Cotes, 5 March 1712/13 . . 386
980 Cotes *to* Newton, 8 March 1712/13 387
981 The Mint *to* Oxford, 9 March 1712/13 388
982 T. Harley *to* Newton and Phelipps, 9 March 1712/13 . 388
983 Cotes *to* Bentley, 10 March 1712/13 389
984 Bentley *to* Cotes, 12 March [1712/13] 390
985 Cotes *to* Newton, 18 [March] 1712/13 391
986 H. Scott *to* Newton, 21 March 1712/13 394
987 Treasury Reference *to* The Mint, 21 March 1712/13 . 396
988 Newton *to* Cotes, 28 March 1713 396
989 Newton *to* Cotes, 31 March 1713 400
990 Newton *to* Oxford, [31 March 1713] 401
991 T. Harley *to* The Mint, 2 April 1713 403
992 The Mint *to* Oxford, 10 April 1713 403
993 Jones *to* Cotes, 29 April 1718 404
994 T. Harley *to* The Mint, 3 May 1713 404
995 Cotes *to* Jones, 3 May 1713 405
996 Newton at the Treasury, 8 May 1713 405
996*a* Newton *to* Oxford, [? February 1712/13] 407
997 Newton's Recommendation of Machin, 12 May 1713. 408
998 Newton *to* Oxford, 22 May 1713 409
999 James Newton *to* Newton, [n.d.] 410
1000 Orders of Council, 24 June 1718 411
1001 Cotes *to* Samuel Clarke, 25 June 1713 412
1002 Bentley *to* Newton, [30 June 1713] 413
1003 Memorandum Concerning a Copper Coinage, [1713] 415
Appendix: Bentley's Accounts for the Printing of the *Principia* 417
Index 419

PREFACE

It is now more than ten years since the editor of the last volume of this *Correspondence* to be published, the late Dr J. F. Scott, began preparations for the present volume. Dr Scott's progress was interrupted by a singular series of untoward events and impeded by the advance of age. After his death in August 1971 the Librarian of the Royal Society took steps to secure the large mass of Newtonian material that Dr Scott had accumulated; when the present editors came to examine this material (in February 1972) it proved to contain a large number of photocopies and rough, annotated transcripts of the letters that Dr Scott had proposed to publish in Volume v, extending to the spring of 1713, when the second editing of Newton's *Principia* was complete. Further examination revealed that not all the material that Dr Scott had intended to publish was easily traceable among his papers and that which we found was far from ready for the printer. Accordingly, we decided to make a fresh beginning while still utilizing what we could of Dr Scott's work. We reviewed the extant correspondence from the summer of 1709 to the spring of 1713, and discovered much additional material, especially relating to the Mint; we carefully verified all transcripts, and made many new ones; we have rewritten every note. A completely new and greatly enlarged typescript was prepared for the press. Hence the faults of the present volume must be ascribed to the present editors.

In general we have followed the practices and style of our predecessors. We have tried to resist the temptation to bring in documents throwing light on Newton's life or thought unless they are directly related to his extant correspondence. The printing of such documents may well be welcome to scholars at first, whatever the context, but an inconvenience is very likely to arise in time if unpublished scientific papers or notes, for example, are included among lettters so that they become separated from other collections of similar material. On the other hand we have tried to be as complete as possible, including every document of an epistolary character, and in addition many that are closely related to the letters. We have omitted no part of the Mint business that was conducted by letter or memorandum, even though a good deal of it is of a more or less routine character. If nothing else, the Mint correspondence allows one to reconstruct a fairly detailed picture of Newton's variety of occupations day by day, week by week, month by month.

Newton created a formidable problem for all who seek to edit his words in the multiplicity of drafts that he wrote. As is usual (and reasonable) we have included here a number of drafts of letters that are not known in any other form; indeed, in some cases we have stated our belief that the letter under draft was never despatched to its putative recipient at all. Where both a transmitted letter and drafts are extant obviously the former is the prime authority; in general, in this volume, variations between transmitted letters (whether written by Newton himself or by a clerk) and Newton's holograph drafts are too insignificant to call for any particular attention to be paid to the draft(s). We have, however, noted variants here and there that seemed of particular interest, and we have (in recording the manuscripts transcribed) noted the location of drafts also, except in the case of the drafts of letters from Newton to Cotes which are all in the University Library, Cambridge, Add. 3984. The application of a rigorous diplomatic treatment to Newton's multiple drafts can rarely be either rational or useful, or so we judge of the documents in the present volume. However, we have on occasion found it useful to combine passages from several drafts to form the text printed here; the locations are noted.

We have tried in print to follow the orthography and form of the documents, though (as before) raised letters have been dropped to the line, producing some forms at first a little strange to the eye. Such forms as 'o^{r}' and 'yor' have been written out in full. Certain minutes written by Treasury clerks are very full of contractions and abbreviations; these we have silently expanded: 'm̄' we have replaced by 'mm', 'annū' by 'annum' and so forth, merely to avoid a complexity of typography that serves no useful purpose. All dates are given in Old Style (N.S. – 11 days), for the sake of homogeneity, except where otherwise indicated and when transcribing correspondents' letters dated in New Style. The figures, unless otherwise indicated, are copied from the original documents.

We have not translated every Latin word or sentence into English, believing that (particularly in the correspondence between Newton and Cotes) to do so would be tiresome for all parties, and needless. As in earlier volumes, documents written wholly in Latin are translated; we have translated also passages from the *Principia*, or drafted for the *Principia*, where these differ significantly from what was printed in the third edition, and then later translated into English. We have not translated passages that can be found in an English version of the *Mathematical Principles* from the references given in the *Correspondence*, either exactly or with obvious alterations in numerical values. Our policy has been to impose as little compulsory Latinity on the reader as possible, while assuming that any reader following the Newton–Cotes exchanges must have some Latin vocabulary, and that he will have a *Principia* at his elbow. It would have been possible to transcribe into our notes much of the first or second editions of the *Principia*, but to do so would have been *actum agere* since the publication of the multi-reading edition edited by the late Alexander Koyré and I. Bernard Cohen. This renders the Newton–Cotes correspondence much more easy to follow than it was before, and we often refer to it in our notes. However, we have thought it useful to print some large sections of the 'copy' sent by Newton to Cotes for the second edition and discussed in the letters, where this 'copy' does not appear in the printed book and is not considered by Messrs Koyré and Cohen.

The editors will welcome notice of errors in or omissions from this and other volumes of the Correspondence since it is our intention to conclude it with a sizeable *addenda* and *corrigenda*. Our next task is to prepare a census of the extant correspondence for the remaining fourteen years of Newton's life. Again, notes of the location of letters will be welcomed.

Our sincere thanks are extended to private owners who have permitted or facilitated the publication of this correspondence, among them the Marquess of Bath and Viscount Parker. We wish to thank the staffs of public institutions, including the British Museum, the Ministry of Defence Library, the Public Record Office, and the Scottish Record Office, who have handled our inquiries with promptness and cordiality. The Libraries of the Universities of Oxford, Cambridge and Dublin have provided many documents, as have the Libraries of King's College and Trinity College in Cambridge; we are very grateful for help received. Mr P. S. Laurie of the Royal Greenwich Observatory has put the Flamsteed archives at our disposal, and the staff of the Library of the Royal Society itself has given willing service at every point. We also wish to express our appreciation of the kindnesses done us by Miss B. M. Austin, Professor J. L. Axtell, Professor I. Bernard Cohen, Dr Eric Forbes, Mr P. J. Gautrey, Dr F. Gyorgyey, Mrs V. Harrison, Mr P. I. King, Mr Robert E. Kohler, Jr, Mr James D. Mack, Miss Winifred Myers, Mr C. A. Potts, Professor C. S. Smith, Mr C. G. Stableforth and Mrs M. A. Welsh.

In encouraging the present editors to undertake this work, Dr D. T. Whiteside generously promised his assistance; all who are acquainted with him will know how well he has fulfilled this promise and how much this volume owes to him.

Finally, our thanks go to Mrs K. H. Fraser and Mrs Frances Couch for their care in typing this volume.

April 1973

A.R.H.
L.T.

A NOTE ON THE MANUSCRIPTS USED IN THIS VOLUME

The letters printed in this volume are drawn largely from four collections. That in the University Library, Cambridge, a part of the former Portsmouth Collection, is by far the most voluminous: it contains (Add. 3983) Newton's original letters from Cotes (all but a few), some drafts of his letters to Cotes (Add. 3984) and the letters he received from John Keill (Add. 3985). Many other letters received by Newton are preserved, scattered, elsewhere because he used them as scrap-paper.

The Library of Trinity College, Cambridge, contains (MS. R.16.38) Newton's original letters to Cotes, and drafts—or in some cases probably rather personal record copies—of Cotes' letters to Newton. There are also a number of original letters from Cotes to Newton (which would otherwise be in U.L.C. Add. 3983); these were borrowed from John Conduitt, to whom Newton's papers passed, by Robert Smith, who was Cotes' literary executor, and kept by him with Cotes' own papers which ultimately came (through Edward Howkins) to Trinity College. These letters were all printed by Joseph Edleston in his *Correspondence*, but it will be understood that Edleston had no access to the materials now accessible to us in the University Library; to that extent only our publication is more complete and accurate than his highly scholarly work.

Besides revised 'copy' for the second edition of the *Principia*, some of which is printed here, the Library of Trinity College also possesses Richard Bentley's correspondence (R.4.47), which we have also used.

As is well known, the retained portion of the Portsmouth Collection was dispersed by sale in 1936. We have used some documents then purchased by Lord Keynes and given to King's College; another large group, relating to Newton's career at the Mint, was bought and presented to the Royal Mint whence (with all other early Mint Records) it has passed to the Public Record Office. These Mint Papers were (rather inadequately) sorted by John Conduitt into subject groupings and assembled in three huge volumes, which now (handsomely bound) are shelfmarked as Mint/19, I–III. A great many papers written or signed by Newton or referred to him are also to be found in the normal Treasury and Mint records.

SHORT TITLES AND ABBREVIATIONS FOR PUBLISHED WORKS AND MANUSCRIPTS

(a) PUBLISHED WORKS

Baily, *Flamsteed*	Francis Baily, *An Account of the Revd. John Flamsteed... & Supplement to the Account*. London, 1835–7; reprinted, Dawsons, London, 1966.
Birch, *History*	Thomas Birch, *History of the Royal Society of London*. London, 1756–7; reprinted, Johnson, New York and London, 1968.
Brewster, *Memoirs*	Sir David Brewster, *Memoirs of the Life, Writings and Discoveries of Sir Isaac Newton*. Edinburgh, 1855; reprinted, Johnson, New York and London, 1965.
C.S.P.D.	*Calendar of State Papers, Domestic Series*.
Cal. Treas. Books	William A. Shaw (ed.), *Calendar of Treasury Books preserved in the Public Record Office* (London H.M.S.O.). (There are two parts in this series for each calendar year, 1709–13. The second part is invariably referred to in this *Correspondence*.)
Cal. Treas. Papers	J. Redlington (ed.), *Calendar of Treasury Papers preserved in Her Majesty's Public Record Office*. London, Longman & Co., 1868–89.
Cohen, *Introduction*	I. Bernard Cohen, *Introduction to Newton's 'Principia'*. Cambridge, 1971.
Commercium Epistolicum	*Commercium Epistolicum D.* Johannis Collins, *et aliorum de Analysis promota: jussu Societatis Regiæ in lucem editum*. London, 1712.
Craig, *Newton*	Sir John Craig, *Newton at the Mint*. Cambridge, 1946.
Cudworth, *Life*	William Cudworth, *Life and Correspondence of Abraham Sharp*. London, 1889.
Edleston, *Correspondence*	Joseph Edleston, *Correspondence of Sir Isaac Newton and Professor Cotes including Letters of other Eminent Men*. London, 1851; reprinted, Cass, London, 1969.
Gerhardt, *Briefwechsel*	C. I. Gerhardt (ed.), *Der Briefwechsel von Gottfried Wilhelm Leibniz mit Mathematikern*. Berlin, 1899; reprinted, Georg Olms, Hildesheim, 1962.
Gerhardt, *Leibniz: Mathematische Schriften*	C. I. Gerhardt (ed.), *G. W. Leibniz Mathematische Schriften*, 7 vols. Halle etc., 1849–63; reprinted, Georg Olms, Hildesheim, 1962.

Hall & Hall, *Oldenburg*	A. Rupert Hall and Marie Boas Hall, *The Correspondence of Henry Oldenburg*. University of Wisconsin Press, vol. I, 1965, to vol. IX, 1973, in progress.
Hall & Hall *Unpublished Scientific Papers*	A. Rupert Hall and Marie Boas Hall, *Unpublished Scientific Papers of Isaac Newton*. Cambridge University Press, 1962.
Hofmann	J. E. Hofmann, *Die Entwicklungsgeschichte der Leibnizschen Mathematik während des Aufenthaltes in Paris (1672–76)*. München, 1949. See also *Leibniz in Paris, 1672–76*, Cambridge University Press 1974.
Koyré and Cohen, *Principia*	*Isaac Newton's Philosophiæ Naturalis Principia Mathematica. The Third Edition (1726) with variant readings assembled and edited by Alexander Koyré and I. Bernard Cohen with the assistance of Anne Whitman*. Cambridge, 1972, 2 vols.
Lyons, *History*	Sir Henry Lyons, *The Royal Society, 1660–1940*. Cambridge, 1944.
MacPike	Eugene Fairfield MacPike, *Correspondence and Papers of Edmond Halley*. London, O.U.P., 1932; Taylor and Francis, 1937.
More	Louis Trenchard More, *Isaac Newton, a Biography*. New York and London, 1934.
Rigaud, *Correspondence*	S. P. and S. J. Rigaud, *Correspondence of Scientific Men of the Seventeenth Century...in the Collections of the Earl of Macclesfield*. Oxford, 1841; reprinted, Georg Olms, Hildesheim, 1965.
Shaw	William A. Shaw, *Select Tracts and Documents Illustrative of English Monetary History, 1626–1730*. London, 1896.
Sotheby Catalogue	*Catalogue of the Newton Papers Sold by Order of the Viscount Lymington...which will be sold by Auction by Messrs. Sotheby and Co.* [on 13 and 14 July, 1936].
Ward, *Lives*	John Ward, *The Lives of the Professors of Gresham College*. London, 1740.
Weld, *History*	C. R. Weld, *A History of the Royal Society*. London, 1848.
Whiteside	D. T. Whiteside, (*ed.*) *The Mathematical Papers of Isaac Newton*. Cambridge, 1967 onwards: in progress.
Wollenschläger, *De Moivre*	Karl Wollenschläger, 'Der mathematische Briefwechsel zwischen Johann [I] Bernoulli und Abraham de Moivre', *Verhandlungen der Naturforschenden Gesellschaft in Basel*, Band XLIII, 1931–2, pp. 151–317.

(*b*) MANUSCRIPTS

Mint Papers	Newton's private file of papers concerning Mint business, sold at Sotheby's in 1936, now bound in three volumes in the Public Record Office (Mint/19, I–III).
P.R.O.	Manuscripts in the Public Record Office.

Sharp Letters	A volume containing 'A collection of Original Letters addressed to Mr Abraham Sharp...by Mr John Flamsteed', placed on permanent loan to the Royal Society by Mr F. S. Edward Bardsley-Powell.
U.L.C.	Manuscripts in the University Library, Cambridge (Portsmouth Collection).

INTRODUCTION

From a bibliographical point of view the period of almost four years, of which the surviving correspondence is published in this book, was the most important period in Newton's life, since it was that in which, with Roger Cotes, he revised his *Philosophiae Naturalis Principia Mathematica* and saw it through the press. Considered as a single group of letters, the Newton–Cotes correspondence is the largest and most important section of Newton's scientific correspondence that we have; nowhere else can one witness Newton in a detailed debate about scientific argument and scientific conclusions—a debate from which he did not always emerge victorious. Nowhere else does Newton write in detail about the text of the *Principia*. And all scholars would agree that this text which was hammered out between Cotes and Newton was the most important of all the versions, printed and unprinted; this was (to all intents and purposes) the *Principia* of subsequent history. So that though the prime of Newton's age for invention was long past, and no great scientific or mathematical innovation was to be expected from the end of his seventh decade, this was nevertheless a time of considerable literary achievement—towards which the collaboration of Cotes was of no slight importance, as we shall see.

In this time too Newton's relationship with Leibniz passed through a new critical phase, which included the publication of the *Commercium Epistolicum*; this volume ends just as Johann Bernoulli was about to occupy a prominent rôle in this lamentable drama. Considering its importance and the gallons of ink subsequently devoted to it, the drama does not figure largely in this volume. There are two reasons for this. First, there is the fact that Newton simply did not carry on a large correspondence about Leibniz, the *Commercium Epistolicum* and so forth at this time; or if he did, it has now vanished. In writing to Cotes Newton might have blamed some of his delays on this distraction, but he did not, so that before 1713 there is hardly any mention of the Leibniz affair between them. Second, and consequent upon this, our policy to exclude drafts not of an epistolary character or clearly related to the letters has led us to exclude from this volume many repetitive pages in which Newton stated his case against Leibniz. For of course the fact that Newton was not (so far as we are aware) writing letters about Leibniz does not mean that he was not writing about Leibniz at all. As is well known, there are a great number of drafts in which Newton sought to demonstrate the priority of his mathematical invention over Leibniz's, and the actual derivation of Leibniz's differential method from his own. Many of these drafts are very alike, and almost all are without date. Some belong to the period of the preparation of the *Commercium Epistolicum* in 1711–12, but many more belong to a later period when Newton had been further embittered by the publication of the *Charta Volans* and other incidents. As we are engaged in publishing Newton's correspondence, we must leave the further study of such drafts to some future historian of Newton's dispute with Leibniz.

Nevertheless, one must recognize that there was here a demand on Newton's time, and a source of strain upon his emotions and character, that were more serious than the letters alone might suggest. It is hardly conceivable that, after the spring of 1711, the Leibniz affair did not weaken Newton's concentration upon the revision of the *Principia*, and it seems likely that it was a factor increasing his impatience with Cotes, for all the latter's diplomacy.

It is possible that but for John Keill these mischiefs would have been at least postponed. It was Keill who had publicly revived the charge of plagiarism against Leibniz in the *Philosophical Transactions* for 1708, Keill who advised Newton of the fresh Leibnizian imputations against his honour (Letter 830), and Keill who prepared the manifesto stating Newton's case at length (Letter 843*a*). Newton, who at first was genuinely reluctant to fan the fires of controversy (Letters 888, 918), might have let the whole business rest as he had for many years. In Keill's defence it must be said that he was honest and sincere in his championship of Newton; and looked at from Newton's point of view, with access to Newton's impressive historical records of youthful performance (but not, of course, to those of Leibniz), Newton stood in need of a champion. Nor was Keill alone in his sense of the injustice; Cotes, Jones, Raphson, Freind and others felt much as he did. On the other side Leibniz and his friends were far from discreet, despite the philosopher's protestations (Letter 822), since in anonymous reviews and in private correspondence (later published—Letter 918) they spread derision of the Newtonian philosophy, gradually adding new dimensions to the original quarrel.(1)

Leibniz's exculpatory claim (Letter 822) that he had never heard of the name of *calculus of fluxions* nor seen the 'symbolism that Mr. Newton has employed' before he saw Wallis's *Opera Mathematica* in the 1690s—some seven or eight years after the appearance of his own first calculus paper—was innocent-sounding and, literally, quite true. But of course it slid over the deeper questions, and by no means put him out of court. For it could—again, up to a point, with plausibility and reason—be claimed on behalf of Newton that *his* phase of mathematical innovation occurred at a still earlier date, at a time when Leibniz really knew no mathematics; and to avoid the retort that it was Newton's own fault if he had published nothing on his new mathematics before the

1687 *Principia* (in which there are neither fluxions nor dotted letters) the Newtonians turned to the evidence of the communication of Newton's discoveries in the 1670s and developed (as we now know, with great exaggeration) the hypothesis that the communication had seminally extended to Leibniz himself. Leibniz's apologia, heightened by the counter-charge that even in 1687 (three years after Leibniz's first paper!) Newton had known very little of the new calculus, forced Newton into trying to prove not only the reality of his mathematical achievement in the years before 1676, but into vastly over-emphasizing the value to Leibniz of the letters from England that Leibniz had received.

Hence the character of Keill's manifesto (Letter 843*a*), and hence the *Commercium Epistolicum* which it foreshadowed. In 1711, after the receipt of Leibniz's Letter 822 complaining of Keill's charges against himself, it was for Newton only a question of historical research and quasi-legal argument; what precisely had Leibniz learnt from him, what did Leibniz's own letters reveal of his attainments at the time when they were less than Newton's? For such an attempt to overwhelm Leibniz by sheer documentation (to which, indeed, there was no adequate rejoinder of the same kind before the nineteenth century) a precedent had been provided by Wallis in his *Opera Mathematica*, and more recently by William Jones (Letter 821). Jones in his *Analysis per quantitatum series, fluxiones ac differentias* (1711) was exceedingly discreet; his edition of Newton's mathematical opuscula, effected with their author's willing cooperation, was completed before Letter 822 was written; nevertheless his acquaintance with Jones can only have confirmed Newton's conviction that no Briton, at any rate, with the evidence fairly before him, could deny Newton's priority and the blackguardly conduct of Leibniz.

When Leibniz's complaint was read at a meeting of the Royal Society on 22 March 1710/11, 'Dr Sloane was ordered to write an Answer to him' (Journal Book of the Royal Society, XI, p. 209). At the meeting of 5 April, however, this quiet tone abruptly changed; Keill, who had not long returned from New England, whither he had escorted a party of Protestant refugees from the war-scarred Palatinate, threw back the challenge, pointing out that

> in the Lipsick Acta Eruditorum for ye year 1705, there is an unfair Account given of Sir Isaac Newton's Discourse of Quadratures, asserting ye Method of Demonstration by him there made use of to Mr. Leibniz &c: Upon which the President gave a short account of that matter, with ye particular time of his first mentioning or discovering [i.e. revealing] his Invention, referring to some Letters published by Dr. Wallis. Upon which Mr. Keill was desired to draw up an account of ye Matter in dispute, and set it in a just light, and also to vindicate himself from a Particular Reflection [i.e. aspersion] in a Letter from Mr. Leibniz to Dr Sloane. (Journal Book of the Royal Society, XI, pp. 210–1)

Presumably Newton had received Keill's Letter 830 and so had been able to prepare his vindication before the meeting. It is clear (if one may rely on the precise wording of the minutes) that at this stage Leibniz's guilt was virtually taken for granted; no one would have expected Keill to make an impartial adjudication between the claims of the two mathematicians; rather his task was to prove the justice of Newton's case. Newton assisted him further the next week, for when the minutes of the last meeting were read they

gave occasion to further discourse of ye Matter mentioned in ye Leipsic Acta: The President was pleased to mention his Letters [written] many years ago to Mr. Collins about his Method of treating Curves &c: and Mr Keill being present was again desired to draw up a Paper to assert the President's right in this matter. (p. 213)

Hence Keill's task was clear, and he completed it (not without private reference to Newton) by 24 May:

A Letter from Mr John Keill to Dr Sloane was produced and read, relating to the Dispute concerning the Priority of the Invention of the Arithmetick of Fluxions, &c: between Sir Isaac Newton and Mr Libnitz [*sic*]...A Copy of this Letter was order'd to be sent to Mr Leibnitz, and Dr Sloane desired to draw up a Letter to accompany it, before it was made publique in ye Transactions, which should not be 'till after ye receipt of Mr Leibnitz's Answer. (p. 224)

The results may be seen in Letters 843 and 843*a*.

More than six months elapsed before Leibniz's response, in predictable terms (Letter 884), became available; it was read at a meeting of the Royal Society on 31 January 1711/12, and 'deliver'd to the President to consider of the Contents thereof' (Journal Book of the Royal Society, XI, p. 267). What may have been the immediate result of Newton's consideration appears as Letter 888, though it probably never went beyond a rough draft. At the next meeting (7 February 1711/12) Newton failed to take the Chair, and nothing further happened for some weeks, or rather nothing relative to Leibniz is known to have happened; Newton carried on active correspondence with Cotes and had some Mint business. However, Newton prepared what is clearly the draft of a speech asserting his own rights, for delivery on 14 February 1711/12 (the silence of the minutes being no good evidence that he did not in fact speak in the following manner):[2]

Gentlemen

The Letter of Mr Leibnitz wch was read before you when I was last here relating to me as well as to Mr Keill I have considered it, & can acquaint you that I did not see the papers in the Acta Leipsica till the last summer & therefore had no hand in beginning this controversy.[3] The controversy is between the author of those papers & Mr Keil. And I have as much reason to complain of that author for questioning my candor & to desire that Mr Leibnitz would set the matter right without engaging me in a dispute wth that author as Mr Leibnitz has to complain of Mr Keil for questioning his candor & to desire that I would set the matter right without engaging him in a controversy with Mr Keil. For if that author in giving an account of my book of Quadratures gave every man his own, as Mr Leibnitz affirms, he has taxed me with borrowing from other men & thereby opposed my candor as much as Mr Keil has opposed the candor of Mr Leibnitz & so was the aggressor. Mr Leibnitz & his friends allow that I was the inventor of the method of fluxions: & claim that he was the inventor of the differential method. Both may be true because the same thing is often invented by several men. For the two methods are one & ye same method variously explained & no men could invent the method of fluxions without knowing first how to work in the augmenta momentanea of fluent quantities wch augmenta Mr Leibnits calls differences. Dr Barrow & Mr Gregory drew tangents by the differential method before the year 1669. I applied it to abstracted æquations before that year & thereby made it generall. Mr Leibnits might do the like about the same time; but I heard nothing of his having the method before the year 1677. When and how he found it must come from himself. By putting the fluxions of quantities to be in the first ratios of ye augmenta momentanea I demonstrated the method & thence called it the method of fluxions: Mr Leibnitz uses it without a Demonstration. I [*ends*]

Certainly something occurred at the Society's meetings that cannot be traced in the Journal Book, since abruptly the minute for 6 March reads:

Upon account of Mr Leibnitz's Letter to Dr. Sloane concerning the Dispute formerly mentioned between him and Mr Keill, a Committee was appointed by the Society to inspect the Letters and Papers relating thereto; viz. Dr Arbuthnot, Mr Hill, Dr Halley, Mr Jones, Mr Machen, and Mr Burnet, who were to make their Report to the Society. (Journal Book of the Royal Society, XI, p. 275)

On the 20th Mr Robartes was added to the Committee, on the 27th 'Mr Bonet, the King of Prussia's Minister', and on 17 April Abraham de Moivre, Francis Aston and Brook Taylor were also nominated to serve. The last three at least can have done little of the work, for their report was presented to the Royal Society only a week later, on 24 April 1712.(4)

One might wonder, perhaps, that such a committee could assemble and sift through so large a body of material as was mentioned in the report, and reach conclusions of so important and (to Leibniz) damaging a kind in the short period available, even though guided by Keill's earlier evaluation. Some, at least, of the mathematical letters demanded scrupulous consideration. The answer is, of course, that the report—of which the Royal Society does not at present possess the original—was prepared by Newton himself, who did not scruple to be judge in his own cause. His own, rather shorter rough draft of the report, with his emendations bringing it closer to the final form, exists still in private possession. In the text of the report transcribed below from the Journal

Book of the Royal Society we have italicized the passages also be to found in Newton's draft,[5] though sometimes with trivial differences in phrasing:

We have consulted the Letters and Letter-Books in the Custody of the R. Society, and those found among the Papers of Mr John Collins dated between the years 1669 and 1677 inclusive, and shewed them to such as knew and avouched *the Hands of Mr Barrow, Mr Collins, Mr Oldenburg and Mr Leibnitz,* and compared those of Mr Gregory with one another and with Copies of some of them taken in the hand of Mr Collins: *And* have *Extracted from them what relates to the Matter* referred to us: All which Extracts herewith delivered to You we believe to be genuine and authentic, and *by these Letters and Papers we find*

1st *That Mr Leibnitz was in London in the beginning of the year 1673 and went thence* in or about March *to Paris, where he kept a Correspondence with Mr Collins by means of Mr Oldenburg till about September 1676 and then returned by London and Amsterdam to Hannover;*[6] *and that Mr Collins was very free in Communicating* to able Mathematicians *what he had received from Mr Newton and Mr Gregory.*

2ly *That when Mr Leibnitz was the first time in London he contended for the* Invention of another *Differential Method* properly so called; and notwithstanding that he was shown by Dr Pell that it was Mouton's Method, he persisted in maintaining it to be his own Invention, by reason that he found it by himself without knowing what Mouton had done before, and had much improved it.[7] *And we find no Mention of his having any other Differential Method than Mouton's, before his Letter of* the 21st of *June 1677,*[8] *which was a Year after a* Copy of *Mr Newton's Letter* of the 10th of December 1672,[9] *had been sent to Paris to be communicated to him,*[10] *and above 4 years after Mr Collins began to communicate that Letter to his Correspondents: In which Letter the Method of Fluxions was sufficiently described to any intelligent Person.*[11]

3ly *That by Mr Newton's Letter of ye 13th of June 1676,*[12] *it appears, that he had the Method of Fluxions above five years before the writing of that Letter; and by his Analysis per æquationes numero terminorum infinitas communicated by Dr Barrow to Mr Collins in July 1669.*[13] *We find that he had invented the Method before that time.*

4ly *That the Differential Method is one and the same with the Method of Fluxions, excepting the Name and Mode of Notation, Mr Leibnitz calling those Quantities Differences, which Mr Newton calls Moments or Fluxions, and marking them with the Letter* d, *a Mark not used by Mr Newton. And* therefore *We take the proper Question to be not who invented this or that Method, but who was the first Inventor of the Method; And we believe that those who have reputed Mr Leibnitz the first Inventor knew little or nothing of his Correspondence with Mr Collins and Mr Oldenburg* long before, *nor of Mr Newton's having that Method above 15 years before Mr Leibnitz began to publish it in the Acta Eruditorum of Leipsick.*

For which reasons we reckon Mr Newton the first Inventor; and are of opinion that Mr Keill in asserting the same has been noways injurious to Mr Leibnitz. And we submit to the Judgement of the Society whether the Extract of Letters and Papers now presented togather with what is Extant to the same purpose in Dr Wallis's 3d Volume, may not Deserve to be made publick.

To which Report the Society agreed nemine contradicente, and ordered that the whole Matter, from the Beginning, with the Extracts of all the Letters relating thereto, and Mr Keill's and Mr Leibnitz's Letters, be published with all convenient speed may be, together withe the Report of the said Committee.

Ordered that Dr Halley, Mr Jones, and Mr Machin, be desired to take care of the said Impression (which they promised) and Mr Jones to make an Estimate of the Charges against the next Meeting.

Mr Keill said he would draw up an Answer to Mr Leibnitz's last Letter, it relating cheifly to himself; which he was also desired to do, and that it should be read at a Meeting of the Royal Society. (pp. 287–9)

It would obviously be naive in the extreme to suppose that the Royal Society Committee acted impartially or that the *Commercium Epistolicum* whose publication resulted from its recommendation was other than a dedicated proclamation of the case for Newton's priority. The documents are fairly and accurately quoted in the *Commercium* but they are by no means always impartially used. In any event, countless drafts testify to Newton's deep involvement in its preparation; one such draft is reproduced by Dr Whiteside in his *Mathematical Papers*(14) with the comment that the Committee was a mere 'blind front' for Newton himself, who prepared the documentation, commentary and footnotes and asked the Committee to do little more than lend their names.

Clearly then, Newton's dispute with Leibniz was a constant element in his thoughts from the spring of 1711 onwards, and we know that all through the summer of 1712 he was busy with the compilation of the *Commercium Epistolicum* although very little of this activity is reflected in his surviving letters. This is not the only instance of an almost inevitable imbalance in the present volume. As it happened, because Newton was excessively cautious, the records of his State business are preserved in great detail. Few would claim that they inform us of the most interesting aspect of Newton's mind. On the other hand there were correspondences of far higher intrinsic merit, in which Newton and the Newtonian philosophy are deeply concerned but in which Newton appears nowhere as a correspondent. Within the present period Halley, Abraham de Moivre and William Burnet were all in correspondence with Johann [I]. Bernoulli. Of the extant remains of these exchanges only the correspondence of De Moivre has been published (see Letter 951*a*). There can be little doubt that behind these English mathematicians stands the invisible figure of Newton, and that as in Keill's case here one must take other men's words as representing the work of Newton's still active mind. We hope that it may be possible to attend more carefully to these letters in the final volumes of this series.

Nevertheless, it was the preparation of the second edition of his *Principia* that furnished Newton with his major intellectual preoccupation during the four years the book was in the press. Professor I. Bernard Cohen in his *Introduction to Newton's 'Principia'* has already dealt in great detail with the antecedents of this edition, including Newton's own preparations, the frustrated aspirations of Nicholas Fatio de Duillier and David Gregory, the various emended copies of the first edition still extant, and so forth, none of which need be repeated here. By June 1708 (Letter 742, vol. IV) Richard Bentley, Master of Trinity College, had emerged as the successful man-midwife and specimen pages had been printed off, but little more happened for another year. As nothing is known about the arrangement Bentley made with Newton (save for Conduitt's story that Newton chose to gratify Bentley's love of money—and that Bentley did gain from Cotes' labours may be seen from the Appendix to this volume)—it is impossible to guess whether this lack of progress was due to some renewed hesitancy on Newton's part, or to his business at the Mint, or even to Bentley's failure to bring forward Roger Cotes as editor, while he himself remained, in effect, as publisher of the new edition. Once Cotes was steadily at work from October 1709 onwards Bentley disappeared from the scene until the new edition was ready to go before the world, although he did on a few occasions act as an intermediary between Cotes and Newton; Cotes himself never seems to have visited Newton in London during the whole period of three and a half years, while Bentley undertook the journey quite frequently.

That Bentley's choice should fall on Cotes for the technical work of seeing Newton's book through the press, examining the old and new matter for consistency and accuracy, was natural enough. Cotes was young (he was born on 10 July 1682), he came of a poor but scholarly family and so had his way to make in the world, he was dependent on Bentley's patronage, and last but not least he was a brilliant mathematician. Son of a Leicestershire parson, he had gone to St Paul's School where, at sixteen, he was already precociously advanced in mathematics and astronomy (see Edleston, *Correspondence*, pp. 190–2). He entered Trinity in 1699—another letter printed by Edleston shows him handling fluxions readily after his second year at Cambridge—and became a junior Fellow on 3 October 1705. Two years later he became the first Plumian Professor of Astronomy, being strongly favoured by the Lucasian Professor, William Whiston (who said he 'was but a child [in mathematics] to Mr Cotes'; *Memoirs*, 1749, p. 133), and by Richard Bentley, much to the annoyance of Flamsteed, who had recommended his own assistant, John Witty (Baily, p. 258; Edleston, *Correspondence*, p. lxxiv, note 158). Newton's reaction to Bentley's (or perhaps Whiston's) suggestion that Cotes should be his coadjutor is unknown, and so is the degree of acquaintance between the two men, forty years apart in age. Early in 1709 when Cotes, as Plumian Professor, was busy setting up the observatory (constructed over the Great Gate of Trinity), he wrote a letter that indicates a slight acquaintance with Newton; he must have been aware that he owed his successful career in part to Newton's goodwill.

In this letter to his uncle, Cotes made no mention of the task he was to undertake for Newton. Yet in the following month (if the dates are correct)[15] Bentley was already writing to Cotes warning him to expect from Newton a copy of the *Principia* prepared for the press; in this he was far too hasty, however, for in spite of a visit by Cotes to Newton himself in July 1709 no copy of the new edition was received in Cambridge until September or early October (Letter 766). Then the printing really began at last. The reader may once again be referred to Professor Cohen's *Introduction* for a review of Cotes' rôle as editor and the progress of the work, portrayed in detail in the letters printed below.

At first the printing went swiftly. Cotes went all through the first Book of the *Principia* and well into the second—that is through about half of the whole work—before (so far as we know) he had to refer to Newton over a considerable difficulty that he did not care to resolve himself (Letter 765).[16] He took the opportunity to apologize to Newton for making 'little alterations' in the text necessary for elegance, clarity or truth, since it would cause 'great inconvenience to the work and uneasiness to yourself' if every detail were submitted to Newton for agreement. Indeed, Cotes' responsibilities were great; each sheet (eight pages) of the book was set, then proofed by Cotes, and when correct the necessary number of copies (750) were pulled and the type distributed. As Letter 775 makes clear, Newton only saw a printed page (in general) at this stage when any further correction by him would have entailed reprinting all 750 sheets. So Cotes had to make sure each sheet was right. At the beginning (Letter 766) Newton had advised Cotes not to try to examine every mathematical argument in turn, but to be satisfied if the printed page corresponded accurately with the copy, amending any slips noted as he went through the book; he repeated that advice in Letter 789. In the earlier stages, therefore, Newton did not expect a great deal of Cotes, and was prepared to take the responsibility for the correctness of the various propositions upon himself—which was reasonable enough. But as time went on, and the inherent difficulty of the propositions increased, Cotes found it necessary to draw Newton's attention to more and more problems in the copy, and to himself draft revisions which, in his view, would state the matter correctly. At this stage—in the early summer of 1710—Cotes was more concerned to satisfy himself that changes introduced by Newton from the first edition were justified and that (whether in the old or the new form) Newton maintained consistency. He did not attempt to verify every one of Newton's conclusions independently, though in Letter 783 he does just this, and satisfies himself that Newton was right. Thus he missed the famous error in Book II, Proposition 10 about this time (see Letters 951 and 951*a*) because there was no obvious inconsistency or slip to catch his eye; even when Newton knew that the 1687 form of this proposition was erroneous it took him several days to locate and correct his precise mistake. Cotes had no reason to look for it. Again, Cotes insists at more than one point (Letter 786, for example) that Newton must be *clear*; he points out cases where Newton has confused quantities or terms (Letter 788), he shows him where a correction has been only partly carried through (Letter 779). Obviously all these minor emendations contributed to the high quality of the new edition, without making any very basic difference to it.

By the end of June 1710 the printer had reached Book II, Proposition 31 and the *Scholium Generale* (in which Newton justified the preceding theory of resisted motion by the report of actual experiments, the number of which he had now much increased); Cotes sent two sheets (Oo and Pp) to Newton for him to check finally before they were printed, then went off on holiday (Letter 793). All Newton's prepared copy had now been printed. When, in September, Cotes examined the next batch sent by Newton, he at once raised a substantial difficulty about the velocity of efflux of fluids, where (Proposition 36) Newton had altered the first edition to produce in the second the result which is (approximately) correct. Cotes, not realising—as Newton soon did—that the issuing jet of water under pressure is not necessarily of the same diameter as the aperture forming it, objected that the *weight* of water issuing in a given time seemed in better conformity with the first edition result (Letter 805). Cotes left it to Newton to clear the difficulty, which was not done till the following Spring (Letter 826).

The press had not proceeded far before Cotes declared himself dissatisfied with Book II, Proposition 47 (on the harmonic motion of particles transmitting wave motion; Letter 854). His objections to Newton's treatment are not easy to understand and were presumably ill-founded, for though he argued against Newton's first attempt to explain himself, when Newton returned to the question a second time (Letter 889) Cotes conceded that 'as to the business of sounds I do entirely agree with you upon considering that matter over again' (Letter 890). A great deal of time had been lost to no purpose, for it was now February 1712, nearly two years since the project began.

Now, at last, they could start on the third and final Book of the *Principia*. Here Cotes did Newton much minor service in saving him from errors of computation and all kinds of minor inconsistencies, errors and omissions; for many numbers were to be worked out by theory and compared (rather too precisely) with other numbers reduced from observations. Newton and Cotes strove together to make the 'fit' as exact as possible; Cotes was almost too exacting—

did it matter if Newton in one place gave the length of a degree of latitude as 57230 French fathoms, and in another as 57220 (Letter 893)? For it is very doubtful whether the measurement was accurate to some sixty feet in seventy miles. There were many tiresome details concerning the best observational values to adopt—for the precession of the equinoxes, for example—or the most appropriate number to derive theoretically, and Cotes, in the same letter, still had energy to argue that Newton was not justified in supposing that even God could not make two equal spheres, each perfectly filled with matter, differ in the quantity of matter they contained. A fair part of Book III, debated in this way, was rewritten yet again before it reached its final form.

Of all the problems in Book III the theory of the Moon was the most intractable; Cotes did an enormous amount of detailed work tidying up in Newton's wake, and stimulating him to rewrite his treatment in order to remove ambiguities and contradictions (see Numbers 912*a*, 917*a*, 937*a*). But Cotes in no sense modified the structure of Newton's lunar theory, and several of the points of criticism he raised proved insubstantial when Newton set out his thoughts plainly.

It may be useful to recall that, neglecting motion of the Moon's orbit as a whole (an important topic to which Newton devoted several propositions in the *Principia*) four inequalities of the Moon's motion within its orbit were known before Newton published his own theory of the Moon. Hipparchos had realized that (to a first approximation) the lunar orbit, like that of the Sun, could be represented by a circle eccentric to the Earth. For the eccentric circle an epicycle on a concentric deferent could be substituted or (much later) an ellipse. Ptolemy discovered that Hipparchos' eccentric was adequate when its apogee was in quadrature with the Sun but in error when the apogee was at intermediate positions. To correct this second inequality, the *evection*, Ptolemy introduced another epicycle; the same result can be attained by supposing the orbit to vary semiannually in its eccentricity so that it is twice a year rounder, and twice a year longer, the area remaining constant. Jeremiah Horrox, who brought the Keplerian ellipse into lunar theory to deal with the first inequality (*c.* 1640), made the empty focus of the ellipse rotate semiannually upon an epicycle in order to deal with the evection. The magnitude of the first inequality (the *equation of the centre*) is about 6° 17′; that of the evection 1° 16′.

The next two inequalities to be defined were both effectively discovered by Tycho Brahe.(17) Both are due, like the evection, to gravitational accelerations caused by the varying distance between the Sun and the Earth–Moon system. The third inequality, known as the *variation*, amounts to about 40′ in the octants of the Moon's orbit; Tycho added yet another epicycle to the lunar system to account for it. The fourth inequality, the *annual equation*, is less still—about 12′—

and was handled by all astronomers before Newton as a component in the equation of time (thereby creating a distinction between the equation of time as applied to the Sun, and that applied to the Moon). In other words astronomers were aware that this correction had to be tabulated but could divine no mechanism for it.

In the first edition of the *Principia* Newton investigated thoroughly the motion of the Moon's nodes and her motion in latitude, but of the orbital inequalities considered only the *equation of the centre* (Proposition 28) and the *variation* (Proposition 29). Further, in a Scholium to Proposition 35 (suppressed in the later editions) he reported computations of the varying motion of the lunar apogee, whose results he regarded as more valuable than those tabulated by Flamsteed from observations. Hence he had not produced a complete theory of the Moon.[18] In 1702 he allowed David Gregory to publish in his *Astronomiæ Physicæ et Geometricæ Elementa* the theory of the Moon that he had developed since 1687 with the assistance of Flamsteed's observations. This consisted of a brief statement of the mean motions involved, followed by an in-principle account of the seven inequalities to be allowed for in computing from them; an in-principle account because although Newton gave maximum values for these inequalities the tables required for their use were not present in the book. In this *Theory* Newton neither explained how the magnitudes assigned to each of the inequalities were derived, nor accounted for them by the theory of gravitation. Of the seven inequalities, three were new discoveries of his own.

In preparing the second edition of the *Principia*, Newton revised the text of the first to accommodate the new numbers which he had derived with Flamsteed's aid and printed in the *Theory*, but was otherwise content to record his new work briefly in the new Scholium to Proposition 35. In the treatment of the lunar theory, the differences between the two editions are, up to this point, largely numerical.

Newton began with a general discussion of the Sun's perturbing effect on the Moon (Propositions 22 and 25),[19] discussed the shape of the Moon's orbit, and then established the magnitude of the *variation* (Proposition 29). The next group of propositions dealt with the motion of the Moon's nodes—essential to the determination of eclipses—and then he turned to the Moon's motion in latitude (Propositions 34, 35). The Scholium treats of the *evection* (by Horrox's method) and the new inequalities.

No one could plausibly argue that Newton's treatment of the lunar motions is easy to follow; for example, he made no attempt to adjust it to the way in which his predecessors had handled the four major inequalities known to them, barely even employing the technical names for them that these astronomers had introduced. Although Newton stated the magnitudes of the new as well as the old inequalities, he did not attempt to compute the tables from which alone a future position of the Moon could be predicted. Nevertheless, it is obvious that Newton completely transformed the problem. In the words of Lalande: 'all the small inequalities...would have remained unsuspected but for the idea of attraction, and they have been determined only by comparing this theory with a great number of good observations; it was reserved for Newton to take the greatest step forward in the theory of the Moon as in all the rest. Guided by the principle of universal gravitation and aided by the observations of Flamsteed, he determined the magnitude of several new equations with their epochs and mean motions' (*Astronomia* II, Paris, 1771, pp. 221–2). Tycho, in discovering the annual equation, had revealed the least inequality detectable by naked-eye observation; the subsequent refinement of observation down to the time of Flamsteed had revealed no more. As a result, the best theory of the Moon based on four inequalities alone—none properly explicable, and one at least misunderstood—was seriously inadequate. Newton was able to effect further improvement by deducing from a general review of the dynamical interaction of Sun, Earth and Moon first the gravitational explanation of the four known inequalities, then the situations in which other, lesser accelerations of the Moon's motion might be expected to occur. To handle the perturbations thus deduced, however, Newton was compelled to resort to crude geometrical models and, in order to make these work, select appropriate numbers in a highly arbitrary fashion.

Accordingly, Newton in print gave no general theory of the Moon, either in the traditional sense of a complex geometrical model or in the more modern sense of a system of algebraic equations whose solutions yield the desired parameters. Consequently each inequality is obtained separately—only in a weak sense are all seven of them derivable from the theory of universal gravitation. Proposition 22 (strangely interpolated between discussions of the shape of the Earth and of the tides) is the only one in the whole of Book III in which Newton attempted to show how the origin of the lunar inequalities (considered as a class of phenomena in themselves) may be found in a single cause—the action of the Sun's gravitation on the behaviour of the Earth–Moon system; and this proposition is general to the point of vagueness (it was in fact copied almost word-for-word from the earlier non-mathematical draft of Book III now called *The System of the World*). It seems perhaps a little extraordinary that when for the second edition of the *Principia*, Newton came to deal with the *evection* and the periodic motions of the lunar apogee (in the new Scholium to Proposition 35)[20] he could only point out qualitatively that the correlations of these motions with the changing eccentricity of the lunar orbit and its inclination to and nodal intersections with the ecliptic, were in accord with the dynamics of gravitation; to express them he turned to the arbitrary geometrical models of Horrox and Halley, with the parameters derived from observation. However, Newton was able to derive a value for the *variation* from a dynamically derived evaluation of the solar acceleration of the Moon in its orbit (Propositions 26 and 27) and so obtain in Proposition 29 a theoretical value for this inequality that could be compared with observation, the final arbiter. On the other hand the treatment of the lesser inequalities, though completely original, turned the draft revised version of the Scholium to Proposition 35 into a strange medley with which Cotes was far from satisfied. Newton repeatedly employed the phrase 'By the theory of gravity...', and indeed asserted (without details) that by theory the large annual equation of the Moon (Tycho's) was 11′ 49″ (or 11′ 52″) whereas by observation it was 11′ 50″; but though the second lesser inequality, the semiannual, depended on the position of the Moon's apogee in relation to the Sun its value of 3′ 45″ was derived from observation. Similarly, the equation of the apogee depended on the orientation of the lunar to the terrestrial orbit, but its values (about 15′ or 20′) 'must be determined from the eclipses' and to express this inequality Newton again resorted to an epicycle; so that now the empty focus of the lunar orbit was carried on an epicycle rotating on an epicycle rotating on an epicycle in a truly Ptolemaic fashion. It is apparent from these discussions that though the intuitive value of the theory of gravitation was extremely high for Newton, as he contemplated with his extraordinary comprehension of interrelated details the geometrical permutations arising from the motion of the Moon in its ellipse while the major axis of the ellipse revolved with respect to the major axis of the Earth's orbit, in the last resort final details of the inequalities that the theory of gravitation enabled him to perceive could not be expressed without the aid of a geometrical model whose parameters were painfully defined by the ancient method of noting discrepant observations.

Once the Moon was out of the way Newton clearly hoped that the *Principia* would proceed smoothly and swiftly to its end, through the new material to be added to the theory of comets and a new *Scholium Generale* to conclude the book. It is true that Newton had not yet (in early September 1712) decisively rejected the idea of expanding the *Principia* by adding to it his thoughts 'about the attraction of the small particles of bodies' and also a new, short treatise on fluxions (see Letters 914, 929 and 977);[21] these, in preliminary form at least, he had by him already and they could, perhaps, have been printed while the final operations of making the body of the *Principia* ready for market were gone through. Then—almost at the end of this long toil—a crushing blow fell on Newton. The unremarked error in Proposition 10, Book II, was brought to his notice by Nikolaus Bernoulli, nephew of Johann who had discovered its general existence by faulting a particular instance, but not the exact nature of the mistake in Newton's demonstration (see Letters 951 and 951*a*). Admittedly Johann Bernoulli was not yet so decidedly and painfully linked with the cause of Leibniz as he was to be a year later; in fact he had for some years been in friendly correspondence with Abraham de Moivre, who furnished a link between British and continental mathematicians. Nikolaus was kindly received on his visit to London (September–October 1712), and Johann was complimented; on 1 December 1712 he was elected F.R.S. on the proposal of Newton himself. Nevertheless, apart from Newton's chagrin at learning of a fairly obvious absurdity following from the first form of Proposition 10, the revelation was awkward in that he feared the 'continental mathematicians' might charge up such an error in the *Principia* as evidence that he could not have been an independent inventor of the calculus: both Johann and Nikolaus Bernoulli were to make just this sort of capital out of the faulty Proposition 10 in a few months' time. (Further, Johann had imparted the discovery to Leibniz two years before he informed Newton though he, like everyone else, had long known that a second edition of the *Principia* was in preparation.) It was inevitable that all Europe would soon know that the author of the *Principia* had been saved from a great blunder for which act Newton, properly speaking, ought to make some public expression of gratitude (but he did not).

The trouble extended immediately to Cotes, who received news of the mistake (and consequential reprinting) calmly enough, merely remarking that he would not seek out the cause of the error himself, but would await Newton's revised version (Letter 953). Even when he received the new version of Proposition 10 Cotes merely observed that the only change was to increase the resistance in the proportion of three to two. Seemingly Cotes—who by now was well aware of Newton's fallibility—was insensitive to the delicacy of the situation; he was probably (up to this point) not so well acquainted as some other men were with Newton's touchiness; nor was he directly involved in the great quarrel with Leibniz brewing since April 1711 and the consequent preparation of the *Commercium Epistolicum*, which must have been nearing completion when Nikolaus Bernoulli arrived in London.

The relationship between the great man and the young don had begun auspiciously. Newton's anxiety not to impose matched Cotes' deference. Cotes was firm, indeed at times insistent; he did not yield easily to his author's authority, but he apologized for the freedom of his criticisms and left his difficulties to Newton's final decision. In return, Newton sends Cotes 'many thanks...for your trouble in correcting this edition' (Letter 794) or remarks 'The corrections you have made are very well and I thank you for them' (Letter 807) or writes in apology for his long delaying the progress of the book:

'I send you at length the paper for which I have made you stay this half year. I beg your pardon for so long a delay' (Letter 826). Only a few days later (31 March 1711) Cotes has occasion to thank Newton for a public compliment to himself that Newton had put into his new copy (Letter 829). There can be no doubt that at this stage Newton was well satisfied with his editor and meant in more than one place to let the public know that Cotes' activity had saved him from mistakes. After this the Leibniz quarrel broke out, and Newton's temper to Cotes worsened. Newton's letter of 18 June 1711 (Letter 851) opens in hitherto familiar style: 'I have read over and considered your alterations and like them very well and return you thanks'—for the last time. Newton was hardly ever thereafter to thank Cotes again; his replies to him became curt, almost impatient. He did not react well to Cotes' criticism of Book II, Proposition 48 and his next letter (Letter 860) sounds an ominous note: 'ever since I received yours of June 23 I have been so taken up with other affairs that I have had no time to think of Mathematics'. Reasonable as the excuse might be—for Mint affairs were pressing—this was a clear storm-warning and when Cotes, ignoring it, renewed his criticisms and then pressed politely for an answer (Letter 871) Newton ignored him for six months, although he does not seem to have been very busy that autumn. When he did condescend to return to the *Principia* Newton offered no apology but merely hoped that the press would halt no more (Letter 889). There is no sign of friendliness, no exchange of gossip, in subsequent letters, until in Letter 903 Newton does thank Cotes again (a little sardonically?) for explaining his objection against Book III, Proposition 6, Corollary 3. If there is some greater warmth of tone in subsequent letters, Newton was still very casual about *Logometria* which Cotes submitted to him for criticism; when Cotes begged for a word Newton merely replied that he had found nothing that needed correction.

Despite his coldness, it seems certain that into the summer of 1712 Newton still meant to pay Cotes the tribute he had every right to expect—the sort of tribute he did pay to his next editor, Henry Pemberton, many years later. It is impossible to be absolutely confident of the date of his draft Preface to Cotes' edition (Number 829*a*) but one would hardly suppose Newton to have embarked on it before the end was in sight—that is, before the summer of 1712—and this period seems to fit well with the document.(22) In this draft Newton's tribute to Cotes is generous enough, and points out that he has both *removed errors* and caused Newton profitably to *reconsider many points*. But it was never printed.

What happened can only be guessed. Perhaps Newton had already deleted the compliments to Cotes which, as we have seen, were once in the text but have now vanished from it, thinking that one compliment in the preface would suffice. One cannot say. What may very plausibly be supposed is that Newton decided against his complimentary draft preface in the late autumn of 1712, after Nikolaus Bernoulli's visit. For he could not now thank Cotes in the manner he had proposed without also thanking Johann Bernoulli for exactly the same reasons—for it was he who, precisely where Cotes had let him down, had caused Newton to reconsider, and so remove a blemish.

Newton's irritation is manifest in Letter 946, before Cotes knows anything of the trouble. Let Cotes stop fiddling, Newton seems to say; the alterations in the lunar scholium were made advisedly (that is, he knows better than Cotes). Certainly the lunar theory requires more observations to be made by those who have leisure, but meanwhile the public must take it as it is. Only in the next letter (by which time, one may be sure, Newton had the new demonstration of Proposition 10 safely in his pocket) did he curtly tell Cotes what was wrong; within a week he wrote again to complain of Cotes' silence. The subsequent correspondence is brief, terse, frigid. After the new *Principia* was out Newton was to send Cotes, probably in December 1713, a long list of trivial mistakes which astonished Cotes considerably. No kind word of Newton's towards him is recorded until after Cotes was dead.(23)

Whether (as Dr Whiteside has suggested to us) Newton attached to Cotes the blame for failing to detect the error found by Bernoulli, or whether his change of attitude towards Cotes in late 1712 was an exaggeration of earlier coolness produced by Newton's irritation and his reluctance to acknowledge gratitude to Johann Bernoulli is not a matter of great significance. At any rate the result may be read in Letter 989—clearly Newton could go so far as to confess that there *had* been mistakes in the first edition but he would thank no man for his help in amending them. If Newton *did* blame Cotes for not discovering in a few weeks an error that he himself (and not wholly incompetent friends) had failed to detect in several years, he was surely unjust, and—with six months to ponder the question—he ought to have appreciated that an immediate feeling of annoyance at his editor's incompetence was out of place. It may be the case that Roger Cotes, who had been pushed along as something of an infant prodigy but cannot be credited with any tremendous achievement in his thirty-five years, was no more than a superb second-class mathematician, not approaching the quality of Johann Bernoulli, and that he excelled in detail rather than in grasping the bigger issues (where his attempts in the correspondence with Newton were not highly successful); it may be that in going through those difficult propositions on motion in a resisting medium he followed Newton somewhat blindly. Still, his blindness was no greater than Newton's own. And he was certainly denied a due tribute—denied in fact all recognition as editor of the *Principia* and not simply author of a preface to it—

in order that Newton could make himself appear in as good a light as possible.

The Preface, Cotes' only visible memorial in the new *Principia*, was once regarded as a defiant statement of Newtonianism.(24) Now that Newton's own thoughts on Descartes, gravity, Leibniz and so forth have been more plainly revealed and more profoundly studied(25) and that, moreover, Newton's own share in formulating other documents (which therefore can be taken as representing his authentic ideas) is better understood, Cotes' preface seems the less telling precisely because Newton's own expressions (whether direct or indirect) are really more interesting to us than Cotes'. For we have every reason to believe that when Newton declared his lack of foreknowledge of this Preface he was speaking the truth (see Letter 989 and his *Ad Lectorem* to the *Commercium Epistolicum*.) Inevitably Cotes sought directions or at least hints for the essay he now found that he was expected to compose (Letters 983–5). But his attempt to interest Newton in a long summary of what he intended to say failed; for once, Newton refused secretly to ghost-write for his champion. The only obvious reason for his refusal, apart from an unusual desire to be genuinely uninvolved, must be a lack of sympathy that he now felt towards Cotes.

Cotes took two months over his first draft, dating the Preface 12 May 1713. It bears a generic if not a close resemblance to the plan outlined in Letter 985. Cotes does say something 'concerning the manner of philosophizing made use of,' contrasting Newton's experimental philosophy—'incomparably the best way of philosophizing'—with the hypothetical model-building of the early corpuscular philosophers, Descartes above all. Then he proceeds to 'a short deduction of the principle of gravity from the phenomena of nature in a popular way', though not quite in the manner proposed in the letter; he seemingly got over the rather absurd difficulty he had raised about the 'invisible Hand'.(26) Next, as he proposed in the letter, he attempts (not very successfully) to defend Newton against the Leibnizian charge that the Newtonian conception of gravity renders it 'occult'. At the opening of the Preface Cotes wrote that Aristotle's philosophy was *occult* because it ascribed phenomena to the peculiar natures of bodies without saying whence those natures derive; he now says that Newtonian gravity should not be called *occult* merely because its cause is unknown: it is a simple (that is, primitive) cause which cannot be further explained. The difference, in Cotes' eyes, then, between Aristotle's treatment of physical causation and Newton's seems to be that Aristotle appealed directly to an unexplained nature (which was occult) whereas Newton descended from phenomena to a primitive cause by means of a causal chain (and so the primitive cause was not occult). However, in avoiding the Scylla of occultism Cotes embraced the Charybdis of materialism. 'Either gravity will have a place among the primitive qualities of all bodies universally,' he wrote, 'or extension, mobility and impenetrability will not.'[27] Now it had long been understood that at least the two last of these attributes were essential to matter (that is, one could not conceive of matter without them) and whether or not one conceived also of extension as the prime attribute of matter in the Cartesian manner, a Newtonian could not suppose matter to be non-extended (though this was the notion later developed by Boscovich).[28] Hence Cotes is proposing that we really ought not to try to conceive of matter without gravity, any more than we can conceive of matter that is penetrable or immobile. Indeed, he must have proposed something worse than this, since in Letter 1001 to Samuel Clarke he assured Clarke that some phrase greatly offensive to Clarke on the score of its materialism had been struck out. Yet what Cotes did print was of course contrary not only to the doctrine of Aristotle, but to that of Descartes, of Leibniz, and of Newton himself. Nor does Cotes' letter to Clarke clear him of the charge of adopting materialism in order to make gravity comprehensible; for the best Cotes can retort is that he did not mean to make gravity essential to matter, but only to claim that we do not know what the essential properties of matter are. By which phrase, of course, he bounces himself back neatly into occultism.

It is impossible to believe that these passages can have pleased Newton, who was beyond all doubt a better philosopher than Cotes. Avoiding pure mathematics altogether and the decision of the *Commercium Epistolicum* committee which he had in his outline proposed to recite, Cotes went on to a long attack upon the theory of vortices, which was by now no less a Leibnizian than a Cartesian theory, and so (like Newton himself) to expatiate upon the universe as the creation of 'the most excellent counsel and supreme dominion of the all-wise and almighty Being'. Then an enthusiastic tribute to Newton's achievement leads him to conclude with words on Richard Bentley 'outstanding alike in modesty[!] and learning' who had produced the edition at his own expense [!] and had 'imposed on me, as was within his rights, the not unwelcome task of seeing that it was as correct as I could make it'.

Perhaps Newton smiled a little as he read these words.

The dispute with Leibniz, clouding Newton's mind from 1711 onwards, broke upon him just as it seemed that his troubles within the Royal Society itself were over. It may be that Newton's biographers have exaggerated the drama of the year 1710, for Sir Henry Lyons, in his administrative history of the Society, has pointed out that this was a time of solid if minor improvements in organization, during which the number of Fellows was tending slowly upwards and the finances of the Society were improving. To Newton events were perhaps less dramatic than they may seem to a posterity forming its image of Newton's presidency from highly coloured sources. In an age when satire wa a profitable branch of literature, when the public was denied solid news and even intelligent speculation, and when the antics of 'society' constituted a large fraction of popular entertainment, insignificant quarrels were fanned by the breath of rumour and blown up into pamphlets. One sees in Flamsteed's letters how eagerly he garnered and retailed the shallowest tittle-tattle about the 'establishment' of the Royal Society. One sees in the oft-quoted pamphlet about the Society's move to Crane Court in 1710 (from which we have, with some hesitation, included selected passages as Number 812) its complement of falsehood, innuendo, and malice. Such documents as these are clearly not to be treated with serious naivety as historical sources, and it need not be surprising that someone as tough in character as Newton or Sloane could learn to live amid such rumour and gossip.

In the Woodward affair (Letters 774, 784) Sloane was, of course, the object of attack, rather than Newton himself. Presumably professional rivalry had more than a little to do with the antipathy between Woodward and the Secretary. And perhaps, in the over-excited atmosphere surrounding all the concerns of the Royal Society at this time (in depicting this the documents surely are not far astray), Hans Sloane really *did* make faces at the angry geologist behind Newton's back. Just as Newton constantly pursued a hard line against currency offenders without any sign of softness (Letter 806) so it did not trouble his conscience to act firmly in holding the Royal Society to what he believed to be its proper course. But was Newton right in upholding Sloane for so long, and indulging what was, presumably, his own personal partiality towards the Secretary? Sloane held the office for twenty years, and although at the moment of his retirement (30 November 1713) he was no more than fifty-three years old, it may well be doubted whether he had given of his best to the Society for a long time before. His profession and his collections made heavy demands on his time and energy. To his biographers Sloane is a paragon, but Lyons, admiring Sloane's abilities in administration (mostly manifest during his own presidency), confesses also his want of tact. One may question too whether Sloane was the best man to manage the Society's scientific activities: a good deal of the material he brought before the meetings was dreadfully dull, and but for the energy of Hauksbee, Papin and Desaguliers the meetings would have been sterile. There was justification in the pressure upon the 'establishment' for a change of secretary. It is interesting to note that when all was over, John Woodward and Newton seem to have been on at least civil terms, for the geologist presented Newton with a copy of his *Naturalis Historia Telluris* (1714; *penes* Dr V. A. Eyles).

On the other debated issue of 1710, the Royal Society's move to Crane Court from Gresham College, the 'establishment' policy was surely correct (see Letter 802). Bishopsgate was outside the fashionable orbit; the crumbling buildings had their disadvantages for meetings, the Gresham Committee was anxious to oust its distinguished but unlucrative tenant, and the only persons who really were likely to suffer from the move were the Gresham Professors (of whom, of course, Woodward was one). The time had come when a divorce should have been profitable to both parties; if Gresham College did not later profit as much as it might have done from its regained independence, that was not the Society's fault. In any case, the separation had never been actively sought by the Newtonian 'establishment' (see Lyons, *History*, pp. 131–7 and Number 802*a*), for whom the Crane Court house was something of a last refuge under a threat of being put into the street.(29) Again, the storm was soon over; the generosity of a number of Fellows (besides Newton himself) in financing the Society's purchase of its house indicates its success.

The fact that Newton's relations with Flamsteed grew steadily more unfriendly is more serious and more painful to contemplate. As was seen in the previous volume of this *Correspondence*, no long period elapsed after Newton became president of the Royal Society before his personal dissatisfaction had been converted into an official dispute between the Royal Society and Flamsteed; moreover, after the issue of the Royal Warrant of 12 December 1710 (Number 814) it was no longer the case that Flamsteed was charged with failing Newton, he was blamed for failing the nation. This is no place for debating the morality of what happened, but it must at least be recorded that there was no precedent for the new claim that the Astronomer Royal at Greenwich was not responsible directly to the Crown which had brought his office into being, but to the Crown through the Royal Society. There was indeed no precedent for an interpretation of the Royal Society's Charter (at this time) which would make it the Crown's chief instrument in managing the scientific affairs of the British people. On the contrary, it had once been specifically declared that navigation 'being a state concern was not proper to be managed by the Society'.(30) Moreover, the Society had a long-standing tradition of making no pronouncements *as a corporate body* on the merits of any particular theory, or argument, or mathematical demonstration, or upon the accuracy of any individual experiment or observation. Flamsteed—a man of sixty-four years in 1710—might well feel astonished that Newton now proposed, not only to publish Flamsteed's observations in his (Newton's) own way, but to direct him to make this observation or that, to equip his Observatory with new instruments provided by the Crown or demand purchase of those furnished by Flamsteed and his friends long before, sending down instrument-makers to remedy their supposed defects.

With the financial support of Prince George of Denmark, the printing of

Flamsteed's astronomical observations had begun in 1706. By October 1707 the first volume (or rather 98 sheets of it, to which Flamsteed wanted to add seven more—see vol. IV, p. 525) had been printed off, and the work ceased. In March 1708 Flamsteed was summoned to meet the Referees who had charge of the publication, presumably to find out what more he had ready for the press; he produced 175 manuscript pages of further observations made between 1698 and 1705—according to his own account, these were left with the Referees—but Newton refused to proceed with printing this material in the second volume until the star-catalogue had been printed. A preliminary, incomplete form of this catalogue had been in Newton's hands since 1705; Flamsteed absolutely refused to let it go to press 'before it be as complete as I can render it at present'. Neither man would give way, nor (it must be added) did Flamsteed make it clear that he was perfecting the catalogue of stars with reasonable speed.(31)

After the death of Prince George and much lapse of time, Newton seems to have concluded that the *impasse* could only be resolved by his own strong, independent action. And possibly if he had not done this, or else completely trusted Flamsteed's goodwill, the observations would have remained unpublished so long as both men lived, for Flamsteed was immovable. Accordingly Newton procured his Royal Warrant (Number 814)—no doubt after drafting it himself. Thereafter followed the exchange of letters between Flamsteed and Arbuthnot, whose rôle in the affair was absurd, and the editing of Flamsteed's material by Halley. It was, of course, certain that the Astronomer Royal would never agree to cooperate with the Royal Society 'establishment' once it was clear that the resumed printing was to proceed on their terms, not his. Hence the *Historia Cœlestis* of 1712.(32)

Again, it is pointless to attempt a moral verdict as between Newton and Flamsteed. The astronomer was old and ailing; his best assistant and old friend, Abraham Sharp, had long retired to Yorkshire. He had reached the time of life when (since few men ever retired voluntarily in the eighteenth century) an office-holder expected to be left in peace, his duties performed by deputy (this was already true, to some extent, of Newton himself at the Royal Mint). He certainly did not expect to be hectored and hurried, or to have his life's work edited by a rival astronomer whose competence he (wrongly) despised. Much more might be said on Flamsteed's behalf; on the other hand, it must be added that he was not a man to meet trouble half way—he rushed to embrace it. His suspicion, if anything, exceeded Newton's own. His consciousness of rectitude and innate superiority was devastating. And he had no intention of entering into any compromise with an 'establishment' that had done nothing to assist his work, and therefore should not pretend the right to lay its result before the world. Of course there was a strong sense in which Newton, who had by now a fair experience of government service and knew that politicians were in the last resort his masters, saw the situation in a more 'modern' way than Flamsteed did. The Reverend John Flamsteed thought of himself as a private gentleman, as (maltster's son though he was) he had considered himself throughout his life; when he added the letters 'MR' after his name it was to own a distinction, not a badge of servitude. He recognised no one as his master in the execution and expression of his professional scientific responsibilities. To say (as one might) that he was the first scientific civil servant under the British crown would have surely struck him as incongruous and improper.

Very little evidence of Newton's personal involvement in the series of moves that led to the publication of the 'official' *Historia Cœlestis* in 1712 now remains in Newton's correspondence. Since the edition was largely destroyed by Flamsteed himself in his rage against Newton and Halley, even the bibliography of the book seems uncertain. One may be sure, however, that neither Arbuthnot nor Halley took a more than routine step without consulting Newton: hence Flamsteed was surely right (however limited his psychological portraiture) in seeing Newton constantly at work, moving invisibly, steadfastly towards his objective. And if Newton's objective was the publication of Flamsteed's observations before a possibly sudden (but hardly, now, premature) death carried him away, it was surely laudable. Once again, the humiliation of Flamsteed was not a misfortune that would deter Newton from pursuing a course that he believed to be incumbent upon him as a loyal servant of the Crown and President of the Royal Society.

Newton's strong sense of responsibility, rendering him overbearing at times, and his compulsion to take all administrative work into his own hands, appear no less obviously in his conduct towards his own relatives. It is not unusual, when a man rises so high in the world as Newton did, for his relatives to be left far behind, poor and undistinguished; by the ethical standards of the time it was incumbent on the successful man to assist his less fortunate connections, and Newton did not avoid this obligation (Letters 906 and 955). As is well known, he was particularly solicitous for the children of his half-sister Hannah Barton. He made financial provision for all of them (Letter 878*a*), promoted the military career of Hannah's son Robert, and when Lt. Col. Barton was drowned in 1711 took successful measures to secure a pension for his widow, Katherine Barton (Letter 949). It is obvious that the fortunes of the whole family hung upon Newton, and that the family sank into oblivion after his death when, within a few years, Woolsthorpe itself passed into other hands.

Among the younger generation there was one bright star, Robert's sister

Catherine, the niece who lived with Newton for many (but indeterminate) years, and who in 1717, at the then mature age of thirty-eight, married John Conduitt. Throughout the period of this volume this Catherine, we must presume, was in some sense the protégée of Charles Montague, Baron Halifax. Such she was for considerably longer than nine years before Halifax's death on 19 May 1715, for already in a first codicil to his will (dated 12 April 1706)[33] Halifax bequeathed to her 'all the Jewels I have at the Time of my Death; and likewise three thousand pounds, as a small token of the great Love and Affection I have long had for her'.[34] Catherine Barton's relation to Halifax is an enigma which no new discovery is likely to illuminate. Facts and dates seem to make it clear that Halifax's friend, Newton's household companion, Conduitt's wife, and Swift's hostess were all one and the same person, Robert Barton's sister and not his wife. The only contemporary narrative distinctly connecting Halifax and Mrs Barton is that of Halifax's anonymous biographer:

I am likewise to account for another Omission in the Course of this History, which is that of the Death of the Lord Halifax's Lady;[35] upon whose Decease, his Lordship took a Resolution of living single thence forward, and cast his Eye upon the Widow of one Colonel Barton, and Neice to the famous Sir Isaac Newton, to be Super-intendant of his domestick Affairs. But as this Lady was young, beautiful, and gay, so those that were given to censure, pass'd a Judgement upon her which she no Ways merited, since she was a Woman of strict Honour and Virtue; and tho' she might be agreeable to his Lordship in every particular, that noble Peer's Complaisance to her, proceeded wholly from the great Esteem he had for her Wit and most exquisite Understanding, as will appear from what relates to her in his Will at the Close of these Memoirs.[36]

The writer was clearly mistaken in supposing that Halifax's Catherine was Colonel Barton's widow (the fact that Col. Barton was still living when Halifax drew up his first codicil renders this impossible), yet his mistake was a natural one owing to the identity of names, which—together with much else in his story—proves the biographer's familiarity with, if not accurate mastery of, the true state of affairs. Swift's *Journal to Stella* shows Newton's niece through the years of this volume as an active member of the smart, slightly raffish society in which Swift was at home, but Newton was not. Nor, one may suppose, was Halifax, very much the stiff grandee; the relation between him and Mrs Barton is never touched on by Swift, and was possibly unknown to him. (At this time, of course, Halifax and Swift were hotly opposed in politics; Mrs Barton remained, like Halifax and her uncle, a Whig.) Catherine could relate to Swift a story about the paucity of virgins that her uncle would not have cared for (3 April 1711). She was accessible, perhaps even notorious. It is likely that she was still or had been once Halifax's mistress; virtually impossible that she was (as De Morgan supposed) Halifax's legal but unacknowledged wife. Whatever the truth, it is certainly clear that at least up to the time of Halifax's death in 1715 he was not the known lover of Newton's niece—the tale that Voltaire caught was only spread later. This is proved by the entirely different testimonies of Swift and Newton himself; and one can only conclude that in the eyes of her family and of Society, Mistress Barton was perfectly respectable.

She may, indeed, have been living with her uncle through these years. Naturally her relatives supposed that she was doing so, and Newton never indicates otherwise. Swift's tales of visits to Mrs Barton at her lodgings do not contradict this supposition because, when Swift was living in Leicester Fields he wrote that Mrs Barton was his near neighbour (9 and 25 October 1711) which, of course, she would have been in her uncle's house. Presumably Catherine would have had separate apartments and her own servants, and therefore Swift's allusions need not be taken as implying her completely independent existence; Swift's failure to mention her relationship to Newton (surely known to Stella already) is balanced by his equal failure to mention Halifax.

Yet Catherine Barton certainly remained in Halifax's thoughts, for on 1 February 1712 he added another lengthy codicil to his will, at the same time nullifying his former bequest to Miss Barton. In the new codicil he presented Newton with one hundred pounds (the same sum had been bequeathed to others for mourning) 'as a Mark of the great Honour and Esteem I have for so Great a Man'; then he proceeded to bequeath to Catherine both a cash sum of £5000 and 'the Rangership and Lodge of Bushy-Park, together with all the Household-Goods and Furniture belonging to the House, Gardens, and Park' as also, for cash income, 'my Manour of Apscourt, in the County of Surrey together with all the Rents, Profits, and Advantages thereunto belonging'. The codicil then goes on to explain that these gifts and legacies are tokens of 'the sincere Love, Affection and Esteem I have long had for her Person, and as a small Recompence for the Pleasure and Happiness I have had in her Conversation'. Halifax then enjoined his executor and residual legatee, his nephew George Montague, 'to give all Aid, Help and Assistance to her' in obtaining these bequests and also in the transfer to her of an annuity of £200 per annum 'purchased in Sir Isaac Newton's Name, which I hold for her in trust'.

To conclude this story (though its end lies outside the proper limits of this volume) it is certain that after Halifax's death (on 19 May 1715) Catherine Barton did *not* officially enjoy all that Halifax intended for her. In April 1716 his heir, George, now in turn Earl of Halifax, received under the Great Seal all offices granted to his uncle on 3 June 1709, including the Rangership of Bushey Park.(37) It seems unlikely that Catherine inherited Apscourt Manor either.(38 What seems certain is, that just as Isaac Newton was involved in the Halifax–

Barton affairs before Halifax's death, so he took much trouble in defending his niece's interest after it. Drafts in the University Library and the Library of King's College at Cambridge, as well as others at Lehigh University, prove his deep involvement. Newton drafted in legalistic phraseology a document reciting the terms of the late Earl's will, by which the new Earl agreed to assign and transfer to Catherine Barton a government annuity for £200 due under an order 'number three thousand eight hundred and six with all respective Tallies thereunto belonging'.[39] Another document (at Lehigh) shows Newton elaborately noting amendments to some legal agreement between George, Earl of Halifax, and Catherine Barton, thus:

Pag. 5. lin. 20. Blot out the words interlined namely [and her said estate in the said severall offices & Lodge house hereby granted shall so long continue

. . . .

Pag. 7 lin. 10. read [enjoyed in good & substantial repair & good Order & will yearly & every year &c] & omit the sentence [& in the same form & Order...hereby granted.]

. . . .

Pag 8 l. 3 Omit the word *Conditions* & lin 10 & sequ. omit the clause within the brackets, & instead thereof insert a convenant for dividing the Parks [at Hampton Court] in these words. And the said George Earle of Halifax for himself & his heirs doth covenant grant & agree to & with the said Catherine Barton that he the said Earle of Halifax shall & will divide the said south & north Parks with a pale or fence between them in such manner as they were heretofore divided, she the said Katherine Barton bearing one half of the charges thereof.[40]

All this is perfectly familiar to readers of Newton's papers; it is only the object of his attention that seems strange. The document certainly belongs to late 1715 or early 1716, as is borne out by the draft on its other side;[41] enquiry into the final outcome of these negotiations we must leave to others. No doubt the business of Halifax's estate had been effectively if not legally disposed of before Catherine married John Conduitt some two years after Halifax's death.

At all events, the meaning of these documents for Newton's own life is unmistakable: throughout all the last years of Charles, Earl of Halifax, his own former patron, Newton was deeply enmeshed in Halifax's relationship with Catherine Barton. Whatever that relationship was, little of its nature can have escaped Newton; and yet he did not at any stage disapprove of Halifax's wish to make a generous financial provision for his own niece, but on the contrary strove rather to give his best advice in order to ensure that she should enjoy the fruits of Halifax's intentions.

It was Halifax who in 1696 had brought Newton to the Mint.[42] Little of a general nature need be written here about Newton's Mint business during these years since Sir John Craig has published an account of it;[43] Craig's account is shown by the documents printed here to be incorrect or incomplete in some details, but its outline is sound enough. After the completion of the Scottish recoinage in the spring of 1709 there was a period of quiet at the Mint, though the final settlement of the Edinburgh accounts was to drag on for a further five years and involve Newton in occasional correspondence. But there came a change with the downfall of the Whig administration and especially after the appointment of Robert Harley (soon to be raised to an earldom) as Lord High Treasurer on 19/29 March 1710/11. Oxford is usually supposed to have been an exceedingly idle Minister; such a reputation is not borne out by the volume of his correspondence with the Queen's Mint. The new government began by creating much extra work through its scheme for increasing the flow of silver plate into the Mint for coinage; it constantly demanded advice about the values of foreign coins (in which the large British forces overseas were paid), and finally in 1712 it allowed the news to get about that a new issue of copper coins was in prospect. (None had been made since the reign of William III and Harley presumably saw the possibility of winning popularity by, effectively, a mild inflation.) At once every copper-smelter and dealer in the South of England tried to seize the opportunity to secure a profitable contract. Newton was sceptical of the need for new halfpence and farthings, sceptical of the would-be contractors' motives, and sceptical of the metal they proposed to employ. His Fabian tactics and obfuscation of the technical issues concerning the purity of copper prevented any decisive measures being taken before the Tories' fall, so that in fact new coppers were only minted in 1717.

Meanwhile, the normal tasks of the Mint went on: receipt of tin from the smelters in Cornwall and its resale to merchants, an activity that brought all the Mint staff a welcome bonus (Letter 808); the prosecution of counterfeiters and clippers of the money (Letter 806); the regular production of silver and gold coin and its periodic check at the Trials of the Pyx (in August 1710 and 1713).

The former trial caused Newton great trouble (Letter 816 and Number 816*a*) because the apparent result of the comparative assay of his gold coins with the standard trial-plate newly made after the Union with Scotland in 1707 was to reveal the coins as slightly below standard, though in assay by weight they proved good enough. Newton's answer was to claim that the 1707 plate was above the standard, and he seems to have made his point since the 1707 plate was never used again. It has often been suggested—partly on the basis of his chemical experimentation both genuine and hypothetical—that Newton was highly skilled in the assay of precious metals. There is not the least evidence in this volume for the unlikely supposition that he conducted official assays himself, but there is much in favour of the more probable view that he did not. His writings on the subject (for example, Number 816*a*) do not exhibit any extraordinary learning in this field, beyond what he could readily have gathered from books and watching the Mint assayers at work; there is little reason to imagine that Newton had acquired manual dexterity in this delicate art.

Of Newton's excruciatingly conscientious attention to every detail of government business there is all too much evidence. He wrote hundreds of draft reports or letters, often many times over; he covered scores of pages with tables of currency values; he carefully garnered copies of indentures, warrants and every other sort of government document; he prepared précis in his own hand of Mint records going far back into the Middle Ages. Many letters to Oxford were fair-written by Newton himself, no doubt from his home. He virtually ran every aspect of the Mint's business.

These years at the Mint provided the setting for two well-known stories concerning Newton. According to one (related by William Derham to Conduitt on 18 July 1733) Newton was offered a bribe of £6000 to favour one of the copper-coinage contractors.(44) According to the other, which seems to originate with Conduitt himself, Newton was offered a pension of £2000 annually by Bolingbroke (in 1713?) if he would retire.(45) It is needless to say that Newton reputedly refused both offers, and that no trace of either remains in his Mint Papers.

NOTES

(1) Leibniz had constantly claimed for himself the first discovery of a new differential calculus since his first paper of 1684 on the subject, paying little or no tribute to such mathematicians as Barrow and James Gregory besides Newton, whose work—of which he had knowledge—had certainly preceded his own however little influence it had upon his own development.

(2) U.L.C. Add. 3965(8), fo. 77v. We have omitted the many deletions and corrections made by Newton, of which the most significant is perhaps the phrase 'this day fortnight' for which he substituted 'when I was last here'. One might, alternatively, suppose that this draft was prepared for the meeting of either 5 or 12 April 1711, but we do not find it fits the proceedings of these dates so well.

(3) Compare Letter 830.

(4) Moreover Abraham Hill was 77 years old and Francis Aston 67; neither they (nor perhaps Arbuthnot) could have been supposed to be active.

(5) We are very much indebted to Dr D. T. Whiteside for providing a transcript of this document. See also Michael Hoskin in *The Listener*, 19 October 1961, p. 599.

(6) This statement is not quite correct. Leibniz's correspondence with Oldenburg had begun in July 1670 (see Hall & Hall, *Oldenburg*, VII, pp. 64–8) and continued throughout the following years. It was by no means wholly mathematical.

(7) See Hall & Hall, *Oldenburg*, IX, pp. 438–47 and Hofmann, § 3.

(8) This is Letter 210 in the Newton *Correspondence* (vol. II, pp. 231–4). Gabriel Mouton had published with his *Observationes diametrorum solis et lunæ apparentium* (Lyons, 1670), as the work of his friend François Regnaud, tables of the successive differences terminating in a constant final difference arising from subtraction of the successive terms of arithmetic progressions. Leibniz thought (mistakenly) that he had first found the general rule for the formation of the differences in the arithmetic triangle, which was in fact given correctly by both Pascal and Mouton.

(9) Printed above in Letter 98, vol. I.

(10) Apparently this was a mistake, though one to which Newton always clung. Leibniz certainly saw and annotated Letter 98 during his second visit to London (October 1676) when Collins allowed him to examine thoroughly his 'Historiola' (extracts of letters from Gregory, Newton and others relating to British mathematics); see Hofmann, § 20.

(11) In fact, however, Newton had there conveyed only a tangent rule with which Leibniz in 1676 was already thoroughly familiar, and therefore the letter had for him by no means the importance Newton attributed to it.

(12) Above, Letter 165, vol. II.

(13) Above, Letters 6 and 7, vol. I.

(14) Whiteside, *Mathematical Papers*, III, p. 20 and note (1).

(15) The Cotes–Smith letter is dated 'Febr 10. 1708', presumably to be read as Old Style, i.e. 1708/9; the Bentley letter 31 March 1706 (to be read as 1709). See Edleston, *Correspondence*, p. xv, notes 197–200.

(16) Professors Koyré and Cohen have provided (*Principia*, II, pp. 819–22) a serially ordered list of the various sections and propositions in the *Principia* to which reference is made in the Newton–Cotes correspondence.

(17) B. R. Goldstein has shown (*Centaurus*, **16** (1972), 257–84) that Tycho was anticipated about 1320 by Levi ben Gerson, whose lunar model embraces the variation.

(18) The computations mentioned by Newton are to be found in U.L.C. Add. 3966 (12), fo. 102–111; Compare Cohen, *Introduction*, pp. 350–1.

(19) In Proposition 25 Newton also shows how the magnitude of the perturbing solar force on the Moon may be computed for any relative position of the three bodies given the distances and angles of Sun and Moon from the Earth.

(20) Neither Horrox's model nor Halley's deals with the mean secular advance of the lunar apogee—they simply recorded this motion 'by observation'.

(21) Newton's design was to some extent realised in the second (1723) Amsterdam reprint of the 1713 *Principia*, to which was appended a reprint of William Jones' little book comprising four of Newton's mathematical tracts (see Letter 821, note (2) and Cohen, *Introduction*, p. 257). Possibly Newton had originally intended that the mathematical addendum to the *Principia* should consist of a much expanded version of *De Quadratura* entitled 'Analysis per quantitates fluentes et earum momenta' (see Whiteside, *History of Science*, **1** (1962), 23; the MS. of this tract is still extant and will be published in Whiteside, *Mathematical Papers*, VIII.

(22) Further, since it follows a *Commercium Epistolicum* draft on the same sheet, it must be later than April 1712.

(23) This is the story related by Cotes' nephew, Robert Smith. However, it should be noted that Cotes' letters to Newton of 1715 are amicable and in the following year Newton tried to do him a good turn (Edleston, *Correspondence*, pp. 179–84 and 228–9).

(24) Of all the annotation in his version of the Motte *Principia* translation, Florian Cajori devoted one-seventh to the Preface. Yet it sinks into relative insignificance in Professor Cohen's *Introduction.*

(25) Alexandre Koyré, 'Newton and Descartes' in his *Newtonian Studies*, Cambridge, Mass., 1965, pp. 53–114. The text is in Hall & Hall, *Unpublished Scientific Papers*, pp. 90–156.

(26) See Koyré, 'Attraction, Newton, and Cotes' (*loc. cit.*, note (25), pp. 273–82). He interpreted Cotes as objecting against Newton that the Third Law would no longer be true of the motion of two bodies (apparently caused by their mutual gravity) if the motion were in fact the result of external (e.g. aetherial) action. However, an objection in this form is only valid if, in the motion, only one body is moved, the other remaining at rest; for if the aetherial action be supposed to affect both bodies the phenomena remain as before and the Third Law's account of the phenomena remains accurate. Moreover, Newton could argue that in his terms the Third Law was well confirmed by the phenomena of the tides and astronomy. The only really strong objection that Cotes could have raised was that the Third Law was not confirmed by experimental evidence (except that relating to the collision of bodies).

(27) *Præfatio*, signature c *verso*.

(28) We may compare, though Cotes could not, Newton's early discussion of these issues in Hall & Hall, *Unpublished Scientific Papers*, pp. 90–156.

(29) What appears to be Newton's final draft—it carries one modification—of the Royal Society's unsuccessful petition to the Crown in 1705 for a grant of land in the Royal Mews, Westminster, because the Gresham 'Trustees and Professors are about pulling down and rebuilding the said [Gresham] College in a new form which will afford your petitioners no convenient accommodation' is in U.L.C. Add. 4006, no. 36. Weld (*History*, I, p. 388) printed the petition from the *House of Commons Journal*.

(30) Birch, *History*, I, p. 249. The same doctrine was repeated twenty years later (1683), *ibid.*, IV, p. 396. See G. N. Clark, *Science and Social Welfare in the Age of Newton*, second edition, Oxford 1949, p. 16.

(31) See his letters to Wren of 19 July 1708 (Letter 747, vol. IV) and to Sharp of 15 May 1711 (Letter 841).

(32) It is unjust to describe this as 'Halley's pirated edition' as was done in the previous volume. A pirated publication is unlawful and may be prevented by appeal to the Courts. Flamsteed sought no legal remedy but later took the law into his own hands. Flamsteed claimed that the actions of Newton and the other Referees were dishonourable, not that they were criminal. In any case Halley was not the prime mover; he merely did the hard technical work (see Letter 892).

(33) The will is dated two days earlier.

(34) *The Works and Life of the Right Honourable Charles late Earl of Halifax* (London, printed for E. Curll, etc.) Contrary to Brewster's aspersions (*Memoirs*, II, p. 273 note) this is a sober production extolling Halifax's literary, public and private virtues. It is mainly filled with his poems and speeches. The will as printed at the end of the book, separately paginated, shows no significant difference from the copy in the Public Record Office, 1715, June no. 113, Carolus Comes de Halifax.

(35) He had married the Dowager Countess of Manchester, who died in 1698. They had no children.

(36) *Works and Life*, pp. 195–6.

(37) *Cal. Treas. Books*, XXX (Part II), 1716, p. 185.

(38) The rather full account in the *Victoria County History*, *Sussex* is marred by loss of the records relating to the Halifax period of possession of the manor.

(39) King's College, Keynes MS. 127; see also U.L.C. Add. 3968 (13), fo. 139v.

(40) MS. 731, Lehigh University.

(41) See A. Rupert Hall and Marie Boas Hall in *Isis*, **52** (1961), 583–5.

(42) Vol. IV, Letter 545.

(43) *Newton at the Mint*, Cambridge, 1946, pp. 76–96.

(44) More, *Newton*, pp. 451–2; *Sotheby Catalogue*, p. 57.

(45) The intermediaries are said to have been Jonathan Swift and Catherine Barton; this suggests that the story is a fabrication.

THE CORRESPONDENCE

763 NEWTON TO [LAUDERDALE]
[? JULY/AUGUST 1709]

From a draft in the Mint Papers.(1)
Perhaps in reply to Letter 759

My Lord

I understand that the Barons of the Exchequer of North Britain(2) are calling in the money due to your Mint from the Collectors of ye Bullion, & I hope there is good progress made in that matter. And as that money can be got in I presume my Ld Treasurer will order it for paying off ye arrears of ye late recoinage & ye other charges of your mint. And if any thing relating to this matter be referred by my Ld Treasurer to ye Officers of the Mint in [the] Tower I shall be ready to promote the affairs of your Mint as much as I can.

Since the Union of ye two kingdoms there has been a Duty collected in North Britain upon the same foot with the coinage Duty in England(3) & the money collected (wch is but a small summ) lies in her Majts Excheqr. I presume my Ld Treasurer will give order about it next winter & also about the 12 hundred pounds applicable (by the new Act of Parliamt) to ye payment of Salaries & charges of repairs. But during this vacation time, his Lordp being seldom at London & not yet having an account of the old arrears in the hands of the collectors of bullion,(4) it will be difficult to get these matters settled. I wrote to Mr Allardes about two or three months ago to sollicit the Barons of the Exchequer to call in those arrears,(5) & as soon as a full state of that matter can be laid before his Lordp, wch I hope may be before winter, I am humbly of opinion that it may be proper to draw up a memorial representing the arrears due to the Mint, & also moving his Lordp for ye paymt thereof & for setling the paymt of such further summs from time to time as shall be requisite for the support of your Mint for ye future, & I believe it will be most proper to get the Secretary of State to lay the Memorial before his Lordp.

NOTES

(1) Mint Papers III, fo. 166. The document is in Newton's hand throughout. It seems likely from internal evidence that it was written in the vacation (summer) of 1709, since Allardes died on 5 October of that year.

(2) They were at this time: John Smith (Lord Chief Baron), John Scroop, Alexander Maitland, George Dalrymple.

(3) From the Treaty of Union in 1704 until 1710 Scotland was liable to the payment of certain taxes on imported liquors which had previously been collected in England for the support of the Mint. They were regularly evaded.

(4) Presumably the reference is to the executors of Daniel Stuart, Collector of Bullion at the Edinburgh Mint, who had died in March 1709. Settlement was deferred until 1710 (Mint Papers, III, fo. 131).

(5) See Letter 753, vol. IV, p. 535; possibly this letter was wrongly dated.

764 THE MINT TO GODOLPHIN

10 AUGUST 1709

From the original in the Scottish Record Office, Edinburgh[1]

To the right honble ye Earl of Godolphin
Lord high Treasr of Great Brittain

May it please your Lordship

In obedience to your Lordships order of reference of the 3d June last upon the annex'd peticon of the provost of the Monyers wherein he sets forth that by her Majties Signe Manual some of the Monyers were sent to the Mint at Edinburgh to assist in the Recoinage of the Monyes of North Brittain[2] with an Allowance of 9*d* per £wt for Coinage and £16 to each man for his Journey backwards and forwards and three shillings per diem to each man for his maintainance there whenever there should not be 1000 £wt coined in one week; We have considered the said peticon and by the Monyer's Book of Accounts signed by the Generall,[3] the Master,[4] and other Officers of the said Mint we find that there was coyned by the said Monyers 103346 £wt of silver monyes. The Coynage of which at 9*d* per £wt amounts to 3875£: 09*s*: 06*d* whereof the said Monyers have reced of Mr. Allardes the Master of the said Mint the Summe of £1409: 6*s*: 2½*d* and there remains still due to them from the said Master the Summe of £2446: 3*s*. 3½*d* as the petition sets forth which Summe is not yet paid by reason that the said Master hath not yet reced money from the Governmt sufficient for defraying the Charge of the said Coynage And whereas the Master of her Majties Mint in the Tower payes 9*d* per £wt to the Monyers whereof 8*d* is out of his own allowance of 10½ per £wt for Coynage and the other penny is placed to her Majties Acct. We are humbly of oppinion that the Master of her Majties Mint at Edinburgh should in like manner pay 8*d* of ye 9*d* out of his own allowance for Coynage and be allowed the other penny by her Majtie in his accts. We find also by a Certificate signed by the Warden of the said Mint[5] that there is further due from her Majtie to the said Monyers for the Journyes of five of them to Edinburgh and four of them back the Summe of £72 and upon their Allowance of 3 Shillings per diem the Summe of £182: 2*s*: All which Summes her Majtie has appointed by her said Warrant to be paid in such Manner as your Lordsp shall think fit.(6)

All which is humbly submitted to your Lopps great Wisdom

C: PEYTON
IS: NEWTON
JN: ELLIS

Mint Office
August ye 10 1709

NOTES

(1) E. 411/10/14. This document in a copyist's hand bears the signature of the Warden, Master and Comptroller.

(2) The recoinage began in Scotland in November 1707 and was substantially completed by December 1708: £320372 12*s* was coined. The moneyers returned to London in March 1709 and the Edinburgh Mint was closed on 4 August 1710. (See Craig: *The Mint, A History of the London Mint from A.D. 287 to 1948*, Cambridge, 1953, p. 210).

(3) John Maitland (*c.* 1655–1710), fifth Earl of Lauderdale. See vol. IV, p. 523, note (4).

(4) George Allardes, see vol. IV, p. 522, note (2).

(5) William Drumond, appointed 9 March 1705.

(6) The deficit was finally disposed of by resolution of the House of Commons; see below Number 883.

765 COTES TO NEWTON

18 AUGUST 1709

From the original in Trinity College Library, Cambridge.(1)
For the answer see Letter 766

Cambridge August 18*th* 1709

Sr.

The earnest desire I have to see a new Edition of Yr Princip. makes me somewhat impatient 'till we receive Yr Copy of it which You was pleased to promise me, about the middle of the last Month, You would send down in about a Fourtnights time. I hope You will pardon me for this uneasiness from which I cannot free my self & for giveing You this Trouble to let You know it. I have been so much obliged to You by Yrself & by Yr Book yt (I desire You to beleive me) I think my self bound in gratitude to take all the Care I possibly can that it shall be correct. Some days ago I was examining the 2d Cor: of Prop 91 Lib I(2) & found it to be true by ye Quadratures of ye 1st & 2d Curves of ye 8th Form of ye second Table in Yr Treatise *De Quadrat.*(3) At the same time I went over ye whole Seventh & Eighth Forms which agreed with my

Computation excepting ye First of ye Seventh & Fourth of ye Eighth which were as follows

$$\text{Form: 7. 1.}\quad \frac{4de\frac{\Upsilon^3}{\xi}-2df\frac{v^3}{x}-8\,dee\sigma+4dfgs}{4\eta eg-\eta ff}=t.$$

$$\text{Form: 8. 4.}\quad \frac{\begin{array}{cccc} +36defg & +8degg & -28defg & -16deeg \\ s & xxv & xv & v \\ -15df^3 & -2dffg & +10df^3 & +10deff \end{array}}{24\eta eg^3-6\eta ffgg}=t.$$

I take this Oportunity to return You my most hearty Thanks for Yr many Favours & Civilitys to me who am(4)

Yr most Obliged humble Servant
ROGER COTES

For Sr Isaac Newton at his House
in Jermin Street near St James's
Church Westminster

NOTES

For the writer of this letter see vol. IV, p. 472, note (7). Richard Bentley's concern to bring out a second edition of the *Principia* (Letter 742, vol. IV) had induced him to invite the cooperation of Cotes, who was presumably recommended by his skill in mathematics and astronomy, to join in the enterprise. Probably in March 1709, one copy or other of the first edition containing emendations by Newton was placed in Cotes' hands (Edleston, *Correspondence*, p. xvi; Cohen, *Introduction*, p. 227). This was not to be the basis of the second edition, however, for certainly on 21 May Bentley wrote to Cotes from London that Newton expected him to call and collect from him 'one part of his Book corrected for ye press' (Edleston, *Correspondence*, p. 1). Cotes was indeed in London in July 1709 (see below, Letter 821) and called on Newton, with the result described in this letter.

(1) R.16.38, no. 2; printed in Edleston, *Correspondence*, pp. 3–4.

(2) The proposition reads: 'To find the attraction of a particle situated upon the axis of a solid, to the several points of which tend centripetal forces decreasing in any proportion to the distances whatever.' Then Corollary 2 specifies: 'Thus also the force with which a spheroid *AGBcD* attracts any body *P*, situated outside it along the axis *AB*, may be known.' A paper written by Cotes bearing on this investigation is in Trinity College Library, Cambridge (R.16.38, nos. 24, 25).

(3) Newton published his *Tractatus de quadratura curvarum* with *Opticks* in 1704. The table to which Cotes refers appears as a scholium and is referred to in its tenth proposition; this table was revised from the 1671 'Tractatus de methodis serierum et fluxionum' where it appears as a 'Catalogus curvarum aliquot ad conicas sectiones relatarum ope Prob. 8 constructus' (Form 7:1 is there Ordo Sextus. I). The error in the 'integral' from Form 7.1 arose from a computational mistake on Newton's part—two terms are missing from the numerator—the other mistake was one of transcription only (see Whiteside, *Mathematical Papers*, III, pp. 250–3, 257–8, note (564)). The forms given by Cotes are correct.

(4) The extent of the earlier acquaintance between Newton and Cotes—Fellows of the same

College and Professors in the same University—appears to be unknown at present. However, in a letter from Cotes to his uncle (John Smith, who had taught him mathematics), dated 10 February 1708/9, he writes of a visit to London: 'Whilst I was in Town Sr Isaac Newton gave order for ye makeing of a Pendulum Clock which he designs as a present to our new Observatory... I believe Sr Isaac's clock can cost him no less yn 50 *lbs*.' (Trinity MS. R.16.38, no. 33; Edleston, *Correspondence*, p. 198).

766 NEWTON TO COTES

11 OCTOBER 1709

From the original in Trinity College Library, Cambridge.[1]

Reply to Letter 765

Sr

I sent you by Mr Whiston[2] the greatest part [3] of ye copy of my Principia in order to a new edition. I then forgot to correct an error in the first sheet pag 3 lin 20, 21 & to write *plusquam duplo* for *quasi quadruplo* & *plusquam decuplo* for *quasi centuplo*.[4]

I forgot also to add the following Note to the end of Corol. 1 pag. 55 lin 6. Nam datis umbilico et puncto contactus & positione tangentis, describi potest Sectio conica quæ curvaturam datam ad punctum illud habebit. Datur autem curvatura ex data vi centripeta: et Orbes duo se mutuo tangentes eadem vi describi non possunt.[5]

I thank you for your Letter & the corrections of ye two Theorems in ye treatise de Quadratura. I would not have you be at the trouble of examining all the Demonstrations in the Principia. Its impossible to print the book wthout some faults & if you print by the copy sent you, correcting only such faults as occurr in reading over the sheets to correct them as they are printed off, you will have labour more then it's fit to give you.

Mr. Livebody is a composer[6] (I mean Mr Livebody who made the wooden cutts) & he thinks that he can sett the cutts better for printing off then other composers can, & offers to come down to Cambridge & assist in composing if it be thought fit. When you have printed off one or two sheets, if you please to send me a copy of them I will send you a further supply of wooden cutts.[7] I am

Your most humble & faithful servant

Is. Newton

London. Octob. 11. 1709.

For Mr. Cotes Professor of Astronomy
in the university of Cambridge at his
Chamber in Trinity College

NOTES

(1) R.16.38, no. 3; printed in Edleston, *Correspondence*, pp. 4–5.

(2) William Whiston; see vol. IV, p. 472, note (8).

(3) So far as the end of Section 6 in Book II (*Principia* 1687, p. 316).

(4) Newton indicates his additional discussion of Definition 5 on p. 3 of the specimen of the opening of the revised *Principia* already prepared by Bentley; this passage is not found in the first edition. (See Koyré and Cohen, *Principia*, I, p. 42.) Suppose a projectile fired from the top of a mountain, horizontally, with such a velocity that its range is two miles. Double the velocity and its range will be *more than twice* (because of the curvature of the Earth). Multiply the velocity by ten and the range will be *more than ten times* as great (neglecting air resistance)—not *about four times* and *about one hundred times*. By the time Newton's letter reached Cambridge, Bentley had already made the University Press print off up to page fifty (see Letter 767), hence the correction came too late; fortunately Bentley (or more likely Cotes) had already caught the mistake (cf. Cohen, *Introduction*, pp. 221–2).

(5) Coroll. 1 to Prop. 13 reads (translating): 'It is a consequence of the last three propositions that if any body P moves away from the point P along any straight line PR, with any velocity whatever, and is at the same time acted upon by a centripetal force reciprocally proportional to the square of the distance from the centre, this body will move in some conic section having its focus at the centre of force, and conversely.' To this he now adds (*Principia*, 1713, p. 53): 'For if the focus, and the point of contact, and the position of the tangent are given, a conic section may be described which shall have a given curvature at that point. But the curvature is given from the given centripetal force and two orbits mutually tangent cannot be described by the same force'.

This is Newton's first direct proof of the inverse problem of motion when the force is reciprocally as the square of the distance; the force may be deduced from the curve, since the curve must coincide with that generated from a known force.

(6) Compositor; though what is described seems rather part of the press-work.

(7) Following this letter there is a long gap in the correspondence. An annotation by Edward Howkins—who gave the Cotes papers to Trinity College—records the absence from them of any copy of a letter from Cotes to Newton dated 9 April 1710, concerning Book II, Proposition 9. No letter of this date is to be found among the originals in the University Library either. Moreover, the opening of Cotes' next letter (Letter 775) makes it plain that the Press had printed off many sheets without Cotes' referring to Newton at all and that he had taken much liberty in altering Newton's 'copy'. Furthermore, if a letter of 9 April be lost, it follows that Newton's answer (in original and draft) is lost also. Hence Howkins, for all his care, may have been mistaken. In the same note recording this putative lost letter Edleston (*Correspondence*, pp. 6 and 7) gives his reasons for supposing that other letters exchanged during this long blank period have vanished, again without any trace of replies from Newton. It is certainly plausible to conjecture that such letters may have been exchanged in late 1709 (though there must nevertheless have been a long gap before Letter 775), yet there is no strong evidence indicating the loss of letters.

767 BENTLEY TO NEWTON

20 OCTOBER 1709

From the original in Trinity College Library, Cambridge.[1]

Reply to Letter 766

Trin: Col. Octob. 20. 1709

Dear Sir,

Mr Cotes, who had been in ye Country for about a month returned hither ye very day, yt Dr Clark[2] brought your letter. In which, I perceive, you think we have not yet begun your book; but I must acquaint you yt five sheets are finely printed off already: and had we not staid for 2 Cuts yt Rowley[3] carried to Town to be mended by Lightbody,[4] wch we have not yet receivd, you had had sent you six sheets by this time. I am sure you'l be pleasd with them, when you see them. Besides ye General Running Title at ye head of every leaf PHILOSOPHIÆ NATURALIS PRINCIPIA MATHEMATICA; I have added the subdivisions of ye Book (like Huygenij de Oscillatione)[5] first DEFINITIONES then AXIOMATA SIVE LEGES MOTUS, then DE MOTU CORPORUM LIBER PRIMUS; next will come—SECUNDUS; and lastly, DE MUNDI SYSTEMATE LIBER TERTIUS. All these stand in ye Top of ye Margin of ye several Leaves. Your new Corollary, which you would have inserted, came just in time; for we had printed to the 50th page of your former Edition; & yt very place, where the Insertion is to be, was in the Compositors hand. The correction in the first sheet, wch you would have, *plusquam duplo*, & *plusquam decuplo*, was provided for before: for we printed it as *quasi duplo* and *quasi decuplo*, which you know amounts to ye same thing: for *Quasi* denotes either the Excess or ye Defect: & in my opinion, since in yt place you add no reason why it will be plusquam, tis neater to put it *quasi*, undetermind, and leave ye reader to find it out. In the old Edition p. 34,[6] line 20 & 21, for *infinite major*, you had twice mended it *minor*: this we thought you did in hast; for it was right before, & so we have printed it *major*:[7] I proposed to our Master Printer to have Lightbody come down & compose, which at first he agreed to; but the next day he had a character of his being a mere sot, & having plaid such pranks yt no body will take him into any Printhouse in London or Oxford; & so he fears he'll debauch all his Men. So we must let Him alone; and I dare say we shall adjust the Cuts very well without him. You need not be so shy of giving Mr Cotes too much trouble: he has more esteem for you, & obligations to you, than to think yt trouble too grievous: but however he does it at my Orders, to whom he owes more than yt. And so pray you be easy as to yt; we will take care yt no little slip in a Calculation shall pass this fine Edition.

Dr. Clark tells me you are thinking for Chelsea(8) where I wish you all satisfaction. I hope my Picture at Thornhills(9) will have your last Sitting, before you leave the Town: the time you set under your Hand is already lapsed. When the 2 Cutts are sent us, we shall print faster than you are aware of; therefore pray take care to be ready for us. I am, Sir,

Your very obedient humble servant

RI. BENTLEY

For Sir Isaac Newton
at his House in Jermin Street
near St James's Church
London

NOTES

(1) R.4.47, fo. 20; printed in Brewster, *Memoirs*, II, pp. 250–2.

(2) Presumably Samuel Clarke (1675–1729), B.A. of Caius College, Cambridge, 1695; D.D., 1710. He had been translator of Newton's *Optice*, 1706.

(3) Possibly John Rowley (d. 1728), mathematical instrument maker, who was at this time supplying instruments to the observatory at Trinity College; or the Cambridge carrier.

(4) 'Livebody' in Letter 766.

(5) Christiaan Huygens, *Horologium oscillatorium* (Paris, 1673).

(6) The sentences read: 'And if DB becomes successively as AD^2, $AD^{\frac{3}{2}}$, $AD^{\frac{4}{3}}$, $AD^{\frac{5}{4}}$, $AD^{\frac{6}{5}}$ etc. there will be formed another infinite series of angles of contact, of which the first is of the same kind as [those of] circles, the second infinitely greater, and any succeeding ones infinitely greater than the first.' (See Koyré and Cohen, *Principia*, I, p. 86.)

(7) So it is, on p. 32 of the second edition.

(8) Newton moved from Jermyn Street to Chelsea in the present month.

(9) The portrait of Newton commissioned by Bentley and painted by Sir James Thornhill (1675–1734) still hangs in the Master's Lodge at Trinity. He painted other versions which have remained with Newton's collateral descendants.

768 FLAMSTEED TO SHARP

25 OCTOBER 1709

From the original in the Royal Society of London(1)

The Observatory Octob. 25. 1709

Sr

This comes to informe you that on Thursday last my Servant delivered to Mr Stamfeild(2) at Mr Knaps all the printed sheets of my Historia Cœlestis from page 101 to ye conclusion, except the reprinted copy of the first sheet of ye Maculæ which I could not yet procure nor so much as set my eyes upon. this is Sr I N[ewton']s return for all my obligeing Civilitys & Kindnesses of which you have sometimes tho many yeares agone been witness. He is now removeing to Chelsea & has been lately much talkt of but not much to his advantage. Our Society is ruined by his close politick & cunning forecast, I fear past retreiveing for our Transactions have been twice burlesqt publickly & now we have had none published I think this four moneths.(3)

... ...

I am apt to think I shall not have occasion to write to yu hereafter so often as formerly except yu find some new causes pray let me hear however of your health sometimes & what you are doeing, & when any thing happens that I thinke may be acceptable to yu I take care to acquaint you with it that yu may see yt I am still & ever will be Sr

Your affectionate freind & Servant
JOHN FLAMSTEED M.R.(4)

For Mr Abraham Sharp at
little Horton near
Bradford in Yorkshire

NOTES

For the recipient of this letter see vol. IV, p. 464, note (1).

(1) Sharp Letters, no. 71.

(2) Abraham Sharp's nephew. Compare Sharp to Flamsteed, 12 July 1710 (Cudworth, *Life*, p. 101): 'The bearer hereof, my nephew (Robert Stansfield), proposes to be in London Tuesday the 18th inst., and...may be heard of during that time at Mr Knapp's in Basinghall Street.'

(3) William King (1663–1712), D.C.L. Oxon, 1692, published at least four pieces satirizing science and scientists, such as *A Journey to England, with some account of the manners and customs of that nation*, London, 1700, a satire on Martin Lister's *Journey to Paris*; the two Flamsteed had in mind were, probably, *The Transactioneer, with some of his philosophical fancies*, London, 1700, aimed at Hans Sloane; and *Useful Transactions in Philosophy and other sorts of Learning*, London, 1709. All these writings were anonymous. *The Transactioneer* at least caused the Society some official embarrassment, partly because Sloane suspected John Woodward of writing it.

(4) Mathematicus Regius.

769 LOWNDES TO THE MINT

9 NOVEMBER 1709

From Newton's copy in the Mint Papers.(1)
For the answer see Letter 770

Gentlemen

My Lord Treasurer(2) Commands me to transmit to you the inclosed Memorial of the Tynners in Cornwall, which his Lordp has lately received from the Lord Warden of the stannarys.(3) My Lord directs you to peruse the same &

report to his Lordp what you think fit to be considered in any future contract made with the Tynners,(4) if her Majty should so think fit to direct particularly as to the Quantity to be taken and the price to be paid for the same, so that her Maty may not be a looser [*sic*] thereby.

I am

Gentlemen

Your most humble Servant
WILLIAM LOWNDES

Treasury Chambers
9 *November* 1709

[The Enclosed Memorial]
To the Honble Hugh Boscowen Esqr
Ld Warden the Stannaries of Cornwall & Devon

We the Gentlemen owners of Tyn Lands & Tyn bounds adventurers for Tyn & others concerned in that Commodity in the County of Cornwall

Humbly shew

That we cannot but with the utmost gratitude contemplate her Majts goodness to us in making & performing the present contract for our Tyn. And being sensible now the said contract is so neare expiring, of the ill consequences that may ensue to her Maty & ourselves by its determination before the commencement of another, take this opportunity to apply to your Honour praying

That considering our present circumstances your Honour will be pleased to intercede with her Maty that a Convocation or Parliament of Tynners of the said County may with all convenient speed be held, wherein we hope the Gentlemen that shall be chosen to represent us will consent to such a price as that her Maty may be no looser thereby. But at the same time desiring your Honours endeavours that in such Convocation your Honour may be impowered to agree for such farther quantity beyond the present stipulation as may prevent the inconveniencies We now labour under on that Acct

And we shall ever pray

NOTES

For the writer of this letter see vol. IV, p. 248, note (3).

(1) Mint Papers III, fos. 486v–487. Both these documents are copies in Newton's hand. See also *Cal. Treas. Books*, XXIII (Part II), 1709, p. 416.

(2) Godolphin.

(3) Hugh Boscawen, later first Viscount Falmouth (d. 1734); he was, besides, M.P. for several Cornish constituencies and holder of other lesser offices.

(4) Responsibility for the tin trade was assumed by the Mint after Newton, in 1703, under-

took to handle the output of the Stannaries. The disposal of tin, in competition with supplies from abroad, was a continual concern of Newton's until the death of Queen Anne when it was transferred, with the Duchy of Cornwall as a whole, to the Prince of Wales (see Craig, *Newton*, pp. 7–61 and *The Mint*, pp. 208–9).

770 THE MINT TO GODOLPHIN

16 NOVEMBER 1709

From a holograph draft in the Mint Papers.(1)
Reply to Letter 769

In Obedience to your Lordps Orders of the 9th Instant &c We humbly represent to your Lordp that by the course of the sale of the Tyn this & the four last years there has been sold in Cornwall & London at a Medium about 1560 Tunns per annum stannary wt.(2) Which being deducted from 1600 Tunns received annually from Cornwall, & allowing 40 Tunns more for Devonshire, there has remained about 80 Tunns yearly unsold: So that 1520 Tunns of Cornish Tynn has been sold yearly, more or less. And if the same course of sale continues, there will still remain at the end of the present contract, so much Tyn unsold, as at the Rate of 3£. 16*s* per hundred Averdupois will produce about 180000£.(3) And so long as the same course of sale continues, if six per cent (the interest of the moneys now advanced to the owners of the Tyn lands) be allowed upon the said dead stock of 180000£, & the Coynage duty & post groats(4) be paid & 1400£ be accounted sufficient to answer accidents & unforseen charges; her Maty may without losing by the contract, give 3£. 9*s*. 6*d* per hundred stannary wt for 1520 Tunns of Tynn the quantity annually consumed, as will appear by the following Accot.(5)

	£	*s*	*d*
1520 Tunns of tynn at 3. 9. 6 per [centum] stan wt	105640.	0.	0
Freight of the same to London	2035.	14.	3
Salaries in London	1350.	0.	0
Incident charges in London, as Porters &c	200.	0.	0
Passing Accots &c	94.	0.	0
Salaries in Cornwal & Truroe	1540.	0.	0
Incidents in Cornwall	700.	0.	0
Interest of 180000£ at 6 per cent	10800.	0.	0
	122359.	14.	3
Insurance & other unforseen accidents	1411.	14.	4
Produced annually by sale of Tyn at 3£ 16*s* per [centum]	123771.	8.	7

And if it be supposed that the consumption may carry off annually 1600 or 1800 Tunns stannary, it will appear by the like recconing upon that supposition, that her Maty may give 3£ 10s in ye first case & 3£ 11s in the second & so in proportion to a greater or less consumption. But every 100 Tunns purchased above what the consumption will carry off, makes a dead stock annually increasing, the interest of which at 6 per cent in the end of seven years will amount to about 10000£ & the loss by the falling of the price of the Tyn at the end of the contract will amount unto about 18 or 20 thousand pounds more

All wch is most humbly submitted to your Lordps great wisdome

C. PEYTON.
I. NEWTON.
J. ELLIS.[6]

Mint Office
Nov. 16. 1709.

NOTES

(1) Mint Papers III, fo. 487. For a similar calculation by Newton see *ibid.* fo. 520. There is a fuller draft of the letter, signed by Newton alone, in *ibid.* fo. 568.

(2) There were 20 cwt., each of 120 lb. avoirdupois, to a stannary ton, which was therefore 2400 lbs. av.

(3) The unsold tin of course represented money unavailable to the Crown; Newton evidently estimated the surplus tin at the termination of the contract as amounting to nearly 2,400 tons av. The quantity of tin bought by the Crown had increased from 860 tons in 1705.

(4) In the Middle Ages for a brief period surplus tin was stamped in 'post coinages', paying an extra charge of 4*d.* on each cwt., known as the 'fine of tinners', or 'post groats'. This charge was revived after 1507. In 1710 post groats amounted to £10 merely. (See G. R. Lewis, *The Stannaries*, Cambridge, Mass., 1924, pp. 152–3 and p. 156.)

(5) Newton adjusts the price of the 1520 tons stannary bought, so that when sold at the market rate the account will almost balance. He seems not to provide for the excess purchase of tin in earlier years, nor for disposal of the stockpile.

(6) All the signatures are in Newton's hand. For John Ellis, Comptroller, see vol. IV, p. 391, note (9), and for Craven Peyton, Warden, *ibid.* p. 521, note (3).

771 JOHN TAYLOUR TO THE MINT

28 JANUARY 1709/10

From the copy in the Public Record Office.[1]
For the answer see Letter 772

Gentlemen

It having been Represented to My Lord Treasurer that the Accompts of the later Master and Worker of the Mint at Edinburgh[2] do stop for want of a Tryal of the pix:[3] his Lordship Directs You to Consider of that Matter and to

Attend his Lordship here next Tuesday at five of the Clock in the Afternoon with Your Report in what Manner, where, and when, you conceive the same may most properly be done; And whether (for the avoiding of Expence and Trouble as much as may be) the pix in the Mint in the Tower may not be Tryed at the same Time. I am

Gentlemen

Your most Humble Servt

J. TAYLOUR

Tre[asu]ry Ch[ambe]rs
20 *January* 1709
Officers of the Mint

NOTES

(1) Mint/1, 8, 166. The writer was one of the four First Clerks of the Treasury.
(2) George Allardes.
(3) The pyx (L. *pyxis*) was a chest into which sample coins, chosen at random, were placed each day while minting was going on. Periodically the pyx was opened and the coins tested for fineness and weight by a jury of the Goldsmiths' Company summoned by the Lord Chancellor, together with representative members of the Privy Council and certain legal officers. (A manuscript by Newton dealing with the trial of the pyx is printed in vol. IV, pp. 371–3.) Allardes' accounts could not be cleared till it was certain that the coins for which he had been responsible were up to standard.

772 THE MINT TO GODOLPHIN

31 JANUARY 1709/10

From the original in the Public Record Office.(1)
Reply to Letter 771

To the most Honble the Ld High Treasurer of great Britain

May it please your Lordp

In obedience to your Lordps Order signified to us by Mr Taylour's Letter of ye 28th instant, concerning the manner time & place of trying the Pix of her Majties Mint at Edinburgh & whether the Pix of the Mint in the Tower should be tried at the same time: We humbly represent that in our opinion it is convenient that the Pixes of both Mints be tried together at Westminster in one of her Majties houses after the usual manner specified in the Indenture of her Majties Mints at such a time as shall be appointed by her Majty in Council, so that the Officers of the Mint at Edinburgh may have timely notice to send or bring up their Pix to London. The most convenient time is when days are long, suppose in May or June, that the whole trial may be dispatched in one day for avoiding of unnecessary charge & trouble. And because the trial ought to be in one of her Majesties houses, we humbly conceive that a new furnace should be erected in one of them where it may be safely performed, the expence thereof being inconsiderable.

All which is most humbly submitted to your
Lordps great wisdome

C: PEYTON
IS. NEWTON
JN ELLIS

Mint Office
31*th Jan* 1709

NOTE

(1) T/1, 120, no. 30. This was written by Newton and signed by all three Principal Officers. Compare Letter 798. There is a draft by Newton in U.L.C. Add. 3965 (12), fo. 216. The letter is endorsed: 'read 31 Janry 1709. My Lord will speak to my Ld Chancellor to appoint the day in June or July. 6 May 1710. My Lord will speak to my Lord Chancellor to appoint some day in the beginning of August.'

773 NEWTON TO GODOLPHIN

16 FEBRUARY 1709/10

From the holograph original in the Public Record Office[1]

To the most Honble the Earl of Godolphin Lord
High Treasurer of great Britain.

May it please your Lordp

In obedience to your Lordps verbal Order I humbly lay before your Lordp the state of ye Question about the allowance to be made in ye Accounts of Mr Allardes for the late coinage of silver moneys at Edinburgh.[2]

By the Indenture of the Mint in the Tower the Master & Worker is allowed sixteen pence half penny upon the pound weight Troy for the coinage of silver moneys. The words of the Indenture are herunto annexed.[3]

By her Majties Warrant of June 20th, 1707 for putting the Act of union in Execution, the Officers of her Majties Mint at Edinburgh were directed to observe the rules of coinage set down in the said Indenture with respect to their several Offices. But I do not remember that any alteration in the allowance for coinage or in the salaries was then under consideration. The words of ye Warrant are hereunto annexed.[3]

By an Act of Parliament made in Scotland A.C. 1690 the Master of that Mint was allowed twenty pounds Scots upon the stone weight Scots for coinage of silver moneys, & this allowance was in use till the Union.(4) The words of the Act are hereunto annexed.(3) Three pounds Scots are worth four shillings sixpence halfpenny English at their just value. But the nation of Scotland valued thirteen shillings Scots at twelve pence English before the Union while the said Act of Parliament was in force, & at thirteen pence English in distributing the Equivalent. The allowance for coinage comes in the first case to almost seventeen pence halfpenny upon the pound weight Troy, in the second to something more than seventeen pence halfpenny, in the third to something more than nineteen pence. There hath been coined 104227£wt 10 oz Troy, & the whole allowance for the coinage thereof comes in the first case to 7533£. 13*s*. 2*d*, in the second to 7655£. 18*s*. 10*d*, in the third to 8293£. 18*s*. 9*d*; & by the Indenture of the Mint it comes only to 7165£. 12*s*. 3*d*. And for paying all the Accounts relating to that coinage, there is a deficiency in the funds of about two or three thound [*sic*] pounds.

The Question is, what shall be allowed in the Accounts of Mr Allardes upon the pound weight Troy for the said coinage.

Which is most humbly submitted to your Lordships great wisdome

Is. Newton

Mint Office
16 *Feb*. 170$\frac{9}{10}$.

NOTES

(1) T1/120/44, fos. 163–5; printed in Shaw, pp. 169–71.
(2) Georges Allardes, Master of the Edinburgh Mint, had died in October 1709.
(3) See Number 773*a*.
(4) The stone weight Scots was nearly 17·2 lb avoirdupois or about 20·8 lb Troy.

773*a* EXTRACTS ENCLOSED WITH THE PREVIOUS LETTER

The Clause in the Indenture of the Mint
appointing the allowance per pound weight for coinage.

And the said Isaac Newton shall have & receive ye summe of one shilling & four pence half penny to be by him taken for the coinage of every pound weight Troy of silver moneys for the paying bearing & susteining all manner of wasts, provisions, necessaries & charges coming arising & growing in or about the coining of her Majties Crowns, half crowns, shillings & sixpences of silver moneys by the mill & press out of the moneys to be paid & payable unto him as is herein after expressed. And the said Master shall out of the one shilling & four pence half penny allowed to him as aforesaid & received by him for every pound weight Troy of the moneys of silver, from time to time pay unto the Moneyers the summ of eight pence for the making of every pound weight Troy of the said silver moneys by the mill & press, according to the undertaking & agreement of the said Moneyers with the said Master & Worker.

The clause in her Majties Warrant directing the Officers of her Mint at Edinburgh to observe the Rules of coinage set down in the said Indenture.

It is our Will & pleasure, & We do hereby authorise & require you & every of you, that in the coinage of such Gold & Silver as shall be imported into Our Mint at Edinburgh, you act under & observe the Rules of coinage wch respect your several Offices & are conteined & exprest in the copie of the Indenture herewith sent, attested by Sr John Stanley Warden of our Mint in the Tower of London,(1) Sr Isaac Newton Master & Worker & John Ellis Esq Comptroller of our said Mint: wch Indenture was made in the first year of our reign with the said Sr Isaac Newton &c.

To the General, Master, Warden, Counter-warden,
Essey-master, & other Officers of our Mint at
Edinburgh

The clause of the Act of Parliament made in Scotland 1690, entituled an Act anent an humble offer to his Majesty for an imposition upon certain commodities for defraying the expence of a free coinage.

It is ordeined that considering that by the Act of Parliament 1686 anent a free coinage there is only allowed eighteen pounds Scots upon the stone weight of silver for defraying the whole charge wast expences & loss upon its coinage, wch allowance is found by experience to be too small & insufficient: therefore their Majesties with the advice & consent of the said estates do hereby rescind in all time coming that clause of the aforesaid Act, & further statute & ordein that the Master of the Mint have allowed to him in time coming twenty pounds Scots instead of eighteen pounds Scots upon the stone weight of silver as the just & reasonable allowance for defraying the said charge expence & loss upon its coynage in manner provided in the said Act.

NOTE

(1) See vol. IV, p. 357, note (6).

774 —— TO NEWTON

28 MARCH 1710

From the original in King's College Library, Cambridge[1]

Sr

The Subject of the Entertainment of the Council last Wednesday was a Surprize to me, & one or two more, who were not let into the Secret before.[2] We think you did wisely to adjourn that Affair: & that 'twill be no less Wisdom, with Respect both to your Self, & to the Society, if not finaly to wave it, & let the two Contendants decide their own Differences, at least to be Equal & Impartial to Each. Dr Harris,[3] who sate all the while faceing Dr Sloane, interposed in the Dispute, & can set you to Rights in the whole, is at Rochester, upon the Duty of the Passion-Week. If you adiourn the Society for that, & the Holyday-Week, the Dr will be back ready to execute the Office of Secretary if Dr Sloane is pleas'd as he gives out, to desert it; but that is taken only for one of his usual Arts. He made a Pretence of quitting it hertofore: & Dr Harris has been ever since ready to attend & act, whenever the Task should be devolved upon him, either alone, as Dr Sloane has done for several Years, or jointly with any other that ye Council, or the Society hereafter shall appoint. This he has declared both to you, & others: & he has, in Readyness, Materials, for Entertainment of you and the Society, that are proper, & suitable to our Institution: & not such as have been, now for several Years past, wont to be produced. There is likewise a Disposition to setle a general Correspondence, to make Experiments & Observations, & set on foot such Methods as may serve both to promote the Design, and advance the Honour of the Society; whenever you shall please to put the Revenue of the Society upon a right Foot: & all these Gentlemen will be as forward to second you in these Designs, as you know, Dr. Sloane, & his Junto of Non-Solvents,[4] by which he is supported, are to impede & defeat you. This once setled, you will see, (I tell you no more than I am sure of) some of the greatest Men in the Nation come in, & be proud to act under you, who do not think it at all for their Honour to venture in such a Bottom as Things now have been so long. If Mr Halley would be as forward in that, as it seems, for Reasons that he best knows, he is, in some other Things that assuredly are not so much for yours or the Society'es Interest, he will tell you how loud Men of Sense all over the Town are in their Complaints, & Men of Wit, in their Railery of the Society on Account of Dr Sloanes Management & Philosophical Transactions.[5] Twill not easyly be imagin'd how greatly the Reputation of the Society, without doors, suffers upon that Account. Tis in your Power to retrieve all: & as you have Ability, so likewise we know you have

Inclination to do it. They intended you an Opportunity of that when they made choice of Dr Harris: & he will serve you in it both assiduously & faithfully if incouraged. If, in your Absence, Mr Roberts,[6] Mr Foley[7] Mr Hill,[8] Mr Halley, or such others as you shall judge fit, were appointed Vice-Presidents, it would conduce greatly to this good End: but no Man is capable to carry on these noble Purposes equaly with your self, if you please to attend the Meetings.

As to the Dispute depending, after due Examination, & hearing the Representations on all Hands, I find the state of it to be as follows. Dr Sloane read a Translation out of the French Memoi[r]es: & I confess I cannot but think a Translation out of a printed French Book a very low, mean, unfit Entertainment for the Royal Society. But he must entertain us with what he can get: & this is an old Practise with him. The Position there maintained, *that the Bezoar is a Gall-Stone*, is neither true, nor so much as probable.[9] Dr Sloanes Assertion, yt *the Stones in the Gall Bladder are the Cause of the Colic*, carrys with it still less Truth & Probability. I do not suppose he is to sit there as a perpetual Dictator: & that nobody is to refuse Assent to every Thing he offers. What Dr Woodward advanced was both pertinent & civil; till Dr Sloane, not able to maintain what he had asserted by Words, had Recourse to Grimaces very strange & surprizing, & such as were enough to provoke any ingenuous sensible Man to a Warmth at least equal to that which Dr Woodward us'd. His Words were *no Man that understands Anatomy can assert that the Stones in the Gall-Bladder are the Cause of the Colic*. When Dr Sloane averr'd that all Medical Writers were of that Opinion; Dr Woodward replyed, *none unless the Writer of the History of Jamaica*;[10] challenging him to assign any one more, which he did not. But appealing to Dr Mead,[11] which was only a small mean Shift, the Dr was forced to give it against him. Those recited were the very Words Dr Woodward used; & whether they were unfit, you are a proper Judge. That they were not spoken till after Dr Sloane had made his Grimaces, twice or thrice, you were assur'd by Mr Clavel:[12] & Mr Knight[13] is ready to confirm the same, if you please to ask him. He is a Gentleman as modest, impartial, & creditable: & indeed with Mr Clavel, & Dr Harris, sate so fronting Dr Sloane as to be able to see his Face & Grimaces. The rest, which were but few, sate out of fair View. In particular, Mr Moreland,[14] that with so much solemn Formality made Asseveration that, to the very best of his Memory, the Words preceded the Grimaces, sate directly behind Dr Sloane; so that he neither did nor possibly could see one of those Grimaces.

As People that are in the Wrong are usualy very noisy, Dr Sloane & his Agents make mighty Remonstrances & Complaints every where; which has caus'd much Talk & Inquiry into the Matter: & several have been so free as to declare upon the whole, there was no real Cause of Convening the Council, especialy in so much Hurry & Hast. They say such Treatment may come to be their own Lot, in Case there be such a Readyness at any of Dr Sloanes Calls, to lead up his Mermidons, & fight his Battels. Tis known he has the Knack of packing Councils: & can get Instruments there enow to act for him; thô Care will be taken to prevent that Practice for the Future. The Man who does not see that at the late Meeting there was a pretty odd Design, deeply laid, & well concerted, is assuredly not over quick-Sighted. Then they observe the Summons were not deliverd till the very Day before; that Dr Woodward or his Friends might have no Time to enquire into the Design; & indeed they had little Cause to suspect any such. Tis reasonable, needful, & indeed the Custom of all the Courts in the World, to give a Person accused due Notice thereof: & Time to make his Defence. I perceive Dr Woodward sincerely believes you his Friend: & by his Demeanour on all Occasions, he has shew'd himself heartily yours. But others greatly wonder why this Design should be kept so very private, as, upon Inquiry, I find it was: & that neither he, nor any of his Friends, should have the least Notice given them of it. You have declar'd, of some Time, to several, again & again, that Notice should be given to the Council beforehand of all Matters to be lay'd before them. You had complain'd of Dr Sloanes Artifices in surprizing you with Things at the Council, frequently very unfit, without having given you any previous Account. As upon others, you had declar'd to more then one Friend how *little qualified he was for the Post of Secretary*, so, upon these Occasions, you as freely declar'd him to be *a Tricking Fellow*: nay *a Villain*, & *Rascal*, for his deceitful and ill Usage of you, especialy in the Affair of Dr Wall.(15) Such Expressions do not fall forth of the Mouth of a Gentleman of your truely good Sense & Breeding without Cause. Indeed all allow you had very great & just Cause &, thô Dr Woodward has not used any such Expressions, he has had Causes as great & just, long & often; of which I have heard the Particulars, but shall not trouble you with them here. Dr Sloane assuredly does not take you to be over fervent in his Interest: & you may depend upon it he is as little fervent in yours. Tis generaly thought, upon the whole, the main of his Drift, in this Affair, was, by thus in his usual Manner surprizing you, to draw you in to be the Instrument of his Revenge upon Dr Woodward; thereby to involve you: & by that means realy sacrifice both, & kill two Birds with one Stone. 'Twas known Dr Harris had been absent, & out of Town, for some Time; &, if Mr Clavel, & Mr Loutharp,(16) by meer Chance, had not been there, the Design had probably succeeded; which would have been very unhappy on many Accounts. As to Dr Woodward in particular, he is not only a Man of Publick Character, & a Professor in Gresham College, but has as good a right to be there as any Man has to be in his own House: &

for the Council, who have themselves no Right to be there, but by his Courtesye, to go about without any just Cause to disturb him, which tis judg'd was the Design of that Meeting, was very Poetical indeed.(17) The Justice & Generosity of it had been likewise very extraordinary. The Lord Mayor & Court of Aldermen, a few years agoe, with very little Ceremony, sent an abrupt warning to the Royal Society to get speedily out of Gresham College. Dr Woodward interpos'd, having first obtain'd the Consent of his Brethren, the other Professors, in whose Power that was: &, in Opposition to that Warning, hath maintain'd the Royal Society quietly & commodiously there ever since. What Returns have been made him, & what Sort of Treatment he hath met with ever since, you can very well judge. This I believe may be granted that had a Slurr been thus surreptitiously put upon him, if it is thought he was so Tame as to acquiesce, which there's no great Reason to expect, his Brethren there, & his Friends in the World, would not suffer it: & had this come to be made publick you will easyly see how far that would have comported with your Honour and that of the Society.

The real Cause of all this Spleen to Dr Woodward is, Dr Sloane imagines he & his Friends were the Means of choosing Dr Harris Secretary; which has terribly broke in upon the old Schemes and Practices. If Dr Woodward could be slurrd, and Dr Harris cast out again, all would be set to rights: & Dr Sloane still sovereign of the whole. This indeed is the very Thing design'd to be brought about by this Bustle. I should be as sorry to see such a Design succeed, as I should to do Dr Sloane any Injury: or to obstruct him in any Thing that is good & honorable. If you resolve to persue the Affair complaind of to the Bottom: & find that Dr Sloane has endeavourd to deceive you, and is realy himself the Aggressor, tis hoped you will be as forward & zealous in your Marks of Resentment towards him, as tis apprehended you were towards the other.

Sr As you live at Distance from me, which deprives me of the Pleasure of Converse with you: & as I, & some others, how well soever our Wishes may be to ye Good & Honour of the Society, care not to enter into such a Fray openly at the Council so freely as Dr Sloanes Partizans have done, I take the Liberty to lay my Thoughts thus cursylaryly before you in Writing. I can tell you with great Truth tis wrote with real Respect to you: & faithly design'd for your Service. If what is offer'd, which is really my Judgment, appears reasonable, & proves of any Light & Use to you, tis what I aim at, & I have my End: if otherwise, no farther Harm is done than by Loss of the Time you spend in the Perusal, & you are still at Liberty to act according to your own Inclination. When this Affair is over, & the Days grown longer, I will take some Opportunity of Discoursing [with] you upon some other Affairs, that deserve the Consideration of a Person that wishes so well to the Society as you do. I confess

I think it, at present, very much off the Hinges: & that nothing will restore it but those Means that first rais'd it to so great Reputation; the settling of a Correspondence, & the making useful Experiments & Observations. If Dr Sloane will concurr in this, tis well; no body intends him any Molestation. Or if he can sit still, & suffer others to do it. But tis judged his Consciousness of his own Incapacity to joyn in that Way, & his Hopes to defeat it, are the true Motives to all this Pother, whatever else may be pretended.

Postscript

Since what I wrote above, there have been with me two of the Council; & two or three of the Fellows. I shall only just mention the Heads of what past.

1. That, at which this Contest happend, was not, according to the Charter & Statutes, a Meeting; there being neither President nor Vice-President in the Chair: & but 7 or 8 Fellows present.(18)

2. Mr Thorp, (19) as he is a private Fellow, ought not to appear at the Council. Were he Clerk indeed he might act there, by Permission. But the Statutes expresly declare that no Fellow shall be Clerk.(20) This, tis said, was the Reason why Dr Cockburn(21) offerd, on Wednesday, to officiate, in absence of the Secretaryes.

3. [That] The Dissolving the Council before the Minutes, & what the Secretary has wrote, be examind, rightly stated, & approv'd of by the Council present, is in some Measure a putting the Government of the Society soley into his Hands. Nay thô the Council approve of the Minutes, since Dr Sloane writes them upon loose Paper, & they are not fair Transcrib'd ratify'd or sign'd by any one else present, they are liable to be changed defalcated or inlarged. There have been long & often private Complaints made of such Practices: & several have wished, for that Reason, a Comitte were appointed to inspect the Books & Registers for some Years past.

4. It seems there is, & has been of late, much Flourishing and Menacing of Expelling Persons out of the Society that shall contradict Dr Sloane.(22) The President & Council have no power but by Charter & Statutes: and they assign onely one Cause of Expulsion,(23) viz: the going about to Injure & defame the Society. Now surely Dr Sloane can never be interpreted ye Society: nor indeed pass for a Person so infallable that no Man may presume to dispute any Thing that he says.

5. Such indirect Methods as have been, & are, taken by him, and his Cabal, for sinking the Reputation & Honour of a Gentleman, meerly because he will not fall into their Measures, is very barbarous: and may justly give Cause of Umbrage & Apprehension to every other Fellow of the Society. Tis not doubted but such violent irregular & arbitrary [*sic*] Proceedings will, by your prudent

Care, Vigilance, & firm Adhesion to the Statutes, be prevented for the Future; or the Royal Society, which in Truth has sufferd but too much from them already, will soon be brought to an End.

March: ye 28 1710
To
Sr Isaac Newton President of
the Royal Society

NOTES

(1) KMS. 151. The document is in an amanuensis hand and there is no direct clue to its authorship. It was clearly written by an opponent of Dr Hans Sloane, Secretary of the Royal Society since 1693 (compare vol. IV, p. 239, note (7); Sloane was created baronet only in 1716), and by a friend to Dr John Woodward, who may have inspired it. The letter was clearly intended to influence the Council Meeting of 29 March 1710.

(2) At this time, as Flamsteed has already noted (Letter 768) there was much resentment against Sloane's autocratic management of the Society, under Newton's tolerant eye. An impending move from Gresham College, though no fault of the officers, was resented by a number of Fellows, among them John Woodward (1665–1728), geologist and physician, who had been Professor of Physic at Gresham College since 1692, and became F.R.S. the following year. He was several times elected to the Society's Council, but his hot temper was notorious. He had fought a duel on a professional point with Richard Mead, and in 1706 had been admonished by the Council (with Newton in the Chair) for 'unjust reflections cast upon some of the Members' (*Council Minutes*, 15 May 1706; compare Letter 785 of 30 May below). Nothing contentious is recorded as occurring at the Council meeting of 22 March 1709/10, though the fact that Newton summoned another Council for the following week is evidence of some unrecorded discussion. At this next Council (on 29 March 1710) Woodward 'desired and insisted that the difference now depending might be put off 'till Dr Harris be present at the Councell' (see note (3) below). His motion did not succeed, and proceedings very uncomfortable for Woodward took place. His quarrel with Sloane began at the Society's meeting of 8 March 1709/10 when Sloane, as Secretary, read a paper 'translated from the History of the french Academy' [which we have failed to identify] concerned with gall stones: 'It was thought [records the minute tactfully] that Gall Stones & Bezoar stones were very different Bodyes'. Woodward's offensive words to Sloane were, according to the Council proceedings of 29 March: 'Speak Sense or English and we shall understand you' and 'If you understood Anatomy you would know better, or to that purpose'; however, it was alleged in mitigation that Sloane had previously 'made Grimaces, with a Laughter, and holding up his hands at Dr. Woodward before the reflecting Words above-mentioned were spoken'. The Council resolved that Sloane had not thereby provoked Woodward, and that Woodward's verbal aspersions 'tended to the Detriment of the Royal Society'.

(3) John Harris (? 1667–1719), D.D., F.R.S. 1696, and Secretary since 1709, best known for his *Lexicon Technicum* (London, 1704 etc.). He was friendly towards Woodward, whose *Essay towards a Natural History of the Earth* London, 1695 he had defended in *Remarks on some late Papers, relating to the Universal Deluge*... London, 1697.

(4) *Junto* at this time had a decidedly pejorative meaning: 'domineering faction'. The opprobrious epithet 'non-Solvents' (cf. 'non-jurors') glances at the insolvent condition of the

Society at this time; its purchase of the house in Crane Court was made possible only by special contributions from some Fellows and a heavy mortgage. An attempt had been made recently to penalize Fellows who failed to pay their dues to the Society, but this had been defeated on grounds of unconstitutionality by another group of Fellows.

(5) Compare Letter 768, note (3).

(6) Francis Robartes: see vol. IV, p. 431, note (6).

(7) Thomas Foley (d. 1733), great grandson of the Worcestershire ironmaster, had been elected F.R.S. in 1696 and was actually appointed Vice-President in 1709. He was raised to the peerage in 1712.

(8) Abraham Hill (1635–1721), one of the surviving Original Fellows of the Royal Society. He had served as Secretary long before, and was one of the Board of Visitors appointed in December 1710 to report on the Royal Observatory.

(9) The bezoar-stone is a concretion formed in the stomachs or intestines of some animals, usually ruminants. It was at this time still valued in the materia medica.

(10) Sloane had published *Catalogus plantarum quæ in insula Jamaica sponte proveniunt* in 1696, and the first volume of *The Natural History of Jamaica* in 1707. It is perhaps worth noting that the irrepressible William King parodied Sloane again in *The Present State of Physic in the Island of Cajamai*, London, ? 1710.

(11) Richard Mead (1673–1754), the leading physician of the day, F.R.S. since 1703.

(12) Walter Clavell, F.R.S. 1704, and now a member of the Council. He had studied for the Bar *c.* 1700 but was never admitted.

(13) John Knight (d. 1736), F.R.S. 1706, possibly a barrister.

(14) Joseph Morland (d. 1716), M.D. Leiden 1699, F.R.S. 1703. He was presumably related to Sir Samuel since, besides publishing a medical essay in 1713, he edited *Hydrostaticks or Instructions concerning Waterworks collected out of the Papers of Sir Samuel Morland*, London, 1697.

(15) We have not found the explanation of this remark. One Dr Wall published a paper on electricity in *Phil. Trans.*, **26** (1708–9), 69. Mottelay (*Bibliog. Hist. Elect. and Magn.*, London 1922, p. 152) identifies this writer with William Wall (1647–1728), divine and biblical scholar, without giving any reason for this identification. Dr Wall presented his experiments before the Royal Society on 2 June 1708 and was proposed Fellow on 30 June, but never elected.

(16) John Lowthorp (1659–1724), see vol. IV, p. 332, note (1).

(17) As the writer goes on to say, the Royal Society's tenure at Gresham College had been uncertain for many years. In 1705 a letter from the Mercers' Company—one of the controlling bodies of the College—had indicated an intention of ejecting the Society altogether (see Weld, *History*, I, p. 387; Lyons, *History*, pp. 131–3).

(18) This assertion is incorrect. Neither the Charters nor Statutes defined a meeting of the Society, nor contained any requirement that the President or Vice-President should occupy the Chair, nor required a quorum. Very small meetings had been minuted as such in the past, and on occasion a Fellow (not President or Deputy) had been voted into the Chair.

(19) John Thorpe (1682–1750), antiquary, F.R.S. 1705, M.D. 1710. He had served as Clerk to the Society since November 1706, when Humfrey Wanley resigned, though not appointed officially till February 1707/8. At this point he suspended his Fellowship (according to Lyons, *History*, p. 143) which he resumed when he resigned the Clerkship in 1713. (He certainly was not *re-elected* in 1713, and was obviously reputed a Fellow at the present time.)

(20) This was untrue. However, the Council on 27 January 1685/6 (Birch, *History*, IV, p. 453) had resolved that any Fellow elected as Clerk should resign his Fellowship. At a

meeting of the Society on the same day Edmond Halley was elected Clerk 'to be assistant to the Secretaries'. He did not resign his Fellowship.

(21) William Cockburn (1669–1739), M.D. Leiden 1691, F.R.S. 1696, a successful physician.

(22) See Letter 785 of 30 May for the relevance of this to Woodward.

(23) This is not correct (see *Record of the Royal Society*, fourth edition, 1940, p. 301) for a Fellow might be ejected, according to Statute, for disobedience to Statutes or Orders; for defamation of the Society; or for 'advisedly, and maliciously, do[ing] any thing to the damage and detriment thereof'—a wide and loose proviso.

775 COTES TO NEWTON

15 APRIL 1710

From the original in the University Library, Cambridge.[1]
For the answer see Letter 778

Sr.

We have printed so much of ye Copy you sent us yt I must now beg of You to think of finishing the remaining part, assoon as You can with convenience. The last sheet yt is printed off ends at page 251 of the old Edition & page 224 of ye new Edition.[2] The whole yt is finished shall be sent You by ye first oportunity. I have ventured to make some little alterations my self whilst I was correcting ye Press, such as I thought either Elegancy or Perspicuity or Truth sometimes required. I hope I shall have Yr pardon if I be found to have trusted perhaps too much to my own Judgment, it not being possible for me without great inconvenience to the Work & uneasiness to Yrself to have Yr approbation in every particular. The Pages which are next to be printed being somewhat more than usually intricate I have been looking over them before-hand. Pag: 270 Reg. 1[3] I think should begin thus—Si servetur tum Medij densitas in A tum velocitas quacum corpus projicitur, & mutetur—I must confess I cannot be certain yt I understand ye design of Reg: 4[4] & ye last part of Reg: 7 & therefore dare not venture to make any alteration without acquainting You wth it. I take it thus, yt in ye 4th Rule You are shewing how to find a Mean among all ye Densitys through which the Projectile passes, not an Arithmeticall Mean between ye two extream Densitys ye greatest & least, but a Mean of all ye Densitys considered togather, which will be somewhat greater yn that Arithmeticall Mean ye Number of Densitys which are greater yn it being somewhat more yn ye Number of Densitys which are lesser yn ye same. If this be Yr design I would thus alter ye 4th Rule wth Yr consent. Quoniam densitas Medij prope verticem Hyperbolæ major est quam in loco A, ut habeatur densitas mediocris, debet ratio minimæ tangentium GT ad tangentem AH

inveniri, & densitas in A augeri in ratione paulo majore quam semisummæ harum tangentium ad minimam tangentium GT. The latter part of ye 7th Rule[5] I understand thus. Simili methodo ex assumptis pluribus longitudinibus AH invenienda sunt plura puncta N & per omnia agenda Curva linea regularis $NNXN$ secans rectam $SMMM$ in X. Assumatur demum AH æqualis abscissæ SX & inde denuo inveniatur longitudo AK; & longitudines, quæ sunt ad assumptam longitudinem AI & hanc ultimam AH, ut longitudo AK per Experimentum cognita ad ultimo inventam longitudinem AK, erunt veræ longitudines AI & AH quas invenire oportuit. Hisce vero datis dabitur & resistentia Medij in loco A, quippe quæ sit ad vim gravitatis ut AH ad $2AI$; augenda est autem densitas Medij per Reg: 4, & resistentia modo inventa, in eadem ratione aucta, fiat accuratior. About ye end of the 8th Rule are these words quorum minor eligendus est which I would either leave out or print thus quorum minor potius eligendus est.[6] Pag: 274, lin: 2 should be $\frac{2TGq}{\overline{nn-n}\times VG}$.[7] There are others like this which I will not trouble you wth. Prop. XIV Prob. IV should be Prop. XIV. Theor. XI. Two lines lower are these words est ut summa vel differentia areæ per quam—I would leave out summa vel.[8] Corol. pag: 281 is restrained to ye 1st Case of Prop. 13, I would alter it and print it thus.[9] Igitur si longitudo aliqua V sumatur in ea ratione ad duplum longitudinis M, quæ oritur applicando aream DET ad BD, quam habet linea DA ad lineam DE spatium quod corpus ascensu vel descensu toto in Medio resistente describit, erit ad spatium quod in Medio non resistente eodem tempore describere posset, ut arearum illarum differentia ad $\frac{BD\times V^2}{4AB}$, ideoque ex dato tempore datur. Nam spatium in Medio non resistente est in duplicata ratione temporis, sive ut V^2, & ob datas BD & AB, ut $\frac{BD\times V^2}{4AB}$. Momentum hujus areæ sive huic æqualis $\frac{DAq\times BD\times M^2}{DEq\times AB}$, est ad momentum differentiæ arearum DET & $AbNK$ ut $\frac{DAq\times BD\times 2M\times m}{DEq\times AB}$ ad $\frac{AP\times BD\times m}{AB}$, hoc est, ut $\frac{DAq\times BD\times M}{DEq}$ ad $\frac{1}{2}AP\times BD$ sive ut $\frac{DAq}{DEq}\times DET$ ad DAP, adeoque ubi areæ DET & DAP quam minimæ sunt, in ratione æqualitatis. Æqualis igitur—&c. Page 286. 1 in. 5 must be thus corrected[10]—Rr & TQ seu ut $\frac{\frac{1}{2}VQ\times PQ}{SQ}$ & $\frac{\frac{1}{2}PQq}{SP}$ quas simul generant, hoc est, ut VQ & PQ seu OS & OP. This Corollary being thus corrected, the following must begin thus. Corol. 4.[11] Corpus itaque gyrari nequit in hac Spirali, nisi ubi vis resistentiæ minor est quam vis centripeta. Fiat resistentia æqualis vi centripetæ & Spiralis conveniet cum linea recta *PS*, inque hac recta—&c. Tis evident (by Corol. 1) yt ye descent along ye line *PS* cannot be made wth an uniform Velocity. Tis as evident I think yt it must be wth an uniform Velocity because ye Resistance & force of Gravity, being equall, mutually destroy each others effect, & consequently no Acceleration or Retardation of motion can be produced. I cannot at present see how to account for this difficulty & I choose rather to own my ignorance to You yn to run the hazard of leaving a Blemish in a Book I so much esteem.[12] Cor. 6. lin. ult. I would print thus—ut $\frac{OP}{OS}$, id est, ut secans anguli ejusdem, vel etiam reciproce ut Medij densitas.[13] If I mistake not ye design of ye 8th Corol: I would alter it thus[14]—Centro *S* intervallis continue proportionalibus *SA*, *SB*, *SC* &c describe circulos quotcunque, & statue tempus revolutionum omnium inter perimetros duorum quorumvis ex his circulis, in Medio de quo egimus, esse ad tempus revolutionum omnium inter eosdem in Medio proposito, ut Medij propositi densitas mediocris inter hos circulos ad Medij de quo egimus densitatem mediocrem inter eosdem quam proxime; Sed & in eadem quoque ratione esse secantem anguli quo Spiralis præfinita in Medio de quo egimus secat radium *AS* ad secantem anguli quo Spiralis nova secat radium eundem in Medio proposito: Atque etiam ut sunt eorundem angulorum tangentes ita esse numerum revolutionum inter circulos eosdem duos quam proxime. Si hæc passim—Prop. 16 must be altered; for by my reckoning, if ye centripetal force be as $\frac{1}{SP^{n+1}}$, the velocity will be as $\frac{1}{SP^{\frac{1}{2}n}}$, the Resistance as $\frac{\overline{1-\frac{1}{2}n,\ OS}}{OP\times SP^{n+1}}$ and consequently the Density as $\frac{\overline{1-\frac{1}{2}n,\ OS}}{OP\times SP}$.[15] With Yr consent I would add this Corollary. Si vis centripeta sit ut $\frac{1}{SP^{\text{cub}}}$, erit $1-\frac{1}{2}n=0$, adeoque Resistentia & Densitas Medij nulla erit, ut in Propositione Nona Libri primi. Another Corollary might be added to show in what cases ye Resistance is affirmative & in what cases Negative.[16] I beg of You to pardon my freedom in this Letter.

Your most humble & obliged
Servant
ROGER COTES

Cambridge Apr. 15*th* 1710

For Sr Isaac Newton
at his House
near ye College in Chelsey
By London.

NOTES

(1) Add. 3983, no. 1. There is a draft of the letter, used by Edleston, in Trinity College, Cambridge, MS. R.16.38, no. 37.

(2) That is, the press was stopped after printing sheet Ff 2. Here begins Book II, Section II, Lemma 2, the fluxions lemma.

(3) Newton has just traversed the complex argument of Proposition 10, dealing with a medium resisting in proportion to its density and the square of the velocity of a body moving through it. In the subsequent Scholium (*Principia* 1687; p. 269; Koyré and Cohen, *Principia*, I, p. 388) he sets out a series of rules for determining approximately the trajectory of a projectile in a uniformly resisting medium, starting from the assumption that this trajectory will not be analogous to the parabola described *in vacuo* but rather to the hyperbolas already defined in Proposition 10. In Rule I Cotes proposes to state specifically that both the density of the medium at point *A and the velocity of projection* remain unchanged; this addition was adopted.

(4) Newton has demonstrated (*Principia* 1687; p. 268) that (under the conditions stated) for the projectile to move hyperbolically the medium density along its path (from *A*, the origin) must vary inversely as a certain length *XY*, which diminishes as the projectile moves forward to a minimum value at the vertex. Hence the density must be supposed to *increase* (not diminish, as Newton had written) from ground level to the top of the trajectory. Since the density of air decreases with height, a correction must be applied, which is in the opposite sense to that originally proposed. Cotes' wording was adopted.

(5) Continuing his approximation of the trajectory, in the first edition Newton shows how, by graphical methods, the ranges *AK*, *Ak* may be found which are to be obtained by shooting off from *A* along the lines *AH*, *Ah* (at equal velocities). (See Koyré and Cohen, *Principia*, II, pp. 391–2, referring to the diagrams there.) The method involves guessing the length *AH*; the accuracy of the initial guess is confirmed by comparing the ratio *AK*/*Ak* for two experimentally measured ranges at this velocity. Since the guessed length *AH* is likely to be wrong, Newton purports to show how, by a second graphical method, its correct value *AH'* may be obtained by interpolation from a series of incorrect guesses. However, Newton's account of this interpolation method was incomplete for although it brought *AK* into correct ratio with *Ak* (as determined by the experiment) it did not also bring *AH'* into correct ratio with *AI* (as required by the theory). Hence Cotes' correct addition: 'and the lengths which are to the assumed length *AI* and this last determined [length] *AH* as the length *AK* (as known by experiment) to the lately found length *AK* will be the true lengths *AI* and *AK*, which were to be found. But when these lengths are known, the resistance of the medium at the point *A* [is known] also, since this is to the force of gravity as *AH* to 2*AI*.'

(6) They were omitted.

(7) The first edition has a root sign before this expression which was probably a printer's error, and an upper case *X* in place of the multiplication sign, which certainly was. Cotes clearly took great trouble over his work. However, this part of the *Principia* was by no means correct yet (see below, Letter 951).

(8) Cotes is right in saying that the words 'the sum or' have no place here, and they were omitted.

(9) Newton had already redrafted this argument, but without altogether freeing it from errors. These are now corrected by Cotes, and the corollary was printed in this form.

(10) Coroll. 3 to Prop. 15. Newton had both the resistance (*Rr*) and the centripetal force

(TQ) too great by a factor of two; hence his expression of the ratio between these two magnitudes ($\frac{1}{2}OS:OP$) was correct. Cotes has observed the second of Newton's errors only, and therefore now proposes the ratio $OS:OP$. In his next letter he draws attention to the first error also.

(11) This correction proved unnecessary for the reason just explained.

(12) The *immediate* difficulty arises from the same source as indicated in note (10); it is discussed fruitlessly in subsequent letters. The printed text (*Principia* 1713; p. 256; Koyré and Cohen, *Principia*, I, p. 411) states that since the resistance must be half the centripetal force to produce the spiral motion under consideration, the acceleration is likewise halved, and so the velocities are decreased in the ratio of $1:\sqrt{2}$, and the times correspondingly increased.

(13) This was adopted.

(14) This rewording was adopted.

(15) The error in the 1687 version of Proposition 16 was considerable enough to attract the attention of Johann Bernoulli (see *Mémoires de l'Académie des Sciences*, vol. for 1711 (Paris, 1714), p. 47, 'Extrait d'une lettre de Monsieur Bernoulli, Ecrite de Basle le 10 Janvier 1711'). The major correction that Cotes makes here to the formulae meant that Newton eventually had to reword the enunciation of the Proposition. Given a centripetal force inversely proportional to some power (say $n+1$) of the distance, in order to produce spiral motion the density of the medium must be inversely proportional to the distance from the centre of force, *not* inversely proportional to the nth power of the distance, as Newton had stated in the first edition. (See Koyré and Cohen, *Principia*, I, p. 415.) Newton did not at first note the full force of Cotes' correction, for in his next Letter (778) he notes Cotes' correction of $\frac{1}{2}n$ to $1-\frac{1}{2}n$, but not the more significant change of SP^n to SP. Cotes points out the latter correction again in Letter 779, and emphasizes that Newton will have to alter the wording of the Proposition. Newton does so in Letter 780.

(16) Newton approved these suggestions in his Letters 778 and 780 and the book was printed accordingly (*Principia* 1713; p. 258).

776 LOWNDES TO THE MINT

28 APRIL 1710

From the copy in the Public Record Office.[1]

For the answer see Letter 782

Officers of the Mint about a Petition of the Pewterers' [Company] of London

Gentlemen,

The Queen having been pleas'd by Order of Councill to refer to my Lord Trea[su]rers Consideration a Petition of the Pewterers of London relating to Her Mats. Preemption of Tynn His Lop. has Commanded me, to transmitt the said Order and Petition to You to Consider thereof and Report to his Lop. a true State of the Petitioners Case wth. Yr. opinion thereupon

I am &c: 28th *Aprill* 1710

WM. LOWNDES

NOTE

(1) T/27, 19, 180. The petition from the Pewterers' Company is not now with this letter copy.

777 COTES TO NEWTON

29 APRIL 1710

From the original in the University Library, Cambridge.(1)

For the answer see Letters 778 and 780

Cambridge Aprill 29th 1710

Sr.

I suppose Mr Crownfield(2) our Printer has delivered to You all the sheets that are allready printed off. I desired him to wait upon You before he return'd to Cambridge that I might have Yr answer to my former Letter or at least to the first part of it. The difficulty which I proposed to You concerning the 4th Corollary of Prop XV I have since removed.(3) Upon examination of that Proposition I think I have observed another mistake in ye 3d Corollary which ballances that I before mentioned to You in the same Corollary.(4) For if I be not deceived the force of Resistance is to ye Centripetall force as $\frac{1}{2}Rr$ to TQ not as Rr to TQ. You will see my reasons in the following alterations which I propose to You. Pag. 284. l: 6(5) Ponantur quæ in superiore Lemmate, & producatur SQ ad V, ut sit SV æqualis SP. Tempore quovis, in Medio resistente, describat corpus arcum quam minimum PQ & tempore duplo arcum quam minimum PR; & decrementa horum arcuum ex resistentia oriunda, sive defectus ab arcubus qui in Medio non resistente ijsdem temporibus describerentur, erunt ad invicem ut quadrata temporum in quibus generantur: est itaque decrementum arcus PQ pars quarta decrementi arcus PR. Postquam vero descriptus est arcus PQ in Medio resistente, si areæ PSQ æqualis capiatur area QSr, erit Qr arcus quem tempore reliquo corpus describet absque ulteriore resistentia arcuumque QR, Qr differentia Rr dupla erit decrementi arcus PQ; adeoque vis resistentiæ & vis centripeta sunt ad invicem, ut lineolæ $\frac{1}{2}Rr$ & TQ, quas simul generant. Quoniam vis centripeta, qua corpus—Pag: 285. l: 4 —$\frac{1}{2}VQ$ fit æqualitatis. Quoniam decrementum Arcus PQ, ex resistentia oriundum, sive hujus duplum Rr est ut resistentia & quadratum temporis conjunctim; erit Resistentia ut $\frac{Rr}{PQq \times SP}$. Erat autem PQ ad—Pag: 286: l: 4.(6) Nam vires illæ sunt ut $\frac{1}{2}Rr$ & TQ sive ut $\frac{\frac{1}{4}VQ \times PQ}{SQ}$ & $\frac{\frac{1}{2}PQq}{SP}$, hoc est, ut $\frac{1}{2}VQ$ &

PQ seu $\frac{1}{2}OS$ & *OP*—I satisfied my self more fully that I am not mistaken in my reasoning after this manner. If ye Centripetal force be as $\frac{1}{SP^{n+1}}$, the force of Resistance will be to ye centripetal force as $\frac{1}{2}Rr$ to *TQ* ie, as $\overline{1-\frac{1}{2}n}$, *OS* to *OP*. Put ye Centripetall force as $\frac{1}{SP}$; & You will have $n=0$, & consequently $\overline{1-\frac{1}{2}n}$, *OS* to *OP* as *OS* to *OP*. Therefore when ye Spirall coincides with the line *PS* the resistance will be equall to ye Centripetal force, & ye Body will descend wth an uniform velocity as it ought to do by Cor. 1 Prop. XV and Cor. 5. Prop. IV Lib I compared togather, and also upon this consideration that ye Velocity in ye Spiral of Prop: XVI is as $\frac{1}{SP^{\frac{1}{2}n}}$ ie as $\frac{1}{SP^0}$. I have some things further to propose to You about ye remaining part of Yr Copy which I will not trouble You wth 'till I have Yr answer to my former Letter

Yr most obliged & humble Servant
ROGER COTES.

For Sr Isaac Newton
at his House near the College
in Chelsey

By London

NOTES

(1) Add. 3983, no. 2. There is a copy of this letter in Trinity College Library, Cambridge (R.16.38, no. 37). It is dated 30 April 1710 and was printed in Edleston, *Correspondence*, pp. 12–14.

(2) Cornelius Crownfield [? Cronfelt], described as a 'Dutchman who had been a soldier', was appointed Inspector (effectively, manager) of the Cambridge Press in 1698. For the reformation of printing in Cambridge effected by Richard Bentley and Crownfield see S. C. Roberts, *A History of the Cambridge University Press, 1521–1921*, Cambridge, 1921, Ch. v. He died in 1743.

(3) See Letter 775, notes (10)–(12).

(4) See Letter 775, note (10).

(5) This is Proposition 15; the second edition (pp. 254–6) adopts Cotes' wording without any changes of mathematical significance. Cotes has now perceived that his proposals in his last letter for rewording p. 286 were partially incorrect, because of the formerly unnoticed error occurring on p. 284 (first edition). He now modifies parts of the whole argument of the proposition so that the correct expression for the ratio of the resistance to the centripetal force is logically justified. (See Koyré and Cohen, *Principia*, I, pp. 409–11.)

(6) See Letter 775, note (15).

778 NEWTON TO COTES

1 AND 2 MAY 1710

From the original in Trinity College Library, Cambridge.(1)
Reply to Letters 775 and 777; for the answer see Letter 779

Chelsea near London May 1st 1710

Sr

I thank you for your letter with your remarks upon the papers now in the Press under your care. As soon as I could get some time to think on things of this kind, from wch I have of late years disused myself, I examined them, & all your corrections may stand till you come at page 287. In page 286 lin 4 for $\frac{1}{2}OS$ read OS.(2) In the same page let Corol. 4 stand thus. Corpus itaque gyrari nequit in hac spirali nisi ubi vis resistentiæ minor est quam vis centripeta. Fiat resistentia æqualis vi centripetæ, et spiralis conveniet cum linea recta *PS*, et motus corporis cessabit.(3) In page 287 & 288 the 8th Corollary may remain as in the Copy I sent you. In page 289 let the 16th Proposition end thus et resistentia in *P* ut $\frac{Rr}{PQ^q\times SP^n}$, sive ut $\frac{1-\frac{1}{2}n,\ VQ}{PQ\times SP^n\times SQ}$, adeoque ut $\frac{1-\frac{1}{2}n,\ OS}{OP\times SP^{n+1}}$, hoc est $\left(\text{ob datum } \frac{1-\frac{1}{2}n,\ OS}{OP}\right)$, reciproce ut SP^{n+1}. Et propterea densitas in *P* est reciproce ut SP^n.(4)

Corol. 1. Si vis centripeta sit reciproce ut SP^{cub}, erit $1-\frac{1}{2}n=0$, adeoque resistentia et densitas Medij nulla erit ut in Propositione nona Libri primi.

Corol. 2. Si vis centripeta sit reciproce ut radij *SP* dignitas aliqua cujus index est major numero 3, resistentia affirmativa in negativam mutabitur.(5)

When you sent me the sheets last printed off, I happened to be away from home, but a[t] night found them left at my house, & thank you for them. I am going to finish the next part of the copy I am to send you, & I hope to have it ready in due time if some experiments succeed.(6) I thank you once more for your corrections & for your care of the edition. I am

Sr

Your most humble
& most obedient servant
IS. NEWTON

P.S. After the writing of this Letter I received your second Letter dated Apr. 29. In the alterations you propose to be made in Prop. XV you say.(7) Postquam vero descriptus est arcus *PQ* in Medio resistente, si areæ *PSQ* æqualis capiatur area *QSr*, erit *Qr* arcus quem tempore reliquo corpus describet absque ulteriore resistentia. And this would be true if the velocity of the body at Q were the same as when the arch PQ is described in the same time in Medio non resistente. But the velocity at Q being less in Medio resistente then in non resistente, the arch Qr will be less in the same proportion & thereby reduce Rr to half the bigness, & make the resistance to the centripetal force as Rr to TQ. I hope therefore that what I have written on the other page of this Letter is right & that your difficulty will be removed by the words & motus corporis cessabit.(8)

I am Yours
I.N.

May 2d

For the Rnd(9) Mr Roger Cotes Professor
of Mathematicks and Fellow of Trinity
College in Cambridge

NOTES

(1) R.16.38, no. 3, printed in Edleston, *Correspondence*, pp. 14–16. There are two drafts of this letter by Newton in U.L.C. Add. 3984, nos. 1 and 2, showing a number of variants.

(2) Newton here accepts Cotes' 'correction' in Letter 775, which (as we have seen) was mistaken.

(3) Again, Newton follows Cotes' 'correction' and removes the factor of one-half; the reading of the first edition will be restored later. Compare Letter 775, notes (10) to (12).

(4) This is still not right, as the phrase 'the density in P is reciprocally as SP^n' has been copied from the first edition. It should read 'as SP' (as Newton agreed on 13 May), this being the crucial change for the sense of the proposition. In the second of the drafts noted above Newton has deleted the index n.

(5) As worded here, these two new corollaries suggested by Cotes passed into the second edition.

(6) Newton's thoughts are certainly on Prop. 40, Prob. 10 (*Principia* 1687; p. 338); in 1713 it became Prob. 9 (p. 317); see also Koyré and Cohen, *Principia*, I, p. 495. Not only did he reconstruct the proposition itself, but he completely altered the empirical content of the following scholium. In the first edition he had drawn data from the oscillations of pendulums in various media; he was now to introduce some direct experiments on the descent of heavy bodies in water, apparently made by himself, and others on the descent of bodies in air which were to be made by Francis Hauksbee in early June, 1710. Newton may refer to his own experiments here or (more likely) to those of Hauksbee.

(7) 'And after the arc PQ has actually been described in a resisting medium, if the area QSr be taken equal to the area PSQ, the arc Qr will be that which the body in the remaining time will describe without further resistance'. Cotes countered Newton's objection to this argument, which he confessed to be well-founded, in his letter of 7 May (Letter 779), when he explained that the objectionable words had been added as a (mistaken) afterthought and quoted his originally intended formulation for Proposition 15, which was ultimately printed.

(8) This was, of course, incorrect as Newton admitted on 13 May (Letter 780).

(9) Cotes was ordained deacon on 29 May 1713, and priest on the following day. Newton simply assumed that he was in orders already.

779 COTES TO NEWTON

7 MAY 1710

From the original in the University Library, Cambridge.[1]
Reply to Letter 778; for the answer see Letter 780

Cambridge May 7 1710

Sr

I received Yr Letter by the last Post. I am not satisfied yt Yr words [& motus corporis cessabit] will remove the difficulty proposed. They cannot in my opinion be reconciled with Cor. 1. I acknowledge Yr objection to be just against those words of mine [erit *Qr* arcus quem tempore reliquo corpus describet absque ulteriore resistentia] I remember that I inserted them into my Letter as I was hastily transcribing yt passage from another paper, & was myself sensible of the mistake soon after my Letter was gone from me. The alteration which I proposed, as it stood in that Paper, was thus.[2] Ponantur quæ in superiore Lemmate & producatur *SQ* ad *V* ut sit *SV* æqualis *SP*. Tempore quovis, in Medio resistente, describat corpus arcum quam minimum *PQ*, & tempore duplo arcum quam minimum *PR*; & decrementa horum arcuum ex resistentia oriunda, sive defectus ab arcubus qui in Medio non resistente ijsdem temporibus describerentur, erunt ad invicem ut quadrata temporum in quibus generantur: est itaque decrementum arcus *PQ* pars quarta decrementi arcus *PR*. Unde etiam, si areæ *PSQ* æqualis capiatur area *QSr*, erit decrementum arcus *PQ* æquale dimidio lineolæ *Rr*; adeoque vis resistentiæ & vis centripeta sunt ad invicem ut lineolæ $\frac{1}{2}Rr$ & *TQ* quas simul generant. I am yet of opinion that this alteration is just & that the resistance is to the Centripetal force as $\frac{1}{2}Rr$ to *TQ*; Yr own objection does, I think, if carefully considered, prove it to be so. To avoid further misunderstanding I will put down my Demonstration more at large, thus[3]

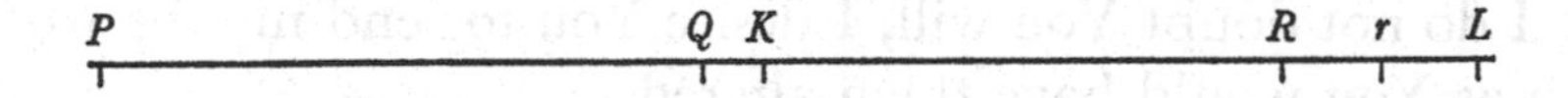

Tempore quovis, in Medio resistente, describat corpus arcum quam minimum *PQ* & tempore duplo arcum quam minimum *PR*; & decrementa horum arcuum ex resistentia oriunda, sive defectus [*QK*, *RL*] ab arcubus [*PK*, *PL*] qui in Medio non resistente ijsdem temporibus describerentur, erunt ad invicem ut quadrata temporum in quibus generantur: Est itaque decrementum [*QK*] arcus *PQ* pars quarta decrementi [*RL*] arcus *PR*. Unde etiam, si areæ *PSQ*

æqualis capiatur area QSr, erit decrementum [QK] arcus PQ æquale dimidio lineolæ Rr. [Nam ut SQ ad SP ita PK ad KL ita PQ ad Qr ita, dividendo, QK ad $KL-Qr$; ergo componendo PK ad PL ut QK ad ($QK+KL-Qr$ sive) rL, unde $rL=2QK$, sed erat $RL=4QK$, itaque $Rr=2QK$] adeoque vis resistentiæ & vis centripeta sunt ad invicem ut lineolæ QK vel $\frac{1}{2}Rr$ & TQ quas simul generant. This I take for a direct demonstration of ye truth of what I proposed, & if You will be pleased to consider what I offered at the end of my second Letter You will find yt also to amount to a Demonstratio per absurdum. I did there assume ye proportion of the resistance to the centripetal force to be as $\frac{1}{2}Rr$ to TQ, & from yt assumption I deduced a consequence whose truth is very evident upon other considerations. But if You take ye proportion to be as Rr to TQ or any other way different from that of $\frac{1}{2}Rr$ to TQ the consequence will be as evidently false. Therefore the proportion can be no other than yt of $\frac{1}{2}Rr$ to TQ. You say in Yr Letter that ye 8th Cor. may remain as in Yr copy, but in Yr copy there are no alterations of ye first Edition. That You may see the reason I had for the alteration I proposed I will put N for ye Number of Revolutions, T for ye Time of those Revolutions, D for ye Density of ye Medium, t for ye Tangent of the angle, s for ye secant of the same. Now in the 6th Corollary You put N as t, T as $\frac{1}{D}$ or s; but in Cor. 8 You put N as $\frac{1}{D}$ or t, and T as s. The alteration which I proposed was to make ye 8th Corollary agree with the 6th, for I am satisfied of ye Truth of ye 6th. In my first Letter I took notice of two mistakes in Prop XVI. You have consented yt one of them may be amended by putting $1-\frac{1}{2}n$ for $\frac{1}{2}n$. The other You seem not to have observed, which was, yt the Density is not reciprocally as SP^n, but reciprocally as SP:(4) For ye resistance in P being as $\frac{\overline{1-\frac{1}{2}n,\ OS}}{OP\times SP^{n+1}}$ and the Velocity in P as $\frac{1}{SP^{\frac{1}{2}n}}$, it follows that the Density in P is as $\frac{\overline{1-\frac{1}{2}n,\ OS}}{OP\times SP}$ not as $\frac{\overline{1-\frac{1}{2}n,\ OS}}{OP\times SP^n}$ the Density being as ye Resistance directly & ye square of the Velocity inversly. If you consent to this correction, as I do not doubt You will, I desire You to send me the words of the Proposition as You would have them altered.

Yr most humble & most

obedient servant

ROGER COTES

It seems to me not improper to add somewhere in this XVI Prop or in a Corollary to it That ye force of resistance is to ye Centripetall force as $\overline{1-\frac{1}{2}n}$, OS to OP

For Sr Isaac Newton
at his House near ye College
in Chelsea
near
London

NOTES

(1) Add. 3983, no. 23. Printed from Cotes' copy in Edleston, *Correspondence*, pp. 16–18.

(2) The text following was in fact printed in the second edition, p. 254. (Koyré and Cohen, *Principia*, I, p. 409). Edleston points out that the draft of Letter 777 contains the words printed here but partially struck out and replaced by those we have printed on p. 29.

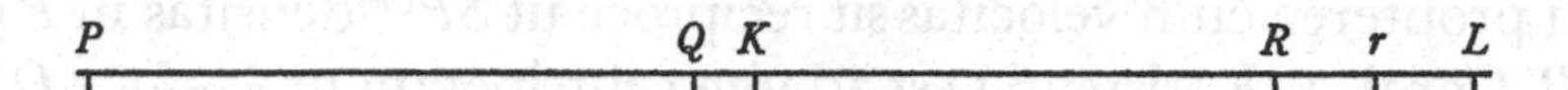

(3) 'Let a body describe in any arbitrary time in a resisting medium the least possible arc PQ, and in a double time the least possible arc PR; and the decrements of these arcs arising from the resistance, that is to say the deficiencies [QK, RL] from the arcs [PK, PL] which they would have described in the same times in a medium without resistance will be to each other as the squares of the times in which they are generated. Accordingly the decrement [QK] of the arc PQ is one fourth part of the decrement RL of the arc PR. Whence also if the area QSr be taken equal to the area PSQ, the decrement [QK] of the arc PQ will be equal to half the little line Rr. [For as SQ to SP, so is PK to KL and PQ to Qr, and so by division QK to $(KL-Qr)$; Therefore *componendo* PK to PL is as QK to $(QK+KL-Qr)$ or rL, whence $rL=2QK$. But $RL=4QK$, hence $Rr=2QK$]; and so the force of resistance and the centripetal force are to each other as the little lines QK or $\frac{1}{2}Rr$ and TQ which they generate at the same time.'

(4) See Letter 775, note (15).

780 NEWTON TO COTES

13 MAY 1710

From the original in Trinity College Library, Cambridge.(1)
Reply to Letters 777 and 779; for the answer see Letter 783

Chelsea. 13 *May*. 1710.

Mr Professor

I have reconsidered the 15th Proposition with its Corollaries & they may stand as you have put them in your Letters. But in pag. 285 lin. 13 after the word *coincident* add the words, *et angulus PSV fit rectus*.(2)

Let the 16th Proposition stand thus

Prop XVI. Theor. XII.[3]

Si Medij densitas in locis singulis sit reciproce ut distantia locorum a centro immobili, sitque vis centripeta reciproce ut dignitas quælibet ejusdem distantiæ: dico quod corpus gyrari potest in spirali quæ radios omnes a centro illo ductos intersecat in angulo dato.

Demonstratur eadem methodo cum Propositione superiore. Nam si vis centripeta in P sit reciproce ut distantiæ SP dignitas quælibet SP^{n+1} cujus index est $n+1$; colligetur ut supra, quod tempus quo corpus describit arcum quemvis PQ erit ut $PQ \times SP^n$,[4] et resistentia in P ut $\frac{Rr}{PQ^q \times SP^n}$, sive ut $\frac{1-\frac{1}{2}n,\ VQ}{PQ \times SP^n \times SQ}$, adeoque ut $\frac{1-\frac{1}{2}n,\ OS}{OP \times SP^{n+1}}$, hoc est, ob datum $\frac{1-\frac{1}{2}n,\ OS}{OP}$, reciproce ut SP^{n+1}. Et propterea cum velocitas sit reciproce ut $SP^{\frac{1}{2}n}$ densitas in P erit reciproce ut SP. Corol. 1. Resistentia est ad vim centripetam ut $1-\frac{1}{2}n \times OS$ ad OP. Corol. 2. Si vis centripeta sit reciproce ut SP^{cub}, erit $1-\frac{1}{2}n=0$, adeoque resistentia et densitas Medij nulla erit, ut in Propositione nona Libri primi.
Corol. 3. Si vis centripeta sit reciproce ut dignitas aliqua radii SP cujus index est major numero 3, resistentia affirmativa in negativam mutabitur.

Pag. 289, lin. 14. For *data lege* read *data velocitatis lege*.

Your most humble servant
IS. NEWTON

NOTES

(1) R.16.38, no. 45, printed in Edleston, *Correspondence*, pp. 19–20. There are again two (longer) drafts in U.L.C. Add. 3984, nos. 3 and 4.

(2) The catchword should be *coincidunt* (in both editions) and the angle (as Cotes marked on the letter and printed) PVQ.

(3) It becomes Theorem 13 in the second edition. At last accepting in full Cotes' corrections of Letter 775, Newton is forced to reword the conditions of the proposition. This form, with further amendments, was adopted in print.

(4) As Cotes noted, this should read $SP^{\frac{1}{2}n}$ (as in the first edition).

781 COTES TO NEWTON

17 MAY 1710

From the copy in Trinity College Library, Cambridge(1)

Cambridge May 17th 1710

Sr.

After I had received Yr Letter I wrote to you again about a week ago,(2) about some difficultys which still remain with me. The Compositor is now at a stand, & I dare not let him go on till you shall be pleased to send me Yr answer.

Yr most Obedient and Faithfull
Servt.
ROGER COTES

NOTES

(1) R.16.38, no. 46.
(2) See Letter 779.

782 THE MINT TO GODOLPHIN

17 MAY 1710

From the original in the Public Record Office.(1) Reply to Letter 776

To the Most Honorable the Lord High Treasurer
of Great Brittain

May it please your Lordship

In Obedience to your Lordship's Order of Reference to us upon the annexed Petition of the Pewterers of London to her Majesty in Councill,(2) We humbly represent to your Lordship that we have considered the same, and upon examining the Allegations thereof do find that the Petitioners are by law oblidged to make their pewter perfectly fine (which we believe they have complyed with) and that on the contrary forreign Nations Manufacture their[s] by mixing lead at pleasure with Tin, whereby they may undersell the Petitioners, as we Judge by the several tryalls and experiments they have made before us for sundry sorts of forreign pewter, the best of which has appeared to be considerably inferior in goodness to that Manufactured in England.

We do likewise find that the Pewterers have some times purchased 500000 £wt. of Tin per Annum of the Stannarys to be totally manufactured by them, paying £3. per hundred to the Owners of said Tin, and 18sh. per hundred to the King or Prince of Wales, for the right of preemption, and covenanting that no other Tin should be solld to be manufacted in England, nor disposed of to the Merchants at a lower rate then to themselves, and when other subjects have farmed the Tin, they have sometimes abated 18sh per hundred to the Pewterers for 100000£wt. of tin per Annum, which abatement we humbly conceive to be the price of the preemption: But the Petitioners have not made it appear to us that the Crown upon farming the Tin hath sold it at a lower rate to the Pewterers then to the Merchants.

We Alsoe humbly certifie to your Lordship that the Dutys of three shillings per hundred upon Tin, and two shillings per hundred upon Pewter exported do determine the 1st of August next, as the petition setts forth, and that it appeares to us by Certificate we have from the Office of the Inspector General at the Customehouse that the pewter exported for Nine years last past hath amounted at a Medium to about Two hundred Tun per Annum.

And further we humbly represent to your Lordships that the Petitioners do not now claim an abatement of the price of Tin as a matter of right, but submitt their case, as to the Encouragement of the exportation of pewter, to her Majes. Grace and pleasure.(3)

All which is most humbly submitted to your Lordships great Wisdom

Mint Office the 17*th May* 1710

CRAV: PEYTON
IS. NEWTON
JN. ELLIS

NOTES

(1) T/1, 122, no. 17A. The report is in a clerical hand, signed by the Officers.

(2) Again, the petition is not in the file, though the Pewterers' rejoinder is. The drift of it is clear enough: the Pewterers, competing with foreign manufacturers using only about 75% English tin to 25% lead, wished to be able to buy the Queen's tin at less than the fixed market price.

(3) On 13 June 1710 the Pewterers submitted to Oxford a criticism of this report, which they argued did not do justice to their case, and again asked 'to partake in some proportion of her Majesty's Royal Bounty'.

783 COTES TO NEWTON

20 MAY 1710

From the original in the University Library, Cambridge.(1)

Reply to Letter 780; for the answer see Letter 784

Cambridge May 20th 1710

Sr.

I thank You for Yr last Letter which came very seasonably. I now beg leave to propose some few alterations in ye remaining part of Yr Copy. Pag. 293 l: 1 I would read thus—secunda *BFK* (per Prop XIX) pro mensura sua æqualiter premuntur. lin. 4 Hac pressione, pro mensura sua, & insuper(2)—Pag 303. l: 6.—nisi forte per particulas intermedias virtute illa auctas—I think these words were better to be left out, as I apprehend it, they alter ye case of the Proposition.(3) l: 11 Ut si particula unaquæque—quadrato-cubi densitatis. I think also yt this whole Period might be omitted, the two Propositions contained in it seeming to me to be erroneous, unless I mistake the sense of Yr words.(4) Pag. 304 Corol 5 & 6 for quadratum temporis directe You have substituted in Yr Copy quadrato-quadratum temporis directe. I find written in the Margin of Yr book by a different hand quadr. quadratum temporis (credo). This Marginal note I beleive was the only occasion of Yr making this alteration.(5) Page 308 l: 10 I would omit the words si verbi gratia arcus sit altero duplo major.(6) With Yr leave I would begin ye 311 page thus.(7) Est itaque incrementum velocitatis ut $V-R$ & particula illa temporis in qua factum est conjunctim, sed & velocitas ipsa est ut incrementum contemporaneum Spatii descripti directe & particula eadem temporis inverse. Unde cum resistentia (per Hypothesin) sit ut quadratum velocitatis, incrementum resistentiæ (per Lem: II) erit ut velocitas & incrementum velocitatis conjunctim, id est, ut momentum Spatij & $V-R$ conjunctim, atque adeo si momentum—In my opinion this alteration is necessary to make the Demonstration accurate. When I first look'd over this passage, upon account of it, I thought the whole Construction was erroneous. I therefore set my self, after the following manner, to examine how it ought to be, which I here put down for a further use I have of it.(8) Taking x, z, v for quantitys analogous to ye Force arising from ye Gravity of ye Pendulous body, ye Force of resistance, & ye velocity in D, tis evident yt the Arch CD will also be as x, & ye Fluxion of the Space BD already described will be as $-\dot{x}$. If therefore $\dot{t}$ be put for the Moment of Time in which ye Fluxion of ye Space $-\dot{x}$, ye Fluxion of ye Velocity $\dot{v}$, ye Fluxion of ye Resistance $\dot{z}$ are generated; You will have $v \parallel -\frac{\dot{x}}{\dot{t}}$, $\dot{v} \parallel \overline{x-z} \times \dot{t}$.

But $z \| v\dot{v}$ & therefore $\dot{z} \| v\dot{v} \| -\frac{\dot{x}}{t} \times \overline{x-z} \times t \| z\dot{x} - x\dot{x}$.

Assuming therefore the determinate quantity (a) of a just magnitude, You will have this Æquation $a\dot{z} = z\dot{x} - x\dot{x}$. To construct this æquation I introduced another indeterminate quantity (y) putting $z = p + qx + ry$ & $\dot{z} = q\dot{x} + r\dot{y}$ which values of z & $\dot{z}$ being substituted in the former aequation I obtained this other

$$aq\dot{x} + ar\dot{y} = p\dot{x} + qx\dot{x} + ry\dot{x} - x\dot{x}.$$

Then putting $q = 1$, $p = a$, I had ye two following æquations

$$\frac{a\dot{y}}{y} = \dot{x}, \quad z = a + x + ry,$$

& the construction of these two æquations agreed intirely with Yr own Solution of the Problem. Being satisfied by this Analysis of the truth of Yr conclusion I easily saw yt my former difficulty lay in the Ambiguity of ye word data in lin. 1. & 5, and ye word detur in lin: 6. which I think may be remedied by the alteration which I propose. Page 312. l: 21. I would leave out ye word quamproxime.[9] Page 313. l: 29 I would conclude ye Demonstration thus. & ex æquo perturbate *Fh* seu *MN* ad *Dd* ut *DK* ad *CF* seu *CM*: ideoque summa omnium $MN \times CM$, id est, $\frac{1}{2}CAq - \frac{1}{2}Caq$ seu $Aa \times \frac{1}{2}aB$, æqualis erit summæ omnium $Dd \times DK$, id est, areæ *BKkVTa*, quam rectangula omnia $Dd \times DK$ seu *DKkd* componunt.[10] Q.E.D. I was further satisfied that there is no mistake in ye Proposition or in this way of concluding it, thus. Taking x for *CD*, & z for *DK*, by ye above mentioned æquation $a\dot{z} = z\dot{x} - x\dot{x}$ it appeares yt $az + \frac{1}{2}xx$ is equall to ye Fluent of $z\dot{x}$; whence I conclude, if *CL* be taken on ye other side ye point *C* equall to *Ca* & ye Ordinate *LQ* be erected, yt ye indeterminate Area *DKVTa* is equall to $\frac{DK}{LQ} \times LQTa + \frac{1}{2}CDq - \frac{1}{2}Caq$ & ye whole Area *BKVTa* is equall to $\frac{1}{2}CBq - \frac{1}{2}Caq$, or $Aa \times \frac{1}{2}aB$. That ye Curve *BRVSa* may still remaine in ye Scheme I propose ye 3 following alterations. Pag. 315. l. 7—& Ellipsis *aBRVS*, centro 0, semiaxibus—l. 22—Nam cum Ellipsis vel Parabola *aBRVS* congruat —l: 24—alterutram *BRV* vel *VSa* excedit figuram —[11] Lin. penult. I would leave out quamproxime. Pag. 319. l: 13 You say—Cum distantiæ particularum Systematis unius sint ad distantias correspondentes &c. The same thing is implyed in ye Demonstration of Prop. 32. I think it ought also to be expressed in the words of Prop. XXXII.[12]

Yr most obliged & faithfull Servant
ROGER COTES

For Sr Isaac Newton
at his House near ye College
in Chelsea
Near London.

NOTES

(1) Add. 3983, no. 47, printed in Edleston, *Correspondence*, pp. 20–3 from the Trinity College manuscript.

(2) In Prop. 20, Theorem 14 (Theorem 15 of the second edition, p. 262, where these words appear); the additions do not change the sense.

(3) These words were excised from the scholium to Proposition 23 (*Principia* 1713; p. 271, Koyré and Cohen, *Principia*, I, p. 431). In this proposition and the one preceding it Newton is concerned with the properties of fluids composed of mutually repulsive particles; however, to simplify the argument, Newton has specified that each particle repels only its immediate neighbours, its action on remote particles being neglected. The offending phrase ('unless perhaps by means of intermediate particles empowered by that virtue') certainly was out of place.

(4) They were omitted. Newton had asserted (in 1687) that if a particle acted not only upon its neighbours, but upon *all* particles in the fluid, then, given a repulsive force inversely proportional to the separation, for constant density the pressure would be as the square of the diameter of the vessel, and for constant volume as the inverse five-thirds power of the density. (In his earlier argument, where all interactions except those between neighbouring particles had been ignored, pressure was of course independent of the size of the vessel.)

(5) The reading of the first edition was restored. I. Bernard Cohen (*Introduction*, p. 214) presents good reasons for supposing that the mistaken *quadrato* was introduced by David Gregory.

(6) The words were omitted as misleading (see Koyré and Cohen, *Principia*, I, p. 436), for Newton did not restrict the following sentence to the case where one arc is twice as big as the other.

(7) This is Proposition 29, Problem 7 (*Principia* 1713; Prob. 6) dealing with the resistance experienced at each point of its swing by a pendulum oscillating in a medium resisting as the square of the velocity. Cotes' wording was printed (*Principia* 1713; p. 279). What Cotes has done is to insert twice the relation of the increment of velocity to the quantity $(V-R)$; the net result is the same as before.

(8) In the argument which follows (where proportionality is expressed by a pair of short vertical lines) Cotes is starting from two conditions $z \propto v^2$ (from the proposition) and $x \propto s$ (if s is the arc CD and the pendulum is cycloidal). If we write instead $z = kv^2$ and $x = cs$, and take the mass of the pendulum bob as m, and follow Cotes' argument in more familiar notation (considering a downward swing with force and velocity both positive towards the bottom of the swing), we have

$$v = -(1/c) . dx/dt, \quad m . dv/dt = x - z.$$

But $z = kv^2$, therefore $$dz/dt = 2kv . dv/dt = -(2k/mc)(x-z) . dx/dt.$$

Hence putting $a = mc/2k$, we obtain $a . dz/dx = z - x$, as Cotes does. Cotes introduces an auxiliary function $y = y(x)$, and writes the solution as $dy/dx = y/a$ and $z = a + x + ry$ (where r is constant). He goes no further towards showing how this agrees with Newton's complicated geometrical solution. If we complete the solution and write $z = a + x + Be^{x/a}$, where B is a constant depending on the boundary conditions of the problem, it becomes easier to see the similarity between this and Newton's solution (see Koyré and Cohen, *Principia*, I, p. 438, referring to the figure there). In Newton's diagram, areas under the hyperbola are proportional to x, whence the abscissae themselves are proportional to $e^{x/a}$. (See also Whiteside, *Mathematical Papers*, VI, p. 444, note (9).)

Newton neither adopted for publication, nor commented upon, Cotes' suggested derivation, but Cotes himself incorporated it into his *Logometria*, the manuscript of which he sent to Newton with Letter 921. The work was published in *Phil. Trans.* **9** (1714), 5–45; see pp. 40–2 for the relevant passages.

(9) This was done. (Koyré and Cohen, *Principia*, I, p. 442.)

(10) Cotes here proposes to cut short Newton's very lengthy conclusion of Proposition 30, Theorem 23 (*Principia* 1713; Theor. 24), by adopting a much more direct argument to the same result. Newton feared that Cotes' method was too obscure, and in the end an expanded version of it was printed (see Letter 786).

(11) Newton's previous references to the points *R* and *S* in the main demonstration having been removed by Cotes' shortening of it, he now simply defines clearly the ellipse mentioned in the corollary only.

(12) In fact no change was made in this passage.

784 NEWTON TO COTES

30 MAY 1710

From the original in Trinity College Library, Cambridge.(1)
Reply to Letter 783; for answers see Letters 786 and 788

Chelsea. May 30. 1710

Sr

The corrections wch you have sent me in your Letter of May 20 are right. But I fear least that wch relates to Prop. XXX may render the Demonstration thereof too obscure. And therefore I think that the Proposition with its Demonstration may stand, & in the end of it, after the words et sic eidem æquabitur quam proxime, may be added these two sentences. Quinimo eidem æquabitur accurate, ideoque conclusiones prædictæ sunt accuratæ. Nam si ad alteras partes puncti *C* capiatur *CL* æqualis ipsi *Ca*, et erigatur normaliter *LQ* ad Curvam *aTVKB* terminata, et pro Curvæ hujus area indeterminata *aTVQL* ad ordinatam *LQ* applicata scribatur litera *M*; area indeterminata *aTVKD* æqualis invenietur quantitati *M*, $DK+\frac{1}{2}CDq-\frac{1}{2}Caq$, et area tota *aTVKB* quantitati $\frac{1}{2}CBq-\frac{1}{2}Caq$, seu $Aa\times\frac{1}{2}aB$.(2)

The *Scholium Generale* wch in the former edition was printed in the end of the seventh Section, I would have printed in the end of the sixt section next after Prop. XXXI.(3) But it wants the following corrections

Pag. 339 lin 21, 22, 23 &c read

Scholium Generale

Ex his Propositionibus per oscillationes Pendulorum in Medijs quibuscunque, invenire licet resistentiam Mediorum. Aeris vero resistentiam investigavi per Experimenta sequentia. Globum ligneum pondere unciarum Romanorum $57\frac{7}{22}$, diametro digitorum Londinensium $6\frac{7}{8}$ fab[r]icatum, filo tenui &c.(4)

Pag. 340. lin 24, 25, blot out, omnino ut in Corollarijs Propositionis XXXII demonstratum est.[5]

Pag 341 lin. 18 for resistentia read resistentiæ

Pag 342 lin 21 blot out, Unde cum corpus tempore, & what follows to the end of the words, longitudinem duplam 30,556 digitorum.[6]

Pag. 343 lin 6 for pedum read digitorum. Ib lin 8 read vis resistentiæ eodem tempore uniformiter continuata. Ib lin 12 read posset.[7]

Pag 344 lin 13, 14 for prima, secunda, tertia read tertia quinta septima & for $\frac{1}{193}$ read $\frac{\frac{1}{2}}{193}$.[8]

Pag. 345 lin 7, 25 for dimidiata read subduplicata. Ib lin. 8 read Nam ratio $7\frac{1}{2}-\frac{1}{3}$ ad $1-\frac{1}{3}$ seu $10\frac{3}{4}$ ad 1, non longe.[9]

Pag. 349 blot out the lines 17, 18, 19, 20, 21, 22, 23, 24, 25, 26, 27[10]

Pag. 350 lin. 32 blot out Quare cum globus aqueus in aere movendo & what follows to the end of the words, probe tamen cum præcedentibus congruebat.[11]

Pag 354 blot out the lines 11, 12, 13, 14, 15.[12]

In the beginning of Sect VII pag. 317 lin. 5 after the words similes sint, insert the words & proportionales

I am

Your most humble servant

Is. Newton

For the Rnd Mr Cotes, Professor of Astronomy, & Fellow of Trinity College in Cambridge.

NOTES

(1) R.16.38, no. 51, printed in Edleston, *Correspondence*, pp. 24–5.

(2) See Letter 786. This phrasing was not printed because Cotes' form of the termination to Prop. 30 took its place.

(3) This was done (*Principia* 1713; p. 284. Koyré and Cohen, *Principia*, I, pp. 448–63).

(4) There is no change of substance here, the experimental details remaining as before.

(5) Prop. 32 is now to follow, in Section VII.

(6) The deleted sentence justified that preceding it by recalling that Huygens had found by experiment that the acceleration due to gravity was $30\frac{1}{6}$ Paris ft. sec^{-2}, or 395 English in. sec^{-2}.

(7) All these were adopted; the first correction was necessitated by a mere slip of the pen.

(8) More corrections of slips.

(9) For *subduplicata* read *duplicata,* as printed in *Principia* 1713; the former means the same as *dimidiata.* The second alteration was struck out later at Newton's request.

(10) Newton struck out an interesting paragraph in which he had pointed out that though in the experiments he had just described the resistance was never in proportion to more than the square of the velocity, when bodies moved very swiftly one would expect the resistance to increase in a higher proportion because of the vacuum created behind the body. (See Koyré and Cohen, *Principia*, I, p. 459.)

(11) This is rather more than a page of the 1687 text: Newton in the first paragraph had again pointed out that the resistance would be much increased if the velocity were sufficient to produce cavitation; then he had briefly mentioned experiments on conical pendulums, whose resistance he wrote agreed with what he had discovered for swinging pendulums. (See Koyré and Cohen, *Principia*, I, p. 460.)

(12) In the deleted final paragraph of the Scholium Newton had written that the resistance of other shapes might be ascertained by the same method that he had used for spheres; 'and thus the different shapes of ships, constructed as small models, might be compared among themselves, so that that which is most suitable for navigation might be searched for at small expense'. This prescient (if idealistic) suggestion of Newton's was not adopted by naval designers until the nineteenth century—notably by William Froude (1810–79); it has proved to be far from cheap. (See Koyré and Cohen, *Principia*, I, p. 463.)

785 WOODWARD TO NEWTON

30 MAY 1710

From the original in the University Library, Cambridge[1]

Sr.

I find it necessary to have a Copy of ye Minutes of your Proceedings in Relation to me May 15. 1706:[2] as also of those at ye 5 or 6 last Councils.[3] I desire yt you will order Copyes to be forthwith deliverd to me: &, if there be any Fees due, I am ready to pay them. Or if you direct yt I may have Access to ye Originals, I will my Self cause a Copy to be taken of them.

I also desire Liberty of access to ye Statute Book.[4] I am

Sr.

Your humble Servant

WOODWARD

Gresh. Coll.
30. *May*. 1710.

That I may not anticipate you, I have forborn makeing Presents of ye Books sent me by Dr. Scheuchser.[5] I wish I had a List of those you intend to present yours unto, yt I may not give any to ye same Persons. The Dr. sent 12 Exemplars to my Lord Arch-Bishop.[6]

To
Sr Isaac Newton
at Chelsea

NOTES

(1) Add. 3965, fo. 288; the paper has been thriftily re-used for a draft relating to the *Principia*, Book III, Prop. 37.

(2) 'Att a Councell May: 15: 1706. It was ordered upon complaint of some irregular proceedings att the Meetings of the Society & unjust reflections cast upon some of the Members particularly by Dr Woodward, that the Members of the Society be admonisht when he is present that if any member of the Society shall hereafter cast reflections on the Society or any of the fellows thereof, the Statutes concerning Ejection shall be taken into consideration, and the Statute about Ejections shall be read this afternoon before such admonition.' Compare Letter 774.

(3) At the Council on 3 May 1710—its first meeting after 29 March, for which see Letter 774—with Woodward himself present, it was ordered that the President or a Vice-President might order the withdrawal of any Fellow who should 'speak words reflecting upon any other' Fellow from either a Council or an ordinary meeting of the Society; a Fellow so ordered to withdraw being instructed not to return to Council or Meetings until readmitted by order of the Council. At a later Council, on 24 May 1710, after Sloane had declared that he meant no affront to Woodward by his grimaces, it was ordered that the latter should apologize to Sloane for his 'reflecting Words'. This Woodward refused to do, and he was then (by a vote of the Council) ordered to remove himself. The Council then thanked Sloane formally 'for his pains and fidelity in serving the Society as Secretary'.

(4) At the Council of 17 June, Newton reported that Woodward had sought a writ from the Queen's Bench in order to be restored to his place on the Council. A committee was appointed to undertake the Society's defence in the event of a prosecution.

(5) Joannes Jacobus Scheuchzerus M.D. (1672–1733), admitted Fellow on 30 November 1703, who visited England in April or May 1710 (Wollenschläger, *De Moivre*, pp. 258, 269). He practised medicine in his native city of Zurich and also taught mathematics in the *Gymnasium*. He was one of the first Swiss to study geology and natural history and was an early supporter of Newton. He was a prolific writer. The book may have been his *Herbarium diluvianum* (Zurich, 1709); Woodward had presented two copies of it to the Royal Society (which still possesses one) on 17 May.

(6) Thomas Tenison (1636–1715), who had once corresponded with Henry Oldenburg.

786 COTES TO NEWTON

1 JUNE 1710

From the original in the University Library, Cambridge.(1)
Reply to Letter 784; for the answer see Letter 787

Cambridge June 1. 1710.

Sr.

I received Yr Letter last night, by which You give Yr consent to ye other alterations which I proposed, but seem to fear least yt which relates to Prop. XXX may render the Demonstration thereof too obscure, & therefore at ye end of ye Corollary after ye words [& sic eidem æquabitur quam proxime] you add Quinimo eidem æquabitur accurate, ideoque &c. I believe you designed those two sentences to be inserted p. 314. l: 18 after ye words [erit etiam æquale areæ *BKTa* quam proxime] & yt by inadvertency in Yr Letter You ordered them to be placed in page 315. l: 25 after the words [eidem æquabitur quam proxime].[2] For though ye Proposition it self & the first part of the Corollary ending with the words [omnino ut in Propositione XXVIII demonstratum est] be accurate, yet as I understand it the remaining part of ye Corollary is still but an Approximation, the Ellipsis & Parabola mentioned in ye latter part of ye Corollary not agreeing perfectly with ye Figure *BKVTa*; but by placing those two sentences as in Yr Letter, even ye latter part of ye Corollary is declared to be accurate. I beg leave of you to express my sense freely, I fear it will be look'd upon as a Blemish in Yr book, first to demonstrate yt ye Proposition is true quamproxime and afterwards to assert it to be true accurate. I am of Opinion yt ye alteration which I proposed p. 313. l: 29 does make ye Demonstration compleat to an intelligent Reader.[3] If you think good, it may be put down more at Large, some such way as this which follows[4]—& ex æquo perturbate (*Fh* seu) *MN* ad *Dd* ut *DK* ad (*CF* seu) *CM*: ideoque summa omnium $MN \times CM$ æqualis erit summæ omnium $Dd \times DK$. Ad punctum mobile *M* erigi Semper intelligatur Ordinata rectangula æqualis indeterminatæ *CM*, quæ motu continuo ducatur in totam longitudinem *Aa*; & trapezium ex illo motu descriptum sive huic æquale rectangulum $Aa \times \frac{1}{2}aB$ æquabitur summæ omnium $MN \times CM$ adeoque summæ omnium $Dd \times DK$, id est, areæ *BKkVTa*. Q.E.D. Or, if you think the demonstration will even this way be too obscure, a new Scheme may be cut with the addition of ye lines here drawn & the demonstration may end thus—et ex æquo perturbate (*Fh* seu) *MN* ad *Dd* ut *DK* ad (*CF* seu) *CM*: ideoque $MN \times CM$ æquabitur $Dd \times DK$. Erigantur normales *AX*, *aZ* æquales ipsis *AC*, *aC* & jungatur *XZ* occurrens normalibus *MY*, *NI* in *Y* & *I*; & erit *YM* æqualis ipsi *CM* atque adeo $MN \times YM$ æquale $Dd \times DK$ & summa omnium $MN \times YM$, id est, trapezium *AaZX* sive huic æquale rectangulum $Aa \times \frac{1}{2}aB$ æquabitur summæ omnium $Dd \times DK$, hoc est, areæ *BKkVTa*.

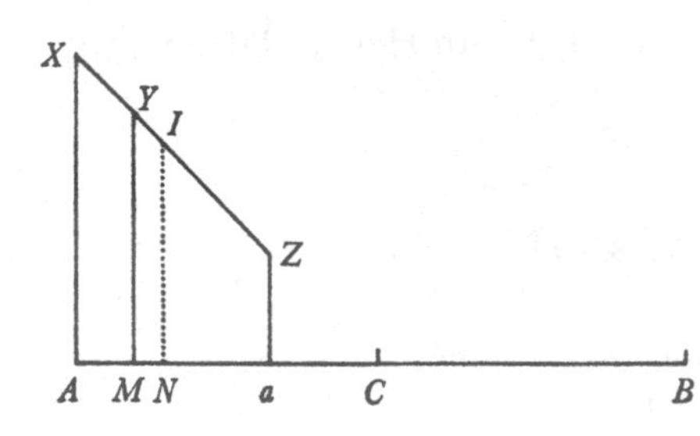

Q.E.D. I think ye first of these two ways sufficiently clear; but will wait for Yr resolution.

Yr most Humble Servant
ROGER COTES

For Sr Isaac Newton
at his House near the College
in Chelsea
near London

NOTES

(1) Add. 3983, no. 5, printed in Edleston, *Correspondence* pp. 25–7, from the Trinity College MSS.

(2) Cotes is of course right. Finally, the passage on p. 315 (*Principia* 1713; p. 283) was left alone, and the passage on p. 314 (*ibid.* p. 281) was modified in the first of the ways Cotes is about to propose.

(3) See Letter 783, and its note (9).

(4) Beginning, as before, at *Principia* 1687; p. 313, l. 29; the passage is printed at *Principia* 1713; p. 281, l. 28. (See Koyré and Cohen, *Principia*, I, p. 444.)

787 NEWTON TO COTES

8 JUNE 1710

From the original in Trinity College Library, Cambridge.(1)
Reply to Letter 786

Sr

I have reconsidered your emendation of the XXXth Proposition with the Demonstration & approve it after the manner you propose in the first of the two ways set down in your Letter of June 1st. In my last letter,(2) as I was sending it away, I crossed out four lines & should have struck out also these words relating to them [Ib. lin: 8, read, Nam ratio $7\frac{1}{2}-\frac{1}{3}$ ad $1-\frac{1}{3}$ seu $10\frac{3}{4}$ ad 1, non longe](3) I am

Your most humble Servant
Is. Newton

Chelsea Jun 8.
1710.

I thank you for mending the Proposition

For the Revnd Mr. Cotes Professor of
Astronomy and Fellow of Trinity
College in Cambridge

NOTES

(1) R.16.38, no. 52, printed in Edleston, *Correspondence*, p. 27.

(2) Letter 784 and its note (9).

(3) Newton is here (*Principia* 1687; p. 345, ll. 8–9; *Principia* 1713, p. 288, ll. 6–7; Koyré and Cohen, *Principia*, I, p. 454) comparing the diameter of a pendulum bob with the resistance it offers to motion, making allowance for the constant resistance of the thread. In the first edition was printed: $(7\frac{1}{3}-\frac{1}{3}):(1-\frac{1}{3})=7:\frac{3}{2}=10\frac{1}{2}:1$. By omitting the second ratio the printer's error was eliminated.

788 COTES TO NEWTON
11 JUNE 1710

From the original in the University Library, Cambridge.[1]
Reply to Letter 784; for the answer see Letter 789

June. 11. 1710 *Cambridge*

Sr.

I received Yr Letter of May 30th. In that which I wrote to You by ye next Post[2] instead of the alteration in pag 313 l: 29 which You thought too obscure, I proposed the following—& ex æquo perturbate Fh seu MN ad Dd ut Dk ad CF seu CM; ideoque summa omnium $MN \times CM$ æqualis erit summæ omnium $Dd \times DK$. Ad punctum mobile M erigi semper intelligatur Ordinata rectangula æqualis indeterminatæ CM, quæ motu continuo ducatur in totam longitudinem Aa; & trapezium ex illo motu descriptum sive huic æquale rectangulum $Aa \times \frac{1}{2}aB$ æquabitur summæ omnium $MN \times CM$ adeoque summæ omnium $Dd \times DK$, id est, areæ $BKkVta$. Q.E.D. We are now at a stand expecting Yr resolution. You gave me orders in Yr Letter to print ye *Scholium Generale* after the Sixth Section & sent me Yr corrections of it. I have not yet had leasure since I received Yr Letter to examine all the Calculations of yt Scholium, being at this time engaged in a Course of Experiments & having some other Buisness upon my hands; but I have read it over & considered the reasoning of it. Page 345. l: 26[3] You say—differentia 0,4475 diminueretur in ratione velocitatis, adeoque evaderet 0,4412. I do'nt see any reason for this diminution, but think it ought to remaine 0,4475 notwithstanding yt the length of ye Pendulum is increased in ye ratio of 126 to $122\frac{1}{2}$, & thereby ye time increased & ye velocity diminished in the subduplicate ratio of 126 to $122\frac{1}{2}$. You will see my reasons in what follows.[4] Quæ tradita sunt in Prop. XXXI & ejus corollariis obtinent ubi Oscillationes sunt Isochronæ. At si Oscillationum tempus quoque mutetur, differentia inter arcum descensu descriptum & arcum subsequente ascensu descriptum erit ut resistentia & quadratum temporis conjunctim: Nam totius retardationis particulæ singulæ, ex quibus differentia illa componitur, sunt in hac ratione, per Lem. X Libri. I

Unde si detur longitudo arcus descripti & resistentia sit ut quadratum velocitatis; manebit differentia, utcunque mutetur tempus atque adeo velocitas corporis oscillantis. Nam ob datam longitudinem arcus descripti, tempus erit ut velocitas inverse; adeoque differentia illa, cum sit ut resistentia & quadratum temporis, erit ut quadratum velocitatis directe & quadratum velocitatis inverse, ac proinde magnitudinem datam habebit.

Idem aliter (vid. fig. prop. XXX)

Manente longitudine arcus descripti *aB* augeatur longitudo Penduli. Si mutata longitudine Penduli maneret resistentia, maneret quoque ratio resistentiæ ad vim gravitatis atque huic æqualis ratio Ordinatæ *DK* ad longitudinem Penduli; adeoque augenda esset Ordinata *DK* in ratione longitudinis Penduli. Verum, ob auctam Penduli longitudinem, augetur quoque tempus in ratione ejus subduplicata adeoque diminuitur velocitas in eadem ratione subduplicata, & resistentia atque huic proportionalis Ordinata *DK* diminuitur in ratione integra. Itaque Ordinata *DK* diminuitur in eadem ratione qua prius augebatur, ac proinde manet ejusdem longitudinis, manetque adeo magnitudo areæ *BKVTa* atque huic æquale rectangulum $Aa \times \frac{1}{2}aB$, & differentia illa *Aa*. If you admit of this reasoning, it will not only affect this place in page 345 but also pag. 348. l: 1 and pag. 353. l: 27 and pag. 341. l: 16. In Page 346. l: 23 You cite ye Corollarys of Prop XL which are now to come after the Scholium. There being no alteration of this place amongst the corrections You sent me I do not know whether You took notice of it, & have therefore mentioned it to you. Pag 348. l: 7 &c. You seem to confound ye *Differentia arcuum* with the *Resistentia Globi*; the former is represented by $AV+CV^2$ & ye latter ought I think to be represented by $\frac{7}{11}AV+\frac{3}{4}CV^2$.(5) I desire Yr answer to this Letter; when I receive it I will examine & alter ye Calculations if there be occasion, according to Yr directions.

Yr most obliged Humble Servant
ROGER COTES.

For Sr Isaac Newton
at his House near ye College
in Chelsea
By London

NOTES

(1) Add. 3983, no. 6; printed in Edleston, *Correspondence*, pp. 28–30, from the Trinity College MSS.

(2) Letter 786.

(3) The question at issue is this: if the length of a pendulum be diminished in the ratio of $122\frac{1}{2}$ to 126, the difference between the length of a descending semi-arc and that of the next ascending semi-arc of swing being 0·4475 ins for the original pendulum, will the same difference hold for the shorter pendulum (given the same initial amplitude of swing), or should the difference be reduced in the ratio of the velocities, i.e. to $0{\cdot}4475\sqrt{(122\frac{1}{2}/126)} = 0{\cdot}4412$? The latter had been Newton's contention in the first edition. However, it seems that Cotes was right (for Newton's argument here is independent of the length of the pendulum) and it was accepted as such by Newton. (See *Principia* 1713; p. 288, and Koyré and Cohen, *Principia*, p. 454.)

(4) 'What is stated in Prop. 31 and its corollaries holds when the oscillations are isochronous. But if the time of the oscillation is changed too, the difference between the arc described

in the descent and the arc described in the subsequent ascent will be as the resistance and the square of the time together; for the several parts of the whole retardation, from which parts the whole of that difference is composed, are (by Lemma 10, Book I) in this ratio. Whence if the length of the arc described be given, and the resistance be as the square of the velocity, the difference will remain [the same], however the time be changed and consequently the velocity of the swinging body. For, since the length of the arc described is given, the time will be in inverse ratio to the velocity; and so that difference (being as the resistance and the square of the time) will be as the square of the velocity directly and the square of the velocity inversely, and hence will possess the [same] magnitude given.

'The same another way (see the figure for Prop. 30): Keeping [the same] length for the arc aB described, let the length of the pendulum be increased. If the resistance were to remain [the same] when the length of the pendulum was changed the ratio of the resistance to the force of gravity would remain [the same] also, and, what is equal to this, the ratio of the ordinate DK to the length of the pendulum. And so the ordinate DK would have to be increased in proportion to the length of the pendulum. However, because the length of the pendulum has been increased the time too is increased as the square root of that ratio and hence the velocity is diminished as the square root of that ratio, and the resistance (with what is proportional to it, the ordinate DK) is diminished in that whole proportion. Thus the ordinate DK is diminished in that same ratio by which formerly it was increased, and hence remains of the same length; thus the size of the area $BKVTa$ will remain, and what is equal to it, the rectangle $Aa \times \frac{1}{2}aB$, and that difference Aa.'

(5) See Whiteside, *Mathematical Papers*, VI, pp. 449–50.

789 NEWTON TO COTES

15 JUNE 1710

From the original in Trinity College Library, Cambridge.[1]
Reply to Letter 788; for the answer see Letter 793

Sr

I sent you a letter the last week[2] in wch I approved your correction of Prop XXX wth its demonstration according to the first of the two ways wch you sent me in your Letter of June 1st & have now repeated in yours of June 11th wch I received last tuesday morning. I thank you for that correction. In my last letter but one[3] I crossed out four corrections wch I had wrote down in it, & should have crossed out a fift wch related to those four & was in these words. Pag. 345 lin. 8 lege, Nam ratio $7\frac{1}{2}-\frac{1}{3}$ ad $1-\frac{1}{3}$ seu $10\frac{3}{4}$ ad 1.[4]

The correction in the Scholium p. 345 lin 26, sent me in your last, is right, & I beg the favour that you would alter the calculations accordingly

In pag. 346 lin 23 strike out the words et propterea (per corollaria Prop XL Libri hujus) resistentia quam Globi majores & velociores in aere movendo sentiunt & so on to the end of the sentence[5]

In pag. 348 lin 7, 14, 15, 16 for A & C put other letters suppose F & G, writing, Designet jam $FV+GV^2$ resistentiam Globi &c because $AV+CV^2$ was used before for the differentia arcuum.(6)

You need not give your self the trouble of examining all the calculations of the Scholium. Such errors as do not depend upon wrong reasoning can be of no great consequence & may be corrected by the Reader.

I am wth many thanks

Sr

Your most humble servant

IS. NEWTON.

Chelsea

June 15*th* 1710

For the Rnd Mr Cotes Professor of Astronomy and Fellow of Trinity College in Cambridge

NOTES

(1) R.16.38, no. 53; printed in Edleston, *Correspondence*, pp. 30–1.

(2) Letter 787.

(3) Letter 785.

(4) See Letter 787, note (3).

(5) In response to Cotes' admonition, Newton simply cuts out (see *Principia* 1713; p. 289 and Koyré and Cohen, *Principia*, I, p. 456) the allusion to Prop. 40—yet to come in the second edition—in which he had remarked that all mediums which resist bodies less than air does, must be less dense than air.

(6) This suggestion was not adopted by Cotes, who preferred to print (*Principia* 1713; p. 290, and Koyré and Cohen, *Principia*, I, p. 457) the corrected expression suggested at the close of Letter 788, and consequentially modify the numbers in the working that follows. No important quantitative result was at stake (Newton's emendation, by introducing two new unknown quantities, would have made it impossible to derive any numbers at all).

790 NEWTON TO [LAUDERDALE]

22 JUNE 1710

From the holograph original in the Scottish Record Office(1)

London June 22
1710.

May it please your Lo[rdshi]p

I received your Lordps Letter yesterday about the tryall of the Pix & this day waited upon my Lord Chancellour(2) to know his sense upon that matter, & his Lordp desired me to signify it to Your Lordp with relation to the two difficulties wch your Lordp proposed in your Letter to him. His Lordp thinks the Order of Council for the triall,(3) a sufficient Warrant for conveying the Pixis to London, & something more then a Warrant because it commands the doing it: but the manner of doing it most safely is left to the prudence of your Lordp & the other Officers of the Mint, as it was lately left to the prudence of the officers of the five country Mints in England to convey their Pixes to London.(4) If it be conveyed safely to the Mint in the Tower, we will take care that it be safely carried thence with our own Pixis to the place of triall. And as to the other Officers wch are to come up hither to the tryall, his Lordp thinks three sufficient, your Lordship, the Master & the Warden of the Mint.(5) For in the triall of the Pix of the Mint in the Tower, three Officers only attend, the Warden the Master & the Comptroller, the rest of the Officers being of no use in the triall. If your Lordship & the Master & Warden think it convenient that the Counter-Warden come to London with you, it will not be found fault with here: but if you excuse him, he will be excused here, his power of acting in the triall being included in your Lordships. I am

My Lord

Your Lordps most humble
and most obedient Servant
IS. NEWTON

NOTES

(1) AD 248/571/2/59. Although this letter was among the Seafield manuscripts and was calendared in H.M.C., 14th Report, Appendix III, p. 223 as being addressed to Seafield, the context and the content make it perfectly clear that the recipient was General of the Edinburgh Mint. Compare the next two letters; obviously Lauderdale wrote directly to Newton about the time that he wrote to Godolphin (17 June) and Newton replied at once—in much the same terms that would later reach Lauderdale through the Treasury.

(2) Presumably the Earl of Seafield. But it is far more likely that Newton, by an easy slip of mind, wrote 'Chancellour' for 'Treasurer'; that he had been to see Godolphin (as Lauderdale had requested) and been shown Lauderdale's letter to Godolphin of 17 June. Thus private negotiations anticipated the formal channels.

(3) See Letters 771 and 772, and Letter 795, note (2). The Order in Council had been made on 14 May.

(4) The five Mints were those set up at Bristol, Chester, Exeter, Norwich and York for the great recoinage of 1696–8.

(5) Compare Letter 796, and its note (5).

791 LOWNDES TO THE MINT

28 JUNE 1710

From the copy in the Public Record Office.(1)

For the answer see Letter 792

Gentlemen

My Lord Trea[su]rer having received the Inclosed Letter from the Earl of Lauderdale General of Her Mats. Mint at Edinburgh touching the Tryal of the pix in that Mint,(2) His Lordship directs You to peruse the said Letter and to Report to Him as soon as You can, what Directions You think may be proper to be given for Transporting the said pix hither, what Officers will be necessary to Attend at the Tryal thereof, and in what manner the Charge of Transporting it may be defrayed together with such other matters as You shall think proper to Observe to His Lordship thereupon. I am

Gentlemen

Your most Humble Servt

Wm. Lowndes

Tre[asu]ry Chrs.
28 *June* 1710
Officers of the Mint.

NOTES

(1) Mint/1, 8, 167.

(2) Lauderdale's letter of 17 June 1710, copied on the previous page of the Record Book, expressed ignorance of the procedure to be followed in attending for the Trial of the Edinburgh Pyx at Westminster on 21 August, and asked for detailed instructions.

792 THE MINT TO GODOLPHIN

29 JUNE 1710

From the copy in the Public Record Office.(1)

Reply to Letter 791

To the Most Honble: the Lord High Trea[su]rer of Great Britain

May it please your Lordship

In Obedience to your Lops Order signified to Us by Mr Lowndes Letter of the 28 Instant We have persued the Annexed Letter of the Earl of Lauderdale General of her Mats: Mint at Edinburgh touching the Tryal of the pix of that Mint,(2) and are humbly of Opinion that Her Mats: Order in Council for trying the same is a sufficient Warrt, for transporting it to London, especially if a Duplicate of the said Order with Your Lops signification thereupon be transmitted to the General, Master & Warden[3] of the said Mint, as is usually done to the Officers of the Mint here, and that the manner of transporting it safely be left to the prudence of the General and other Officers of the said Mint (no particular Order being given to the Officers of the late five Country Mints[4] about the manner of Conveying their pixes to the place of Tryal) and that the charges thereof be placed in the next Accompt of the Master and Worker as being to be born by Her Majesty.

And We are further humbly of Opinion that the persons who are Intrusted with the Keeping of the Keys of the pix should bring them up themselves, & be present at the Opening of it, & by reason one of them has usually been kept by the Treasury or Treasurer; as the Generals Letter setts forth, that the same be sent or brought up in such a manner as Your Lordship shall be pleased to Direct & that the Master & Worker who is to appear in behalf of the late Master[5] do also Attend at the Tryal and that the Coming up of the Counter warden be left to the Discretion of the Officers of that Mint.

All wch is humbly submitted to Your Lops: great wisdom

[C PEYTON
IS. NEWTON
JN. ELLIS][6]

Mint Office the 29 *June*
1710

NOTES

(1) Mint/1, 8, 167; there is another copy in the Scottish Record Office (E/105/1); see also P.R.O. T/17, II, 188.

(2) See Letter 791, note (2), and Letter 772.

(3) That is, Lauderdale himself, John Montgomerie, and William Drummond.

(4) See Letter 790, note (4).

(5) Allardes.

(6) The record copy is naturally not signed. The Scottish copy carries the signatures shown.

793 COTES TO NEWTON

30 JUNE 1710

From the original in the University Library, Cambridge.[1]
Reply to Letter 789; for the answer see Letter 794

Cambridge June 30*th* 1710

Sr.

Wee have now finished all Yr Copy & the Scholium Generale.[2] I received Your Letter of June 15th, in which You consent to the alteration yt I proposed in that Scholium. I have examined the whole Calculation, & done it anew where I thought it necessary. The discourse it self is also a little altered in those places which I mentioned in my last, as You will perceive by the 2 inclosed Sheets. They are not yet printed off, but will stay for Your corrections if You shall think fit to make any. I could wish You would be pleased to look 'em over, for I fear I may possibly have injured You. The Press being now at a Stand I will take this opportunity to visit my Relations in Lincoln-shire & Leicester-shire.(3) I hope I shall come back again to College in 5 or 6 weeks. When I return I will write to You to desire the remaining part of Yr Copy.

I am Your most Obedient
& Humble Servant
ROGER COTES.

For Sr Isaac Newton
at his House near the College
in Chelsea
By London

NOTES

(1) Add. 3983, no. 7; printed in Edleston, *Correspondence*, p. 31, from the Trinity College MSS. According to an endorsement it was sent 'by Peeney post'.

(2) That is, to the end of signature Pp (p. 296) in the new printing, as Newton's next letter confirms.

(3) Roger Cotes was born at Burbage, Leics., where his father was rector; his mother's family came from Barwell in the same county. His uncle, the Reverend John Smith, with whom he corresponded regularly over many years, lived near Gainsborough, Lincs.

794 NEWTON TO COTES

1 JULY 1710

From the original in Trinity College Library, Cambridge.(1)
Reply to Letter 793

Chelsea June 31. 1710.

Sr

I received yours of June 30 this noon with the two inclosed proof sheets, & I have perused them without observing any faults except in the last page of the second sheet lin 28 where vires autem motrices should be vires autem acceleratrices. And in the preceding page (pag. 295) upon reconsidering the words of Prop. XXXIII,(2) I think the words will be better understood if they run as in the former edition, vizt Iisdem positis, dico quod Systematum partes majores resistuntur in ratione composita &c. The remaining part of the copy will be ready against your return from the visit you are going to make to your friends.

I am wth my humble service to your Master & many thanks to your self for your trouble in correcting this edition(3)

Sr

Your most humble servant

Is. Newton.

For the Rnd Mr Cotes Professor of Astronomy and Fellow of Trinity College in Cambridge

NOTES

(1) R.16.38, no. 60; printed in Edleston, *Correspondence*, p. 32.

(2) This was done.

(3) Edleston supposed (*ibid.*, p. 32) that there was missing from the copies of Cotes' letters at Trinity a letter dated 11 July 1710, following here another annotation left by Edward Howkins (see Letter 766, note (7)). However, there is no letter of this date among the originals in U.L.C. Add. 3983 either. If Howkins was correct, it may have been that the missing letter was a very brief acknowledgement.

795 GODOLPHIN TO THE MINT

4 JULY 1710

From the copy in the Public Record Office(1)

Officers of the Mint to Attend at the Tryall of the Pix 21st. Augt. 1710

[Recites the Order in Council of 14 May 1710](2)

Let the Warden, Master and Worker and Comptroller of Her Mats. Mint in the Tower of London take Care that Her Mats. Pleasure signifyed by the within written Order of Councill be duely Complyed with so far as Appertaines unto them Whitehall Trea[su]ry Chambers *4th July* 1710

Godolphin

NOTES

(1) T/17, 2, 189. This is a record copy.

(2) On 14 May an Order in Council was made that 'the Lords of her Mats. Most honble. Privy Councill do meet at the house inhabited by the Usher within the Receipt of her Mats. Excheqr. at Westmr. on Munday the 21st. day of August next at Nine of the Clock in the Morning for the Tryal of her Mats. Coins in the Pixs of the Sevll. Mints within this Kingdom...' (T/17, 2, 188) and in preparation for this event the Principal Officers of the Mint were summoned to the Treasury on 17 August to make the final arrangements (*Cal. Treas.*

Books, XXIV, 1710 (Part II), p. 350). On 4 July warrants were sent to the Tower and Edinburgh Mints for their attendance at this trial, together with copies of the Order in Council and (to Edinburgh) of the Tower Mint's letter of 29 June concerning procedure. The Officers of the Scottish Mint participating were: William Drummond, Warden; Walter Boswell, Counterwarden; Patrick Scott, clerk to the Master; James Penman, Assaymaster; Robert Miller, Clerk; and William Bowles who brought the General's Keys to the Pyx. £252 6*s* 0*d* was paid them for the expenses of their trip on 13 September (*ibid.* p. 450) but Penman and Miller made much fuss later about what they received. This Trial, owing to an excess of 2·7 parts per 1000 in the gold test plate used in trying the gold coins, brought Newton much trouble, since it caused his coins to appear below standard (Craig, *Newton*, pp. 77–9). See Letter 816.

796 NEWTON TO MONTGOMERIE

[JULY 1710]

From a draft in the Mint Papers[1]

Sr

Upon the receipt of your Letter[2] I went to my Ld Chancellour's[3] to speak wth him about it, but he was gone into ye Country for some days. Then I discoursed the matter wth one of my fellow Officers, & it being observed that the duplicate of the Order of Council wch was sent you for attending the trial of the Pix, was expresly directed to ye General the Master & the Warden of your Mint, we apprehended that none of those three Officers can be excused from obeying the summons but by the Council it self. Mr Scot[4] being Deputy both to ye late Master & to your self & having coyned the money to be tried is best able to answer any questions wch may arise about it & might, on that account, be a proper person to attend the trial in your room if it were done with ye consent of all parties concerned. But to get the approbation of her Majty or Council will be difficult.[5] I am wth all respect

Sr

Mr Montgomerie

NOTES

(1) Mint Papers III, fo. 49. The draft is in Newton's hand throughout. It must have been written between 4 July (when the Order in Council for the Trial of the Pyx was sent out) and 25 July.

(2) Not found, but compare Letter 790.

(3) See Letter 790, note (3).

(4) Patrick Scott (d. 1712), deputy-Master to Allardes and Montgomerie; compare Letter 798.

(5) This forecast was pessimistic. On 25 July, Lowndes wrote to Lauderdale and Seafield excusing the new Master, Montgomerie, from attendance as none of the coins to be tried had been minted while he held office (*Cal. Treas. Books*, XXIV (Part II), 1710, p. 383).

797 LOWNDES TO THE MINT

25 JULY 1710

From the copy in the Public Record Office[1]

Officers of the Mint in England

Gentlemen

My Lord Treasurer directs You forthwith to prepare and transmit to his Lordp. such a Draft of an Indenture[2] for the Mint at Edinburgh as you shall think most agreeable to the Indenture of her Mats. Mint in the Tower and so as the same do not any waies interfere or disagree with the Articles of Union, to the end directions may be given for passing the same under the proper Seales in Scotland. *25th July* 1710.

WM. LOWNDES

NOTES

(1) T/17, 2, 193.

(2) John Montgomerie had been appointed Master of the Edinburgh Mint in succession to George Allardes on 22 June 1710 (there is a copy of the Order in Council appointing him in Mint/1, 8, no. 168). There is a draft or copy of the Indenture between the Queen and Montgomerie in the Mint Papers, III, nos. 81–93. This is in a clerical hand and is not annotated by Newton. Draft suggestions for the wording of the Indenture written by Newton himself may be found in *ibid.* III, 48 and 60.

The salaries stated in the Indenture were: General, £300; Master and Worker, £200; Warden, £150; Counterwarden or Comptroller, £60; Assay-master, £100; Queen's Ingraver, £50; Queen's Smith, £30; total, £890.

798 SEAFIELD TO NEWTON

2 AUGUST 1710

From the original in the Mint Papers[1]

Edinburgh August 2d. 1710

Sir

The Officers of the Mint here concerned in the late coynage are to sett out this day with the pixis for London. Mr Scott who was the late Master's Deputy goes as one of them by vertue of a Deputation from the present Master, he has the honour to be knowen to you and is a very honest man & very capable to act his part at this tryall & he is the only person among the Officers that has any concern to advert to the interest of Mr Allardice,[2] You have been at a great deal of trouble in directing this coynage you have done most justly to all concerned in it and I hope you will continue to assist in directing him in what concerns this tryall and in any thing els that may prevent Mr. Allardice his Gaine for haveing so seasonably served the government and the proprietors of the old money; and who did so readily give his Credit & bestow his pains and attendance at the earnest desyre of My Lord Treasurer and the government. Mr Scott will inform you what further many be needfull and I entreat that you may be assisting to him in procureing the payt of the Ballance due to Allardice; & the allowance of his Charges for this journey with some encouragement for his bypast Services in the late recoynage pardon me for this trouble & I am with the greatest respect

Sr

Your most humble & OBedient Servt

SEAFIELD

NOTES

(1) Mint Papers III, fo. 40; compare the preceding letter.

(2) Another spelling of the name Allardes, closer to its pronunciation.

799 JOHN CHAMBERLAYNE TO NEWTON

21 AUGUST 1710

From the original in King's College Library, Cambridge(1)

Petty France Westmr

21 *Aug:* 1710

Honor'd Sr

Signor Bianchi the worthy Resident of Venice(2) (whom I proposed to your Society for Member) will I fear leave England before he can be fully chosen;(3) however it wil be some satisfaction to him to learn from your own Mouth that he is admitted or approv'd by the Council, but it wil be a much greater as he himself says, to see the renown'd President before he goes, wherefore I have promist him to ask an Audience of you one quarter of an hour either in a morning or afternoon (excepting Wednesday or Thursday of this week) as shal be most for your convenience wch I beg you wil signify by the Bearer to

Hond Sir

your most humble servant

JOHN CHAMBERLAYNE

NOTES

(1) Keynes MS. 99C. On the verso are theological notes in Newton's hand concerning Photinus and Marcellus, fourth century bishops of the Greek Church. John Chamberlayne,

born in 1666, was elected F.R.S. in 1702. He held several court offices, including that of Gentleman of the Privy Chamber. He did much translating and popular writing, besides contributing three papers to the *Philosophical Transactions*. He was an active member of the S.P.C.K. and died in 1723.

(2) Vendramino Bianchi (d. 1738) was a Venetian professional diplomat, formerly Secretary of the Council of Ten and also Resident at Milan. He took part later in the Peace of Passarowitz (1718). He wrote a book about this and another about Switzerland.

(3) He was elected F.R.S. on 8 November 1710. The Society observed the summer Long Vacation.

800 COTES TO NEWTON

4 SEPTEMBER 1710

From the original in Trinity College Library, Cambridge.(1)

For the answer see Letter 803

Monday Sept. 4th 1710.

Sr.

I hope to be at Cambridge again on Wednesday next. I have been somewhat longer in ye Country yn I at first intended, I hope you will excuse Me. For the future I shall, I hope, be ready without any further intermission to attend upon ye Edition of Yr Principia. I desire You to send me the remaining part of Yr Copy assoon as You can.

Yr most Humble Servant

ROGER COTES

For Sr Isaac Newton
at his House
near the College
in Chelsea
near London

NOTE

(1) R.16.38, no. 61, printed in Edleston *Correspondence*, p. 33.

801 LOWNDES TO THE MINT

5 SEPTEMBER 1710

From the copy in the Public Record Office.(1)

For the answer see Letter 804

Whitehall Treasury Chambers 5th: *Sept*: 1710

The Lords Commissioners of Her Mats: Treasury are pleased to Refer this Petition to the Warden, Master and Worker and Comptroller of Her Mats:

Mint who are to Consider the same and Certifye their Lordships a true State of the Matters therein Contained together with their Opinion what is fit to be done therein.

WM. LOWNDES

Penman Ref.: to Officers Mint.

NOTE

(1) Mint/1, 8, 171. The enclosed petition from James Penman, assay-master of the Edinburgh Mint (immediately preceding the above letter) asks a gratification because of the hard work and expense involved in the Scottish recoinage, in which he worked as assistant assay-master from September 1707 to May 1708, and as assay-master (after his predecessor's death) thence until March 1709.

802 NEWTON TO SLOANE

[13 SEPTEMBER 1710]

From the original in the British Museum[1]

Wednesday noon.[2]

[I] am glad Sr Christopher & Mr Wren[3] like the house[4] [&] hope they like the price also. I have inclosed a Note [to] Mr Hunt[5] to call a Council on Saturday next[6] at [twe]lve a clock, & beg the favour that you would send [it] to him by the Porter who brings you this. I am

Sr

Your most humble Servant

IS. NEWTON

NOTES

(1) Sloane 4060, fo. 70; printed in John Nichols: *Illustrations of the Literary History of the Eighteenth Century*, London, 1822, IV, p. 60. The left edge of the paper is imperfect.

(2) Almost certainly the Wednesday preceding Saturday 16 September, when a Council was held; compare Edleston, *Correspondence*, p. xxxviii.

(3) Christopher Wren (1675–1747), son of the great architect, and compiler of *Parentalia*. He had been elected F.R.S. in 1693, and to the Council in this present year.

(4) The difficulties that the Royal Society had faced for several years arose from the fact that the Mercers' Company—one of the Gresham Trustee bodies—wished to rebuild the unsuitable Elizabethan town house which was Gresham College, selling some of its valuable site in the process. They intended to offer the Royal Society space in their new building. In 1701 the Gresham Trustees promoted a bill in Parliament to enable them to achieve these objects, but it was rejected in the House of Lords on the petition of Robert Hooke, who alone of the Gresham Professors opposed the rebuilding (see Lyons, *History*, p. 131). Negotiations continued, but the uncertainty was such that a Committee of the Council was appointed to consider the

purchase of a new meeting-house elsewhere. Meanwhile, Sir Christopher Wren had, by request, prepared plans for accommodating the Society in a building to be erected at the Gresham Trustees' expense upon the old site, plans which Wren's consciousness of the 'very great Quality' of certain Fellows rendered grandiose (see *ibid*., p. 132). Wren's plans proved too much for the Gresham Committee, which at Mercers' Hall on 13 December 1704 resolved neither to meet the Society's demands, nor to revive its Bill for rebuilding Gresham College (see Number 802*a* following). This decision reached the Royal Society in the garbled form that the Mercers had resolved to grant the Society no room at all; however, it is clear that eviction from their existing rooms at the old Gresham College was not intended.

A few months later, on 13 June 1705, Newton laid several proposals before the Society, of which the idea of buying a house in Westminster was most favoured, and other expedients were canvassed such as moving to Cotton House, where the Cottonian Library was. All came to nothing until Newton drew attention to the availability for purchase of a house belonging to the late Dr Edward Browne (d. 28 August 1708), President of the Royal College of Physicians, in Crane Court off Fleet Street, for a price of £1450. In fact, this price also brought a second adjacent house.

(5) Henry Hunt (d. 1713) served Robert Hooke as a boy assistant (*Diary*, 4 and 9 January 1672/3, etc.) and was trained by him. Soon after the death of Richard Shortgrave, the Society's first operator, Hunt was appointed to succeed him (2 November 1676); he also carried on the making of meteorological instruments for Fellows and others that Shortgrave had commenced. On 14 January 1679/80 his salary was raised from £10 to £40 p.a. (the same as Hooke's) in return for his devoting all his time to the Society's affairs. In 1696 he was appointed Keeper of the Library and Repository. By the time of his death the Society owed him £650 advanced by him for the purchase and fitting out of Crane Court.

(6) Besides Newton himself and Sloane (Secretary) only three members of the Council attended this Saturday meeting; it was adjourned to the following Wednesday (20 September) when with the President and thirteen members present it was resolved (with one contrary vote) to buy the Crane Court house.

802*a*

MINUTES OF THE JOINT COMMITTEE FOR GRESHAM AFFAIRS HELD AT MERCER'S HALL ON WEDNESDAY 13 DECEMBER 1704

From the original at Mercers' Hall[1]

Present

For the City	For the Company
Sr: Thomas Rawlinson[2]	Sr: Samuel Moyer Barrt:
Sr: William Withers	Sr: Edmond Harrison Knt:
Mr: Deputy Gardiner	John Morice Esqr
Mr: Thomas Cooper	Mr: Samuel Totton } Wardens
Mr: Deputy Munford	Mr: John Hart } Wardens

Mr: Treasurer Eyre	Mr: Nathaniel Houlton
Mr: Isaac Brand	Mr: Jasper Clotterbooke
Mr: Loveday.	Mr: William Willys.

The Acts of the last Joynt Committee were read.

Order abt the Gresham College Bill. Not to bee presented.

Sr. Francis Child(3) and Sr. Gilbert Heathcot(4) Aldermen of London being now present, Sr. Francis Child acquainted the Committee, hee had hitherto deferr'd offering the Gresham College Bill to bee read in the house of Commons[;] for the Royall Society, who had beene severall yeares accomodated in Gresham College by the Curtesie of the Citty and Company did expect (tho they neither can, nor do, insist on itt as any right) to have some provision made in the Bill for a Hall and other Conveniences for them att the charge of the Citty & Company, and if they were not made easy, itt seemed very plaine that the Bill would bee obstructed by their Interest—That Doctor Newton President Doctor Sloane Secretary of ye sd. Society and Sr. Christopher Wren and Mr. Asky Members thereof had by order of their Society mett Sr. Francis Child Sr. Gilbert Heathcot & the Clarks of this Committee in order to accomodate this matter, and first insisted on forty foot of ground in front and sixty foot in depth in the best part of the intended Square and that the Citty and Company at their charge should erect A building thereon, and that the Society should bee interested therein Gratis at their Freehold, or for nine hundred ninety nine years at a pepper corne Rent—That the sd. Proposition being thought very unreasonable, another meeting was appointed to consider of some other method, which being had, it was there declared by the Gentl[emen] in behalf of the Royall Society, that they would bee content with five and twenty foot of ground fronting the Square and thirty foot depth joyning to the Hall intended for the Lectures, and a house to bee built thereon by the Citty and Company, and alsoe two Storys over the sd. Hall & a Platforme of lead to view the Heavens, to bee intire for the Societys use, and to have an Interest therein Gratis for. 999. yeares at a pepper Corne Rent, but sd. they would fit up the inside of the house for their convenience at their owne charge —Sr. Gilbert Heathcot declared he was of Opinion, it would bee for the Generall advantage of the Citty of London to accomodate the said Society, in regard they were very famous in forreigne parts, and would probably draw a great concourse of people where ever they were,—Whereupon after long debate, and the Will of Sr. Thos. Gresham being read & considered of, the Question was put whether itt is the Sense of this Committee that the Citty and Company should accomodate the Royall Society according to the last exposition above mentioned and itt was unanimously carried in the negative—Then the Question was put whether the Gresham College Bill should be presented this session of Parliament or noe, and twas' carry'd in the negative, & the Committee ordered the Clarkes to wayte on Sr: Francis Child & Sr: Gilbert Heathcot and give them thanks for the Paynes they have taken in this affaire, and accquaint them with the Resolutions of this Committee.

NOTES

(1) Gresham Repertory, 1678–1722, fo. 397. Some normal contractions have been silently expanded in the transcription.

(2) (1647–1708); former Master of the Vintners' Company, Lord Mayor of London, 1705.

(3) (1642–1713); from being a goldsmith became one of the most important London bankers; he had been Lord Mayor and was now M.P.

(4) (? 1651–1733); chief founder of the East India Company, 1693, and major participant in the Bank of England, 1694, he was Sheriff and M.P. for the City; Lord Mayor, 1710–11. Heathcote was a Cambridge graduate.

803 NEWTON TO COTES

13 SEPTEMBER 1710

From the original in Trinity College Library, Cambridge.[1]
Reply to Letter 800; for the answer see Letter 805

Sr

This Letter accompanies the next part[2] of the Principia. I am not certain that you have all ye cutts in wood, but if any be wanting pray send me a draught in paper of what is wanting & I'le get them cut [in] wood. I am

Sr

Your most humble Servant

Is. Newton.

Chelsea. Sept 13
1710.

For the Rnd Mr R. Cotes Professor of
Mathematicks & Fellow of Trinity
College in ye University of Cambridge

NOTES

(1) R.16.38, no. 62, printed in Edleston, *Correspondence*, p. 33.

(2) This made up signatures Qq to Ddd of the new edition.

804 NEWTON AND ELLIS TO THE TREASURY

18 SEPTEMBER 1710

From the copy in the Public Record Office.(1)
Reply to Letter 801

To the Right Honble: the Lords Comrs: of Her Majts: Treasury

May it please your Lordships

In Obedience to your Lordships Order of Reference of the 5th Instant upon the Petition of Mr. James Penman Assay Master to the Mint at Edenborough, we have considered the same and upon Enquiring into the Matter therein contained we beleive that the Busines of the Assay master during the late Recoinage of the moneys in Scotland was very laborious by reason that every Journey of new Money was severally Assayed according to the custom of that Mint, & that the performance thereof required a Clerk and Assistant to the Assay master, for the Maintenance of whom during the Recoinage which lasted a Year and Six Months, We are humbly of Opinion that there may be allowed to the Petitioner for Himself & Mr Borthwick(2) a Summe after the Rate of Forty Pounds per Annum in all Sixty pounds to be divided between them in proportion to the time they severally acted as Assaymasters during that Service, there being the like Allowance made to some other new Clerks of that Mint.

All which is most humbly submitted to your Lordships great Wisdom.

Is. Newton
Mint Office Jn. Ellis.
Sept: 18: 1710.

NOTES

(1) Mint/1, 8, 171.
(2) Borthwick was the previous assay-master who died in May 1708.

805 COTES TO NEWTON

21 SEPTEMBER 1710

From the original in the University Library, Cambridge.(1)
Reply to Letter 803; for the answer see Letter 807

Cambridge Sept. 21*st* 1710

Sr.

I have received the second part of Yr Copy; there are wanting only two wooden Cutts, which I can get made at Cambridge. I have read over what relates to ye Resistance of Fluids, I thank You for the satisfaction I have received in seeing yt Theory so perfectly compleated.(2) I confess I was not a little Surprized upon ye first reading of Prop: 36,(3) but I now begin to be better reconciled to it.(4) One of my greatest difficultys was an Experiment of Monsr. Marriotte which he says (pag. 245 Traite du Mouvm: des Eaux) he often repeated wth great care.(5) By his Experiment I concluded yt the Velocity of the effluent water was equall to yt gotten by a heavy Body falling but from half the height of the Vessel, as You had determined in ye former Edition of Yr Book. He tells us yt 14 Paris Pints of water were evacuated in a Minute of time through a circular aperture of $\frac{1}{4}$ Inch diameter, ye Altitude of ye Vessel being 13 feet. He describes ye Paris pint to be ye 35th part of ye Cube of ye Paris foot. Therefore ye water evacuated in a second was $\frac{14\times 1728}{35\times 60}$ or $\frac{2\times 144}{25}$ [cubic] Inches. The Area of the Aperture was $\frac{11}{14\times 16}$ [square] Inches. Hence ye length of a Cylinder equall in magnitude to ye evacuated water & having ye abovementioned aperture for its Basis is $\frac{2\times 14\times 16\times 144}{11\times 25}$ Inches, & this length is ye space described in a second of time with ye uniform velocity of the Water as it passes through the aperture. The Space described in a second of time wth ye uniform velocity acquired by a falling Body in the same time is $2\times 15\frac{1}{12}$ feet, or 2×181 Inches. The Velocity therefore which a falling Body acquires in a Second, is to ye Velocity of ye water as 1 to $\frac{16\times 14\times 144}{11\times 25\times 181}$. Hence ye Altitude of 181 Inches, is to ye altitude A from which an heavy Body must fall to acquire ye velocity of ye water as 1 to $\overline{\frac{16\times 14\times 144}{11\times 25\times 181}}\Big|^2$ Therefore $A=\overline{\frac{16\times 14\times 144}{11\times 25}}\Big|^2 \times\frac{1}{181}=76$ Inches, that is, nearly half 13 feet. I wish the Experiment were made in an intire Cylindrical vessel such as You describe in Prop: 36. I think I have reason to suspect yt the figure of his Vessel may contribute very much to the diminution of ye Velocity of the Water.(6) With Yr leave I would alter the latter part of Cor. 2. Prop 38 thus(7)

Nam Globus tempore casus sui, cum velocitate cadendo acquisita describet spatium quod erit ad octo tertias diametri suæ ut densitas Globi ad densitatem Fluidi; et vis ponderis motum hunc generans erit ad vim quæ motum eundem generare possit quo tempore Globus octo tertias diametri suæ eadem velocitate describit ut densitas Fluidi ad densitatem Globi: ideoque per hanc Propositionem vis ponderis æqualis erit vi resistentiæ & propterea Globum accelerare non potest.

I propose the following alteration in Prop. 40[8]
Demittatur Globus ut pondere suo B in Fluido descendat; & sit P tempus cadendi, idque in minutis secundis si tempus G in minutis secundis habeatur. Inveniatur numerus absolutus N qui congruit Logarithmo $0{,}4342944819 \times \frac{2P}{G}$, sitque L Logarithmus numeri $\frac{N+1}{N}$ & velocitas cadendo acquisita erit $\frac{N-1}{N+1}H$, altitudo autem descripta erit $\frac{2PF}{G} - 1{,}3862943613F + 4{,}605170186LF$. Si Fluidum satis profundum sit, negligi potest terminus $4{,}605170186LF$, & erit $\frac{2PF}{G} - 1{,}3862943613F$ altitudo descripta quamproxime. Patent hæc per Prop. IX Lib. II & ejus Corollaria, ex Hypothesi quod Globus nullam aliam Patiatur resistentiam nisi quæ oritur ab inertia materiæ. Si vero aliam insuper resistentiam patiatur, descensus erit tardior, & ex retardatione innotescet quantitas hujus resistentiæ

Ut corporis in Fluido cadentis velocitates & descensus facilius innotescant composui Tabulam sequentem, cujus columna—&c.—Numeri in quarta columna sunt $\frac{2P}{G}$, & subducendo numerum $1{,}3862944 - 4{,}6051702L$, inveniuntur numeri in tertia columna & multiplicandi sunt hi numeri per spatium F ut habeantur spatia cadendo descripta. I think this alteration or some other to the same effect (which You may be pleased to send me) will make the Theory more easy. I have not as yet had time to go over the Calculation of the Scholium which is annexed to this Proposition.

Yr much obliged & humble Servant
ROGER COTES

NOTES

(1) Add. 3983, no. 8; partially printed by Edleston, *Correspondence*, pp. 34–5, from the incomplete draft in Trinity College Library, MS. R.16.38.

(2) Presumably Cotes means to embrace here the remainder of Book II, Section VII, that is Props. 36 to 39.

(3) Prop. 37 in the first edition. Newton here argued (p. 331) that the velocity of water flowing out through a hole, under a head of height h, is $\sqrt{(gh)}$; hence, ideally, the jet would be capable of ascending to a height $h/2$. As Cotes goes on to point out, measurements of the rate of efflux of water appeared to confirm this result.

(4) Experimental trials, of which Torricelli's were the first, suggested that Newton's result was incorrect in that the height $\frac{1}{2}h$ could be exceeded; in some experiments made by Halley before the Royal Society on 18 and 25 March 1691 the jet spouted 'far above the middle of the height' of the head of water, 'whence it is to be noted that there is a mistake in the 37th Prop.

of Mr Newton's 2nd Book, whereof it was ordered that Mr Newton should be certified' (Edleston, *Correspondence*, p. 35). Hence Newton re-examined his argument and (as Cotes makes clear) demonstrated in the new copy for Prop. 36, that $\sqrt{(2gh)}$ gives the correct value of the velocity (under ideal conditions). But this (as Cotes now argues) was *greater* than the velocity found by experiment, just as $\sqrt{(gh)}$ was *less*.

(5) Evangelista Torricelli was probably the first to suggest that actual experiments justified the $\sqrt{(2gh)}$ velocity rule, though he had proposed this rule as an axiom in the first instance—that the outflowing water would have the velocity of any drop falling through the height h ('De motu aquarum', in *De motu gravium naturaliter descendentium et proiectorum* (1644); Gino Loria e Giuseppe Vassura, *Opere di E. Torricelli*, Faenza, 1919, II, pp. 185–7). Edmé Mariotte's experiments were reported in his *Traité du Mouvement des Eaux*, Paris, first edition, 1686, second edition, 1700; Cotes used the later edition (compare *Œuvres de Mr Mariotte*, Leiden, 1717, p. 426). Mariotte himself believed that ideally the $\sqrt{(2gh)}$ rule was true, but that friction of the water at the aperture and air-resistance reduced the height of the jet by an amount increasing as the square of the height (*ibid.* pp. 436–8). Newton himself seems to have relied initially on Mariotte's experiments; see U.L.C. Add. 3968(9), fo. 102v.

(6) Newton was impelled by Cotes' remarks to investigate the matter a third time, as will be seen in later letters.

(7) This formulation is printed in the second edition, p. 316. It clarifies what had been printed before.

(8) This revision is printed in the second edition, p. 318, with negligible numerical alterations. Prop. 40 is on the movement of spheres through perfectly fluid mediums of known density, and prepares the way for Newton's verification of his theory by experiments in the following *Scholium*. This section had already been considerably modified by Newton, so that direct comparison of the first edition text with that of the second becomes profitless.

806 NEWTON AND ELLIS TO HENRY ST JOHN

[late SEPTEMBER 1710]

From a draft in the Mint Papers(1)

Sr

On Saturday Sep 16th Instant we received a Letter from Mr Secretary Boyle dated the day before & directed to ye Officers of her Majties Mint signifying that her Majty having been pleased to reprieve Jane Housden & Mary Pitman who were condemned last Sessions at the old Baily for high Treason in counterfeiting the coyn of this kingdom, & that there having been some circumstances represented which might induce her Majty to pardon them, he was commanded to inform himself of the said Officers what objections they might have in relation to their being pardoned.

We the Mr & Controller of the Mint in the absence of the Warden whose buisiness it is to prosecute such Offenders have therefore e[n]quired into that matter & find that Jane Housden was committed to prison by the name of Jane

Newstead in the year 1696 for clipping the coin of this kingdom & that a warrant was also then issued out against one Tho. Newstead her pretended husband for ye same crime & information was soon after given that the said Tho. Newstead used indirect practises to stifle ye said warrant & procure ye liberty of the said Jane Newstead.

About two years after, the said Jane Newstead was again committed to prison by the name of Jane Newsted alias Housden for putting of counterfeit money & upon suspicion of coining the same, & about four pounds of counterfeit money were then taken upon her & three files with some sand found in her house & she confessed that she received the said counterfeit money & ye three files of one Mr Tuck whom she knew to be a coyner & that the sand was left in her custody by another man who had brought her into ye acquaintaince of the said Tuck.

Aftwerds in the years 1702 she was again committed to prison & convicted of counterfeiting the coyn of this kingdom & pardoned by her Maty in order to be transported, & was set at liberty upon giving security to transport herself & therefore being now found in England she is liable to be called down to the barr upon the judgment then given against her

And whereas she & Mary Pitman were accused the last Sessions at the old Baily of coining together & convicted by the finding of the coining Tools & counterfeit money, we observe that this is the second conviction of Jane Housdens; But against Mary Pitman we meet wth nothing antecedent to the fact of which she now stands convicted

These things we pray you to lay before her Majty [in answer to the Letter wch Mr Boyle wrote to us by her Majties order.] We are](2) & remain

Your most humble
& most obedt Servants.

And (3) this present year being again accused of high Treason in counterfeiting the coyn of this kingdom when she was apprehended she dropt a parcel into the Thames wch was found to be a parcel of counterfeit money & the coining Tools were found in ye house of Mary Pitman where these two weomen were said to coine together, & by these circumstances she stands now convicted a second time. But against Mary Pitman we meet with nothing antecedent to the fact of wch she now stands convicted

To the Rt Honble Henry St John Esqr
her Majties Principal Secretary of State

NOTES

Henry St John (1678–1751), to be created Viscount Bolingbroke in 1712, was appointed Principal Secretary of State for the Northern Department in succession to Henry Boyle (d. 1725; cr. Lord Carleton 1714) on 21 September 1710.

(1) Mint Papers I, fos. 466–7. It must have been written in the last days of September 1710. It is a highly corrected holograph draft.

(2) Brackets in original.

(3) This paragraph is written on the facing page; the place of its insertion is not indicated.

807 NEWTON TO COTES

30 SEPTEMBER 1710

From the original in Trinity College Library, Cambridge.(1)
Reply to Letter 805; for the answer see Letter 809

London. Sept. 30. 1710

Sr

Since the receipt of your Letter I have been removing from Chelsea to London,(2) wch has retarded my returning an answer to your last. I have not seen Mariots book concerning the motion of running water, but certainly there is something amiss in his experiment wch you give me an account of. For I have seen this experiment tried & it has been tried also before the Royal Society, that a vessel a foot & an half or two foot high & six or eight inches wide with a hollow place in the side next the bottom & a small hole in the upper side of the hollow, being filled with water; the water wch spouted out of the small hole, rose right up in a small streame as high as the top of the water wch stagnated in the vessel, abating only about half an inch by reason of the resistance of the air. The small hole was made in a thin plate of sheet tin & well polished, that the water might pass th[r]ough it with as little friction as possible. It was about the bigness of a hole made with an ordinary pin.(3)

The corrections you have made are very well & I thank you for them, & am glad that the Theory of the resistance of fluids does not displease you provided the XXXVIth Proposition be true, as I think it is.

Direct your next Letters to me in St Martins Street neare Leicester fields. I am

Your most humble Servant
Is. NEWTON

For the Rnd Mr Cotes Professor of
Astronomy & Fellow of Trinity
College in Cambridge in
Cambridgeshire.

NOTES

(1) R.16.38, no. 64; printed in Edleston, *Correspondence*, pp. 36–7.

(2) Newton had now established himself in St. Martin's Street, Leicester Fields, in a house later numbered 35. This street—running south from Leicester Square to the back of the National Gallery—has been totally rebuilt. Newton moved thence to Kensington in January 1724/25.

(3) See Letter 805, note (4). Mariotte's experiments were not at fault; the apparent discrepancy between theory and practice was caused by the contraction of the jet after leaving the hole forming it, which makes the effective diameter of the jet less than that of the hole forming it (see Letter 826).

808 NEWTON AND ELLIS TO THE TREASURY

4 OCTOBER 1710

From the original in the Public Record Office[1]

To the Right Honorable the Lords Commissioners of her Majestys Treasury.[2]

May it please your Lordships

Your Lordships having been pleased to appoint the Warden, Master and Worker, and Comptroller of her Majes Mint for the Management of the Tin Affaire[3] in London, and having directed them in their Constitution to sell and dispose of the Tin, that shall be consigned to them, according to their best discretion at such price or prices and pursuant to such Orders and directions as they shall from time to time receive from your Lordships, We humbly desire your Lordships Warrant as to the price, which in the former Contract was sett at £3. 16.– –per hundred Merchant weight, and upon which the new one was calculated.

Being likewise Authorised and required by your Lordships to Imploy such porters, Carmen, or others to receive the said Tin from on shipboard, and to carry the same into the Mint, and to deliver out the same when sold at such Salarys, Wages or Allowances as we shall thinck reasonable to be approved from time to time by your Lordships, and to be satisfied out of such Money and in such Manner as your Lordships shall direct, We humbly represent to your Lordships that in the former Contract we were impowered by a subsequent Warrant to issue our Orders from time to time to the Treasurer or Receiver of the Monies arising by the sale of her Majes Tin, to pay out of the proceed thereof both the charges of freight, and all other incident charges proper for her Majesty to bear on account of the bringing in of the said Tin, or the selling or delivering out of the same, or otherwise howsoever.

We farther humbly represent to your Lordships that the following persons were imployed in the Management of the Tin Affaire in the former Contract with these severall Allowances, which if your Lordships do approve of, We desire your Lordship's Warrant for their continuance viz. Ourselves each £150. per Annum £450.–.–

Dr. Fauquier[4] Deputy to the Master and Worker as our Assistant to attend dayly for the selling or otherwise disposing of her Majesty's Tin, with an allowance of One hundred pounds per Annum £100.–.–

Mr. Beresford[5] Deputy to the Comptroller to keep a Leidger book and to enter all orders and Minutes at forty pounds per Annum 40.–.–

The Warden's, and one of the Master & Worker's Clerks to keep books of all Tin received and sold, and attending to the Weighing of the Tin when landed at Thirty pounds per Annum each 60.–.–

Tobias Dixon head porter and Warehouse keeper, who is to attend constantly the Receipts and deliverys, and is to give us £10000. security for the trust reposed in him as Warehouse keeper at One hundred pounds per An: 100.–.–

Mr Edwd Webster who keeps books of entrys both of all papers and Accts. transmitted from the Agents in Cornwall and Devon, and of all Warrts. and orders relating to the Tin affaire at one hundred pounds per Annum 100.–.–

All which is most humbly submitted to your Lordships great Wisdom

IS. NEWTON
JN ELLIS

Mint Office the 4th October
1710.

NOTES

(1) T. 1/125, no. 4. This was written by a copyist and signed by Newton and Ellis.

(2) Lord Godolphin was dismissed from office on 19 August 1710, when the Treasury was put into commission under Earl Paulett, Robert Harley (afterwards Earl of Oxford), Henry Paget, Sir Thomas Mansell and Robert Benson. Harley was appointed Lord Treasurer after the General Election of 1710, on 29 March 1711.

(3) Compare Letter 769. For the various references to the Tin Trade, see Index.

(4) John Francis Fauquier, or Fauquiere (died 1726) was later a Director of the Bank of England. He was appointed Deputy Master by Neale, Newton's predecessor as Master, and was given the salary of £60 per annum, allowed to the Master's Assayer. See Craig: *Newton*, p. 34.

(5) The Beresford (or Berisford) here mentioned served as Deputy Comptroller between 1708 and 1717. See vol. IV, p. 376, note (3).

809 COTES TO NEWTON

5 OCTOBER 1710

From the original in the University Library, Cambridge.(1)

Reply to Letter 807; for the answer see Letter 811

Cambridge Octobr. 5*th* 1710

Sr.

I received yr Letter the first day of this Month. I was satysfied by several Experiments of Marriotte & others & by some Tryals which I had formerly made my self that the Spout of Water when directed upwards would ascend nearly to ye whole height of the Water which stagnates in ye Vessel. But notwithstanding this assurance I had still a Suspicion that the Velocity with which the water flowed out of the Vessel was equall to that which an Heavy body acquires by falling but from half that Height. To satisfy my self in this matter I did Yesterday make the following Experiment.

I procured a Vessel *ABCD* of the Figure here described, the diameter of ye Bottom *AB* & the height of the Vessel *IE* were between 2 & 3 feet. About ye middle of the Bottom, which was $\frac{3}{4}$ of an Inch thick, I made a Conical Hole *EGHF*, the diameter *EF* being about an Inch. Over this Hole I fixed a Plate of Tin having a circular aperture in it whose diameter I measured with all the care I could & found it to be exactly $\frac{1}{2}$ Inch. The Tin was so placed that the centre of the Aperture might coincide with the centre of ye Circle *EF*, & the water was prevented from coming under the Tin by stopping all passage that way with a mixture of Bees-wax & Turpentine. The lower vessel *OKLP* which received the effluent water was a Frustum of a Cone as regular as I could procure.

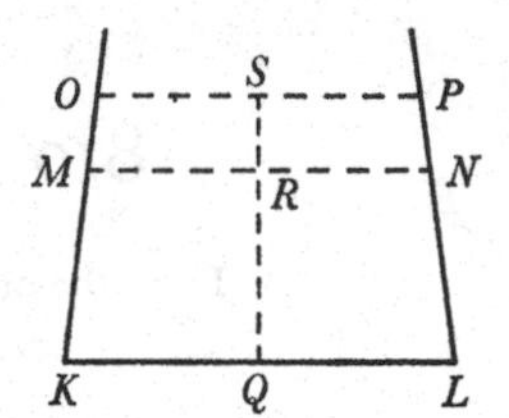

The upper Vessel being fill'd with water so yt its surface *CD* was 32,8 Inches above ye Tin the water was permitted to run out during ye time of 135 Seconds, new water being continually supplied to keep ye surface *CD* all ye while at ye Altitude of 32.8. By this efflux the lower Vessel had received so much water yt the Altitude *RQ* of its surface *MN* was 13,8 Inches.

I repeated ye Experiment & found yt when the time of efflux was 143 Seconds the Altitude *SQ* of the surface *OP* was 14,65 Inches.

The Diameter *KL* was 16,4 Inches, the Diameters *MN*, *OP* were not sensibly different, by my measure they were each equall to 15,85 Inches.

Now by the first Experiment I conclude yt the water evacuated in a Second of time was 20,877 Cubick Inches; by ye second Experiment it was 20,923 Cubick Inches. I took as a Mean 20,9 Cubick Inches, & found yt the Velocity of the effluent water was equall to ye velocity acquired by an Heavy body falling 14,64 Inches, not 32,8.(2)

I was not accommodated with all the conveniences I could have desired to make the Experiment as exact as was possible yet I am satisfied by this Tryal that Monsr Marriotte's Experiment ought not to be disregarded. I wish You would be pleased Yr self to make the Experiment before You resolve to print the 36 Proposition as it now stands.

I am Your most humble servant
ROGER COTES

For Sr Isaac Newton
at His House in St Martin's Street
near Leicester Fields
London

NOTES

(1) Add. 3983, no. 9.

(2) If the flow is nearly 21 cu. in. sec^{-1}, and the aperture $\pi/16$ sq in., then (assuming the column of liquid formed to be the same diameter as the aperture) the velocity of the water in the column is about 107 in. sec^{-1}; hence if $\surd(2gh) = 107$, $h \approx 15$ in.

810 COTES TO NEWTON

26 OCTOBER 1710

From the original in the University Library, Cambridge.(1)
For the answer see Letter 811

Octobr. 26*th* 1710

Sr.

It is now three Weeks since I wrote to You & gave You an account of an Experiment I had made to determine the velocity of the effluent water.(2) I concluded from this Experiment, as I did before from yt of Monsr. Marriotte, that it was equall to the Velocity acquired by an heavy body falling but from half ye height of the water in the Vessel; & I wish'd You would be pleased to make the Experiment Your self before You resolv'd to print the 36th Proposition. Having received no Letter from You since that time, I fear my own was miscarried.

I am Your most Humble
Servant ROG: COTES

NOTES

(1) Add. 3983, no. 10.

(2) Newton acknowledged this letter the following day, but did not answer it until the following March (see Letter 826).

811 NEWTON TO COTES

27 OCTOBER 1710

From the original in Trinity College Library, Cambridge.(1)
Reply to Letters 809 and 810

Sr

I received both your Letters & am sensible that I must try three or four Experiments before I can answer your former.(2) My time has been taken up partly with removing to this house, partly with journeys about purchasing a house for the Royal Society(3) & partly wth settling some matters in the Mint in order to go on wth ye coynage(4) that I have had no time to take these matters into consideration but hope wthin a fortnight to try the experiments & settle the matters in doubt & beg the favour that you will let the press stay till you hear from me again. I am

Your most faithfull friend
& humble Servant
IS. NEWTON

London
Octob. 27. 1710
For the Rnd Mr Cotes Professor of
Astronomy, at his chamber in
Trinity College in Cambridge.

NOTES

(1) R.16.38, no. 65; printed in Edleston, *Correspondence*, pp. 37–8.

(2) For an account of Newton's experiments, see Letter 826.

(3) See Letter 802.

(4) The Mint was not extraordinarily productive in 1710. However, there was a change of ministry in August (when the Treasury was put in commissions) and preparations were probably already afoot for the great deal of work done in 1711.

812 AN ACCOUNT OF THE LATE PROCEEDINGS IN THE COUNCIL OF THE ROYAL SOCIETY, IN ORDER TO REMOVE FROM GRESHAM-COLLEGE INTO CRANE-COURT IN FLEET-STREET.

IN A LETTER TO A FRIEND.

22 NOVEMBER 1710

Extracts from the printed pamphlet in the British Museum[1]

Sir,

You desire an Account of the surprising Convulsions that have lately happen'd in the Royal Society...I fear, the real causes that have produc'd these odd Effects, are bury'd too deep for me to reach. However I have endeavour'd to inform myself of all Particulars, amongst those whom I esteem'd most likely to have a perfect Knowledge of what past (at least *openly*) in their Debates...

Nov. 8. 1710.[2] The President of the Royal Society, acquainted the Council that the Dwelling-house of the late Dr. Brown, in Crane-Court, in Fleet-Street, (being an Inheritance in Fee Simple) was to be Sold; and that he thought it would afford a very convenient Reception to the Society. Upon this Information...a Committee was appointed to view, in order to Purchase it, for their Use.

A few Days after, the President Summon'd another Council to Meet at Gresham College upon Saturday, Sept. 16. at 12 a Clock:[3] And upon Tuesday preceding, at Night, he sent Orders to Summon also as many of the Fellows as were in Town, or could be found, to Meet at the same Time and Place.[4] At this Extraordinary Meeting of the Society he told them, That they were without any Being of their own; that their continuing in Gresham-College was very Precarious; that Dr. Brown's House had been propos'd to 'em, and a Committee had view'd it, and that he thought it very convenient for the Uses of the Society: He added, That he had call'd 'em thither that he might hear what Objections they had to offer against the Proposal, that the Council might consider of them and take their final Resolutions accordingly.

The profound Silence that follow'd, sufficiently exprest a general Surprize: Till the President (after a little while) began the Debate; and addressing himself to some particular Members, ask'd their Objections. They told him, that the very Embryo of the Society had been *form'd* in Gresham-College, and that they kept their Weekly Meeting in that Place some time before they obtain'd the Royal Charter of Incorporation; that the Society had continu'd there almost ever since even in their more flourishing Condition; that they yet enjoy'd the same Freedom and Conveniencies as formerly, without the least Disturbance or Impediment; that the present Professors of Gresham-College, are as willing as their Predecessors were before them, that the Society should long continue to enjoy the same Accommodation; and if any new Privileges could be reasonably desir'd they might be assur'd to meet with a ready Compliance from Gentlemen of so much Learning and Ingenuity. And therefore they hoped to hear the Reasons that induc'd him, and a few others who appear'd as zealous and earnest to remove from thence.(5) Till that Question was debated and determin'd it was out of Season to enquire into the Inconveniencies of the House he had recommended.

The President was not prepar'd (or, perhaps, not instructed) to enter upon that Debate: But freely (tho' methinks not very civilly) reply'd, That he had good Reasons for their Removing, which he did not think proper to be given there.

The Acting Secretary,(6) who had engross'd the whole Management of the Society's Affairs into his own Hands, and despotically Directs the President, as well as every other Member, took upon him to relate a Fact, which he thought would determine every Vote. He told them, that One of the Gresham Committee ask'd Him (not long ago) Why the Royal Society did not remove from Gresham College? Since the City had several times sent them Warning to that Purpose...

[This does not appear to have satisfied the Members, who were of the opinion that there were no grounds for removing.]

They likewise Remonstrated to the President, that the Season of the Year, and the short Notice he had given of this Meeting, made it very improper to determine an Affair of so great Importance to the Society at that time: And therefore they mov'd that the Debate should be adjourn'd to St. Andrew's-Day, or at least to some other Extraordinary Meeting;...This the President would not hear of. They therefore offer'd to give him their Opinion either by Ballotting or Voting *viva voce* [other objections were made] But in vain; his Scruples were unmoveable. So that some of the Gentlemen with warmth enough ask'd him, To what purpose then had he call'd them thither? Upon which the Meeting broke up somewhat abruptly, and not only the Members of the Society, but most of the Council also, left the President with Dr. Sloan, Mr. Waller, and one or two more, to take such Measures at the Councill as they best lik'd.

The next Meeting of the Councill [was] on Wednesday following...Of the 14 Fellows that then appear'd there were Two only who dissented.(7)

...In a few Days another Council was summon'd: and then 'twas resolv'd that towards the Purchase 450*l.* should be paid out of the Treasurer's Hands;

That Sir Godfrey Copley's Legacy of 100*l*. should be apply'd to this Use, and that 900*l*....should be borrow'd upon a Mortgage of the Premises at 6*l*. *per Cent* Interest.(8)...

The President was so elated with this Success, that he presently summon'd another Council to meet at the House in Crane-Court: And at the same time gave notice that the ordinary Meetings of the Society would begin there on Wednesday November 8....

I am,
Sir,
Your Most Faithful
Humble Servant

22 *Nov.*
1710.

NOTES

(1) The pamphlet is of 32 pages; its pressmark is 740c 24(2). It was first used by Weld, *History*, I, pp. 391–4.

(2) *Recte*, 8 September. At this meeting, at which eight members were present besides the President, 'the President acquainted the Councill that he had summoned this Councill on occasion of the late Dr. Brown's House in Crane-Court in Fleet Street, being now to be sold; being in the middle of town, out of noise and might be a proper place to be purchased by the Society for their Meetings'. 8 November was the date of the Society's first meeting in Crane Court (*Council Minute Book*, fo. 171).

(3) Compare Letter 802.

(4) The writer of the pamphlet was certainly misinformed as to these matters, which throws doubt on the whole narrative. The Council meeting on Saturday 16 September, at which only Halley, Moreland and Waller were present besides Newton and Sloane, was adjourned to the following Wednesday (20 September) when the President and 13 members attended. Here Newton reported that he and several other Council members, after visiting the Crane Court house, judged it suitable for the Society's purposes; on the question being put, whether Dr Browne's interest in the house(s) should be bought for £1450, twelve votes were cast in favour and one against. (*Council Minute Book*, fo. 173.) There seems to be no other record of the extraordinary assembly of the Royal Society recorded in the pamphlet; obviously it could only have been called on or after 20 September.

(5) These arguments of the pamphleteer either ignore or deliberately conceal all the difficulties concerned with the management of Gresham College, previously noted.

(6) Sloane is, of course, intended, the word *acting* being used as equivalent to active; the other, or Second Secretary, had little share in the Society's business.

(7) The pamphleteer now returns to the Council of 20 September, with some inaccuracy.

(8) This paragraph gives a garbled version of events at the Council meeting of 26 October (*Council Minute Book*, fo. 174). Newton reported that he had, with the committee of Council appointed for the purpose, contracted for the purchase of the leases in Crane Court; at this point two Council members, Richard Harris (Second Secretary) and Walter Clavell, who were both opponents of Sloane and friends of Woodward, left the meeting. The remaining 11

Council members then agreed unanimously to devote £550 of the Society's funds to the purchase, raising the rest (£900) on mortgage, at 6% interest. Newton himself gave the Society £110 towards the purchase.

The allusion to the Copley Bequest is a misrepresentation. Isaac Pitfeild, the Treasurer, was ordered to receive the £100, the Society accepting its responsibilities under the terms of Sir Godfrey Copley's will. There was no suggestion that this bequest be used for the purchase of Crane Court.

813 THE TREASURY TO THE MINT

29 NOVEMBER 1710

From the minute in the Public Record Office.[1]
For the answer see Letter 815

William Palmes desires a Licence from Her Maty. to Coyn Copper halfe pence & Farthings in Like manner as Sr. Jos. Hearne & others[2] had from their late Majts. K. Wm. & Q: Mary for ye space of 7: years, it being long since that time is expired & that Coyn being pretty well exhausted, Humbly desires ye like Grant for ye space of 14 Years not exceeding in ye Whole 700: Tun & Subject to ye Same Limitations & Covenants as was in ye 7: years patent. 29th Novr. 1710: Ref. to ye Warden, Master & Worker, & Comptroller of Her Mats. Mint to consider & Report &c.

Wm: Lowndes.

NOTES

(1) INDEX 4622, p. 431. Who the petitioner was is not clear; several persons of this name appear in the *Calendar of Treasury Books*. His petition, in the terms recorded here and dated 1 September 1710, is at Longleat House (Portland Papers, IX, fo. 173). The matter was first brought before the Lords of the Treasury on 28 August 1710 (see *Cal. Treas. Books*, XXIV, (Part II), 1710, pp. 522, 582, 601).

(2) See vol. IV, p. 409, note (2).

814 ROYAL WARRANT TO NEWTON

12 DECEMBER 1710

From the original in the University Library, Cambridge[1]

Anne R.

Trusty and Well beloved.

We Greet you well. Whereas We have been given to understand, that it would contribute very much to the Improvement of Astronomy and Navigation if We should appoint constant Visitors of our Royal Observatory at Greenwich, with sufficient Powers for the due Execution of that Trust.(2) We have therefore thought fit in Consideration of the great Learning Experience and other necessary Qualifications of our Royal Society to constitute and appoint, as we do by these presents constitute and appoint you the President, and in your Absence the Vice President of our said Royal Society for the time being, together with such others as the Council of our said Royal Society shall think fitt to join with you to be constant Visitors of Our Said Royal Observatory at Greenwich; Authorising and requiring you to demand of Our Astronomer & Keeper of our Said Observatory for the time being to deliver to you within six months after every Year shall be elapsed a true & fair Copy of the Annual Observations he shall have made.(3) And Our further Will & Pleasure is, that you do likewise from time to time Order and Direct Our said Astronomer & Keeper of Our Said Royal Observatory, to make such Astronomical Observations as you in your Judgement shall think proper. And that you do survey and Inspect Our Instruments in Our said Observatory, and as often as you shall find any of them defective, that you do inform the Principal Officers of Our Ordnance thereof, that so the said Instruments may be either Exchanged or repair'd. And for so doing this shall be Your Warrant.(4) And so We bid you Farewel.

Given at Our Court of St James's the Twelfth day of Decembr. 1710 In the Ninth Year of Our Reign.

By her Majties. Command

H. St. John

Visitors of the Royal Observatory at Greenwich

NOTES

(1) Add. 4006, fo. 29. There is a copy in the Royal Society *Council Minute Book*, II, fo. 180. The original is in a non-clerical hand but is signed by St John, Secretary of State since 21 September. Flamsteed's copy is printed in Baily, *Flamsteed*, pp. 90–1.

(2) The first stage of the dispute between Newton and Flamsteed over the printing of the *Historia Cœlestis Britannica* ended in March 1707/8 (see vol. IV, Number 737). In the same year the death of Prince George of Denmark removed his personal and financial intervention. All was quiet—and the catalogue of fixed stars which lay at the centre of the dispute remained unprinted—until after the change of ministry in 1710. Whether (as Flamsteed supposed: Baily, *Flamsteed*, pp. 89–90) this political upheaval gave Newton a new sense of power, or whether Newton felt the situation had gone too long unresolved, one cannot tell. The emphasis on the state of the Observatory suggests a new grievance.

(3) The Royal Society, like Flamsteed himself, received the Royal Warrant on 14 December (it was brought to Arbuthnot, for whom see below). It was then resolved by the Council that 'the President, Mr. Roberts, Dr. Arbuthnot, Dr. Halley, Dr. Mead, Mr. Hill, Sr. Christopher Wren, Mr. Wren and Dr. Sloane be Ordered a Committee to go to Greenwich, any three of them (of which the President, or the Vice-President to be one) to be of the Quorum,

and to report their Opinion of ye Condition of the Observatory and the Instruments therein; and to take an Inventory of the Instruments' (*Council Minute Book*, fo. 182). This group was later made a Standing Committee of Visitors.

(4) On the same date St John wrote to the Board of Ordnance informing them of their unusual responsibilities; in his letter St John laid it down that if any useful instruments that were not royal property were found at the Observatory they were to be purchased for the Crown, and that the Board 'should have regard to any complaints the said Visitors may make to you of the behaviour of Her Majesty's Astronomer and Keeper of the said Observatory in the execution of his office'. (See Weld, *History*, I, p. 402.)

815 NEWTON TO THE TREASURY

13 DECEMBER 1710

From the copy in the Mint Papers.[1]
Reply to Letter 813

To the Rt. Honble: The Lords Commrs of her Majties Treasury

May it please your Lordsps:

In obedience to your Lordsps: Order of Reference of the 29 of November last upon the Annexed memorial to Mr: Palmes for a New Coynage of 700 Tuns of half pence & farthings of Copper[2] in 14 years after the manner of the patent Granted in the Reigne of their Late Majties King William & Queen Mary to Sr Joseph Hearne and others[3] Wee doe humbly acquaint your Lordsps; that wee have Enquired into all ye Coynages of that sort since ye yeare 1672 & found that in the Reignes of King Charles ye: 2d: King James ye: 2d: & in the beginning of the Reigne of their late Majties King William & Queen Mary The Coinage of half pence & Farthings was performed by commrs: who had money Impressed from ye: Exchеqr: to bey [*sic*] Swedish Copper and Tinn & coined at most at 20*d* per pound averdupoize & accounted upon Oath to ye Governmt: for ye Charge & produce thereof by Tale: And that afterwards upon calling in ye Tinn Farthings & half pence by reason of ye complaints made against them The patent above mencond was grainted [*sic*] to Sr: Josh: Hearne & others who contracted to change ye same & to Enable them to bear yt: charge They were allowed to coin 700 Tuns at 21*d* per pound wt: of English Copper which is Cheaper then ye Swedish without being accountable to the Governmt: for the Tale The reason of wch: allowance now ceasing. Wee are humbly of opinion yt the sd patent be not Drawn into prsident Especially since the money made thereby was light, of bad copper, & ill coyned And as to yt species of moneys being wasted there may be some want thereof in

London & parts adjacent but yt those more remoter are unfurnished does not appear to us—

All which is humbly submitted to your Lsps Great Wisdom[4]
13 *Decr* 1710

NOTES

(1) Mint Papers II, fo. 387. This document is in a clerical hand, and lacks signature. In phraseology it is a repetition, in part, of Newton's letter of similar purport, dated 13 July 1703 (vol. IV, pp. 408–9).

(2) Newton was always strongly opposed to the idea that the country was short of copper coins, and to the idea that they might be issued as of less than intrinsic metallic value. However, in 1713 he was forced to yield and the minting of copper at the Tower Mint commenced, though no issue occurred.

(3) See vol. IV, p. 409, note (2).

(4) This Mint report on Palmes' proposal was read by Oxford on 19 February 1711. Palmes and the Mint Officers were ordered to attend next Saturday week (3 March). The result of the meeting is not recorded, but on 19 September 1711 the Principal Officers were again written to—the letter does not survive—to attend on Tuesday next (25 September) to consider the state of the copper currency (*Cal. Treas. Books*, XXV (Part II), 1711, pp. 96, 624).

816 NEWTON TO THE TREASURY

[*c.* 31 DECEMBER 1710]

From the draft in the Mint Papers[1]

To the Rt Honble the Lords Comm[ission]ers of her Majesties Treasury.

May it please your Lordps

Finding reason to suspect that the present indented trial piece is too fine;[2] I have nicely examined it & find that it is finer then the last trial piece by about a quarter of a grain & that the last trial piece is also something too fine by the assay. Which excesses of fineness being of great consequence, I have further endeavoured to find out the reason thereof that the like accidents in making new triall pieces hereafter may be avoided. And by the assay I am satisfied that there are various degrees of fine gold, some being 24 carats fine by the assay, some a quarter of a grain coarser or finer or above, & that gold may be refined so high as to be almost half a grain finer then 24 carats.[3] And accordingly as the fine gold of wch the standard pieces are made is finer or coarser the standard pieces will be finer or coarser in proportion. By wch means the standard of gold is rendred uncertain And the like for silver

I humbly offer therefore to your Lordps consideration, whether for ascertaining the value of gold & silver there should not be one common standard of gold & one of silver for the money plate & Merchantable Ingots in all great Britain, setled by the assay which is the rule of the market; & whether the standard once setled should not be preserved in the Exchequer for a rule to Juries in makeing trial pieces for the future without varying or the present trial piece remain.(4)

NOTES

(1) Mint Papers I, fo. 275. In Anne's reign the Treasury was in commission only from 11 August 1710 to 28 March 1711, therefore this letter was written some time after the trial of the pyx in August, but before the end of the following March. We place it arbitrarily here, and add a memorandum explaining Newton's views in detail. There is also a later letter on the same topic, written before the next trial of the pyx in 1713.

Newton's concern for refining and assay at this time probably retarded his work on the *Principia*.

(2) At the trial of the pyx on 21 August 1710 the standard gold test plate, with which the sample coins in the pyx were to be compared, was that made as a consequence of the Union with Scotland in 1707, but never hitherto used. This plate contained a shade too much gold—917·1 parts per 1000 instead of 916·6 parts, which was standard—but Newton claimed that by the normal assay it was much finer still, at 920·9 parts per 1000. The money made since 1707 had been made to match the last previous trial plate, that of 1688, which was by Newton's reckoning rather above standard (918·3 parts) and really a little below. The Jury of Goldsmiths at the 1710 trial accordingly found, when comparing the money made with the 1707 test plate, that the money was a quarter of a grain (2·6 parts) below standard—at 914 parts—while Newton (not disputing the deficiency of gold content in the comparison) claimed that the money was above standard at $(920{\cdot}9-2{\cdot}6=)918{\cdot}3$ parts. By modern assay the goldsmiths were more accurate than Newton and the Mint; the money was certainly under 915 parts. Newton was very angry at the slur on the Mint's competence, even though the value of one-quarter of a grain of gold was less than a penny. The assays on which he relied were, in fact, too high by about 3·8 parts in 1000, presumably as a result of impurity in the gold he was using as a standard.

(3) See Newton's memorandum following.

(4) If a given sample of gold were arbitrarily defined as containing n parts per thousand, then other samples could be prepared, of different fineness, by reference to this. But ordinary practice was to assume that *pure* gold ($n=1000$) could be prepared, and start from there; this would cause error should the presumptive 'pure gold' later be shown to contain other metals.

816a MEMORANDUM BY NEWTON

From the original in the Mint Papers(1)

Of the assaying of Gold and Silver, the making of indented Triall-pieces, and trying the moneys in the Pix.

1 of the Assay

Assaying and refining are operations of the same kind. The assayer refines a small peice of any mass of Gold or Silver, and by the decrease of its weight makes his report, and if there be no decrease, that is, if the mass be of the same fineness with the refined assay-peice, he reports (or ought to report) the Gold 24 carats fine and the silver 12 ounces fine, and this is fine Gold and fine Silver in the Sense of the Law. And all Gold and Silver of the same finenesse with the Assay peice is fine Gold and fine Silver in the sense of the Law. And because the Assayer works more exactly to a rule than the Refiner and makes better dispatch, the Assay is made the standing universall rule of Valuing Gold and Silver in all Nations in point of finenesse, and the Law in ordeining that standard Gold shall be 22 carats fine and standard silver 11 ounces two penny weight fine, means by the Assay.(2)

The Assay of Gold ought to be made with two waters and noe more, this being the constant practice of assaying, and the waters ought to be of the usual strength (the Second Water stronger than the first) and to work the usual time and in the usual heat, and the assay peice ought to be hammer'd to the usual thinnesse that the Assays may be uniforme, and the Assays of Silver ought to be made with a due proportion of lead in a due and even heat, and as soon as the lead is blown off and the Silver looks bright and glittering, the Silver must begin to cool without roasting it, and it must cool slowly that it doe not spring, But in refining Gold & Silver in great quantities these niceties are not observed.

Assays are lyable to errors, but the errors are generally very small and seldome exceed a quarter of a grain in Gold and an half penny weight in Silver, and by reason of these little errors the Assayer in single Assays makes his report to noe lesse then a quarter of a grain in Gold and an half penny weight in Silver, But if two or more assays be made of the same peice of Gold or of the same peice of Silver, and the assays agree without any considerable difference and a medium be taken between them, the finenesse of the Gold may be determined to lesse then half a quarter of a graine, and the finenesse of the Silver to lesse than an half penny weight, and this is the exactest way of assaying hitherto in use(3)

2 of making the Indented Trial-peices

The Standard Trial-peices are made by the assay. First a Jury of workmen summoned and sworn by Order of Council procures Gold and silver refin'd by the Refiner, and assays them to see if they be of a just degree of fineness, that is, the Gold just 24 carats fine and the Silver just 12 ounces fine, Then they melt this Gold and Silver severally with allay in due proportion and stir them well together in fusion severall times to mix them very well with the Allay, and pour them off before the allay evaporates, and then assay them severall times to see if they be standard, taking assays from severall places to see if the mixture be uniforme, It must agree therefore with the assay as exactly as is possible least there be two standards, one by the Assay-weights, the other by the Trial-peices.

Refiners find it difficult to refine Gold to the degree of 24 carats. they seldome make it above 23 car. 3 gr. 3 qters fine, and by fine Gold generally understand gold of this degree of finenesse, and if Gold at any time prove finer upon the assay, Assayers out of prejudice do not report it finer, and thence it comes to passe that Goldsmiths are generally of opinion that Gold cannot be above 24 carats fine, not knowing that there are ways of making it finer then by the assay. [But[4] if when they have watered their granulated gold once or twice with Aqua fortis, they should dulcify it & grind it very fine as painters do their colours, & then water it once or twice more with double Aqua fortis in the same degree of heat as before & keep it longer in the water then before stirring it now & then with a wooden stick to make the gold mix wth fresh water: the gold would become finer then by the Assay, & by consequence finer then four & twenty carats. Chymists also tell us that Gold may be made finer by Antimony then by Aqua fortis & by consequence then by the Assay; & gold refined by Antimony is of a better colour then Gold refined by Aqua fortis, & by reason of its fineness will go much further in gilding, as I have heard. But the Refiners of this city know not how to Refine gold by Antimony. And so silver also by being tested at more lead & roasted becomes finer then by the common way of assaying, but not a half penny weight finer [*sic*].[5]

If Refiners should work perfectly in the same manner with Assayers, that is, if they should mix gold with silver in the same proportion & drive it off the test with the same proportion of lead & hammer it to the same thinness & water it with waters of the same strength in the same degree of heat during the same length of time, their gold would become just 24 carats fine. And by imitating the Assayer their silver would become twelve ounces fine. But they work not with so much curiosity & exactness. Their fine gold & fine silver must be assayed to know the just degree of fineness.] Thence also it may have sometime happened that at the making of tryal-peices the Assayer may have reported the fine Gold not soe fine as really it was, and by that means the tryal-peice may have been made too fine. And if the fine gold was but 23 car. 3 gr. 3 qters fine, The Trial-peice may have been made too coarse, and there are other ways of erring, as by assaying with waters too strong or too weak or after any other unusuall manner, or by scattering any part of the allay or of the fine Gold or Suffering a sencible part of the Allay to evaporate or not mixing the Gold with the allay very well, or using a faulty crucible, or roasting the fine silver or suffering it to spring in the assay, And for avoiding these errors the Jury ought to consist of workmen very well skilled and exercised in assaying, refining and allaying of Gold and Silver.

3 of trying the Pix

The tryall of the moneys in the Pix is to be performed by a Jury of Assayers in the presence of the warden Master & comptroller of the Mint after the most just manner that can be made by fire, by water by touch, or by weight, or by all or by any of them, as is described in the Indenture of the Mint. The Pix is opened, and the Jury sworn before the Queen or such of her Councill as her Majesty shall apoint. If the tryal peices be exactly made the tryal thereby is the most expedite and the least lyable to errors or fallacy. But a Tryal peice may happen to be erroneous, and then the other ways of assaying, as they are lawfull, soe alsoe they may be usefull, For the assay by the assay weights exactly performed will discover the error of the Tryal peice if there be any and how great that error is, and the assay by the touch may be also used to see how it agrees with the other assays, tho it be less exact & not to be depended upon alone.(6)

If at any time the Tryal-peice doth not agree with the Assay, either the error must be reported by the Jury or it must not be reported.(7) If it be reported, either the Master of the Mint must be authorized to allow for the error in coining the money by that Trial-peice for the future or a New Tryal-peice must be made. If it must not be reported, the Mint-Master must goe on to coine the money by an erroneous Tryal-peice, and the Goldsmiths will have it in their power to alter the Standard without Controul as often as they are to make a New Trial-peice, and to make a New Standard instead of making a New Trial peice agreeable to the Standard established by Law.(8)

[When(9) I came first to the Mint & for many years before, Importers were allowed almost all the Remedy & the money was coyned unequally some peices being two or three grains too heavy & others as much too light, & the heavy Guineas were called Come-again-Guineas because they were culled out & brought back to ye Mint to be recoined (as was then the common opinion) &

thereby the public Moneys called the Coynage Duty were squandred away to the profit of the Master & Moneyers & Goldsmiths & the new moneys which remained after the heavy pieces were culled out, & was put away by the Goldsmiths among the people, was without the Remedy. The money is now coined equally so that the culling trade is at an end. And the Importers are not allowed the advantage of the Remedy, but the money is coined to the just value.(10)

When the Importers were allowed the advantage of the Remedy there wanted about 30 grains of fine gold in 44½ Guineas & about 34 grains of fine silver in sixty two shillings. There is now the just quantity of Gold & Silver in the moneys, & there wants only about 15 grains of fine Copper in 44½ Guineas or the third part of a grain of copper in a Guinea wch want is of no value or consequence & is occasioned by the too great fineness of ye trial pieces.

The end of the Indenture of the Mint & of all the Rules of Coynage is that the moneys for ye sake of commerce have its just intrinsic value And if the Jury pleases to examin the Moneys in the Pix by the weight & assay together they will find that they are as justly coined to the value as ever they were.]

At the last Trial of the Pix(11) the Gold money was Standard full by the Assay and the Trial-peice a quarter of a grain better than the money & the Jury in their Verdict represented the money a quarter of a grain worse than Standard by the Trial peice. This Trial peice was made upon the Union A: C: 1707 It was made (I think) Wthout an Order of Council(12) and by many assays very carefully made is five twelfths of a grain better than Standard, That of 1688 made by order of K. James II is a sixth part of A grain better than Standard, and that of 1660 made by order of K. Charles II is Standard.(13)

[The (14) Standard pieces should be made by the Queens Order, as ye Indenture of the Mint expresses, but were made the last time (A.C. 1707) by the Lord Treasurers warrant. The Jury did not then refine the Gold & Silver but bought these of my Deputy Mr Carlick(15) They know not how he refined them and wondred that he could make them so fine. But being generally of opinion that no gold or silver could be too fine, they made the trial pieces but absolutely refused to give receipts of them for making their plate thereby.(16) And these trial pieces were never established under the broad seal, as they ought to be. And if they be established the Merchants will thereby lose about 4 or 5*s* in the pound out of all gold imported.]

Quære 1 If upon trying the Pix, the Trial-peice at any time doth not agree with the Assay, are not the Jury to report the Error.

Quære 2 If any doubt arise about the manner of the report or Verdict are not the Jury to make a Special report of the matter of fact & leave it to the Queen and Council to make a Judgment thereupon.

Quære 3 If any doubt arise about the truth of the tale, weight, or assay, are not the Jury (Especially at the motion of the Officers of the Mint) to repeat the operation.[17]

NOTES

(1) Mint Papers I, fo. 109; this is a clerical copy. Newton attached great importance to this paper, of which there are eight drafts (*ibid.*, fos. 243–51, 288, 291–2, 295–8) and several related fragments. Some of the drafts have a different title; the numbered paragraphs are different; and there is variation in the content, but several are virtually identical with this 'final' version. We have added substantial passages from two drafts, as noted below. There is no evidence to show that this paper was ever submitted officially, but the trial-plate of 1707 was never used again (Craig, *Newton*, pp. 77–9). For much of the following explanation we are indebted to Professor Cyril Stanley Smith.

(2) The use of proof-plates to provide legal standards is very old. Fragments of the official sterling silver plate made in 1279 and the gold plate of 1477 are still extant (J. H. Watson, *Ancient Trial Plates*, London, 1962). The mint normally used a secondary trial plate for its running assays, the master plates appearing only for use in the formal Trial of the Pyx.

In discussing the making of trial plates and the assay of coinage Newton was concerned with the fact that the legal and absolute compositions are not necessarily the same. The legal assay is specified on the basis of comparison with a standard; if the standard is actually not true to its nominal composition the reported figures will not represent the correct fraction of gold by weight. Comparative assays are necessary because some gold is lost by volatilization during cupelling and by solution in the parting acid, while some silver remains alloyed with the finally parted gold. All of these effects are small, but they vary with the actual regimen of fire and assay used by the assayer and correction has to be made for them.

A brief description of the assaying process will clarify the problem. Silver is assayed by heating a weighed sample with at least four times its weight of lead on a little thick-bottomed cup (cupel) made of porous bone-ash. All the base metal impurities are oxidized and carried by the molten lead oxide to be absorbed in the cupel, leaving a bead of pure silver which is weighed. Gold alloys are similarly cupelled to remove the base impurities, but any silver that is present is removed by a subsequent parting operation. The gold bead is hammered or rolled into a thin strip and then subjected to the action of nitric acid (the first 'water'). (Since alloys rich in gold are only superficially attacked by the acid it is necessary to add to them, before cupelling, an amount of silver equal to about three times the weight of the gold. This step, called inquartation, is not mentioned by Newton.) The gold remaining after acid treatment still retains about 2·5 parts of silver per thousand. Some but not all of this silver is removed by a second treatment with stronger acid (the second water). After this the gold, which is intact but spongy, is annealed to consolidate and colour it, and then weighed on a sensitive balance.

Now it is obvious that if an assay is corrected on the basis of the yield from a check piece processed along with it, the corrected assay report will be proportionately high if the standard actually contains less than it is supposed to. The early assay books all insist on the care to be taken in preparing the standard gold. It is clearly stated that gold purified by parting is not pure enough, and treatment with sulphur or antimony sulphide is preferable. However, Newton later remarks that the London refiners do not know how to do this (see note (4)).

Newton had probably learned the practice of the Mint mainly by observing the assayers in operation but he could have read the excellent description of Lazarus Ercker (*Beschreibung*

Allerfürnemisten Mineralischen Ertzt Unnd Berckwercksarten, Prague, 1574) in the poor English translation of Sir John Pettus, 1683. (See also the modern annotated translation by A. G. Sisco and C. S. Smith, Chicago, 1951.)

(3) Ercker gives minimum balance weights of one quarter grain, and notes that half a grain was the minimum reported in Mint assay. Using today's sensitive balances reports are made to one part in 10000.

(4) Here between the brackets we have interpolated a paragraph written on Mint Papers I, fo. 291. Newton, arguing that the trial plate made with properly refined gold will give assay results greater than 24 carats (see note (2))here suggests the purification of parted gold by comminution and further acid attack. The resulting gold would still contain more silver than that produced by refining with antimony sulphide and this was probably the reason for the erroneously high assay results that Newton reports later (see note (16)).

(5) On 20 March 1711/12 Newton assured the Royal Society that 'there was a difference in the Weight of fine Gold, arising as he believed from the different Methods of refining it'. (Journal Book of the Royal Society, XI, 279.)

(6) Though legally permitted at the Trial of the Pyx, assay by touchstone is essentially qualitative. The sample to be tested is rubbed against a fine-grained stone (usually a very dark green or black quartzite) to leave a streak beside which other streaks are made from alloys of known composition for comparison. (The roughness of the stone causes multiple reflections which intensify the colour of the alloy. Further information comes from observing the action of nitric acid upon the streak.) An experienced assayer can detect differences of as little as 5 parts per 1000 in gold alloys containing only copper or silver, and the method is still used by jewellers for a rough assay. By far the best assay is by cupellation and parting, with the use of check pieces (note (2)) to correct for inherent error.

(7) Now Newton comes to the heart of the matter. At the pyx trial of 1710 the coins made in the last three years were not found defective by assay, but were slightly below the standard of the 1707 trial-plate. Newton's view was that the difference between the assay and the trial-plate should have been reported because the fault was (in his view) not in the badness of the coin but in the super-excellence of the trial-plate.

(8) As Craig remarked, Newton was making a great to-do about a pennyworth of gold, on the very limit of accuracy in such measurements at that time.

(9) The next three paragraphs, enclosed in brackets, are from Mint Papers I, fo. 250.

(10) The remedy was the permissible variation of any coin tested at random from the prescribed weight and fineness. It was necessary to cover inevitable small irregularities in practice. As practice improved a Mint Master would aim to work close to the lower limit of weight and fineness but there might still remain enough variation between coins to make possible the 'culling' of the better coins for melting down and resale as bullion to the Mint, as described by Newton.

(11) On 21 August 1710.

(12) We have not come across a Warrant for this trial plate, nevertheless, the making of new trial plates common to England and Scotland was strongly recommended by the Mint Officers (including Newton, of course) on 24 March and 2 June 1707 (*Cal. Treas. Books*, XXI (Part II), 1707, pp. 263, 264; compare vol. IV, pp. 490, 491).

(13) The determinations on which Newton relied were erroneous. 'Standard' being 916·6 parts of pure gold per 1000, the 1660 plate contained only 912·9 parts; the 1707 plate had 917·1.

(14) This paragraph is from Mint Papers I, fo. 293.

(15) Or Cartlich.

(16) The lines are much crossed out and interlineated so that it is difficult to make out the word-order Newton intended.

(17) According to Craig (*Newton*, p. 77) at the pyx trial of 1710 the Mint officers 'seem to have protested with so much vigour that they were ejected from the assaying chamber by the Goldsmiths' constituting the Jury. Hence the bitterness of Newton's attitude to the Jury here.

817 WILLIAM KERRIDGE TO NEWTON AND SLOANE

16 JANUARY 1710/11

From the original in the British Museum[1]

May it Please yr Honours

I have had 40 years Experience in all sorts of Water Workes & was Chiefe Engineer to Sir Samll. Mooreland[2] & ordered ye fine Glasse Fountaine at Fox hall.[3]

I have a Modell in Wood that I had of the Prince Reuport & on it all the 4 Motions. bÿ Water bee it over shott or under shaft water wheeles. Horse wind or mans labour

I can order a small Engin to water a garden or to beat ye worme of the wall Trees: so that one man may Water a large peece of Ground in a small tyme:

I can mount Water verie high & a great quantity by the Compression of Are & other secrets

I can so Compose & Order a Water Wheele to Raise ye Water from a Sestern to give a motion to A clocke in a Dyning Roome: to strike or chime

Allso to Order vocall singing of Birds ye Nightingall Black Birds sky larke or others Aparrt or alltogether &c.

As For Fountaines or Grotts above or under ground to be S [*sic*] with lighted candles only so haveing only mentioned in some part what I can doe

I most humbly Begg this Honourable Society to make Tryall of me

Or to Recommend me to some Nobleman as his Servant: to Keepe in Repayre all his Plomers Worke: now finisht; or to Add new Worke; desiering no profit but only a Livelwhod.

As I shall Indevor to meritt
& no otherwise

Jan ye 16*th* 1710

WM. KERRIDGE

I live att Mr Knights brokers in fore street neare the City green yards going to more feilds

To Sr Isaac Newtom
To Dr Slone
These humbly present

NOTES

(1) Sloane 4042, fo. 227. At a meeting of the Royal Society on 17 January 1710/11 this letter was referred to the attention of the Council, who seem to have taken no action upon it. Sloane did acquire or more likely create at the Manor House at Chelsea, which he bought in 1712, a 'Waterwork' valued at £4000 at the time of his death (G. R. de Beer, *Sir Hans Sloane and the British Museum*, London, 1953, pp. 60, 145, 149).

(2) Sir Samuel Morland (1625–95), Master of Mechanicks to Charles II, was well known for his pumps and fire-engines; the chief maker of his devices was Isaac Thompson, but a number of workmen must have been involved.

(3) For Morland's association with Vauxhall House see H. W. Dickinson, *Sir Samuel Morland*, Heffer, Cambridge, 1970, pp. 53–5. The fountain 'played in the room and all the glasses stood under little streams of water'.

818 FLAMSTEED TO SHARP

23 JANUARY 1710/11

Extracts from the original in the Library of the Royal Society[1]

The Observatory Jan 23 ♂ 1711/0

Sr

...Dec 15 last I receaved a letter from ye Sec. of State[2] yt signified to me yt it was her Maties pleasure for the Improvemt of Astronomy to appoint ye Presidt Vicepresidt & such others as ye council of the R.S. should think fit to be the Constant Visitors of ye Observatory. that they should see her Maties Instrumts repaired & purchase those yt were mine. & that I should every yeare give ym an Account of what Observations I made, & make such as they appointed. a like letter was sent to ye Office of ye Ordnance & one to ye R.S. it happens very Unluckly for ye procurer of these letters, you know who he is, that all the Instruments in ye Observatory are either absolutely given to me especially (and not to ye Observatory) by Sr Jonas Moore or else built at my owne expense & Charge[3] And I have neither any Need nor desire to sell them so yt part of ye letter falls. as for ye other. Dr H: Slone is ye sole Vicepresident [.] ye council I am apt to thinke consists of persons not less ingenious yn he so yt I am in little pain about this Visitation. but to obviate ye inconveniencys yt might emerge from that letter God has raised me some freinds, yt I hope will give the Qn. a true state both of ye R: Observatory & R.S. & doubt not but in a little time ye Author & its Pr. & Vice P. will be ashamed of their attempt.[4]

... ...

Your affectionate freind & servant
JOHN FLAMSTEED M.R.

NOTES

(1) R. S. Sharp Letters, no. 75; printed in Baily, *Flamsteed*, p. 279 and in Cudworth, *Life*, p. 99.

(2) A copy of the Royal Warrant (Number 814), together with a letter from St John to himself (also printed in Baily, *Flamsteed*, pp. 91–2).

(3) See vol. II, p. 427, note (15) for Sir Jonas Moore (1617–79), one of Flamsteed's patrons. He presented the astronomer with a seven-foot iron sextant, two great clocks by Tompion (on which see Derek Howse, 'The Tompion Clocks at Greenwich and the Dead-beat Escapement', *Antiquarian Horology*, **7** (1970–1971), and a telescope object glass for his use at the new Greenwich Observatory.

(4) Flamsteed (despite the tolerant contempt for the antics of little men displayed here) had reacted strongly to St John's letters; on the day after receiving them he called on St John in person, and told him that his work would be impeded by the Visitors, that he wanted no new instruments and the Visitors in any case lacked the ability to contrive any, that the instruments and clocks at the Observatory were all his own, and that the Royal Society was trying to steal the fruits of his labours, having none of their own worthy of the public. But St John, according to Flamsteed, 'seemed not to regard what I said, but answered me haughtily, "The Queen would be obeyed"' (Baily, *Flamsteed*, p. 92). He called to the same effect upon the Earl of Rochester (Lawrence Hyde, the Queen's uncle), where he thought his words made more impression. Further, on 29 December 1710 he drafted a petition to Queen Anne (*ibid.*, pp. 278–9) in which he set out his own blamelessness, begged that neither 'the President of the Royal Society, nor any of their Council' should be set over him as Visitors, and asked for the release of the remainder of the money provided by Prince George for printing the *Historia Cœlestis*.

819 NEWTON TO THE TREASURY

23 JANUARY 1710/11

From the original in the Mint Papers[1]

To the Rt Honble the Lords Commrs of her Majts Treasury

May it please your Lo[rdshi]ps

I understand by Mr Ward[2] that the East India Company are willing to give 7000£ for one hundred Tunns of Tin[3] to be paid out of the first moneys they shall receive of your Lordsps for salt peter & in the same moneys & desire the Tin may be speedily delivered if your Lordps approve thereof.[4]

All which is most humbly submitted to your Lordps great wisdome.

Mint Office
Jan. 23. 1710/1

Is. Newton

NOTES

(1) Mint Papers III, fo. 557. On 15 December 1710 the Warden, Craven Peyton, had been to the Treasury to inform the Lords that the East India Company wished to take a 'parcel of tin to carry to the East Indies' at the rate of £3 10*s* 0*d* per hundredweight (six shillings less than the normal price) paying for it in Exchequer Bills. The Lords agreed to 50 tons being sold on these terms (*Cal. Treas. Books*, XXIV (Part II), 1710, p. 111). Further, on 8 January 1711 it was agreed that the Mint might treat with the Company for the export of as much tin as it wished (*ibid.*, XXV (Part II), 1711, p. 3).

(2) John Ward was a director of the East India Company.

(3) That is, seventy shillings per hundredweight as before. Whether this proposal submitted by Newton is a revision of that put up by Peyton, or an additional one, is not clear.

(4) The importation of saltpetre by the Company was strategically its most important contribution to the Nation's trade. It is worth noting again that the sale of tin abroad—directly or indirectly—provided the country with an important source of foreign exchange. A direct sale of 1000 tons of tin by the government might put £76000 of foreign money into its hands for the payment of troops—no trivial contribution to the cost of a campaign. The Treasury was anxious to reduce the dead stock of the metal held at the Mint both for this reason and because the capital invested in it was unproductive (though it could be used as a guarantee for borrowings). In August 1710, for example, the Lords of the Treasury had entered into earnest negotiations with London merchants over the sale of tin and had on two successive days called for the attendance of the Principal Officers of the Mint to assist the deliberations (*Cal. Treas. Books*, XXIV (Part II), 1710, p. 43).

820 THE MINT TO THE TREASURY

2 FEBRUARY 1710/11

From the original, in Newton's hand, in the Public Record Office[(1)]

To the Rt Honble the Lords Commissioners
of her Majties Trea[su]ry.

May it please your Lordps

In obedience to your Lordps order of Reference signified to us by Mr Lowndes his Letter of Jan 29[(2)] upon the annexed proposal of Mr Robt Ball merchant for paying three pounds sixteen shillings per cwt for 200 Tunns of Tin upon his factors receipt thereof at Leghorn we have considered this proposal & humbly represent that Sr Theodore Janssen[(3)] paid 4£ 10*s* per cwt to the late Ld Treasurer[(4)] for Tin runn into barrs barrelled shipt of & delivered at Leghorn & Genoa at her Majties charge

	£	s	d
The price at the Tower is	3.	16.	00.
Charges of melting into barrs barrelling & porters	0.	2.	06.
Shipping off if carried to the Nore	0.	0.	05.
Freight to Leghorn	0.	4.	06.
Insurance 10 per cent	0.	8.	00.

Something is also to be recconed for imbezzlements of Tin by the seamen if carried in her Majesties ships, wch one voyage with another came to about 2*s* 6*d* per cwt. And something also for forbearance of payment till the Tin arrived at Leghorn.

All wch being considered, wee are humbly of opinion that by the Proposal above mentioned her Majesty would be a great loser, but submit it to your Lordps great wisdome

C: PEYTON
IS. NEWTON
JN ELLIS

Mint Office
2 *Feb.* 171$\frac{0}{1}$

NOTES

(1) T/1, 131, no. 37. There is a draft in the Mint Papers, III, fo. 556.
(2) This has not been found.
(3) Compare Letter 845.
(4) Godolphin.

821 COTES TO WILLIAM JONES

15 FEBRUARY 1710/11

From the original in a private collection[1]

Febr. 15. 1711

Sr

I yesterday received Yr most valuable & acceptable Gift,[2] togather with Your very kind Letter. I return You my hearty thanks for 'em both. Not having heard anything of Your Book till I saw it, I received it with the additionall pleasure of a Surprize. You have highly obliged the Mathematicall part of the World by collecting into one Volume those curious & usefull Treatises which were before too much dispersed, but more especially by the Publication of the Analysis per Æquationes Infinitas & the Methodus Differentialis. I could heartily wish yt nothing of Sr Isaac's might be lost. I hope You will endeavour (as You find an oportunity) to persuade him to publish other Papers;

for I believe he has yet many excellent things in Reserve. About a Year & an half ago (when I was last in Town)[3] I acquainted Mr Ralphson[4] that You had some papers of Sir Isaac's in Your hands which were long ago communicated to Mr Collins.[5] I thought they might have been pertinent to his design of writing an History of the Method of Fluxions.[6] I afterwards understood yt you gave him a sight of those Papers, & yt he thought 'em not to be for his purpose, which I do now very much wonder at if his Intention was to do justice to Sr Isaac. If that was not his Intention I think Your Preface has already sufficiently defeated all his attempts.[7] We are now at a Stand as to Sr Isaac's Principia; he designs to make some few Experiments before we proceed any further. The first Book & ye six first Sections of ye Second are printed off...

NOTES

(1) This was printed in Rigaud, *Correspondence*, I, pp. 257–8. William Jones (1675–1749), of Anglesey, was an autodidact mathematician who taught and published in London until he entered the service of Thomas Parker (later Lord Macclesfield and Lord Chancellor); the latter part of his life was spent at Shirburn Castle, Oxfordshire. He formed a fine mathematical library including (from 1708) the books and papers of the mathematical practitioner John Collins (1625–83) who had developed so wide a correspondence. Jones was elected F.R.S. in 1711.

(2) A copy of Jones' *Analysis per Quantitatum Series, Fluxiones, ac Differentias* London, 1711. Jones had become acquainted with Newton in 1706. Within a few years Newton thought so well of him that he allowed Jones to print *De analysi per æquationes numero terminorum infinitas*, *De quadrature curvarum*, *Enumeratio linearum tertii ordinis*, and *Methodus differentialis*, together with extracts from correspondence. Of the opuscula, the second and third had already been published by Newton himself in the 1704 *Opticks*. Jones had found Collins' transcripts of some unpublished papers, but Newton lent him his own manuscripts for greater accuracy. Together with Jones' preface, this publication was a powerful confirmation of Newton's priority in the new mathematics.

(3) This was when Cotes first visited Newton; compare Letter 765.

(4) Joseph Raphson or Ralphson, of Middlesex, was admitted a Fellow Commoner of Jesus College, Cambridge in 1692 and in the same year proceeded M.A. *Literis Regiis*. Before the book discussed here he had published *Analysis æquationum universalis*, London, 1690. He died *ca* 1715.

(5) Besides the letters written to him by Newton, Collins had left transcripts of mathematical papers that Newton had sent him at various times—taken with or without Newton's knowledge. He had also made copies of letters from Newton to Leibniz and others. Cotes does not seem to question Newton's approval of Raphson's use of these documents.

(6) *The History of Fluxions. By the late Mr Joseph Raphson*, was published at London in 1715.

(7) Without mentioning Leibniz or the calculus dispute, Jones in eleven short pages presented powerful evidence of Newton's mathematical originality as far back as 1665. For the first time the testimony of Newton's earliest communications to Barrow and Collins (with allusions to vol. I, Letters 5, 6, 7, 8 and others which were quoted) was set before the public, thus anticipating the fuller documentation attempted in the *Commercium Epistolicum*.

822 LEIBNIZ TO SLOANE

21 FEBRUARY 1710/11

From the copy made by Sloane for Newton in the University Library, Cambridge.(1)
For the answer see Letter 843

Epistola Dni. *Leibnitii* ad D. *Hans Sloane*
Regiæ Societatis Secretarium 4to Martii S.N. 1711 data

Gratias ago quod novissimum Volumen præclari Operis *Transactionum Philosophicarum* ad me misisti; quamvis nunc demum mihi *Berolinum* excurrenti redditum sit.(2) Namque excusabis quod pro munere superioris anni nunc demum gratiæ dudum debitæ redduntur.

Vellem inspectio Operis me non cogeret nunc secunda vice ad Vos querelam deferre. Olim *Nicolaus Fatius Duillierius* me pupugerat in publico scripto, tanquam alienum inventum mihi attribuissem.(3) Ego eum in Actis Eruditorum *Lipsiensibus* meliora docui;(4) et Vos Ipsi, ut ex literis a Secretario Societatis vestræ inclytæ (id est quantum memini a Teipso) scriptis didici, hoc improbastis.(5) Improbavit *Newtonus* ipse Vir excellentissimus (quantum intellexi) præposterum quorundam hac in re erga vestram gentem et se studium. Et tamen D. Keilleus in hoc ipso volumine,(6) Mens. *Sept* & *Octob* 1708 pag 185 renovare ineptissimam accusationem visus est, cum scripsit. *Fluxionem Arithmeticam a Newtono inventam, mutato nomine et notationis modo, a me editam fuisse*. Quæ qui legit et credit non potest non suspicari alterius inventum a me larvatum subdititiis nominibus characteribusque fuisse protrusum. Id quidem quam falsum sit nemo melius ipso *Newtono* novit: certe Ego nec nomen *Calculi Fluxionum* fando audivi, nec Characteres quos adhibuit D. *Newtonus* his oculis vidi, antequam in *Wallisianis Operibus* prodiere.(7) Rem etiam me habuisse multis ante annis quam edidi, ipsæ literæ(8)apud *Wallisium* editæ demonstrant: quomodo ergo aliena mutata edidi quæ ignorabam?

Etsi autem D. *Keillium* (a quo magis præcipiti judicio quam malo animo precatum puto) pro calumniatore non habeam, non possum tamen non ipsam accusationem in me injuriam pro calumnia habere. Et quia verendum est ne sæpe vel ab improbis vel ab imprudentibus repetatur; cogor remedium ab inclyta Vestra *Societate Regia* petere. Nempe æquum esse Vos ipsi credo judicabitis, ut D. *Keillius* testetur publice, non fuisse sibi animum imputandi mihi quod verba insinuare videntur, quasi ab alio hoc quicquid est inventi didicerim, et mihi attribuerim. Ita ille et mihi læso satisfaciet et calumniandi animum a se alienum esse ostendet; et aliis, alias similia aliquando jactaturis, frænum injicietur. Quod superest Vale et fave.

Dabam Berolini 4to Martii 1711 [N.S.]

Translation

A letter from Mr. Leibniz to Mr. Hans Sloane, Secretary of the Royal Society, 4 March 1711 N.S.

I return you thanks for sending me the latest volume of your distinguished production, the *Philosophical Transactions*; although it was now at last delivered to me while on a visit to Berlin.(2) And so you will excuse me for only now returning you due thanks at long last for a gift of last year.

I could wish that an examination of the work did not compel me to make a complaint against your countrymen for the second time. Some time ago Nicholas Fatio de Duillier attacked me in a published paper for having attributed to myself another's discovery.(3) I taught him to know better in the *Acta Eruditorum* of Leipzig,(4) and you [English] yourselves disapproved of this [charge] as I learned from a letter written by the Secretary of your distinguished Society (that is, to the best of my recollection, by yourself).(5) Newton himself, a truly excellent person, disapproved of this misplaced zeal of certain persons on behalf of your nation and himself, as I understand. And yet Mr. Keill in this very volume,(6) in the [*Transactions* for] September and October 1708, page 185, has seen fit to renew this most impertinent accusation when he writes that I have published the arithmetic of fluxions invented by Newton, after altering the name and the style of notation. Whoever has read and believed this could not but suspect that I have given out another's discovery disguised by substitute names and symbolism. But no one knows better than Newton himself how false this is; never did I hear the name *calculus of fluxions* spoken nor see with these eyes the symbolism that Mr. Newton has employed before they appeared in Wallis's *Works*.(7) The very letters published by Wallis(8) prove that I had mastered the subject many years before I gave it out; how then could I have published another's work modified of which I was ignorant? However, although I do not take Mr. Keill to be a slanderer (for I think he is to be blamed rather for hastiness of judgement than for malice) yet I cannot but take that accusation which is injurious to myself as a slander. And because it is to be feared that it may be frequently repeated by imprudent or dishonest people I am driven to seek a remedy from your distinguished Royal Society. For I think you yourself will judge it equitable that Mr. Keill should testify publicly that he did not mean to charge me with that which his words seem to imply, as though I had found out something invented by another person and claimed it as my own. In this way he may give satisfaction for his injury to me, and show that he had no intention of uttering a slander, and a curb will be put on other persons who might at some time give voice to other similar [charges]. For the rest, farewell and thrive. *Berlin*, 4 *March* 1711 [N.S.]

NOTES

(1) Add. 3968(18), fo. 262; the paper, slightly torn, is in Sloane's hand; the reverse is addressed to Newton. 21 February is the Old Style date corresponding to 4 March N.S. The text was first printed in *Commercium Epistolicum*, p. 109. Another copy may be found in the Letter Book of the Royal Society (original), XIV, fos. 273–4.

(2) *Phil. Trans.* **26** (1708) appeared in 1710.

(3) See *Nicolai Fatii Duillierii R.S.S. Lineæ brevissimi descensus investigatio geometrica duplex. Cui addita est investigatio geometrica solidi rotundi in quod minima fiat resistentia* (London, 1699). On p. 18 Fatio wrote:

Quæret forsan Cl. Leibnitius, unde mihi cognitus sit iste Calculus, quo utor? Ejus equidem Fundamenta universa, ac plerasque Regulas, proprio Marte, Anno 1687, circa Mensem Aprilem & sequentes, aliisque deinceps Annis, inveni; quo tempore neminem eo Calculi genere, præter me ipsum, uti putabam. Nec mihi minus cognitus foret, si nondum natus esset Leibnitius. Aliis itaque glorietur Discipulis, me certe non potest. Quod plus satis patebit, si olim Litteræ, quæ, inter Clarissimum Hugenium meque, intercesserunt, publici juris fiant. Newtonum tamen primum, ac pluribus Annis vetustissimum, hujus Calculi Inventorem, ipsa rerum evidentia coactus, agnosco: a quo utrum quicquam mutuatus sit Leibnitius, secundus ejus Inventor, malo eorum, quam meum, sit Judicium, quibus visæ fuerint Newtoni Litteræ, aliique ejusdem Manuscripti Codices. Neque modestioris Newtoni Silentium, aut prona Leibnitii Sedulitas, Inventionem hujus Calculi sibi passim tribuentis, ullis imponet, qui ea pertractarint, quæ ipse evolvi, Instrumenta.

(Perhaps Mr. Leibniz may enquire whence that calculus became known to me, of which I make use? In fact I discovered its universal foundations and most of its rules, by my own exertions, around April and the following months of 1687 and in other years thereafter, at which time no one (as I believed) was employing that sort of calculus besides myself. And it would be no less known to me had Leibniz not yet been born. So let him pride himself upon other disciples, for upon me certainly he cannot. And this will be more than obvious if the letters exchanged between Mr. Huygens and myself should one day be made public. Yet I recognize that Newton was the first and by many years the most senior inventor of the calculus, being driven thereto by the factual evidence on this point; as to whether Leibniz, its second inventor, borrowed anything from him, I prefer to let those judge who have seen Newton's letters and other manuscript papers, not myself. Neither the silence of the more modest Newton nor the eager zeal of Leibniz in ubiquitously attributing the invention of this calculus to himself will impose on any who have perused those documents, which I myself have examined.)

(4) 'G.G.L. Responsio ad Dn. Nic. Fatii Duillierii imputationes: accessit nova artis analyticæ promotio specimine indicata...', *Acta Eruditorum*, May 1700, pp. 198–208; reprinted in *Commercium Epistolicum*, p. 107. Leibniz was also the anonymous reviewer of Fatio's booklet in the previous year (1699; pp. 510–13). The editors of the journal did not allow Fatio to print more than a bare rejoinder to the technical content of this (see *Acta Eruditorum*, March 1701, pp. 134–6).

(5) Leibniz had written to Wallis on 6 August 1699, complaining of Fatio's treatment of him. Wallis passed a transcript of this to Sloane on 22 August, and Sloane sent Wallis a reply (which we have not traced) on 26 August. (See Letterbook of the Royal Society (copy) XII, fos. 142 and 152). Wallis sent Leibniz a transcript (also lost) and his own comments in a letter of 29 August. (See Gerhardt, *Leibniz: Mathematische Schriften*, IV, p. 71.)

(6) See Letter 830, note (5).

(7) Wallis, *Opera Mathematica*, II (1693) pp. 393–6. (See also vol. III, Number 394 and note (1).)

(8) Wallis published a collection of letters under the title 'Epistolarum Quarundam Collectio, Rem Mathematicam Spectantium' in *Opera Mathematica* III (1699), pp. 615–708. These included Leibniz's correspondence with Oldenburg in this period 1674–1677.

823 ARBUTHNOT TO FLAMSTEED

14 MARCH 1710/11

From the original in the Royal Greenwich Observatory.(1)
For the answer see Letter 824

Sir

Her Majesty having commanded me,(2) to take care, that The *Historia Cœlestis*, which was begun at his Royal Highnesses order,(3) and carry'd on at his charge,(4) should be finish'd as soon as possible, & that it should appear in a dress suitable to the honour of such a patron; I should fail in my duty; If I did not acquaint you, That ther remains severall things to be perform'd on your part towards the perfection of so usefull a work;(5) & particularly what retards us at present is, the want of Your most accurate Catalogue of the fixd starrs, which The world has so long wishd to see:(6) The Copy you have hitherto deliver'd is imperfect, wanting the six Northern constellations of Draco, Ursa major & Minor, Cepheus, Cassiopeia & Hercules; Draco & ursa minor wholly, & the rest without longitudes & latitudes & differences; so that for want therof they are disabled to proceed on the Edition; Therefor I desire you would deliver into my hands as soon as possible, a perfect copy of your Catalogue of the fixd starrs, & you shall have a receipt in due form upon the delivery of it, & I can assure you ther shall no pains be wanting, that both the catalogue & the rest of the work be publish'd in as creditable a manner as is fitt for so usefull a work: I am the more fully perswaded you will complye with so reasonable a request, because of the regard you have for the Memory of the prince, as well as for your own reputation, both which are interested somewhat in this performance. I expect your answer by the bearer or as soon as you can, being with all respect

Sir

Your Most humble servant
JO: ARBUTHNOTT

London: March 14*th*
$17\frac{10}{11}$

Dr Arbuthnot in St James's place
Receaved *Monday morning March* 19. 1710(7)

NOTES

(1) Flamsteed MSS, 55, fos. 89–90; printed in Baily, *Flamsteed*, p. 280. For John Arbuthnot (1667–1735) see vol. IV, p. 431, note (7). He was a man of some mathematical knowledge, F.R.S. since 1704, and with influence over the Queen as her physician (1705–14). He was reckoned one of Newton's supporters.

(2) Arbuthnot was one of the 'Gentlemen to whom his Royal Highness the Prince hath referred the care of printing Mr Flamsteeds Astronomical Papers'. (Vol. IV, pp. 430, 524.) It is postulated in the literature bearing on this dispute (e.g. Baily, *Flamsteed*, p. 93 note; Brewster, *Memoirs*, II, p. 238; More, p. 544) that Newton or at any rate the Royal Society had procured an 'order' or 'royal grant' from the Queen to continue the printing of *Historia Cœlestis Britannica*, suspended after the death of Prince George on 28 October 1708 (compare vol. IV, Number 749). No such order has been found by anyone, or by ourselves; Flamsteed himself did not attribute the procurement of any such order to Newton; and this correspondence between Arbuthnot and Flamsteed appears to exclude the necessity for it. We are inclined to believe that Newton was glad to keep himself officially out of the business.

(3) See vol. IV, p. 430 etc.

(4) The account for £350 expended, as made up by Newton on 8 April 1710, is printed in Baily, *Flamsteed*, p. 275.

(5) The resumption of the printing of the *Historia Cœlestis* had been discussed already by the Royal Society (see the meeting of 7 March 1710/11, Journal Book, IX, p. 207), and Flamsteed had been privately advised 'that my catalogue (which I was then working upon to complete it as far as I then could) was in the press': nevertheless Arbuthnot's letter demanding further star places 'that had not been delivered into Sir Isaac Newton's hands when he got the rest into his possession by tricks and pretences' astonished him as being 'one of the boldest things that ever was attempted. None that had less dexterity and boldness and art than the Doctor would have had the confidence to have mentioned such a demand' (Baily, *Flamsteed*, p. 93).

(6) 175 pages of a preliminary version of Flamsteed's star catalogue had been placed in Newton's hands some years before. It had never been possible to reach an agreement on the procedure for checking, printing, proof-reading and publishing them. This partial list Edmond Halley revised and extended with an additional 500 places computed from Flamsteed's own records (vol. IV, p. 529). This 'Halley' version of the star-catalogue was that to be issued later, in 1712. (Compare vol. IV, p. 458 and notes thereon.)

(7) The endorsement is Flamsteed's.

824 FLAMSTEED TO ARBUTHNOT

23/25 MARCH 1710/11

From a copy in the Royal Greenwich Observatory.(1)
Reply to Letter 823; for the answer see Letter 827

The Observatory March ye 23 1711(2)

Sr

Tis no small Satisfaction to me to find by Yours of ye 14th Instant received this week by Mr Hunt,(3) that her Majesty is pleased the Historia Brittanica Cœlestis; yt was begun to be printed at ye Charge of his Royall Highness should be published as soon as possibly it can; and appear in a dress worthy of so great & Excellent a Patron.(4) It has been allwayes my Endeavour, and is still ye same, that it should do so: In order to it as soon as I found ye press at a full stop; I carried on ye Large catalogue of ye fixd Stars with all the dilligence speed and care I could, and compleated it as far as I thought would be necessary till it should come to be printed I had no sooner done this but ye good providence of God (that has hitherto conducted all my Labours; and I doubt not will do so to an happy conclusion) afforded me an occasion of carrying them much beyond those bounds, which I had first proposed to my selfe or could reasonably hope. The great differences I found this and some foregoeing years; betwixt ye planets places in the Heavens derived from my observations, and their places calculated by the best numbers, wth a small intervening accident, put me upon formeing new Tables for one of the Superior planets; my success herein carried me on to a second and third I have now ye fourth under my hands; and a large Stock of Materials ready for ye rest, by what I have done I have found wherein ye faults of ye common numbers be and how they are to be limited & altered. but a great deale more help is requisite, and must be procured to calculate ye new Tables and the planets places therefrom to render the Work compleat. Worthy of the Brittish Nation, the name it bears, her Majestys patronage and to commend the memory of his Royall Highness to posterity; in order to which it is very necessary that I should have a few hours discourse with you, if possibly you can, at ye Observatory; where I can show you ye result of my endeavours; and we might consider togeather how to carry on ye work and keep it free from such hindrance and delayes as have formerly retarded ye progress of it: I will draw up some short Notes for this purpose against you come to Dine with me, but if your necessary affairs & attendance will not allow you to afford me that favour, please to let me know by a short Note at wt hours any day in ye Week, except Monday Morning or Saturday afternoon, I may find you at best leasure; I will wait upon you, and discourse more fully with you concerning this business, and I shall esteem it a favour done to

Sr

Your oblieged and humble Servant

Jo. Flamsteed M.R.

March ye 25 at evening[2]

NOTES

(1) Flamsteed MSS vol. 53, fo. 87, printed in Baily, *Flamsteed*, pp. 280–1.

(2) The copy is in an amanuensis' hand, only the date at the foot being written by Flamsteed.

(3) See Letter 802, note (5).

(4) No doubt Flamsteed expected—as authors so often have—that he would be able to rewrite, enlarge and correct his *magnum opus* in proof. He did not yet realize that it was intended only to publish the imperfect manuscript that he had handed over on 15 March 1705/6.

825 NEWTON TO FLAMSTEED

[c. 24 MARCH 1710/11]

From a draft in the University Library, Cambridge(1)

Sr

By discoursing wth Dr Arbothnot about your book of observations wch is in the Press, I understand that he has wrote to you by her Mats order for such observations as are requisite to complete the catalogue of the fixed stars & you have given an indirect & delatory answer.(2) You know that the Prince had appointed five gentlemen to examin what was fit to be printed at his Highness expence, & to take care that the same should be printed.(3) Their order was only to print what they judged proper for the Princes honour & you undertook under your hand & seal to supply them therewith, & thereupon your observations were put into the press. The observatory was founded to the intent that a complete catologue of the fixt stars should be composed by observations to be made at Greenwich & the duty of your place is to furnish the observations. But you have delivered an imperfect catalogue wthout so much as sending the observations of the stars that are wanting, & I heare that the Press now stops for want of them. You are therefore desired either to send the rest of your cataloge to Dr Arbothnot or at least to send him the observations wch are wanting to complete it, that the press may proceed. And if instead thereof you propose any thing else or make any excuses or unnecessary delays it will be taken for an indirect refusal to comply wth her Majts order. Your speedy & direct answer & compliance is expected.

NOTES

(1) Add. 4006, fo. 38r. This draft evidently expresses Newton's passionate reaction to the immediately preceding letter from Flamsteed to Arbuthnot. It was first printed by Brewster, *Memoirs*, II, p. 489. There is no evidence that Newton in fact addressed Flamsteed in these harsh terms; the absence of any known response on Flamsteed's part makes it unlikely that he did.

(2) See Letters 823 and 824.

(3) See Letter 680 (vol. IV, p. 430). Strictly, only *four* 'gentlemen' (Newton, Francis Robartes, Sir Christopher Wren and Arbuthnot) were named as referees, together with 'others of your Society'. David Gregory and Francis Aston both signed documents as referees later, making six.

826 NEWTON TO COTES
24 MARCH 1710/11

From the original in Trinity College Library, Cambridge.(1)
For the answer see Letter 829

St Martins street by Leicester
Fields Mar. 24*th* 171$\frac{0}{1}$

Sr

I send you at length the Paper(2) for wch I have made you stay this half year. I beg your pardon for so long a delay. I hope you will find the difficulty cleared, but I know not whether I have been able to express my self clearly enough upon this difficult subject, & leave it to you to mend any thing either in the expression or in the sense of what I send you. And if you meet wth any thing wch appears to you either erroneous or dubious, if you please to give me notice of it I will reconsider it. The emendations of Corol. 2 Prop 38 & Prop 40 are your own. You sent them to me in yours of Sept. 21, 1710,(3) & I thank you for them. That you may have a clearer Idea of the experiments in the beginning of the inclosed paper,(4) let *ABCD* represent a vessel full of water perforated in the side with a small hole *EF* made in a very thin plate of sheet tin. And conceive that the water converges towards the hole from all parts of the vessel & passes through the hole with a converging motion & thereby grows into a smaller stream after it is past the hole then it was in the hole. In my trial the hole *EF* was $\frac{5}{8}$ths of an inch in diameter & about half an inch from the hole the diameter of the stream *RS* was but $\frac{21}{40}$ of an inch. And therefore the streame had the same velocity as if it had flowed directly out of a hole but $\frac{21}{40}$ of an inch wide. And so in Marriotts experimt the stream had the same velocity as if it had flowed directly out of a hole but $\frac{21}{100}$ of an inch wide.(5) In computing the velocity of the water wch flows out we are not to take the diameter of the hole for the diameter of the streame, but to measure the diameter of the streame after it is come out of the hole & has formed it self into an eaven & uniform stream. And the velocity thus found will be what a body would get in falling from ye top of the water: as is manifest also by the distance *CG* to which the stream will shoot it self, &

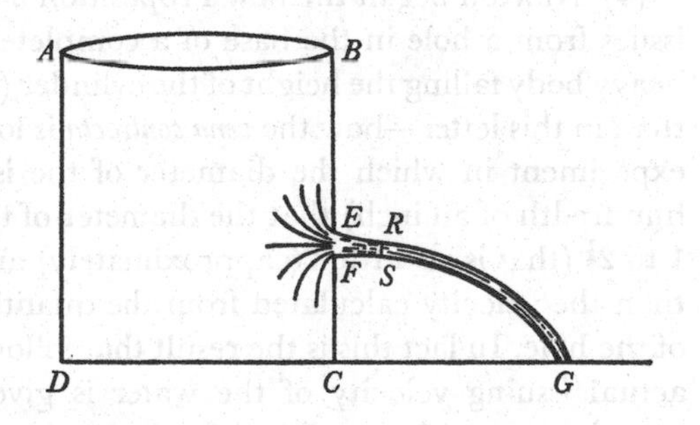

also by the stream's ascending as high as the top of ye water stagnating in the vessel, if the motion be turned upwards.

I am
Your most humble & most obliged Servant
IS. NEWTON

For the Rnd Mr Roger Cotes Professor of Astronomy at his Chamber in Trinity College in the University of Cambridge

NOTES

(1) R.16.38, no. 66, printed in Edleston, *Correspondence*, pp. 38–9; this is Newton's reply to Cotes' last letters of October 1710.

(2) This seems to have been largely lost; Edleston supposed that Newton sent four folio sheets, of which he found only the last (*Correspondence*, p. 40). Certainly the new material covered at least Propositions 36 and 37 of Book II (*Principia* 1713; pp. 303–13), and presumably revised versions of some later passages at least up to Proposition 40; however, it is certain from Cotes' Letter 853 that he received copy at least up to Proposition 48.

(3) See above, pp. 65–8, Letter 805.

(4) Newton began the new Proposition 36 with a 'thought experiment' to show that water issues from a hole in the base of a completely filled cylinder with the velocity obtained by a heavy body falling the height of the cylinder ($v = \sqrt{(2gh)}$). He then explained—more elaborately than in this letter—how the *vena contracta* is formed and claimed (overtly on the basis of a single experiment in which the diameter of the issuing stream was purportedly measured to one hundredth of an inch) that the diameter of the *vena contracta* is to that of the hole forming it as 1 to $2^{\frac{1}{4}}$ (that is, as 21 to 25 approximately) and so the velocity of the actual stream is $\sqrt{2}$ greater than the velocity calculated from the quantity of water emitted and the measured diameter of the hole. In fact this is the result that follows from computation if it is postulated that (i) the actual issuing velocity of the water is given by $\sqrt{(2gh)}$; (ii) the velocity calculated from experiments on the quantity of water emitted and the measured diameter of the hole is $\sqrt{(gh)}$. Then in the latter part of Proposition 36 Newton supported these results by reference to the (rough) equality of the height of a vertical jet of water with the head forming it, and to the projectile-like breadth of the parabola formed when the jet is horizontal.

(5) See Letter 805. Mariotte's experimental hole was $\frac{1}{4}$ inch in diameter: $0{\cdot}25/2^{\frac{1}{4}} = 0{\cdot}21$.

827 ARBUTHNOT TO FLAMSTEED

26 MARCH 1711

From the original in the Royal Greenwich Observatory.[1]
Reply to Letter 824; for the answer see Letter 828

Sir

I receaved yours, and am extremely Glad, at any improvement so noble a Science as Astronomy can receave, & shall be willing as farr as lyes in my power to give my helping hand towards publishing those observations & tables mention'd in your letter; but that being beyond my Commission, (which was only to oversee the publishing of the observations which were giv'n in to the referees befor his Royal Highnesses death) I cannot at present say any thing more to it: Than that when those are printed, I shall be ready to sollicite her Majesty that those may be published as an Appendix to the work, what I desired in my letter was that you would be pleasd to deliver to me those Constellations that are wanting in the Catalogue you have Allready delivered; or such of them as you have compleat; if you have nothing more to adde to the Catalogue, lett me know so much by a line & I shall order the press to proceed with what we have; I beg your positive Answer to this, for the press at present stands still & I am complain'd off, for delays.

I shall be ready to waitt on you any wher in town & at any hour, only sending me a Note in the morning or night befor, I am with all respect

Sir

Your Most humble Servant

JO: ARBUTHNOTT

London: March 26
1711

NOTE

(1) Flamsteed MSS, vol. 55, fos. 97–8; printed in Baily, *Flamsteed*, p. 281.

828 FLAMSTEED TO ARBUTHNOT
28 MARCH 1711

From the copy in the Royal Greenwich Observatory.(1)
Reply to Letter 827

The Observatory March 28 ☿ 1711

Sr

I am oblieged to you for ye favour of yours recd this morning by ye Bearer. & ye more because it expresses your good will so fully to her Majestys Observatory. a small touch of ye Gout yt came upon me last Sunday kept me at home. but I thank God I have no paine, so yt I hope nevertheless yt I may be in London by ye Stage Coach to Morrow by 11 a Clock or soon after, where if yr occasions permit, you will find me at Garraways Coffee house,(2) and I shall inform you of ye State of my Works it being to long to be told you in a Lettr for an Answer to which ye Messenger stays. I am with all due respect & hearty thanks

Your Most oblieged humble Sert
JOHN FLAMSTEED M R

NOTES

(1) Flamsteed MSS, vol. 55, fo. 99, printed in Baily, *Flamsteed*, pp. 281–2.

(2) Flamsteed did in fact meet James Hodgson (husband of his own niece, who assisted him in proof-correcting and other matters), George Clarke (secretary to the late Prince George) and Arbuthnot at Garraway's coffee-house in Cornhill on the 29th. According to Flamsteed's Diary for this day (Baily, *Flamsteed*, p. 226):

He [Arbuthnot] urged to have the catalogue [of fixed stars] made up: I told him I was willing: that Sir I. Newton had *two* imperfect copies in his hands: I desired the latter might be returned, to save me the pains of transcribing, and I would fill it up with all I had finished: but that the variations of Right Ascension were wanting in those stars that were within 30° of the pole, by reason these variations altered so enormously, the tables I used would not seem to find them so exactly as in those above 30° from it: he seemed satisfied: I desired him to come down to see the work, which he neither consented to nor seemed to refuse: I inquired of him, whether the catalogue were printing or not: he assured me '*not a sheet of it was printed*'; though I am assured by others that some sheets are wrought off; I desired that if it were to be printed, I might have the last proof sheets sent to me, to be examined and corrected: he stuck at this; but promised (and *pronounced it with great earnestness*) he would give me £10 for every error or fault from my copy, that should be shown him in the press-work: I presented him with my printed estimate, and written copy of my letter to Sir C. Wren [see vol. IV, pp. 525–7], occasioned by Sir I. Newton's cunning order, or agreement: he said I had spoken ill of Sir I. Newton...after this I told him I was very desirous to proceed, *provided that I might have just, honourable, equitable and civil usage*: which he assured me I should...

829 COTES TO NEWTON

31 MARCH 1711

From the original in the University Library, Cambridge.(1)

Reply to Letter 826

Cambridge March. 31*st.* 1711

Sr.

I have received Your Letter with the inclosed Paper & am very well satisfied with Yr solution of the difficulty which I formerly proposed to You concerning the Velocity of the effluent water. I find that 25 & 21 express the proportion of $\surd\surd 2$ to 1(2) as nearly as it is possible for so small numbers to do it, whence it is probable yt the exact proportion of the diameter of the Hole to ye diameter of ye Stream is that of $\surd\surd 2$ to 1, & then ye proportion of 44 to 37 will be much nearer the truth than yt of 25 to 21. I am sensible of the undeserved honour You do me in one of Yr additions & I return You my thanks for Yr kindness.(3) After I had read Your paper I was extreamly pleased to see the whole Theory (as I thought) so well settled: however I was resolved to read over the Propositions once more, with all the care I could, before I delivered them to the Printer.

It seems to me that, in transcribing the Copy, You have omitted something in the first section of ye 36th Proposition. For, as I have it, I cannot be certain yt I do fully & precisely understand your sense & design.(4) I think Your Idea is this: You imagine that as the water descends freely from *AB* by the force of its own gravity so by some other force, whatever that be, it is moved at the same time with an Horizontal motion towards ye Axis of ye stream which Horizontall motion is supposed so to be adapted to the motion of Descent that it may not anywise accelerate or retard it, but only just suffice to keep the Stream intire and uninterrupted; & by this meanes ye Space *ABNFEM* is always full of water, but ye space around this (vizt. *AMEC*, *BNFD*) is always void of water. If this be Your Idea, You have not express'd it in Your Copy. I am persuaded it is Your Idea, for otherwise I cannot see any argument in this section, or understand the sense of Your words [uti prius] in ye 2d section. But Yr words [defluere in vas & ipsum perpetuo implere] & a little lower [& vas perpetuo plenum manebit](5) lead the Reader to think this is not Your Idea since You seem by these words to suppose the whole Cylinder to be perpetually full. I cannot think You intend in this place to represent the whole Cylinder as perpetually full; for if so, You are got no further in the 3rd Section than You was in the first. I will transcribe the whole first section from Yr Copy that You may supply what is wanting.(6) Cas. 1. Sit *ACDB* vas cylindricum,

AB ejus orificium, *CD* fundum horizonti parallelum, *EF* foramen circulare in medio fundi, *G* centrum foraminis, & *GH* axis cylindri horizonti perpendicularis. Et producatur axis *GH* ad *I* ut sit *IH* ad *IG* in duplicata ratione areæ foraminis *EF* ad aream circuli *AB*, et per punctum *I* ducatur linea Horizontalis *KL* vasi hinc inde occurrens in *K* & *L*. Concipe cylindrum glaciej *APQB* ejusdem esse latitudinis cum cavitate vasis, et uniformi cum motu perpetuo descendere, et partes ejus quamprimum attingunt superficiem *AB* liquescere et in aquam conversas gravitate sua defluere in vas & ipsum perpetuo implere ut constans & uniformis sit aquæ defluxus per foramen *EF*. Sit autem ea glaciej descendentis velocitas quam aqua cadendo & casu suo describendo altitudinem *IH* acquirere potest; & vas perpetuo plenum manebit, et velocitas aquæ per foıamen *EF* effluentis ea erit quam aqua cadendo ab *I* & casu suo describendo altitudinem *IG*, acquirere potest, ideoque per Theoremata Galilæi, erit ad velocitatem aquæ in circulo *AB* in subduplicata ratione *IG* ad *IH*, id est (per constructionem) in simplici ratione circuli *AB* ad aream foraminis *EF*, et propterea transibit per foramen *EF*, & transuendo implebit foramen illud accurate. Nam circulus horizonti parallelus per quem aqua cadens adæquate transit, est reciproce ut velocitas aquæ. De velocitate aquæ horizontem versus hic agitur. Et motus horizonti parallelus quo partes aquæ candentis ad invicem accedunt, cum non oriatur a gravitate, nec motum horizonti perpendicularem a gravitate oriundum mutet, hic non consideratur.

In the 2d Section of the 36th Proposition You have these words [& pondus totum columnæ aquæ *ABNFEM* impendetur in defluxum ejus generandum.][7] It may possibly appear to some Readers that you have committed a mistake in these words whilst they suppose You to assert that ye weight of this Column alone is of it self sufficient to generate the motion of the defluent water. Let ye Hyperbolick figures *EMA* & *FNB* be infinitely produced towards their Asymptote *KL*, & the weight of the whole solid between ye plane *EF* & ye infinite plane *KL*, or ye weight of ye double of ye Cylindr whose base is *EF* & height *GI* will be ye total weight which is the compleat and adæquate cause of ye motion of ye defluent water. I suppose You agree with me in this Assertion & therefore I understand Your words thus: That ye weight of ye Column *ABNFEM* is not ye adæquate cause of ye motion; but of the whole water in the vessel, this column is ye only part which by its weight concurrs to ye effect. To prevent cavils I think it would not be amiss if You should express Your self more determinately in this place.

In the 3d section of ye 37th Proposition are these words [Sed hic Cylindrus *D*, pondere suo cadendo in vacuo & casu suo describendo altitudinem aquæ in vase, velocitatem acquiret qua fundum prædictum ascendit.][8] I cannot bring my self to agree with You in this assertion, nor indeed can I any way at present from Your premises, deduce the conclusion which is delivered in the words of this Proposition; yet I am very well satisfied of ye truth of that Conclusion by the Experiments subjoyned to ye 40th Proposition. I will endeavour to give You my sense of ye matter, as clearly as I can, in what follows. Let *ABCD* represent a Cylindricall vessel, as in Prop: 36; let *EF* be ye Fundum; *CE*, *FD* ye rima annularis in circuitu fundi; *AB* the upper surface of ye water in ye vessel; & let ye solid *EHF* answer to ye glacies in circuitu aquæ defluentis of ye 36th Propos:(9) & supposing ye same construction as in yt Proposition: the

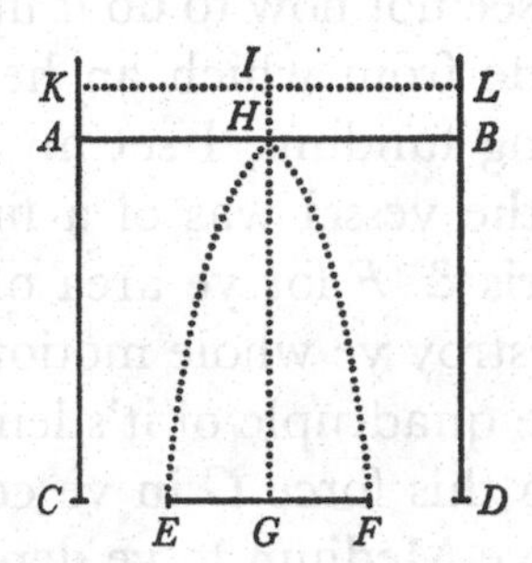

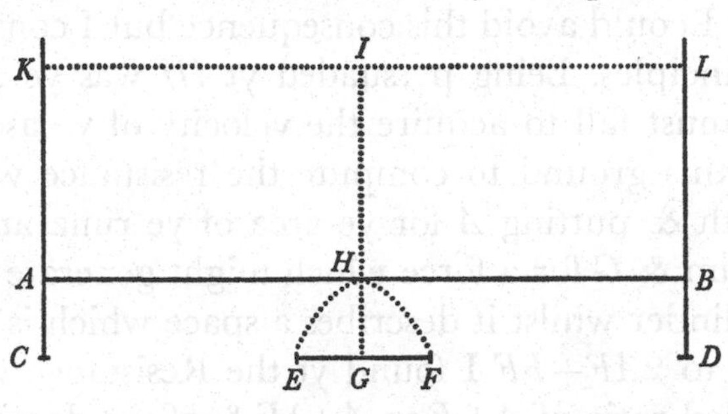

velocity of ye water in *AB* is ye same as by falling from *I* to *H*, and ye velocity of ye water passing through the rima annularis is ye same as by falling from *I* to *G*, & ye weight of ye Solid *EHF* is ye resistance which ye Fundum sustains from ye defluent water. Now if we suppose yt things remain no longer in ye State yt has been described; but imagine a new State, in which ye same resistance shall be caused by ye ascent of ye Fundum; tis evident to me that the Velocity with which ye fundum ascends in this latter state must be equal to ye velocity with which ye water in *AB* descends in ye former state, or to ye velocity acquired by an heavy body in falling from *I* to *H*. This I think You also affirm Your self in the 2d section of ye 37th Proposition, when You say, Et propterea si glacies Cylindrica quiescat & vas cum fundo suo, eadem velocitate qua glacies descendebat, ascendat in glaciem motus omnes aquæ in vase respectu vasis & glaciei, et omnes ejus pressiones eadem erunt ac prius.(10) But notwithstanding this in ye above cited words of ye 3d section you seem to affirm, yt ye velocity with which the fundum ascends is equal to ye velocity acquired by falling (not from *I* to *H*, but) from *H* to *G*. According to my apprehension these assertions cannot be reconciled in any measure; unless it be said, when ye rima annularis & ye breadth of ye vessel become infinite (which is ye case of ye 3d section) yt then *IH* & *HG* become equall: but this I think ought not to be said. For let ye altitude *KC* remain unaltered that ye velocity of ye water in its passage through the rima annularis may remain so likewise; let ye Fundum *EF* also remain of ye same magnitude, but imagine ye rima annularis to be made larger continually: upon these suppositions the proportion of ye circle *AB* to ye rima annularis will continually approach to ye proportion of equality & consequently the proportion of *IG* to *IH* which is duplicate of ye former: so yt ye ultima ratio of *IG* to *IH* will be yt of equality, not yt of 2 to 1, & therefore the ultima ratio of *IH* to *HG* will not be yt of equality, but ye furthest from it yt is possible. I think it hence also appeares yt by infinitely inlarging ye breadth of ye vessel ye resistance of ye Fundum is infinitely diminished, if ye weight of ye solid *EHF* be ye measure of ye resistance: for tis evident yt this solid is infinitely diminished as its altitude *GH* becomes infinitely little. I wish I could avoid this consequence but I confess I see not how to do it upon Yr principles. Being persuaded yt *IH* was ye altitude from which an heavy body must fall to acquire the velocity of ye ascending fundum, I set my self upon this ground to compute the resistance when the vessel was of a finite breadth & putting *A* for ye area of ye rima annularis & *F* for ye area of ye Fundum & *G* for a force which might generate or destroy ye whole motion of ye Cylinder whilst it describes a space which is to the quadruple of it's length as AA to $2AF+FF$ I found yt the Resistance was to this force *G* in ye compounded ratio of $A+F$ to $A+\frac{1}{2}F$ & of the density of ye Medium to ye density of ye Cylinder. Now let ye breadth of ye Vessel become infinite & AA will be to $2AF+FF$ as A to $2F$ or as an infinite to a finite, & ye ratio of $A+F$ to $A+\frac{1}{2}F$ will be a ratio of equality. Consequently in this case the Resistance will be to a force which may generate or destroy ye motion of ye Cylinder whilst it describes an infinite space, that is, to a Force infinitely less yn the gravity of ye Cylinder, as the density of ye Medium to ye density of ye Cylinder. And hence it again appeares yt by infinitely inlarging the breadth of the Vessel the Resistance is infinitely diminished. For my part I must confess I see not how to acquit my self of this difficulty, I hope You will have better success with it. If I have made any mistake or apprehended any thing amiss You will easily perceive it. I have designedly expressed my self more fully than I should otherwise have done yt You may ye better observe my Errors if I have committed any, for I would not willingly give you an unnecessary trouble.[11]

I am Your most obliged humble Servant
ROGER COTES.

For Sr Isaac Newton
at his House
in St Martin's Street
by Leicester feilds
London

NOTES

(1) Add. 3983, no. 11.

(2) Compare Letter 826, note (4).

(3) No compliment to Cotes survived into the printed version of the second edition. The *Auctoris Præfatio* added to it much later (see Letter 989) ignored his work completely. Possibly the compliment of which Cotes writes here remarked on his attention to Proposition 27. Certainly Newton *originally* intended to pay due tribute to Cotes. Professor Cohen has already noted (*Introduction*, p. 247) a phrase in a Newton manuscript related to Letter 784 (U.L.C. Add. 3984, fo. 5v: 'ut me admonuit D. Cotes acutissimus Astronomiæ Professor apud Cantabrigienses'—*this* compliment would have related to Cotes' work on Book II, Proposition 29 (Letter 783). However, this part of the second edition had been printed (without any such phrase) by June 1710. The document following this letter (Number 829*a*) again reveals Newton's early intentions of expressing warm gratitude to Cotes.

(4) Cotes now criticizes Newton's description of his 'thought experiment' justifying the hypothesis that particles of water descend in a cylindrical vessel as though falling freely through the height from the water-surface to the hole from which they emerge. Newton supposed that the descending particles form a contracting column whose top is the water-surface and whose (truncated) point is the hole. But at the same time he wrote of the water 'filling' the cylinder, which is inconsistent with the model.

(5) 'as before...flowing into the vessel and filling it continually...and the vessel will remain perpetually full'. If the cylinder is imagined to be perpetually full, then Newton's cone-model appears quite arbitrary.

(6) This introduction to Proposition 36 (it is not Case 1 in the printed text) was very considerably modified in order to meet Cotes' criticism of this first draft, which reads:

Let *ACDB* be a cylindrical vessel, *AB* its mouth, *CD* its base parallel to the horizon, *EF* a circular hole in the middle of the base, *G* the centre of the hole, and *GH* the axis of the cylinder perpendicular to the horizon. And let the axis *GH* be produced to *I* so that *IH* to *IG* is as the square of the area of the hole *EF* to the square of the area of the circle *AB*, and let the horizontal line *KL* be drawn through *I* meeting the vessel on either side at *K* and *L*. Imagine a cylinder of ice *APQB* to be of the same width as the cavity of the vessel, and to descend perpetually with a uniform motion, its particles melting as soon as they reach the surface *AB* and, turned into water, flowing into the vessel and filling it continually, so that the flow of water through the hole *EF* may be constant and uniform. Let the velocity of the descending ice be that which falling water acquires in descending the distance *IH*, and the vessel will remain perpetually full, and the velocity of the water flowing through *EF* will be that which water acquires in falling from *I* through the altitude *IG*, and so (by Galileo's theorem) it will be to the velocity of the water at the circle *AB* as the square root of *IG* to *IH*, that is (by construction) in the simple ratio of [the area of] the circle *AB* to the area of the hole *EF*, and moreover it will pass through the hole *EF* and in passing through it will fill the hole exactly. For the circle parallel to the horizon through which the water precisely falls, is reciprocal to the velocity of the water. Here the velocity of the water contrary to the horizontal is discussed. And the motion parallel to the horizontal by which the particles of water are made to approach each other mutually in falling, as it does not arise from gravity nor modify the motion perpendicular to the horizon produced by gravity, is not considered here.

This draft of the 'thought-experiment' (which so far does not explain the formation of the *vena contracta*) seems highly arbitrary in its assumptions. All that Newton achieves is a demonstration of hydraulic continuity: that is, since the quantity of water flowing through a system of pipes is constant, there is a constant product formed from cross-sectional area and velocity of flow. Hence *if* the area of the cylinder be A, and that of the hole a, when the water leaves the hole with a velocity of $\sqrt{(2gh)}$ the velocity of descent of its water surface will be $(a/A)\sqrt{(2gh)}$. But this does not tell us what the velocity of descent of the water-surface (or the melting ice) really is.

(7) '...and the whole weight of the column of water *ABNFEM* weighs downwards in producing its flow.'

ABNEFM is the section of a hyperboloid of revolution constituting the volume of descending water (within the cylindrical vessel) in Newton's 'thought experiment'. Despite Cotes' objection, this expression remains in *Principia*, 1713; p. 304, lines 23–5.

(8) 'But here the cylinder *D*, falling because of its weight *in vacuo* and traversing the height of the water in the vessel as it falls, acquires that velocity with which the aforesaid base ascended.' This sentence is modified in *Principia* 1713; p. 311, middle; it seems to make no sense in relation to the argument of Proposition 37 as printed.

(9) '...base...the annular gap running round the base...the ice around the flowing water...'

(10) 'And therefore if the cylinder of ice is at rest and the vessel with its base ascends towards the ice with the same velocity as the ice descended, all the motions of the water with respect to the vessel and the ice and all its pressures will be the same as before.' This sentence does not appear in the printed text. Cotes may have missed a point in Newton's argument here—in the absence of the draft under discussion in these letters one cannot be certain. Newton's argument in *Principia*, 1713, p. 311 concerns not the case Cotes treats here where the base of the vessel *ABCD* moves upwards, instead of the water in the vessel flowing downwards, but the quite separate case of a 'little circle' placed in a cylindrical canal *below* the aperture in the vessel, and the pressure on it of the water issuing from the vessel. Cotes, however, seems mistakenly to have thought that Newton referred still to a 'little circle' ascending within the vessel itself, and assumed the case Newton was now treating was that where the aperture in the base of the vessel was widened to occupy its whole width, but blocked in the middle by the 'little circle'.

(11) Newton delayed answering this letter until (presumably) 2 June; his answer has vanished.

829*a* DRAFT OF A PREFACE TO THE NEW EDITION

[? AUTUMN 1712]

The original is in private possession(1)

Ad Lectorem.

In hac secunda Principiorum editione, nonnulla immutati sunt, nonnull[a] addita. Theoria resistentiæ fluidorum quæ in priore editione ob defectum experimentorum imperfecta prodijt, hic perficitur. Methodus synthetica investigandi vires quibus corpora in sectionibus Conicis revolvi possint, absque omni analysi exponitur. Theoria Lunæ ex causis suis plenius derivatur & corrigitur & Theoria Cometarum [*exacte congruere phænomenis Cometarum aliquot ostenditur*][(2)] per phænomena probatur. Et cum vis gravitatis tam late pateat ut per hanc solam, motus omnes corporum cœlestium regantur, & mare nostrum etiam fluat et refluat, superest ut vires reliquas attractivas, vis scilicet electrica et vis magnetica, examinentur, et earum leges et effectus [*varias*][(2)] ad motus [*minimarum particularum materiæ corporeæ*][(2)] minimorum corporum in dissolutione, fermentatione, vegetatione, [*digestione, præcipitatione, separatione,*][(2)] & similibus operationibus [*observentur*][(2)] inveniantur. Nam effectus harum virium late patent, particulis ferreis compositionem plurimorum corporum ingredientibus & spiritu electrico quantum sentio, per omnia prope dixerim corpora diffuso, sed vires suas ad parvas tantum distantias exercenti nisi ubi per frictionem excitatur. Habent itaque Philosophi campum late patentem in quo vires exerceant ingenij.[(3)]

Analysin [*fluentium*][(2)] qua Propositiones in Libris Principiorum investigavi[*mus*][(2)] visum est jam subjungere ut Lectores eadem instructi Propositiones in his libris traditas facilius examinare possint, & earum numerum inventis novis augere. Hujus An[a]lyseos partes varias chartis sparsis olim cum amicis communicavi[*mus*][(2)], & partes illas hic [*in unum*][(2)] conjunxi[*mus*][(2)] ut methodus tota [*uno intuito*][(2)] simul legatur, facilius & melius intelligatur, et magis prosit. Partium ultima est methodus differentialis, qua utique area figuræ curvilineæ ex paucis ordinatis per earum differentias colligitur quamproxime. Nam inventio arearum ex ordinatis describentibus [*est pars hujus Analyseos non contemnenda*][(2)] ad hanc Analysin omnino pertinent.[(4)]

In his omnibus edendis Vir doctissimus D. Rogerius Cotes Astronomiæ apud Cantabrigienses Professor operam navavit, prioris editionis [*sphalmata*][(2)] errata correxit & me submonuit ut nonnulla ad incudem revocarem. Unde factum est ut hæc editio priore sit emendatior.[(5)]

Translation

To the Reader

Several changes and additions have been made in this second edition of the *Principia*. The theory of resistance of fluids is here made perfect, which in the first edition appeared imperfectly, because of a shortage of experiments. The synthetic method of investigating the forces with which bodies may be revolved in conic sections is expressed without any [use of] analysis. The theory of the moon is derived from its causes in greater detail, and rendered more correct; and the theory of comets is proved by the phenomena. And because the force of gravity is so widespread that all the movements of the heavenly bodies are governed by it, and our sea too is caused to ebb and flow, it remains to investigate the remaining attractive forces, that is to say the electric force and the force of magnetism, and to discover their laws and their effects upon the motions of the least of bodies [as manifested] in solution, fermentation, vegetation and similar processes. For the effects of these forces are widely diffused in that iron particles enter into the composition of most bodies, and the electric force (as I perceive) is extended I might almost say through all bodies; however, (except when excited by friction) its force is effective only at very short distances. Thus philosophers may see a broad area lying before them in which they may exercise their wits.(3)

It has seemed fitting now to annex the analysis by whose aid I investigated the propositions in the books of the *Principia* so that readers acquainted with it may the more easily examine the propositions discussed in these books and increase their number by discovering new ones. The various branches of this analysis were in former times imparted by me to my friends in separate papers and I have conjoined those branches here so that the whole method may be perused at once, may be understood more easily and more thoroughly, and may be of greater benefit. The concluding branch is the differential method, by which the area of curvilinear figures may invariably be established, approximately, from a small number of ordinates by means of the differences between them. For the finding out of areas by describing the ordinates is emphatically a part of this analysis.(4)

In publishing all this the very learned Mr Roger Cotes, Professor of Astronomy at Cambridge, has been my collaborator: he corrected the errors in the former edition and advised me to reconsider many points. Whence it has come about that this edition is more correct than the former one.(5)

NOTES

(1) We are indebted to Dr D. T. Whiteside for providing a transcript of this draft from a manuscript in private possession. The same MS bears a draft in Newton's hand of the opening sentences of the report of the *Commercium Epistolicum* Committee (see the Introduction, p. xxvi). It must be obvious that the draft was written before Nikolaus Bernoulli imparted to Newton news of the error in Book II, Proposition 10 (see Letter 951 *a*).

(2) The italicized words in brackets have been deleted by Newton.

(3) This paragraph was clearly pointing ahead to the (suppressed) longer version of the concluding *Scholium Generale*, in which Newton intended to discuss interparticulate forces at greater length; cf. Letter 977, note (2).

(4) This paragraph paves the way for a treatise entitled *Analysis per Quantitates fluentes et earum Momenta* which Newton at this time (that is, about September 1712) intended to annex to the *Principia*. Compare Letter 914, p. 278.

(5) Newton began his draft with these sentences, in almost identical words, as the opening paragraph; deleted them, and rewrote the paragraph at the end.

830 JOHN KEILL TO NEWTON

3 APRIL 1711

From the original in the University Library, Cambridge[1]

April 3*d*. 1711

Sr

I have here sent you the Acta Lipsiæ[2] where there is an account given of your book, I desire you will read from pag 39 at these words. ceterum autor non attingit focos vel umbilicos curvarum &c to the end.[3]

I have not the Volume in wch Wolfius[4] has answered my letter,[5] but I have sent you his letter transcribed from thence, and also a copy of my letter to him, I wish yould take the pains to read that part of their supplements wherein they give an account of Dr Freinds Book and from thence you may gather how unfairly they deall with you,[6] but realy these things are trifles that are not worth your while since you can spend your time to much better purpose than minding any thing such men can say,[7] however if you would look upon them so far as to let me have your sentiments on that matter you will much oblige

Sr

your most Humble Servant

Jo. Keill

NOTES

(1) Add. 3985, no. 1; published in Brewster, *Memoirs*, II, p. 44, note 1. John Keill (1671–1721) was a mathematician of Scottish origin, a pupil of David Gregory at Edinburgh, and was chiefly remarkable for introducing experimental demonstrations into his lectures on natural philosophy at Oxford in the late 1690s. He was also a popularizer of Newtonian science, and a keen partisan of Newton in person. He was elected F.R.S. in 1708, and Savilian professor of astronomy in 1712.

The occasion for the present letter was as follows: after Wallis and Fatio de Duillier had insisted on Newton's original invention of the calculus and Leibniz's indebtedness to him, there was a period of quiet. Then, in publishing *De quadratura curvarum* in 1704, Newton added the note that he had hit upon the method of fluxions employed in that work in the years 1665 and 1666. This provoked the reaction described in the next note, and Keill's violent response (note (5), below) which reopened the whole dispute. Keill, as this letter shows, was no man of discretion. Leibniz complained to Sloane (Letter 822); Keill then (22 March 1711) read at a meeting of the Royal Society parts of an answer he had drafted. Whether or not Newton felt any doubt about the wisdom of this answer, Keill's letter to him was obviously designed to enhance his bitterness towards Leibniz. And he succeeded.

(2) *The Acta Eruditorum* (a journal with which Leibniz was much concerned) for January, 1705. This issue printed an unsigned review of *De quadratura curvarum* certainly written by Leibniz himself. It contains a passage which may be translated as follows:

So that this may be better comprehended, it should be understood that when any magnitude increases continually, as a line may increase (for example) by the flowing of a point that describes it, those momentary increments are to be called *differences*, signifying the difference between the magnitude that was before and that which is produced by the instantaneous change and from hence arises the *differential calculus* and its inverse, the *integral calculus*; the elementary principles of which were discussed by their inventor, Mr. Gottfried Wilhelm Leibniz, in these *Acta*, and of which the various uses have been displayed both by himself, by the brothers [Jacob and Johann] Bernoulli and by the Marquis de l'Hôpital. Instead of the Leibnizian differences, therefore, Mr. Newton employs and has always employed, fluxions, which are almost the same as the increments of the fluents generated in the least equal portions of time. He has made elegant use of these both in his *Mathematical Principles of Nature* and in all publications since, just as Honoré Fabri in his *Synopsis geometrica* substituted the advance of movements for the method of Cavalieri.

The first and second of these sentences were unexceptionable, except to one already convinced that Leibniz was a plagiarist; but the third was indefensible, insofar as it contained a clear implication that Newton (in his 1687 *Principia* and later) had employed a method of fluxions imitated from Leibniz's calculus, described by him in 1684. But it must be remembered that Leibniz was not the first to make accusations of plagiarism.

(3) The passage signalized by Keill is actually on p. 34: 'moreover the author [Newton] does not touch on the foci of curves of the second degree, still less on those of higher degrees'.

(4) See *Acta Eruditorum*, 1710, p. 78. Christian Wolf (1679–1754), at this time professor of mathematics and philosophy at Halle; expelled in 1723, he taught at Marburg until 1741 when Frederick the Great restored him to Halle. His correspondence with Leibniz (1704) is printed in Gerhardt, *Leibniz: Mathematische Schriften* Supplement-Band (1860).

(5) Into a paper on the laws of centripetal force of 1708 (*Phil. Trans.* **26** (1708), 174) Keill had inserted in rejoinder to Leibniz a strong claim that Newton was the first inventor of fluxions, from which Leibniz's calculus was a mere derivation (p. 185: 'Hæc omnia sequuntur ex celebratissima nunc dierum Fluxionum Arithmetica, quam sine omni dubio Primus Invenit Dominus Newtonus, ut cui libet ejus Epistolas a Wallisio editas legenti, facile constabit, eadem tamen Arithmetica postea mutatis nomine & notationis modo; a Domino Leibnitio in Actis Eruditorum edita est.'). Keill's paper (in the form of a 'letter' to Halley) appeared only in 1710. Wolf rebutted it in the *Acta Eruditorum* for the same year; Leibniz himself replied in Letter 822 to Hans Sloane.

(6) The book was John Freind's *Prælectiones chymicæ* (first published at London in 1709); the English version *Chymical Lectures: in which almost all the Operations of Chymistry are reduced to their True Principles and the Laws of Nature*, London, 1712, contains a reprint of the review of the Amsterdam, 1710 (Latin) edition from the *Acta Eruditorum* of that year (pp. 412–16) together with Freind's rejoinder to its criticisms. The reviewer blamed Freind for following Keill in seeking to explain chemical phenomena in terms of attractive forces between the least particles of matter, an endeavour leading to the reintroduction of occult qualities ('redditur ad Philosophiam quandam phantasticam Scholæ vel etiam Enthusiasticam, qualis Fluddi fuit') and undoing at a stroke the work of Boyle and others in England in explaining phenomena mechanically, that is to say rationally (*Chymical Lectures* 1712; pp. 162–3). This of course indirectly wounded Newton.

(7) Of course Keill must have known how much Newton minded! At a later meeting of the Royal Society (5 April) Keill again referred to the unjust attack upon Newton in the *Acta Eruditorum* for 1705; he was instructed to draw up a true statement of the issues in dispute and also to vindicate himself from the charge of 'reflecting' particularly on Leibniz (who had, of

course, been a Fellow for nearly forty years). This letter, summarized in the Journal Book of the Royal Society for 24 May 1711, was finally approved for sending to Leibniz on 31 May (see below, Letter 843).

830a NEWTON TO [? SLOANE]

[? APRIL 1711]

From the draft in the University Library, Cambridge[1]

Sr

Upon speaking wth Mr Keil about ye complaint of Mr Leibnitz concerning what he had inserted into the Ph. Transactions, he represented to me that what he there said was to obviate the usage which I & my friends met with in the Acta Leipsica, & shewed me some passages in those Acta, to justify what he said. I had not seen those passages before, but upon reading them I found that I have more reason to complain of the collectors of ye mathematical papers in those Acta then Mr Leibnitz hath to complain of Mr. Keil. For the collectors of those papers everywhere insinuate to their readers that ye method of fluxions is the differential method of Mr Leibnitz & do it in such a manner as if he was the true author & I had taken it from him, & give such an account of the Booke of Quadratures as if it was nothing else then an improvement of what had been found out before by Mr Leibnitz Dr Sheen[2] & Mr Craig.[3] Whereas he that compares that book with the Letters wch passed between me & Mr Leibnitz by means of Mr Oldenburg before Mr Leibnitz began to discover his Knowledge of his differential method will see yt the things contained in this book were invented before the writing of those Letters. For the first Propotion is set down in those Letters enigmatically.

NOTES

(1) Add. 3968(30), fo. 440v. This is the fourth of a group of drafts. The first of them (439r) begins:

Sr

Since you shewed me the passage in Mr. Leibnitz letter concerning Mr Keill I have discussed the matter with Mr Keil....

But then Newton made a mark thus // and continued after it

That by your Letter to Mr Collins dated [*blank*] & printed by Dr Wallis he was convinced that you did not then use the methodus differentialis // That the next year in my letter dated [*blank*] & communicated to you by Mr Oldenburg...(etc.)

Here Newton addresses himself directly to Leibniz, launching into a long defence of his priority on the basis of his correspondence and other papers, somewhat in the manner of Letter 843*a*.

It is hardly possible to doubt that these drafts are linked with Letter 830, or that they represent

Newton's reaction to Letter 822 from Leibniz. The fact that Newton cannot recall the relevant dates is indicative of his coming fresh to the business.

(2) Presumably George Cheyne (1671–1743), M.D. Aberdeen 1701, who practised in London and Bath but was chiefly notable as the author of *Philosophical Principles of Religion Natural and Revealed* (London, 1705; completed in the second edition, London, 1715); see Hélène Metzger, *Attraction Universelle et Religion Naturelle chez quelques commentateurs anglais de Newton* (Paris, Hermann, 1938), pp. 140–53. Cheyne had published at the request of his medical teacher Archibald Pitcairne *Fluxionum methodus inversa* (London, 1703) which, according to David Gregory, provoked Newton to publish his *De Quadratura* with *Opticks* in the following year; and also stimulated an attack from Abraham de Moivre (*Animadversiones in D. Georgii Cheynei Tractatum*, 1704). For (adverse) comments by Gregory on Cheyne see W. G. Hiscock, *David Gregory, Isaac Newton and their Circle* (Oxford, 1937), pp. 15–20, with a letter from Cheyne to Gregory about his book, pp. 43–5. See also vol. II, p. 153, note (26) and vol. III, p. 228, note (1).

(3) John Craige (d. 1731), F.R.S. 1711, author of *Methodus figurarum...quadraturas determinandi* (London, 1685) and *Tractatus mathematicus de figurarum curvilinearam quadraturis...* (London, 1693); see vol. II, p. 501, note (1) and vol. III, p. 9, note (5). Craige, but not Cheyne, used the Leibnizian notation although closely associated with Newton. Craige's name is often given as Craig (as in vol. III); but as he himself used Craige, we prefer to use this spelling.

831 ARBUTHNOT TO FLAMSTEED

6 APRIL 1711

From the original in the Royal Greenwich Observatory.(1)
For the answer see Letter 833

Sir

I send you what is left of The duplicate Catalogue in Sir Isaac Newton's Hands(2) The rest having been long since deliverd to your Kinsman Mr Hodgeson, I hope you will find me what is wanting with all speed, being I am call'd upon to end it which makes me the more importunate This is from

Sir

Your Most humble servant

Jo: Arbuthnott

London
Aprile 6*th*
1711

NOTES

(1) Flamsteed MSS, vol. 55, fo. 101; printed in Baily, *Flamsteed*, p. 282. Flamsteed received this letter only on 14 April, when he returned home, and replied to it on the 19th. He added the following explanation: 'came to Greenwich ☿ [*recte* Tuesday] Apr 10 mane when I was goeing to London where I stayd till ♄ [Saturday] noon ye 14 & was to wait on ye Dr twice ♀ 13. twice but found him both times absent'. According to Flamsteed's Diary (Baily, p. 227) he also wrote a note to Arbuthnot on the 12th, of which no copy survives.

(2) See Letter 828, note (2).

832 ARBUTHNOT TO FLAMSTEED

16 APRIL 1711

From the original in the Royal Greenwich Observatory.(1)

For the answer see Letter 833

Sr

I think I undertook & promis'd to you that Your Catalogue of the fixd starrs should be Correctly printed, & that I would take the blame upon me if it was not: how much was printed at that time I really did not know: but I will be faithfull to my promise.(2) You complain that by the Alteration of ptolomys names All the Antient Observations will be render'd useless. as to that I can answer that ptolomys names are religiously adher'd to as farr as is consistent with the order that is observ'd in the Brittanick Catalogue(3) which differs very much from ptolomys & to please you every wher *preceeding* & *following North* & *South upper* & *lower* are putt instead of right & left for you know you were not well satisfyd with ptolomys postures. & to remove the objection I have compard one of the Constellations mentiond with Ptolomys Catalogue & will undertake immediately to find any starr of ptolomy in your Catalogue. so that I cannot see how the antient observations are renderd useless. you know ptolomys order & number too differs much from yours, but the references are as plain as can be. as to some Alterations in the numbers of your Catalogue they are plainly Corrected in this & what to be sure you will stand to because they are slips of the pen or Computation in the Copy which you gave us & which this was printed from. but that I may still proceed wth all Candour in this matter I beg still that you would send me the constellations that are wanting; that they maybe inserted & if after the Catalogue is printed you don't agree to the Correction, upon a just representation of the exceptions; your own shall be printed just as it stands: & you shall correct it your self. If you will not agree to this to send me the Constellations that are wanting we must be contented with what we have & be at the pains to Compute them from your observations, which is a little hard Considering that you can supplye them & have promised so to do. I can say no more to this matter but beg your Answer as soon as possible being with all respect

Sr

Your Most humble servant

Jo. Arbuthnott

London: Aprile $\frac{16}{1711}$

NOTES

(1) Flamsteed MSS., vol. 55, fos. 103–4, printed in Baily, *Flamsteed*, pp. 282–3.

(2) The copy of his catalogue for the latter six signs of the zodiac that Flamsteed found awaiting him at Greenwich on the 14th was obviously his own MS catalogue; but early in April Flamsteed had gained access to the first and third *printed* sheets only of his star catalogue. Thus, as Arbuthnot now admits, the secret was out. Flamsteed was infuriated by the changes introduced into the printed text, and knew that Halley (whom he loathed) was responsible for them. The denotation and nomenclature of the stars now became his prinipal grievance against Newton, Halley, and their friends. Clearly, he had already made Arbuthnot aware of his dissatisfaction—now made specific—possibly in his lost note of 12 April.

(3) That is, the *Historia Cœlestis Britannica*.

833 FLAMSTEED TO ARBUTHNOT

19 APRIL 1711

Extracts from the copy in the Royal Greenwich Observatory.[1]
Reply to Letters 831 and 832; for the answer see Letter 834

Sr

I met with yours of ye 6th Instant[2] at ye Observatory when I returnd from London, on saturday last[3] & with it ye Copy of my Catalogue of ye fixd Stars, for ye six latter Stars[4] of ye Zodiack, ye rest you tell me was left long since with Mr Hodgson, he assures me ye never came to his hands and I am as sure they never came to mine; On ye left hand side of this Copy I had caused Ptolemys Greek names &c be wrote against ye Stars to which ye belong; in my Catalogue; This I am apt to believe is ye true reason yt part is detained, and you are told it was returned; for had you seen them you would have seen wt an outrageous fault Dr Halley has committed in altering ye Ptolemaick names in my Catalogue; I lent a fair Ptolemy to Sr. I: Newton; I believe tis still in his hands; if you please you may do well to send for it & Collate my Translation: & Dr Hs with it for your own satisfaction.

In yrs of ye 16 Instant you acknowledge your promise yt my Catalogue should be correctly printed: I have seen as yet only ye 1st & 3rd Sheets of it in ye 1st I have noted more then forty alterations & deviations from my Copy and as many in ye 3d wth which I am apt to think you hav not been acquainted. I could wish you would order all ye Sheets printed of to be sent me, yt I might give you all ye faults made in my Works by this confident person (all together).

You tell me that Ptolemys names are religiously adhered to. I fear you write only by hear say; for if you please to compare his Greek and my translation with Dr Halleys, you will find every where notorious differences his names being sometimes contrary to Ptolemys, & sometimes not approaching sense...

Ulug Beig follows Ptolemy strictly in his Arab Catalogue, So does Copernicus, Clavius, & Tycho...Hevelius follows Ptolemy to[o] most commonly... Kepler & Bulialdus, Copy Tycho so do ye Catalogues I have seen printed in French and Spanish: Now I believe Ulug: Bieg Copernicus Tycho &c to have been as wise and as Candid persons as any yt have lived since, and therefore I adhære religiously to them...I have now spent 35 years in [the] composeing & Work of my Catalogue which may in time be published for ye use of her Majesty's subjects and Ingenious men all ye world over: I have endured long and painfull distempers by my night watches & Day Labours, I have spent a large sum of Money above my appointment, out of my own Estate to compleate my Catalogue and finish my Astronomical works under my hands: do not tease me with banter by telling me yt these alterations are made to please me when you are sensible nothing can be more displeasing nor injurious, then to be told so.

Make my case your own, & tell me Ingeniously, & sincerely were you in my circumstances, and had been at all my labour charge & trouble, would you like to have your Labours surreptitiously forced out of your hands, convey'd into the hands of your de[c]lared profligate Enemys, printed without your consent, and spoyled as mine are in ye impression? would you suffer your Enemyes to make themselves Judges, of what they really understand not? would you not withdraw your Copy out of their hands, trust no more in theirs and Publish your own Works rather at your own expence, then see them spoyled and your self Laught at for suffering it.

I see no way to prevent ye evill Consequences of Dr Halleys conduct. But this) I have caused my Servant to take a new Copy of my Catalogue, of which I shall cause as much to [be] printed off as Dr Halley has spoyled, and take care of ye correction of ye press my Self, provided you will allow me ye nameing of ye Printer, and yt all ye last proof Sheets may be sent to Greenwich at my Charge, by ye penny post and not printed off, till I have seen a proof without faults, after which I will proceed to print ye remaining part of ye Catalogue as fast as my health and ye small help I have will suffer mee.

But if you like not this I shall print it alone, at my own Charge, on better paper & with fairer types, then those your present printer uses; for I cannot bear to see my own Labours thus spoyled; to ye dishonour of ye Nation Queen & People: If Dr Halley proceed it will be a ref[l]ection on ye Pr. of ye R. So. and your self will suffer in your reputation, for encourageing one, of whome ye wisest of his Companions used to say: yt ye only way to have any buiseness spoyled effectualy was to trust it to his management.

But I hope better things of you and yt you will endeavour to make me easy after all my long painfull & chargeable Labours by affording me your assistance, as occasion shall serve whereby you will ever obliege Sr

Your humble Servant & sincere friend
JOHN FLAMSTEED

NOTES

(1) Flamsteed MSS, vol. 53, fos. 95–6, printed in Baily, *Flamsteed*, pp. 283–4.
(2) Letter 831.
(3) 14 April.
(4) Read: 'Signs'.

834 ARBUTHNOT TO FLAMSTEED

21 APRIL 1711

Extracts from the copy in the Royal Greenwich Obseravtory.[1]
Reply to Letter 833

Sr

I told you nothing in my Letter as to ye Catalogue I sent but wt was told me and Sr Is. Newton does stand to it yt Mr Hodgson did take away some of ye Catalogue, how much he cannot precisely tell, but yt was all yt was left, so yt matter they may clear between them.

I think I told you in my Letter that I had compared some part of Ptolemy's Catalogue, with the translation in ye edition of ye Britannick Catalogue, and found them to agree, bateing ye exceptions yt I had mentioned & it seems a little hard to say after yt that my Information was only upon hear say; for I made use of your book that is in Sir Is Newtons hands...

I can assure you for my self yt I have no design to rob you of ye fruit of your labours but to make ye Catalogue correct, so as it may be fitt to appear in Publick; and if you would have given us a compleat one it should have been done long ago, but since you are not pleased to do so I will not delay any longer but take ye same method to make out ye rest of the Catalogue yt you have done, which is to employ people to Calculate from the observations what is wanting; and why we should not succeed as well in this piece of Journey Work I cannot imagine; and if after all you do not like ye performance you shall be free to print your own. I promised to send you a Copy of the sheets before ye Catalogue is published; and so I will, and wether you send me the remaining part of ye Catalogue or not I will keep my promise; but I cannot but say it is a little hard yt when you can so easily supply wt is wanting you will not so far gratifie those concerned as to let it be printed perfect in this manner, and then it shall be reprinted changed or altered which way you please, I shall give you no further trouble in the matter being

Sr

You most humble servant

JO: ARBUTHNOTT[2]

London Apr 21

1711

NOTES

(1) Flamsteed MSS, vol. 53, fo. 97. The original is in Flamsteed MSS, vol. 55, fos. 107–9, printed in Baily, *Flamsteed*, p. 285.

(2) Flamsteed replied on the 24th with a list of Halley's specific enormities; Arbuthnot acknowledged this on 15 May and enclosed Halley's reply. A week later Flamsteed again wrote to Arbuthnot: 'All the designed mistakes and falsities [Halley] so rudely charges me with, are all entirely the mistakes and falsities of himself, and his excuses trifling...' (Baily, *Flamsteed*, p. 289). Arbuthnot did not reply.

835 THE MINT TO THE TREASURY

25 APRIL 1711

From the copy in the Public Record Office[1]

To the Rt Honble: the Lords Commrs: of Her Majestys Treasury

May it Please Your Lordships

In Obedience to Your Lordships verbal Order concerning the Value of several Standards for Plate we humbly lay before your Lordships[2]

That old Plate of eleven ounces two pennyweight of fine Silver in the pound weight Troy being of the same standard with the money is of the same value wch: is 5*s*: 2*d* per ounce.[3]

That new Plate of eleven ounces ten pennyweight of fine Silver in the pound weight Troy is worth 5*s*: 4*d* & $\frac{26}{111}$ parts of a penny or about 5*s*: $4\frac{1}{4}d$ per ounce.[4]

That Plate of Ten ounces 15 pennyweight of fine Silver in the pound weight Troy[5] would be worth 5*s* & $\frac{5}{111}$ parts of a Penny, or 5*s* and about the sixth part of a farthing per ounce. And if the Fraction which is inconsiderable be neglected, it may be reckoned at 5*s* per ounce.

That Plate of ten ounces four pennyweight of fine Silver in the pound weight

Troy[6] would be worth 4*s*: 9*d*, wanting about the 9th part of a farthing per ounce. And if the fraction wch: is inconsiderable be neglected, it may be reckoned worth 4*s*: 9*d* per ounce.

That Plate of eleven ounces of fine Silver in the pound weight Troy[7] would be worth 5*s*: 1½*d* per ounce abating $\frac{13}{222}$ parts of a penny or a quarter of a Farthing.

And we further take leave to make as near an Estimate as we can of the value of the Plate already made

That old Plate is worth about 5*s* & a farthing per ounce for an abatement must be made of about 6*d*. or 7*d*. per £wt for the worseness of the plate; & 10*d*. or 12*d* per £wt for the worseness of the Soder, & 1*d*. for the Charge of melting it into Ingotts & 2½*d*. or 3*d*. per £wt for the Waste in that melting, that is in all about 21*d* per £wt or seven farthings per ounce.

That new Plate according to the best of our Judgements is worth about 5*s*: 2¼*d* per ounce, for an abatement must be made of about 8*d*. or 10*d*. per £wt for the worseness of the Plate, & 10*d*. or 12*d*. per £wt for the worseness of the Soder and 1*d*. per £wt for the charge of melting it into Ingotts, & 2½*d*. or 3*d* per £wt for the Waste in that melting, that is, in all about 2*s*. per £wt or 2*d*. per ounce.[8]

All wch: is most humbly submitted to your Lordships great Wisdom

C. P[EYTON]. Is: N. I. ELL[IS].

25*th Apl:* 1711.

NOTES

(1) Mint/1, 7, 57, a clerical copy.

(2) This is the first indication of a new policy of encouragement to owners of silver plate to bring it into the Mint to be coined. See Number 837 below, note (2).

(3) That is, sterling silver of 925/1000 fine.

(4) Sterling silver was plate-marked with the lion passant; new standard plate (9584/10000 fine) with the figure of Britannia, whence such plate is usually described as 'of Britannia standard'. The Britannia standard was compulsorily in use only from 1696 to 1720, though it has remained lawful since. The denominator 111 occurs in the odd fractions of a penny because the old standard of silver contained 222 dwt. fine per ounce.

(5) 8958/10000 fine.

(6) 850/1000 fine.

(7) 9166/10000 fine.

(8) As set out in the Warrant of 10 May (Number 837) the prices are about 5*d* per ounce higher than those suggested by Newton as economic; or about threepence above the normal rate for sterling silver of 5*s* 2*d* per ounce. Hence, if the Mint ordinarily coined plate at a working loss approaching twopence an ounce, the prices in the Warrant increased this loss to fivepence per ounce.

836 LOWNDES TO THE MINT

8 MAY 1711

From the original in the Public Record Office.[1]
For the answer see Letter 845

Whitehall Treasury Chambers 8*th. May* 1711

The Lords Commrs. of her Mats Treasury are pleased to referr this Proposall to the Warden Master & Worker & Comptroller of her Mats. Mint who are to consider the same and Report to their Lopps their Opinion what they thinke fit to be done therein.

Wm Lowndes

[The Proposal enclosed]

To the Right honble the Lords Comers. of her
Majesties Treasury

My Lords

If your Lordps do think fitt to send any Tin for the Streights.[2] I humbly propose to take any quantity not exceeding one hundred and fifty Tuns upon the following Terms viz

To pay to your Lordps for the quantity of Tin that shall be deliverd at Leghorne into the hands of Messrs Arundell Bates and Company and at Genoa into the hands of Messrs Guillaume Boisser & Company at the rate of four pounds Ten shillings per hundred weight viz
for $\frac{1}{3}$ part as soon as the Tin is deliverd there
for $\frac{1}{3}$ part three months after and
for $\frac{1}{3}$ part Six months after
provided the quantity deliverd at Genoa doth not exceed one third part of the whole I am

Your Lordps
Most obed. humble Serv
Theod: Janssen

10 *March* $171\frac{0}{1}$

NOTES

(1) T/1, 134, no. 9A. The Treasury letter is written on the reverse of that from Sir Theodore Janssen (1656–1748). Of French birth, he had been naturalized in 1685. He was one of the great merchants, Director of the South Sea Company and closely associated with the government. In a later letter (Mint Papers III, fo. 556) Newton calculated that this contract ended in a loss to the Mint.

(2) The implication is that the tin would be shipped in naval vessels. This had been done before, to the annoyance of the Levant Company.

837 ROYAL WARRANT TO THE MINT

10 MAY 1711

From a copy in the Mint Papers[1]

Anne R

Whereas Our Commons in Parliament assembled Did on the first Instant resolve that for encourageing the bringing Wrought Plate into Our Mint to be Coined,[2] there should be allowed to such persons as should so bring the same after ye rate of five Shillings and five pence per ounce for the Old Standard, & five Shillings & Eight pence per ounce for the New Standard for all Plate,[3] on which the Mark of ye Goldsmiths Company of London or any other City is Set, & for Uncertaine Plate not so Marked (being reduced to Standard) after the rate of five Shillings and Six pence per ounce. And whereas Our said Commons by theire Address of the fifth instant have besought Us to give direcon to the Officers of Our Mint to receive in all such Wrought Plate as should be so brought to them & to give Receipts to such persons as should bring ye same for ye Amount thereof at ye several Rates and Prices aforesaid, And that ye same may be immediatly Coined into Shillings and Sixpences Our Will & Pleasure is and We do hereby Authorize and Command That You the Warden, Master and Worker and Comptroller of Our Mint in the Tower of London Do take and Receive from all persons and Bodys politick or Corporate all such Wrought Plate as they or any of them shall bring to Our said Mint of ye kinds & Standards above menconed And that You do give such Receipts for ye same Respectively as are desired by ye said Address of Our Commons and forthwith cause the same to be melted down and Assayed & That You the Master and Worker of Our said Mint do immediately Coin the same into Shillings and Sixpences and pay the Monies produced into the Receipt of Our Exchequer & take Tallies for Your discharge And for so doing this shall be to You & all others herein concerned a sufficient Warrant. Given at Our Court at St. James this tenth day of May 1711 In the tenth Yeare of Our Reigne

By Her Majts Command

Poulet[4]
J. Paget[5]
T. Mansel

To the Warden, Master & Worker
& Comptr of Our Mint now
and for ye time being

Warrt for takeing in Plate

NOTES

(1) Mint Papers II, fo. 508; another copy is at *ibid.*, fo. 539. The Warrant was drafted by Newton himself (see his memorandum below of 25/26 July) and there are many related papers e.g. *ibid.* fos. 526, 528, 536, 562.

(2) The normal obligation of the Mint was to accept suitable silver, returning to the owners coin minted at the rate of 5*s* 2*d* per fine ounce (925 parts of silver per 1000). But bullion or plate was, in fact, rarely thus offered to the Mint. From 1701 onwards the Mint relied for its supply of silver for minting on the compulsory offer of silver from native lead mines, amounting to a few thousand pounds' worth of coin annually, supplemented by special measures approved by the Government, such as the minting in 1704 of part of the treasure taken at Vigo Bay two years before. In April 1709 the Mint had been granted £6000 in order to enable it to pay 5*s* 4½*d* per ounce for foreign coin, or for plate, brought to the Mint between 20 April and 1 December 1709. Coin to the value of £78811 was minted as a result, little more than half of what had been hoped for.

The policy of the new administration which took office in 1710 was to push the premium offered to those selling plate to the Mint slightly higher; these 'Importers' of plate to the Mint were not to be paid in cash, however, but in certificates which were to be accepted as subscriptions to the proposed new government loan (see the Resolution of the House of Commons, 1 May 1711, and its Address to the Queen of 5 May which was accepted on the 10th; *Journal of the House of Commons*, **16** (1803), 623, 629, 658). The loan for two million pounds was finally provided for by an Act of Parliament which received the Royal Assent on 8 June 1711.

Difficulties were caused to the Mint both by the swift inrush of plate, and the failure of the House of Commons to complete the necessary legislation.

(3) Compare Letter 835.

(4) John Poulett (1663–1743), fourth Baron, created Earl after the Treaty of Union; F.R.S. 1708. He was appointed First Lord of the Treasury when the office of Treasurer was placed in commission after the fall of Godolphin in August 1710.

(5) The Hon. Henry Paget and Sir Thomas Mansell were also members of the Treasury Commission.

838 THE MINT TO THE TREASURY

14 MAY 1711

From the original in the Public Record Office[1]

May it please your Lordps

In obedience to your Lordps Order[2] signified to us by Mr Lowndes that we should prepare an account of what Quantity of Plate has been brought in upon the Resolutions of the house of Commons[3] & the value thereof & lay the same before your Lordps this morning: We humbly represent that on friday[4] when we began to receive & on saturday there was brought in 71584 oz 10 dwt of old Plate & 35767 oz of new. The first at 5*s* 5*d* per ounce comes to 19387£. 9*s*. 4½*d*

The second at 5*s* 8*d* per ounce comes to 10133£. 19*s*. 8*d*. And both together come to twenty nine thousand five hundred twenty one pounds nine shillings. And there lies about two or thre[e] thousand pounds worth of Plate at the Mint to be weighed this morning. We are

My Lords

Your Lordps most humble & most obedient Servants

C. PEYTON

IS. NEWTON

JN ELLIS

Mint Office
May 14 1711

NOTES

(1) T. 1/133, no. 74; this is in Newton's hand, written on the morning of Monday 14 May.
(2) We have not come across this Order.
(3) See the Royal Warrant of 10 May.
(4) 11 May.

839 LOWNDES TO THE MINT

14 MAY 1711

From the original in the Mint Papers[1]

Gentlemen

The Lords Commrs of her Majts Treasury desire you to forbare the taking in any more Plate pursuant to the last Order in that Behalf[2] (the Value whereof is to be accepted as money upon any the funds or Supplys given in this session of Parliamt) unless the persons bringing in the same do declare under their hands that they do not expect the Value of the Plate from henceforth to be tendred by them shall be accepted as part of the first payment of the sum for which there is a Bill now depending in the house of Commons[3]

I am

Gentlemen

14*th May* 1711

Your most humble servt

WM. LOWNDES

NOTES

(1) Mint Papers II, fo. 543; it was written on the evening of the 14th after receipt of the previous letter.
(2) That is, the Warrant of 10 May.
(3) Although this instruction removed the privilege of having the premium value of plate treated as a subscription to the public loan, it did not alter the values of plate set out in the

Warrant. Hence, should further plate be offered to the Mint at these values after 14 May (as proved to be the case) no means existed for making up any difference between what was paid by the Mint for plate, and the actual coin product of the plate bought. Moreover, the Warrant instructed Newton to transfer all the new coin to the Exchequer, hence he had no source of cash for paying the vendors of plate.

840 NEWTON TO THE TREASURY

[15 MAY 1711]

From a draft in the Mint Papers(1)

May it please your Lordps

The plate taken in yesterday, after the rates of 5*s* 5*d* & 5*s* 8*d* per ounce, amounts to about 9760 pounds: wch with 29521 pounds worth taken in on friday & saturday(2) makes the whole value about 39281£, I cannot yet be exact in the summ because our Clerks differed a little in their recconing. The uncertain plate may further amount to about 4000£, & some parcells of markt plate wch were lodged in the Mint but not yet weighed because the parties were absent when it came to their turn, may amount to about 1000£ besides some parcels brought in last night after we had done weighing wch I think do scarce amount to 1200£

What shall be done thereupon is most humbly submitted to your Lor[dshi]ps great wisdome(3)

NOTES

(1) Mint Papers II, fo. 517; Newton's holograph draft.

(2) 11 and 12 May.

(3) The account of what happened is related in Newton's memoranda of 24 and 25/26 July (Numbers 858, 859). For the time being, the Mint simply issued receipt certificates to the importers of plate, pending a decision on the manner of their reimbursement. See further Letter 855.

841 FLAMSTEED TO SHARP

15 MAY 1711

Extracts from the original in the Library of the Royal Society of London(1)

The Observatory May: 15 ♂ 1711

...March ye 19th last. I receaved a letter from Dr Arbuthnot(2) one of ye Queens Physitians signyfieing yt ye Copy of a part of My Catalogue which had been delivered into Sr I.N. hands at his desire sealed up was now in the Doctors. who desired that I would give him 4 Constellations yt were wanteing in it with ye Variations &c for ye rest for when the parte of my Catalogue was put into Sr I Ns hands March ye 15 1705/6 these constellations were not begun & the rest imperfect which tho Sir Isaack knew very well he still persisted to have ye keepeing of it in his hands sealed up that as he said he might have all things in his power. he would not suffer any of my sheets to be printed till he had gained this point & I was forced to yeild it that he might Not pretend & say ye Prince would have printed my works & I hindred it my selfe.

How ye press went on & Sr I.N. hindred its progress by continual shuffles & tricks you have been informed formerly. I shall onely to tell you more that ye press had wrought of a 98 sheets of the first Volume on Octob 21. 1707. that wee met on March 20. 1707/8 & then Sr Isaack had opened the Catalogue & desired me to insert the Magnitudes of ye stars to their places, for they had not allwayes been inserted in it, & of 173 *lb* I had disburst ordered 125 *lb* to be paid me wch with some trouble I got sometime after

But at ye same time he got a second & More compleat Copy of ye Eclipticall Constellations then shewed him into his hands of which Dr Hally returnd but ye 6 latter signes into Mr Hodgsons hands about a moneth agone(3) the other halfe is lost or Dr Hally deteins it with Dr Arbuthnots privity for they are both of one Church.

⊙ March ye 25th last past I was informed by a freind that my Catalogue was in ye press & some sheets of it printed of on ye 29th I met Dr Arbuthnot at Garways who affirmed there was not a sheet printed, but Aprill ye 2d I got ye printed 1st Sheet & soon after ye 3 wherein I found yt Many of ye Names I used which were translated from Ptolemy...were altered...some Names Made Nonsense some stars omitted others inserted in improper places. I learnt further yt Dr Halley lookt after ye press & was ye Author of all this confusion. till I knew this I was willing to have filled up ye Copy of the Catalogue, but perceaving hereby yt Halley was minding to spoyle ye Work & with more views then one or two, I sent to Dr Arbuthnot an Account of his Villa[i]nous outrage, & desired he would permit me to print my own Catalogue at my own Charge

... ...

...this action of Hallys has exposed him to all ye town & they forbear not to say he is an impudent f[ool] & k[nave] Sir Is: turnes of the blame Conningly on Arbuthnot & Hally is very Willing to take it all on himselfe to oblige his Master who made him 15 years agone Controller of ye Mint at Chester,(4) but everyone sees his craft and loaths him for it:...

Ever your freind to love & serve you
JOHN FLAMSTEED MR

NOTES

(1) Sharp Letters, no. 76; printed in Baily, *Flamsteed*, pp. 291–2.

(2) Letter 823.

(3) See Letter 831.

(4) Halley was appointed Deputy Comptroller at the Mint in Chester in 1696—not necessarily owing to Newton's patronage.

842 NEWTON, SLOANE, MEAD TO FLAMSTEED

30 MAY 1711

From the original in the Royal Greenwich Observatory(1)

Sir,

By Virtue of Her Majesties Letter to Us directed, dated the twelveth Day of December 1710.(2) We do hereby order and direct You to observe the Eclipses of the Sun and Moon this Year, and particularly that of the Sun of July ye 4th Ensuing; and We desire You to send such Your Observations to us at Our Meeting at the House of the Royal Society in Crane-Court in Fleet-street. We are

Yr. humble Servts.

IS. NEWTON P.R.S.

HANS SLOANE S.R. Secr.

RI[CHAR]D MEAD

Crane-Court, *May* 30. 1711.

To the Revd

Mr John Flamsteed, at the
Royal Observatory
at Greenwich

These

NOTES

(1) Flamsteed MSS, vol. 55, fo. 115; the order is also entered in the *Royal Society Council Minutes* (1711), p. 189, for 30 May 1711. It was written in pursuance of an order made by Newton, from the Chair, at the meeting of the previous week, 24 May.

(2) The Royal Warrant, Number 814.

843 SLOANE TO LEIBNIZ

[MAY 1711]

From the Letter-Book of the Royal Society.(1)
Reply to Letter 822; for the answer see Letter 884

Clarissime atque Doctissime Vir

Epistolam tuam a Domino Louttono Accepi et cum Societate Regia Communicavi, Interfuit Consessui(2) illi Dominus Keill ex America nuper redux qui palam professus est Actorum Eruditorum Lipsiæ Autores sibi occasionem dedisse ea Scribendi quæ in Transactione illa Philosophica Reperiuntur, et de quibus Conqueritur Illustris Vir, hujus rei rationem eum ut Literis consignaret rogavit Societas Regia quam ad te transmittendam æquum Judicavit.(3) Cæterum me Credas Tibi omnibus modis inservire Semper esse paratissimum.

HANS SLOANE

Translation

Famous and learned Sir,

I received your letter from Mr. Loutton [?] and I communicated it to the Royal Society. Mr Keill was present at the meeting,(2) having recently returned from America; he openly asserted that the authors of the *Acta Eruditorum* of Leipzig had given him grounds for writing what is to be found in that *Philosophical Transaction*, of which complaint is made by yourself, illustrious Sir; the Royal Society asked him to express in a letter his account of this affair, which it was thought fair to send on to you.(3) For the rest believe me always most ready to serve you in everything.

NOTES

(1) LBC 14 fo. 313–14. This letter was read and approved at a meeting on 31 May. The following holograph draft (U.L.C., Add. 3968(30), fo. 442v) seems to represent an attempt by Newton to prime Sloane:

> I shewed your Letter to Sr Is. Newton at [*del*: before] a meeting of ye R. Society [*del*: & upon his discoursing the matter wth Mr Keil,] & Mr Keil shewing [*del*: shewed] some things in the Acta Leipsica wch gave occasion to what he wrote in the Transactions [*del*: & chose rather to write the inclosed answer to yours then to retract & beg pardon] he was hereupon desired by the Society to draw up an Account of that matter, wch account I herewith send you.

A similar draft in U.L.C. Add. 3965(10), fo. 116v, adds after 'Acta Leipsica': '& particularly the account given of his Quadratura Curvarum Ann. 1705'.

(2) See above, Letter 830, note (7), and the Introduction, p. xxiv.

(3) Letter 843*a* which follows was presumably enclosed with the present letter.

843a KEILL TO SLOANE FOR LEIBNIZ

[MAY 1711]

From a clerical copy among Newton's papers in the University Library, Cambridge[1]

Joann: Keill A.M.
ex Æde Christi Oxon. & R.S.S. Epistola ad
Clarissimum & Doctissimum virum Hans Sloane M.D.
Regiæ Societatis Secretarium

Cum Dni. Leibnitii Epistolam[2] mecum vir clarissime communicare dignatus sis; ea etiam quæ mihi visum fuerit rescribere, ne graveris accipere. Sento virum egregium accerrime de me queri, quasi ei injuriam fecerim, & rerum a se inventarum gloriam alio transtulerim; Fateor querelam hanc ideo mihi molestam esse, quod nolim ea sit de me hominum opinio, quasi ego calumniandi studio, cuiquam in rebus Mathematicis versanti, nedum viro in ijsdem versatissimo, obtrectarem; certe nihil ab ingenio meo magis alienum est, quam alterius laboribus quicquam detrahere.

Agnosco me dixisse[3] Fluxionum Arithemeticam a Dno. Newtono inventam fuisse, quæ mutato nomine & notationis modo a Leibnitio edita fuit, sed nollem hæc verba ita accipi, quasi aut nomen quod methodo suæ imposuit Newtonus, aut Notationis formam quam adhibuit, Dno. Leibnitio innotuisse contenderem sed hoc solum innuebam Dnum. Newtonum fuisse primum inventorem Arithmeticæ Fluxionum; seu Calculi Differentialis; cum autem in duabus[a] ad Oldenburgum scriptis Epistolis, & ab illo ad Leibnitium transmissis,[4] indicia dedisse perspicacissimi ingenij viro, satis obvia, unde Leibnitius principia istius Calculi hausit; vel saltem haurire potuit; At cum Loquendi & Notandi formulas, quibus usus est Newtonus, Ratiocinandi assequi nequiret vir illustris; suas imposuit.

Hæc ut scriberem impulerunt Actorum Lipsiensium Editores, qui in ea quam exhibent operis Newtoniani de Fluxionibus seu Quadraturis enarratione[5] diserte affirmant Dominum Leibnitium fuisse istius methodi inventorem, & Newtonum ajunt pro differentiis Leibnitianis, fluxiones adhibere, semperque adhibuisse; id quidem in ijsdem Scriptoribus observatu dignum, quod loquendi & Notandi formam, a Newtono adhibitam, in Leibnitianam passim in eadem enarratione transferunt; de differentiis scilicet & summis, & calculo summatoris loquuntur, de quibus est nullus apud Newtonum sermo; Quasi inventa Newtoni Leibnitianis posteriora fuerint; & a calculo Leibnitij in

(a) Extant in Tertio volumine operum Wallisij

Actis Lipsiensibus Anno 1684 descripto ortum derivarint; Cum revera Newtonus ut ex sequentibus patebit, Fluxionum methodum invenerit, octodecim saltem annos, antequam Leibnitius quicquam de calculo differentiali edidisset Tractatumque[6] de ea re conscripserit, cujus rem specimina[7] quædam Leibnitio ostensa sint; rationi non incongruum est, ea aditum illi ad calculum differentialem aperuisse.

Unde si quid de Leibnitio liberius dixisse videar, id eo Animo feci non ut ei quicquam eriperem, sed ut quod Newtoni esse arbitrabar, auctori suo vindicarem.

Maxima equidem esse Leibnitii in Rempublicam Literariam merita; lubens agnosco; nec eum in reconditiore mathesi scientissimum esse diffitebitur, qui ejus in Actis Lipsiensibus scripta perlegerit; cum autem tantas tamque indubitatas opes, de proprio possideat; certe non video cur spolijs ab alijs detractis, onerandus sit. Quare cum intellexerim populares suos, ita illi favere ut eum laudibus non suis accumulent; haud præposterum in gentem nostram studium esse duxi, si Newtono quod suum est tueri & conservare anniterer. Nam si Lipsiensibus fas fuerit, aliena Leibnitio affingere, Britannis saltem ea quæ a Newtono erepta sunt sine crimine calumniæ reposcere licebit; Itaque cum ad Regiam Societatem appellet vir illustris, meque publice testari velit calumniandi animum a me alienum esse; Ut Calumniandi crimen a me amoveam mihi ostendendum incumbit Dnum. Newtonum verum & primum fuisse Arithmeticæ Fluxionum seu Calculi Differentialis Inventorem; deinde ipsum adeo clara & obvia methodi suæ indicia, Leibnitio dedisse, ut inde ipsi facile fuerit in eandem Methodum incidere.

Sciendum vero primum est, Celeberrimos tunc temporis Geometras Dnos. Franciscum Slusium, Isaacum Barrovium, & Jacobum Gregorium[8] Methodum habuisse qua curvarum tangentes ducebant, quæ a Fluxionum methodo non multum abludebat; & ijsdem principijs innixa fuit. Nam si pro litera o quæ in Jacobi Gregorij Parte Matheseos Universalis[9] quantitatem infinite parvam representat; aut pro literis a vel e quas ad eandem designandam adhibet Barrovius;[10] ponamus $\dot{x}$ vel $\dot{y}$ Newtoni vel dx seu dy Leibnitij in formulas Fluxionum vel Calculi Differentialis incidemus, & regressus quo a data Tangentium proprietate ad naturam Curvæ perveniebant, (quem methodum Tangentium inversam nominabant) eadem plane res erat, ac methodus qua a Fluxionibus ad Fluentes revertitur; Interim suam illam methodum non ultra Fluxiones primas extendebant; neque eandem ad quantitates surdis aut Fractionibus involutas accommodare potuerunt; At priusquam quicquam de hoc argumento, a summis hisce viris publico datum est, Dnus. Newtonus methodum excogitavit, priori quidem non dissimilem, sed multo latius patentem, quæ non substitit ad æquationes eas in quibus una vel utraque quantitas indefinita Radicalibus est involuta, sed absque ullo æquationum apparatu, Tangentem confestim ducere monstrabat; Quæstiones de Maximis & Minimis eodem Artificio tractabat; & speculationes de Quadraturis facilius explicuit; Hæc constant ex Epistola Newtoni ad Dnum. Collinium data Decembris diei 10 Anno 1672 & inter Collinij chartas reperta cujus hæc sunt verba.(11)

'Vehementer Gaudeo Amici nostri Barovij Lectiones Mathematicis extraneis plurimum placuisse; Id etiam haud parum Gaudeo, quod intelligam, eos jam eadem qua ego Methodo in Tangentibus ducendis uti. Qualem ego illorum methodum esse conjicio hoc exemplo facilius disces. Recta *CB* in Curva quavis terminata ad *AB* in quovis dato Angulo applicetur, sitque *AB*, *x*, *BC*, *y*. & relatio inter *x* & *y* quavis æquatione exprimatur, Scil. hac

$$x^3-2x^2y+bx^2-bbx+byy-y^3=0$$

qua curva determinatur. Ad Tangentem ducendam hæc est Regula. Multiplica terminos æquationis per progressionem Arithmeticam, secundum dimensiones ipsius *y*, scil. $\underset{0}{x^3}-\underset{1}{2x^2y}+\underset{0}{bx^2}-\underset{0}{b^2x}+\underset{2}{by^2}-\underset{3}{y^3}=0$ eosdem etiam per progressionem secundum dimensiones ipsius *x* multiplica scil.

$$\underset{3}{x^3}-\underset{2}{2xxy}+\underset{2}{bxx}-\underset{1}{bbx}+\underset{0}{byy}-\underset{0}{y^3}=0$$

prior productus fiet numerator, posterior per *x* divisus fiet denominator Fractionis, quæ exprimit longitudinem *DB* cujus extremitati *D* Tangens *CD* ducenda est. Itaque longitudo *CD*(12) est $\dfrac{-2xxy+2byy-3y^3}{3x^2-4xy+2bx-bb}$ hic est casus particularis, seu potius Corollarium methodi Generalis quæ sine calculi molestia non solum in Tangentibus ducendis, ad curvas omnes sive illæ Geometricæ sunt, sive mechanicæ, sive quacunque ratione ad rectas aut etiam ad alias curvas referantur, sed etiam ad alia magis recondita Problematum genera circa Curvedines, Areas, Longitudines & centra Gravitatum Curvarum se diffundit, Neque hæc Methodus (sicut ea Huddenij de maximis & minimis & consequenter nova illa Tangentium quam Slusium habere suppono)(13) ad eas æquationes restringitur, quæ surdis non implicantur. Hanc methodum alijs immiscui, quibus æquationes ad Series infinitas reducendæ sunt. Memini me cum jam Lectiones

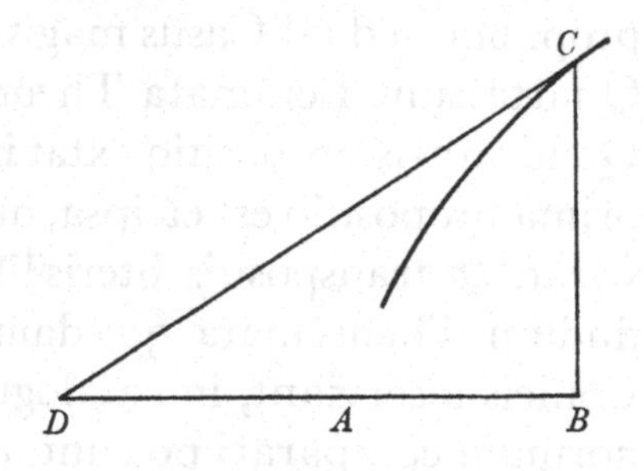

editurus esset Barovius mentionem hujusce methodi fecisse, quam mihi in tangentibus ducendis cognitam esse dixi. Sed sermones nostros nescio quid interpellavit quo minus illam ea Tempestate explicarim.'

Ex hac Epistola, clare constat Dnum. Newtonum methodum Fluxionum habuisse ante annum 1670 eodem nempe quo Barrovij Lectiones editæ sunt.

Anno 1669 misit Newtonus ad Dnum. Collinium Tractatum de Analysi per æquationes infinitas; quem etiam inter schedas Collinij repertum, Dnus. Jones nuper edidit.[14] Sub hujus fine, habetur demonstratio Regulæ pro quadraturis curvarum nata ex proportione augmentorum Nascentium abscissæ & ordinatæ; cum abscissa sit x & ordinata $x^{\frac{p}{n}}$, quæ quidem demonstratio, commune fundamentum est tam Doctrinæ Fluxionum, quam calculi differentialis, ex eo autem Tractatu, non pauca amicis suis communicanda deprompsit Collinius. Unde certum est Dno. Newtono ante illud tempus, Fluxionum Arithemeticam innotuisse. Preterea constat ex posteriore Newtoni ad Oldenburgum Epistola:[15] 'eum suadentibus amicis circa annum 1671 Tractatum de hisce rebus conscripsisse, Quem una cum Theoria Lucis & Colorum, in publicum dare statuerat; scribitque Oldenburgo series infinitas, non magnam ibi obtinuisse partem; seque alia haud pauca congessisse, inter quæ erat methodus ducendi Tangentes quam solertissimus Slusius ante annos duos tresve cum Oldenburgo communicaverit; sed quæ generalior facta, non ad æquationes, quæ surdis aut Fractionibus involutæ sunt hærebat; Et eodem fundamento usum ad Theoremata generalia, Quadraturas curvarum spectantia pervenisse se ait Newtonus; Horum unum Exempli loco in ipsa epistola ponit; Seriem exhibens cujus termini dant Quadraturam curvæ cum abscissa est z & ordinatim applicata $dz^{\theta}\times\overline{e+fz^{n}}|^{\lambda}$.' quæ series abrumpiter & terminis finitis, Curvæ quadraturam comprehendit, quandocunque illa finita æquatione exprimi potest hoc dicit esse primum Theorematum Generaliorum; unde sequitur eum alia ad Casus difficiliores & magis intricatos accommodata habuisse, est autem Theorema illud propositio 5ta. in Tractatu de Quadraturis, eodem etiam spectat ejusdem prop. 6ta. sed ad Casus magis implicatos se extendit.[16] Propositiones Tertia & Quarta sunt Lemmata Theor. hisce demonstrandis præmissa, 2da. autem in Quadraturis propositio extat in Tractatu de Analysi per æquationes infinitas & prima propositio est ea ipsa, quam in dicta epistola fundamentum operationum vocat, & transpositis literis[17] celari tunc voluit. Scribit etiam Newtonus se dudum Theoremata quædam quæ comparationi curvarum cum sectionibus Conicis inserviant, in catalogum retulisse, & ordinatas curvarum quæ ad eam normam comparari possunt, in eadem Epistola describit; quæ profecta eadem plane sunt cum ijs quas Tabula 2da. ad scholium propositionis 10. in Tractatu de Quadraturis exhibet;[18] unde satis liquet Tabulam illam & propositiones

7, 8, 9 & decimam quæ sunt in Tractatu de Quadraturis, (a quibus Tabula pendet) Newtonum dudum invenisse; ante annum 1676 quo scripta est Epistola illa posterior; Cum vero in prima ad Oldenburgum Epistola dicit se ab ejusmodi studijs per Quinquennium abstinuisse,(19) hinc satis clare colligi potest, Propositiones in Tractatu de Quadraturis a Dno. Newtono inventas fuisse Quinquennio saltem antequam Epistolæ illæ ad Oldenburgum scriptæ essent. Totamque illam de Fluxionibus Doctrinam, ante illud Tempus ulterius a Newtono provectam esse; quam ad hunc usque diem a quoquam alio factum est, sub nomine calculi differentialis. Certe Neminem novi qui in hac Provincia peragranda æquis passibus cum Newtono progressus sit, Et pauci sunt, ijque insignes Geometra qui prospicere queant, quo usque ille in eadem Provincia processerit. Præterea in Posteriore illa ad Oldenburgum Epistola modum describit quo in seriem incidit cujus termini Fluxiones seu differentias quantitatum in infinitum exhibent, quæ postquam inventa esset, dicit pestem ingruentem ipsum coegisse hæc studia deserere et alia cogitare,(20) At Pestis illa contigit annis 1665 & 1666; unde patet, etiam ante illud Tempus; Fluxionum Calculum Domino Newtono innotuisse; hoc est duodecim saltem annos antequam calculum suum Oldenburgo communicavit Leibnitius; & novem decem annos antequam vir illustris eandem in Actis Lipsiensibus edidit; & certe ante visas hasce duas Newtoni Epistolas; Leibnitium calculum suum differentialem habuisse, nulla apparent vestigias. His omnibus rite perpensis certissime cuivis constabit, Dominum Newtonum pro vero Inventore Arithmeticæ Fluxionum habendum esse.

Restat Jam ut inquiramus quænam fuere Indicia Leibnitio a Newtono derivata, unde ei facile foret calculum differentialem elicere. Et primo ut dixi nullibi ostendit Leibnitius sibi notum fuisse calculum differentialem, ante visas has duas Newtoni Epistolas, imo ante illud tempus longiore usus est circuitu, cum res facilius multo, & succinctius, ex calculo fluerent differentiali. Hujus rei testis sit Epistola ad Oldenburgum data 18/28 Novembris 1676,(21) quæ in Operum Wallisianorum Tomo 3tio. etiam extat, in qua modum tradit exprimendi rationem subtangentis ad Ordinatam, in terminis quos non ingreditur Ordinata, ubi si loco y & dy ipsarum valores vinculo inclusos posuisset, statim scopum attigisset.

In prima Epistola quæ per Oldenburgum ad Leibnitium transmissa est, Docuit Newtonus methodum qua quantitates in Series infinitas reducendæ sint; i.e. qua Quantitates incrementa exhiberi possunt; In ipso enim initio, seriem ostendit, cujus termini hæc incrementa representant. Sed illa Dnum. Leibnitium prorsus latebat ante visam Newtoni Epistolam qua exponitur.(22)

Sit o incrementum momentaneum quantitatis fluentis x & $\frac{m}{n}$ index dignitatis ejusdem, & si pro x scribatur $x+o$, $x+2o$, $x+3o$, $x+4o$ &c. & Quantitates $(x+o)^{\frac{m}{n}}$, $(x+2o)^{\frac{m}{n}}$, $(x+3o)^{\frac{m}{n}}$, $(x+4o)^{\frac{m}{n}}$ &c. in series infinitas expandantur, habebimus totidem series quarum prima, hæc est quæ sequitur

$$x^{\frac{m}{n}}+\frac{m}{n}ox^{\frac{m-n}{n}}+\frac{m^2-mn}{2n^2}oox^{\frac{m-2n}{n}}+\frac{m^3-2m^2n+2mn^2}{6n^3}ooox^{\frac{m-3n}{n}} \text{ \&c.}$$

In omnibus seriebus primus terminus erit ipsa quantitas fluens $x^{\frac{m}{n}}$ & si prior quælibet series a posteriore auferatur habebimus harum serierum differentias primas. in quibus omnibus primus terminus est seriei primæ terminus primus quem ingreditur quantitas o scilicet $\frac{m}{n}ox^{\frac{m-n}{n}}$. & evanescante o fit ille terminus differentiis hisce primis æqualis; vel quod idem est erit quantitas $\frac{m}{n}ox^{\frac{m-n}{n}}$ fluentis incrementum primum.

Præterea si differentia quælibet prior a posteriori auferatur; deveniemus ad differentias secundas, quarum omnium terminus primus per 2 divisus, idem est cum termino secundo seriei primæ quem ingreditur quantitas o & evanescente o fiunt differentiæ illæ per binarium divisæ singulæ æquales termino illo primo seriei qui est $\frac{m^2-mn}{2n^2}oox^{\frac{m-2n}{n}}$. Et eodem modo inveniemus supra descriptæ seriei terminum $\frac{m^3-2m^2n+2mn^2}{6n^3}ooox^{\frac{m-3n}{n}}$ æqualem esse singulis differentiis tertijs per sex divisis; & quilibet terminus ejusdem seriei ad differentias respectivas semper habebit datam rationem scilicet terminus primus quem ingreditur o æqualis est differentijs primis, 2dus. est differentiarum 2darum. pars media, tertius pars sexta differentiarum tertiarum &c Hasce series quarum[23] termini differentias omnes in infinitum representant, Invenit Newtonus, uti dixi, ante annum 1665, sed illæ ante visam Newtoni Epistolam in qua exponitur, Dnum Leibnitium[(a)] latebant; nam ante illud tempus agnoscit Leibnitius semper ipsi necesse fuisse, transmutare quantitatem irrationalem, in Fractionem rationalem, & deinde dividendo Mercatoris methodo, Fractionem in seriem reducere. Exinde etiam patet seriem hanc differentias continentem non habuisse Dnum Leibnitium; quod postquam ipsi per Oldenburgum ostensa est[(a)] rogat ut Dnus. Newtonus ipsius originem sibi pandat.

Sit jam quantitas quælibet ex constanti & indeterminatis utcunque composita & vinculo inclusa scilicet $(a+bx^c)^{\frac{m}{n}}$ cujus differentia habenda est;[24] constat per Regulam prius traditam quantitatis $a+bx^c$ differentiam esse $cbx^{c-1}o$ (posito quod o sit incrementum momentaneum fluentis x) quare si pro

(a) vide Epistolam Leibnitij ad Oldenburgum 27 Augusti 1676 [N.S.]

$a+bx^c$ scribatur z, & pro cbx^{c-1} scribatur w erit $(a+bx^c+cbx^{c-1}o)^{\frac{m}{n}}=(z+w)^{\frac{m}{n}}$ quæ si per regulam Newtoni in seriem expandatur fit $z^{\frac{m}{n}}+\frac{m}{n}wz^{\frac{m-n}{n}}+$&c cujus seriei terminus $\frac{m}{n}wz^{\frac{m-n}{n}}$ est differentia prima quantitatis $z^{\frac{m}{n}}$ seu $(a+bx^c)^{\frac{m}{n}}$ unde si loco z et w restituantur ipsorum valores, $a+bx^c$ & $cbx^{c-1}o$ habebimus differentiam quantitatis $(a+bx^c)^{\frac{m}{n}}=\frac{m}{n}cbx^{c-1}o\times(a+bx^c)^{\frac{m-n}{n}}$ vel si more Leibnitiano pro o ponatur dx erit quantitatis $(a+bx^c)^{\frac{m}{n}}$ differentia $\frac{m}{n}cbx^{c-1}dx\times(a+bx^c)^{\frac{m-n}{n}}$ ubi vidimus quantitatem differentialem $\frac{m}{n}cbx^{c-1}dx$ extra vinculum semper manere; Atque hinc facile fuit Dno. Leibnitio ope Regulæ Newtonianæ differentias quantitatum omnium exhibere utcunque quantitates fluentes surdis aut fractionibus sint implicatæ. (Quæ ante Epistolicum illud per Oldenburgum cum Newtono commercium ipsi minime nota fuit).

Quamvis hæc per se satis manifesta sunt calculi differentialis indicia; In secunda tamen Epistola quæ per Oldenburgum ad Leibnitium[(25)] missa est; alias adhuc clariores describit Newtonus methodi suæ notas. Dicit enim[(26)] se habuisse methodum ducendi Tangentes quam solertissimus Slusius ante annos duos tresve Oldenburgo impertitus est, ita ut habito suo fundamento nemo posset Tangentes aliter ducere nisi de industria a recto tramite erraret. Quinetiam ibi quoque ostendit 'methodum hanc non hærere ad æquationes quibus una vel utraque quantitas indefinita radicalibus involuta est, sed absque ulla æquationum reductione (quæ opus plerumque redderet immensum) Tangentem confestim ducere, & eodem modo in quæstionibus de Maximis & minimis alijsque quibusdam rem sic se habere. Fundamentum harum Operationum dicit esse satis obvium, quod tamen transpositis literis in illa Epistola, celare voluit; hoc etiam adjicit, hoc Fundamento speculationes de Quadraturis curvarum Simpliciores se reddidisse & ad Theoremata quædam generalia se pervenisse scribit.'

Cum vero methodus Slusiana tunc temporis Leibnitium minime latere potuit; utpote in Actis Philosophicis Lond. publicata;[(27)] Cumque Newtonus dicit eandem & sibi innotuisse, ex fundamento quo habito, non hærebat ad æquationes radicalibus utcunque involutas; (in qua quidem tota rei difficultas posita est). Cumque in priore Epistola seriem descripsit, cujus ope differentiæ haberi possunt, ubi Fluentes surdis aut Fractionibus utcunque implicatæ sunt. cum denique idem Fundamentum ad Quadraturas curvarum se applicuisse dicit, minime dubitandum est hæc omnia facem Leibnitio prætulisse, quo facilius methodum Newtoni perspiceret.

Quod si hæc non suffecisse videantur indicia; etiam Ulterius processit Newtonus, & Exempla Methodi suæ dedit, Et Regulam ostendit qua ex datis quarundem curvarum ordinatis earundem Areæ exhibentur in terminis finitis cum hoc fieri potest, hoc est in stylo Leibnitiano ipsi exempla tradidit quibus a differentijs ad summas pervenitur. Et a simplicioribus orsus, proponit primo[28] Parabolam cujus abscissa est z, & ordinatim applicata $\sqrt{az}=a^{\frac{1}{2}}z^{\frac{1}{2}}$, & curvæ area erit $\frac{2}{3}a^{\frac{1}{2}}z^{\frac{3}{2}}$ hoc est quando differentia Areæ est $dz\times\sqrt{az}$, seu $a^{\frac{1}{2}}z^{\frac{1}{2}}\times dz$ ostendit fore Aream $\frac{2}{3}a^{\frac{1}{2}}z^{\frac{3}{2}}$, unde vicissim concluditur si quantitas differentianda sit $a^{\frac{1}{2}}z^{\frac{3}{2}}$ fore ejus differentiam $\frac{3}{2}a^{\frac{1}{2}}z^{\frac{1}{2}}\,dz$ seu $\frac{3}{2}dz\sqrt{az}$. Exemplum ejus secundum[29] est curva, cujus abscissa est z & ordinatim applicata $\frac{a^4z}{(c^2-z^2)^2}$, ubi ostendit Newtonus curvæ aream fore $\frac{a^4}{2c^2-2z^2}$, hoc est si differentia Areæ sit $\frac{a^4z\,dz}{(c^2-z^2)^2}$, ostendit aream fore $\frac{a^4}{2c^2-2z^2}$. Unde vicissim si quantitas differentiandi sit $\frac{a^4}{2c^2-2z^2}$, concludi potest differentiam fore $\frac{a^4z\times dz}{(c^2-z^2)^2}$ vel si ejusdem curvæ ordinata sic enuncietur $\frac{a^4}{z^3\times(c^2z^{-1}-1)^2}$ erit Area $=\frac{a^4z^2}{2c^4-2c^2z^2}$. Quare & vicissim si quantitas differentiandi sit $\frac{a^4z^2}{2c^4-2c^2z^2}$ erit differentia $\frac{a^4dz}{z^3\times(c^2z^{-1}-1)^2}$.

Hinc ad exempla quædam difficiliora progreditur Newtonus, in ijsque ostendit, quomodo ab ordinatis, hoc est a differentiis ad summas perveniendum sit: ex quibus patebit, Curvam omnem quadrabilem fore, cujus ordinata in differentiam abscissæ ducta, fit quantitatis alicujus differentia; & hinc innumera curvarum Genera assignari possunt etiam Geometrice quadrabilia.

His indicijs atque his adjutum Exemplis, Ingenium vulgare methodum Newtonianam penitus discerneret; ita ut ne suspicari fas sit, eam acerrimum Leibnitij acumen posse latuisse; Quem quidem usum fuisse his ipsis clavibus; ad hæc sua quæ feruntur Inventa, aditum; etiam ex ipsius ore satis elucescit. Nam in Epistola ad Oldenburgum[30] data, post explicatum calculum differentialem exemplum addit, quod coincidere agnoscit, cum Regula Slusiana, & postea addit.[31] 'Sed Methodus ipsa (priore) Nostra longe est amplior, non tantum enim exhiberi potest cum plures sint literæ indeterminatæ quam x & y (quod sæpe fit maximo cum fructu) sed et tunc utilis est, cum interveniunt Irrationales, quippe quæ eam nullo morantur modo, neque ullo modo necesse est irrationales tolli; quod in Regula Slusii necesse est, & calculi difficultatem in immensum auget.' Hæc omnia a Newtono prius in secunda ejus Epistola dicta sunt. Inde Exempla proponit, quorum quidem quod primum est,[32] nescio quo fato, idem prorsus est ac id quod in ea Epistola quam Leibnitio transmiserat Oldenburgus etiam primum protulerit Newtonus.

Mox addit vir illustrissimus,[33] 'Arbitror quæ celare voluit Newtonus de Tangentibus ducendis, ab his non abludere. Quod addit ex hoc Fundamento Quadraturas quoque reddi faciliores, me in hac sententia confirmat; nimirum semper figuræ illæ quadrabiles, quæ sunt ad æquationem differentialem. Æquationem differentialem voco talem qua valor ipsius dx exprimitur, quæque ex alia derivata est, qua valor ipsius x exprimebatur.' Et paulo post, suam de hac re sententiam plenius aperit, dicitque[34] Hanc unicam Regulam pro infinitis figuris quadrandis inservire, diversæ plane naturæ ab ijs quæ hactenus quadrari solebant. Quis est jam qui hæc perpendet & non videbit Indicia & Exempla Newtoni satis a Leibnitio perspecta fuisse; [saltem quoad differentias primas. Nam quoad differentias secundas, Leibnitium methodum Newtonianam tardius intellexisse videtur, quod brevi forsan clarius monstrabo.][35]

Interim facile illustri viro Assentior; & credo eum nec nomen calculi Fluxonium fando audivisse, nec Characteres quos adhibuit Newtonus oculis vidisse, antequam in Wallisianis operibus prodiere.[36] observo enim ipsum Newtonum sæpius mutasse nomen & notationem calculi. In Tractatu de Analysi æquationum per series infinitas, incrementum abscissæ per literam o[37] designat: Et in Principijs Philosophiæ[38] Fluentem quantitatem Genitam vocat, ejusque incrementum momentum[39] appellat; Illam literis majoribus A vel B, hoc minusculis a & b designat.

Id etiam ultra agnosco, inter cætera quæ de re mathematica præclare meritus est Leibnitius, hoc itidem illi deberi, quod primus fuerit qui calculum hunc typis edidit & in publicum produxit,[40] itaque eo saltem nomine magnam apud Matheseos amantes inibit gratiam, quod inventum ita nobile & in multiplices usus deducendum diutius eos noluerit latere.

Habes vir Clarissime quæ de hoc argumento scribenda duxi, unde facile credo percipies, hoc qualecunque fuerit meum in gentem nostram studium, ita parum præposterum fuisse, ut nihil omnino nisi quod Newtoni erat Leibnitio detraxerim; nec dubito quin æqui rerum æstimatores uno ore fateantur me, uti nullo calumniandi animo, ita nec præcipiti Iudicio ea dixisse, quæ tibi tot argumentis luce meridiana clarius comprobavi.

Translation

A letter from John Keill M.A., F.R.S., of Christchurch, Oxford
to the very famous and learned Hans Sloane M.D., Secretary
of the Royal Society

As you have honoured me by imparting to me Mr. Leibniz's letter,(2) famous Sir, be so good as to receive also what I have thought fit to reply. I perceive that that remarkable person has complained bitterly about me, as though I had done him an injury, and transferred to another the glory of things discovered by himself. I confess that this complaint troubles me exceedingly for I would not wish men to form the opinion that out of a desire to broadcast slanders I would disparage anyone who is skilled in mathematics, not to say a person who is most distinguished in that way; nothing certainly is farther from my nature than to depreciate anything in any other man's work.

I admit that I said(3) that the Arithmetic of Fluxions was discovered by Mr. Newton, which was published with a change of name and method of notation by Leibniz but I do not mean these words to be understood as though I were arguing that either the name which Newton gave to his method or the form of notation that he developed were known to Mr. Leibniz; I suggested only this, that Mr. Newton was the first discoverer of the Arithmetic of Fluxions or Differential Calculus; however, as he had in two(a) letters written to Oldenburg (which the latter transmitted to Leibniz)(4) given pretty plain indications to that man of most perceptive intelligence, whence Leibniz derived the principles of that calculus or at least could have derived them; But as that illustrious man did not need for his reasoning the form of speaking and notation which Newton had used, he imposed his own.

I have been impelled to write these lines by the publisher of the *Acta* [*Eruditorum*] of Leipzig, who in the account(5) they have given of Newton's work on fluxions or quadrature expressly affirm that Mr. Leibniz was the discoverer of this method, and they say that Newton employed fluxions instead of Leibniz's differentials, and always has done so; and it is noteworthy in these same authors that in this same account they transpose throughout the forms of speech and notation employed by Newton into the Leibnizian form, for they speak of differences, and sums, and the calculus of summation, of which there is no word in Newton; as though Newton's discoveries were posterior to those of Leibniz and had their origin in the calculus of Leibniz described in the *Acta* of Leipzig for the year 1684. Whereas in truth (as will appear from what follows) Newton had discovered the method of fluxions at least eighteen years before Leibniz had published anything on the differential calculus, and had written a treatise(6) on this topic, [and] as some specimens of it were revealed to Leibniz(7) it is not contrary to reason that these gave him an entrance to the differential calculus.

Whence, if I seem to have spoken pretty freely about Leibniz, I did so not with the

(a) These appear in the third volume of Wallis's works

intention of snatching anything from him but rather in order to vindicate Newton's authorship of what I take to be his own.

Surely the merits of Leibniz in the world of learning are very great; this I freely acknowledge, nor can anyone who has read his contributions to the *Acta* of Leipzig deny that he is most learned in the more obscure parts of mathematics. Since he possesses so many unchallengeable riches of his own, certainly I fail to see why he wishes to load himself with spoils stolen from others. Accordingly when I perceived that his associates were so partial towards him that they heaped undeserved praise upon him, I supposed it no misplaced zeal on behalf of our nation to endeavor to make safe and preserve for Newton what is really his own. For if it was proper for those of Leipzig to pin on Leibniz another's garland, it is proper for Britons to restore to Newton what was snatched from him, without accusations of slander. And so, as the illustrious person appeals to the Royal Society and wishes me to testify publicly that I had no intention of slandering him, to discharge myself of the accusation of slander it falls to me to show that Mr. Newton was the true and first discoverer of the Arithmetic of Fluxions or Differential Calculus, and further that he himself gave such clear and obvious hints of his method to Leibniz that it was easy for the latter to hit upon the same method.

It should first be understood that the most famous geometers of that time, Messrs. François de Sluse, Isaac Barrow, and James Gregory[8] had a method for drawing tangents to curves which did not very much differ from the method of fluxions, and was based on the same principles. For if we substitute the $\dot{x}$ or $\dot{y}$ of Newton or the dx or dy of Leibniz for the letter o which represents an infinitely small quantity in James Gregory's *Pars Matheseos Universalis*,[9] or for the letters a or e which Barrow employs[10] to designate the same, we meet with the formulas of the fluxional or differential calculus and by going backwards from a given property of the tangents they discovered the nature of the curve (and this they called the inverse method of tangents) and this was clearly the same thing as to go from the fluxions to the fluents. However, their method did not reach beyond the first fluxions nor could it be adapted to quantities involving surds or fractions. But before anything of this kind had been published by any of these outstanding men, Mr. Newton had thought out a method not dissimilar from the former but much more broadly conceived, which was not limited to equations in which one or both indefinite quantities are involved in roots, but at once showed how the tangent may be drawn without any apparatus of equations; it dealt by the same trick with problems of maxima and minima and easily unraveled speculations about quadratures. All this is evident from Newton's letter to Mr. Collins dated 10 December 1672, found among Collins' papers, of which the words are:[11]

'...I am heartily glad at the acceptance wch our...friend...Barrow's Lectures finds wth forreign Mathematicians, & it pleased me not a little to understand that they are fallen into the same method of drawing Tangents with me. What I guess their method to be you will apprehend by this example. Suppose CB applyed to AB in any given angle be terminated at any Curve line AC, and calling $AB\ x$ & $BC\ y$ let the relation

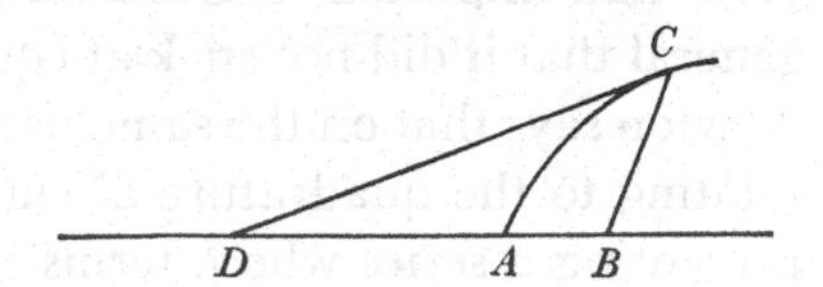

between x and y be expressed by any æquation as $x^3-2x^2y+bx^2-bbx+byy-y^3=0$ whereby the curve is determined. To draw the tangent the rule is this. Multiply the termes of the æquation by any arithmeticall progression according to the dimensions of y suppose thus $\underset{0}{x^3}-\underset{1}{2x^2y}+\underset{0}{bx^2}-\underset{0}{b^2x}+\underset{2}{by^2}-\underset{3}{y^3}=0$ also according to the dimensions of x suppose thus $\underset{3}{x^3}-\underset{2}{2xxy}+\underset{2}{bxx}-\underset{1}{bbx}+\underset{0}{byy}-\underset{0}{y^3}=0$ the first product shall be the Numerator, & the last divided by x the Denominator of a fraction wch expresseth the length of DB to whose end D the tangent CD must be drawn. The length CD(12) therefore is

$$\frac{-2x^2y+2by^2-3y^3}{3x^2-4xy+2bx-b^2}.$$

This is one particular [case], or rather a Corollary of a Generall Method wch extends it selfe wthout any troublesome calculation, not onely to the drawing tangents to all curve lines whether Geometrick or mechanick or how ever related to streight lines or to other curve lines but also to the resolving other abstruser Kinds of Problems about the crookedness, areas, lengths, centers of gravity of curves. Nor is it (as Hudde's method de maximis et minimis & consequently that new method of Tangents which I presume Slusius to possess)(13) limited to æquations wch are free from surd quantities. This method I have interwoven wth that other of working in æquations by reducing them to infinite series. I remember I once...told Dr. Barrow when he was about to publish his Lectures that I had such a method of drawing Tangents but some divertisment or other hindered me from describing it to him [on that occasion].'

It clearly appears from this letter that Mr. Newton had the method of fluxions before the year 1670, that in which Barrow's lectures were published.

In the year 1669 Newton sent the treatise *De analysi per æquationes infinitas* to Mr. Collins, which Mr. Jones published recently after finding it among Collins' papers.(14) Near the end of this there is the demonstration of the rule for the quadrature of curves arising from the proportion of the nascent increases of the abscissæ and ordinates, when the abscissa is x and the ordinate $x^{\underline{p}}$ which demonstration is the common foundation alike of the theory of fluxions and of the differential calculus [and] was extracted by Collins from that treatise for communication to not a few of his friends, whence it is certain that the arithmetic of fluxions was known to Mr. Newton before that time. Moreover it appears from the later epistle of Newton to Oldenburg:(15) 'that he wrote about the year 1671 at the instance of his friends a treatise on these matters which he had resolved to publish together with the Theory of Light and Colours, and he wrote to Oldenburg that infinite series did not figure largely in it; that he had put together not a few other things among which was the method of drawing tangents which the very skilful Sluse had imparted to Oldenburg two or three years before, but made so much more general that it did not stick at equations in which surds or fractions were involved. And Newton says that on the same basis he has arrived at an application to general theorems relating to the quadrature of curves. He gives one example at that place in his letter, presenting a series whose terms give the quadrature of a curve whose abscissa is z and applied ordinate $dz^{\theta} \times (e+fz^{n})^{\lambda}$.' This series is cut short and embraces the quadrature of the curve in finite terms whenever it can be expressed by that finite equation. This he calls the first of the General Theorems whence it follows that he has fitted others to more difficult and more intricate cases; however, that theorem is Proposition 5 in the *Treatise of Quadratures* and Proposition 6 of the same also deals with it but reaches to more involved cases.(16) The third and fourth propositions are lemmas prefixed for the demonstration of these theorems; but the second proposition occurs in the treatise *De analysi per æquationes infinitas* and is there the very first proposition, which in the letter above mentioned he calls the foundation of the operations and meant at that time to conceal beneath transposed letters.(17) Newton also writes that he has long ago reduced to a catalogue certain theorems which permit the comparison of curves with the conic sections and describes in the same letter the ordinates of curves which can be compared with that standard; which surely it is obvious are the same as those which Table 2 to the scholium following Proposition 10 in the Treatise on Quadratures presents.(18) From which it is clear enough that that Table, and Propositions 7, 8, 9 and 10 in the Treatise on Quadratures depending upon that Table, were discovered by Newton long ago, before the year 1676 in which that *Epistola Posterior* was written. And since he says himself in the first letter to Oldenburg that he had given up these investigations for five years,(19) it may be pretty clearly understood from this that the propositions in the Treatise on Quadratures were discovered by Mr. Newton at least five years before those letters were written to Oldenburg. And the whole of that theory of fluxions was further advanced by Newton before that time, than to this very day has been done by anyone else, under the name of differential calculus. Certainly no one I know of who has been active in this field of knowledge has made strides equal to those of Newton. And there are few, and these outstanding geometers who know how to look far ahead, who have advanced as far in this field as he. Moreover, in that later letter to Oldenburg he describes the way in which he hit upon a series whose terms display the fluxions or differences of quantities to infinity; after he had discovered this, he says, the onset of the plague compelled him to leave off such studies and think of other matters.(20) Now that plague befell in 1665 and 1666; whence it appears that even before this time the calculus of fluxions was known to Mr. Newton; that is to say, at least twelve years before Leibniz communicated his calculus to Oldenburg and nineteen years before that illustrious person published it in the *Acta* of Leipzig; and certainly there is no evidence at all that Leibniz had his differential calculus before he had seen these two letters of Newton's. When all these considerations are duly weighed it will most surely appear to any one, that Mr. Newton is to be held the true discoverer of the arithmetic of fluxions.

It now remains for us to inquire what were the hints that Leibniz received from Newton, whence it would be easy for him to extract the differential calculus. And first (as I have said) Leibniz nowhere demonstrated that the calculus of differences was known to him, before he had seen these two letters of Newton; indeed, up to this time he used a roundabout method where results follow much more easily and briefly from the differential calculus. Witness to this is his letter(21) to Oldenburg dated 18/28 November 1676 which may also be found in print in the third tome of Wallis' *Opera*, in which he deals with the way of expressing the ratio of the subtangent to the ordinate, in terms into which the ordinate does not enter, where if instead of y and dy he had put their values enclosed in a bracket he would have reached his goal immediately.

In the first letter sent to Leibniz by Oldenburg Mr. Newton taught the method by which quantities may be reduced to infinite series, that is, by which the increments of flowing quantities may be displayed; at the beginning of it he shows a series whose terms represent these increments. But this was quite hidden from Mr. Leibniz until he saw Newton's letter explaining it.[(22)]

Let o be the momentary increment of the flowing quantity x and m/n the exponent of some power of x, and if for x we write $x+o$, $x+2o$, $x+3o$, $x+4o$ &c and the quantities $(x+o)^{m/n}$, $(x+2o)^{m/n}$, $(x+3o)^{m/n}$, $(x+4o)^{m/n}$ etc. are expanded into infinite series, we shall have the same number of series the first of which is the following:

$$x^{m/n}+\frac{m}{n}\,ox^{(m-n)/n}+\frac{m^2-mn}{2n^2}\,o^2x^{(m-2n)/n}+\frac{m^3-3m^2n+2mn^2}{6n^3}\,o^3x^{(m-3n)/n}\text{ etc.}$$

In all of these series the first term will be the flowing quantity itself $x^{m/n}$, and if any one of these series is subtracted from the next we shall have the first differences of these series, in all of which the first term is the first term of the first series into which the quantity o enters, that is to say $\frac{m}{n}ox^{(m-n)/n}$. And o vanishing this term becomes equal to these first differences, or, what is the same thing, $\frac{m}{n}ox^{(m-n)/n}$ will be the first increment of the flowing quantity.

Moreover, if any one difference is subtracted from the next, we come to the second differences, the first term of which, when divided by two, is the same as the second term of the first series into which the term o enters, and (o vanishing) these differences divided by two become each equal to that term in the first series which is $\frac{m^2-mn}{2n^2}o^2x^{(m-2n)/n}$. And in the same way we may discover that the term $\frac{m^3-3m^2n+2mn^2}{6n^3}o^3x^{(m-3n)/2}$ of the series described above is equal to each of the third differences divided by six; and any term of the same series will always have a given ratio to the respective differences, that is to say the first term into which o enters is equal to the first differences, the second is the half part of the second differences, the third is the sixth part of the third differences &c. Newton discovered these series[(23)] the terms of which represent all the differences to infinity before 1665, as I have said, but they were hidden from Mr. Leibniz until he saw Newton's letter in which they were explained;[(a)] for Leibniz admitted that before that time it was always necessary for him to change an irrational quantity into a rational fraction and then, by division using Mercator's method, reduce the fraction to a series. Thence too it appears that the series containing these differences was not then possessed by Mr. Leibniz; and when it was shown to him by Oldenburg he asked[(a)] Mr. Newton to reveal its origin to him.

(a) See Leibniz's letter to Oldenburg of 27 August 1676

Now let there be any quantity composed of a constant and any indeterminate whatever, enclosed within a bracket thus $(a+bx^c)^{m/n}$ whose differences are to be obtained;[24] from the Rule first stated the difference of the quantity $a+bx^c$ is $cbx^{c-1}o$ (assuming that o is the momentary increment of the fluent quantity x) whence if z be written for $a+bx^c$ and w for $cbx^{c-1}o$, $(a+bx^c+cbx^{c-1}o)^{m/n}=(z+w)^{m/n}$. which becomes, if expanded into a series by Newton's rule, $z^{m/n}+\frac{m}{n}wz^{(m-n)/n}+$ &c in which series the term $\frac{m}{n}wz^{(m-n)/n}$ is the first difference of the quantity $z^{m/n}$ or $(a+bx^c)^{m/n}$ whence if in place of z and w their values be restored, $a+bx^c$ and $cbx^{c-1}o$, we shall have the difference of the quantity $(a+bx^c)^{m/n}$ equal to $\frac{m}{n}cbx^{c-1}o\times(a+bx^c)^{(m-n)/n}$ or if, in Leibniz's fashion we put dx for o the difference of the quantity $(a+bx^c)^{m/n}$ will be $\frac{m}{n}cbx^{c-1}dx\times(a+bx^c)^{(m-n)/n}$ where we see that the differential $\frac{m}{n}cbx^{c-1}dx$ always remains outside the bracket; and thus it was easy for Mr. Leibniz, using Newton's rule, to present the differences of all fluent quantities whatsoever, [even] involving surds and fractions (of which he had no knowledge at all before that correspondence with Newton through Oldenburg).

Although these are by themselves manifest indications of the differential calculus, yet in the second letter sent through Oldenburg to Leibniz[25] Newton describes other still clearer notable points of his method. For he says[26] that he has had the method of drawing tangents which the very skilful Sluse imparted to Oldenburg two or three years before, so that being once possessed of his basis no one could possibly draw tangents in any other manner unless of set purpose he wished to stray from the straight path. Also, since he there shows that 'this method does not stop at equations in whose roots one or more indefinite quantities are involved, but without any reducing of equations (which commonly renders the labour immense) draws the tangent immediately, and the situation is just the same in questions of maxima and minima, and other things. The foundation of these operations is, he says, pretty obvious, yet he meant to conceal it in this letter by transposed letters; yet this he adds too, that upon this basis conjectures about the quadratures of curves are made easier and, he writes, he has arrived at certain general theorems.'

Since the method of Slusius certainly could not be unknown to Leibniz at that time, as it was published in the *Philosophical Transactions*;[27] and since Newton says he was acquainted with the same method too but, from the basis he employed, it did not stick at equations involving any radicals whatever; (in which the whole difficulty of the thing consists) and since in the first letter he has written down a series by means of which the differences can be obtained when the fluents involve any sort of surds or fractions; and since, lastly, he says that he had applied the same basis to the quadrature of curves, there can be no room for doubting that all these things presented Leibniz with a beacon guiding him to the perception of Newton's method.

And if these hints do not seem to have been enough, Newton went yet further and gave examples of his method and showed the rule whereby the areas of any curves are discovered in finite terms from their given ordinates, when this can be done; that is, to put it in Leibniz's terms, he gave examples of the procedure from differentials to integrals. And, starting with the easier cases, he first[28] proposes a parabola whose abscissa is z and applied ordinate $\sqrt{(az)} = a^{1/2}z^{1/2}$, and the area of the curve will be $\frac{2}{3}a^{1/2}z^{3/2}$ that is, when the difference of the area is $dz.\sqrt{(az)}$ or $a^{1/2}z^{1/2}.dz$ he showed that the area will be $\frac{2}{3}a^{1/2}z^{3/2}$ whence conversely it is concluded that if the quantity to be differentiated is $a^{1/2}z^{3/2}$ its difference will be $\frac{3}{2}a^{1/2}z^{1/2}\,dz$ or $\frac{3}{2}dz\sqrt{(az)}$. His second example[29] is the curve whose abscissa is z and the applied ordinate $\frac{a^4z}{(c^2-z^2)^2}$ where Newton showed that the area of the curve will be $\frac{a^4}{2c^2-2z^2}$ that is, if the difference of the area is $\frac{a^4z.dz}{(c^2-z^2)^2}$ he showed that the area will be $\frac{a^4}{2c^2-2z^2}$. Whence conversely if the quantity to be differentiated be $\frac{a^4}{2c^2-2z^2}$ it may be reasoned that the difference will be $\frac{a^4z.dz}{(c^2-z^2)^2}$; or if the ordinate of the same curve may be expressed thus $\frac{a^4}{z^3(c^2z^{-1}-1)^2}$ the area will be equal to $\frac{a^4z^2}{2c^4-2c^2z^2}$. And therefore conversely if the quantity to be differentiated be $\frac{a^4z^2}{2c^4-2c^2z^2}$ the difference will be $\frac{a^4\,dz}{z^3(c^2z^{-1}-1)^2}$.

From this Newton progressed to more difficult examples and in these he showed how one may arrive at the integrals from the ordinates that is to say the differences, from which it will be obvious that any curve will be capable of quadrature whose ordinate multiplied into the difference of the abscissa forms the difference of any quantity, and so innumerable kinds of curves may be enumerated which are capable of quadrature, even geometrically.

Aided by these hints and these examples even an ordinary intellect would see quite through the Newtonian method, so is it not proper to suspect that it could not be concealed from Leibniz's very sharp mind? And that he used these keys to gain an entrance to what are said to be his own discoveries appears plainly enough from his own testimony. For in his letter to Oldenburg[30] after explaining the differential calculus he adds an example which he admits agrees with the rule of Sluse and afterwards he adds:[31] 'But that (the former) method of ours is much more complete, for not only can it be exploited when there are more indeterminate letters than x and y (which often proves most fruitful) but it is useful when irrationals intervene, for these do not hinder it at all nor is it necessary to remove the irrationals as it is with the Slusian rule, which greatly increases the difficulty of calculation.' All these things were first set out by Newton in the second of his letters. Then he proposes examples of his method, of which the very first[32] (decreed by what fate I know not) is the very same as that which Newton had first proposed in that letter which Oldenburg had transmitted to Leibniz.

Soon afterwards the most illustrious [Leibniz] adds:[33] 'I believe that what Newton wanted to conceal about the drawing of tangents does not greatly differ from these [methods]. What he goes on to say, that quadratures too are simplified upon this basis, confirms me in this opinion, in as much as those figures are always capable of quadrature which are [associated] with a differential equation. I call that a differential equation which expresses the value of *dx* and which is derived from another equation expressing the value of *x*.' And a little further on he discloses his opinion of this matter more fully, and he says:[34] This unique rule serves for the quadrature of infinite figures, of a type quite different from those submitted to quadrature hitherto. Who is there who can now consider this and fail to perceive that the hints and examples of Newton were sufficiently understood by Leibniz, [at least as to the first differences; for as to the second differences it seems that Leibniz was rather slow to comprehend the Newtonian method, as perhaps I will show more clearly in a little while].[35]

Meanwhile I can easily agree with that illustrious person, believing that he never heard the name of the calculus of fluxions pronounced, nor saw with his eyes the characters that Newton employed, before they appeared in Wallis's works[36] for I observe that Newton himself changed the name and notation of his calculus pretty often. In the treatise on the analysis of equations by infinite series he denoted the increment of the abscissa by the letter *o*;[37] and in the *Principia philosophiæ*[38] he speaks of the fluent quantity as *generated* and calls its increment a *moment*,[39] denoting the former by upper-case letters (*A*, *B*) and the latter by lower-case (*a*, *b*).

And this much more I acknowledge, among Leibniz's other outstanding merits in mathematics, that this likewise is owing to him: he was the first to publish this calculus in print and to lay it before the public[40] and so for this reason at least he has earned much gratitude from lovers of mathematics, because he was unwilling to leave concealed from them any longer a discovery that was so noble and capable of extension to so many applications.

There you have, famous Sir, what I am induced to write on this subject whence I believe you will easily perceive that this zeal (such as it is) of mine on behalf of our nation was so little out of place that I have detracted not a jot from Leibniz that was not Newton's; and there can be no doubt that candid judges of these things will with one voice allow that I said what I did say out of no slanderous spirit nor haste of judgement, which I have demonstrated to you with so much argument, clearer than the noonday sun.

NOTES

(1) Add 3968, (22); fos. 333r–338r; published in the *Commercium Epistolicum* (1712), pp. 110–17. The footnotes marked (a) in the letter are Keill's own; Newton's emendations are noted below. The letter was read to the Royal Society on 24 May 1711. A much briefer first sketch, hastily written by Keill and not annotated by Newton, is in the University Library, Cambridge (Res 1894(a), Packet 11).

(2) Letter 822.

(3) See Letter 830, note (5).

(4) Keill refers here to the *Epistola Prior* and *Epistola Posterior* of 1676. (See vol. II, Letters 165 and 188.)

(5) See Letter 830, note (2).

(6) Keill refers here to Newton's manuscript *De Analysi* which was sent by Barrow to Collins in 1669 (see vol. I, Letters 5, 6 and 7). The text of the manuscript itself contained only a brief hint of the fluxional method. (See Whiteside, *Mathematical Papers* II, p. 206.)

(7) When Leibniz made his second visit to London in 1676 Collins showed him copies of a number of Newton's early mathematical papers, and allowed him to make a transcript of Newton's letter to Collins dated 10 December 1672 (vol. I, Number 98) containing an account of Newton's method of tangents. It is now known that Leibniz's own mathematical researches had advanced beyond the point where this information, or the hints contained in Newton's letters, could have been of any value to him (see Hofmann, *Entwicklungsgeschichte.*)

(8) For Sluse and his contributions to calculus see vol. I, p. 28, note (1); for Barrow, vol. I, p. 9, note (22); and for Gregory, vol. I, p. 15, note (1).

(9) That is, James Gregory's *Geometriæ pars universalis* Padua (1668), pp. 20–2.

(10) Barrow's method of tangents is described in his *Lectiones Geometricæ*, London, 1670, Lectio x, §14, pp. 80–4.

(11) See vol. I, Letter 98. For our translation of the extract we quote from the original English version (but see notes (12) and (13) below). The figure given here with the extract is considerably less clear than the version in Newton's original letter to Collins. (Vol. I, p. 247.)

(12) Read *BD*, as in Newton's original letter.

(13) We have departed from Newton's English in our translation of the passage in parentheses, for Keill's Latin has a somewhat different meaning.

(14) See Letter 821, note (2).

(15) This is a paraphrase of a passage in the *Epistola Posterior* (see vol. II, Letter 188, pp. 144–5). The meaning is essentially that of the original.

(16) Propositions 5 and 6 of the *De Quadratura curvarum* (1704) were in fact both more general than the example Newton had given in the *Epistola Posterior.*

(17) The proposition concealed in the anagram was 'given an equation involving any number of fluent quantities to find the fluxions, and conversely'. (See vol. II, p. 153, note (25).) In *De Quadratura curvarum* proposition 1 showed how fluxions could be found from fluents (differentiation), and proposition 2 (which Keill refers to here) defined the converse operation (integration or quadrature) as its inverse. The *De Analysi* (see note (6) above) began (Rule I) by stating an algorithm for the simple case of the quadrature of the curve $y = ax^{m/n}$, and this was demonstrated at the end of the treatise.

(18) See vol. II, p. 155, note (43). Newton had, in fact, been constructing tables of integrals since 1665. See Whiteside, *Mathematical Papers,* vol. I, p. 154 *passim,* and Letter 765, note (3).

(19) See vol. II, p. 29 and p. 39.

(20) See vol. II, pp. 111–13 and pp. 130–2.

(21) The original letter has not been found, but Collins' primary transcript of it is in U.L.C. Add.3971 (2), fo. 62. There is also a partial transcript in Collins' letter to Newton of 5 March 1676/7 (see vol. II, Letter 205, p. 198, and Wallis, *Opera,* III, p. 646). Leibniz had shown how to find dy/dx as a function of x only, for the equation

$$ax^2 + by^2 + cyx + dx + ey + f = 0.$$

Presumably Keill intends to imply that, had Leibniz known how to treat irrational functions, he could have differentiated directly after solving the equation as a quadratic in y.

(22) The passage which follows is adapted by Keill not from any of Newton's letters but from the Scholium to the *Tractatus de Quadratura Curvarum* appended to the 1704 *Opticks* (see

also D. T. Whiteside, *The Mathematical Works of Sir Isaac Newton*, I, p. 156). Note that the text printed here wrongly gives $\frac{m^3-2m^2n+2mn^2}{6n^3}$ in the third term of the expansion, in place of $\frac{m^3-3m^2n+2mn^2}{6n^3}$. However, Keill's exposition is considerably clearer than Newton's, and avoids any confusion that might arise between the terms of the binomial expansion itself and successive differentials of $x^{m/n}$.

(23) The singular, 'seriem cujus', has been changed to the plural, 'series quarum', apparently by Newton.

(24) Keill shows here how a knowledge of the binomial expansion makes possible the differentiation of certain cases of a function of a function. Again, the process did not appear explicitly in the *Epistola Prior*, but Keill is implying that Leibniz could easily have inferred it.

(25) The *Epistola Posterior* (see Letter 188, vol. II).

(26) The remainder of the paragraph is a paraphrase of a passage in the *Epistola Posterior* (see vol. II, p. 115). The meaning is essentially the same as it was there.

(27) See 'An Extract of a Letter from the Excellent *Renatus Franciscus Slusius*,...concerning his short and easie *Method of drawing Tangents to all Geometrical Curves*', *Phil. Trans.* 7 (1672), 5143; see also Hall & Hall, *Correspondence of Oldenburg*, IX, Letter 2124 and associated correspondence.

(28) This is Newton's 'Example I' in the *Epistola Posterior.* He did not there give the converse procedure of differentiation, as Keill does here.

(29) Which of the two results given is correct depends, of course, upon the limits of integration, as Newton had pointed out. Keill does not mention this. Keill again gives the converse process of differentiation, which Newton did not do. Nor did Newton use *dz*, which Keill presumably introduces here to show the easy transition between the Newtonian and the Leibnizian methods.

(30) 11 June 1677 (see vol. II, Letter 209, p. 212).

(31) The passage is a near-exact transcription from Leibniz's letter. (See vol. II, p. 214, ll. 9–14.)

(32) Newton's 'Example 1' in the *Epistola Posterior* had been the determination of the area of a parabolic figure; that is, $\int_0^z \sqrt{(az)}\,.\,dz$. Leibniz's first specific example was the *differentiation* of $\sqrt{(az)}$.

(33) This sentence is quoted word for word from Leibniz's letter. (See vol. II, p. 215, ll. 7–12.)

(34) The sentence which follows is an exact quotation. (See vol. II, p. 215, ll. 20–1.)

(35) The passage in square brackets has apparently been inserted by Newton, and was printed. There is nothing in the remainder of the letter about Leibniz's knowledge of second differences.

(36) See Letter 822, note (7).

(37) See note (22) above.

(38) Book II, Lemma 2 (the passage in question), writes D. T. Whiteside: 'manifestly based on Propositions 3–10 of the main version of [Newton's] "Geometria Curvilinea"...occupies a fairly incongruous position in the second book of Newton's published *Principia*' (*Mathematical Papers*, IV, p. 521, note 1). The Lemma demonstrates that: 'The moment of any generated quantity is equal to the moments of each of the generating terms multiplied by the indices of the powers of the same terms, and by their coefficients, continually.' (That is, the moment of

the product AB is $aB+Ab$, where a and b are the moments of A and B.) In it occurs the well-known warning that moments are not finite small magnitudes, but are rather 'to be understood as the just nascent first-beginnings of finite magnitudes'. The word *fluxions* occurs only once, in the sentence: 'Eodem recidit si loco momentorum usurpentur vel velocitates incrementorum ac decrementorum (quas etiam motus, mutationes & fluxiones quantitatum nominare licet) vel finitæ quævis quantitates velocitatibus hisce proportionales.' ('It amounts to the same thing if, instead of moments, we make use either of the velocities of the increments and decrements (which may also be called the motions, changes, and flowings of quantities) or any finite quantities proportional to these velocities.') See on the whole Lemma; Whiteside, *Mathematical Papers*, IV, pp. 521–5.

(39) This word was omitted from Newton's copy and has been inserted in what appears to be his hand.

(40) Leibniz published an outline of his method in the *Acta Eruditorum* for 1684.

844 COTES TO NEWTON

4 JUNE 1711

From the original in the University Library, Cambridge.[1]
For the answer see Letter 846

Cambridge June 4th. 1711

Sr.

On Saturday [2 June] I received Yr Corrections with some wooden cuts & Muys's Elementa Physices.[2] I thank You for yt book & much more for Yr own published by Mr Jones which You was pleased to send me by Dr Bentley some time ago.[3] I have this morning been looking over yr papers & am glad to see ye difficulty removed, as at present I think it is. If upon further consideration I find any material Objection yet remaining I will acquaint You with it. At present I will only observe to You what I think to be amiss in ye Demonstration of Proposition 37.[4] I think the first Paragraph to be needless: Instead of what is there delivered this ought to have been proved; That ye velocity with which ye Water passes through ye Annular Section between ye Circellus & ye Latera Canalis, is equall to yt acquired by falling from I to G & then ye Velocitas aquæ in tota Canali will not be yt which is acquired by falling from I to G but will be lesser than yt.[5] I am sensible You had not as yet in ye first Paragraph placed ye Circellus within ye Canalis & therefore yt Paragraph is not erroneous. However it is needless & will naturally lead Your Reader (as it did me) into a mistake. For I thought You had supposed the Velocitas aquæ in tota Canali to remain ye same after ye Circellus had taken its place as before, till I came to Yr conclusion in ye 4th Paragraph where You say [Velocitas autem Circelli erit ad Velocitatem quam aqua cadendo & casu suo describendo altitudinem IG acquirit, ut Spatium annulare inter Circellum & latera Canalis est ad orificium Canalis,][6] which Conclusion is very true, but not so upon Supposition yt the Velocitas aquæ in tota Canali remaines ye same as in ye first Paragraph. In the 4th Paragraph You have also these words [Augeatur velocitas Circelli in eadem ratione & resistentia ejus augebitur in ratione duplicata][7] In the fifth You have these [Et cum resistentia in Medio densiore vel rariore (ob vim inertiæ partium Medij densitati proportionalem) sit major vel minor in ratione densitatis.][8] As I apprehend it You argue in both these places in a Circle. These are consequences to be drawn from ye Proposition & therefore they must not be supposed in order to prove it. That would be giving up ye cause to ye Patrons of a Plenum. I will not give You ye trouble to make any alterations Yr self unless you have sufficient leasure. I think I can easily rectifie ye whole Demonstration without any assistance. I see there are some things in Muys which relate to this Subject. I intend to read him over & to examine Yr Papers more carefully before I deliver them to ye Printer. What alterations I make I will send You for Yr approbation or amendment.

I am Yr &c.

ROGER COTES

For Sr Isaac Newton
in the St Martin's Street
near Leicester-Fields
London

NOTES

(1) Add. 3983, no. 12. Cotes' last letter to Newton was Letter 829. Newton's reply of 2 June has not been found; it clearly contained another reworking of Propositions 36 and 37.

(2) Wyer Willem Muys, *Elementa physices methodo mathematica demonstrata quibus accedunt dissertationes duæ: prior de causa soliditatis corporum; posterior de causa resistentiæ fluidorum*, Amsterdam, 1711. Muys was a pluralist professor at Franeker.

(3) See Letter 821.

(4) Since Cotes completely redrafted Proposition 37 in his next letter, and this redraft was printed with a few modifications, there is no means of knowing what Newton wrote in the version submitted to Cotes and discussed here.

(5) This is achieved in the printed (or Cotes') version by appealing to Case 5, Case 6 and Corol. 1 of Prop. 36—also rewritten by Cotes.

(6) 'However, the velocity of the little circle will be to the velocity which the water would acquire in falling through the distance IG as the annular space between the little circle and the sides of the channel is to the orifice of the channel.' This sentence was finally printed in the *third* paragraph (*Principia* 1713; p. 311).

(7) 'Let the velocity of the little circle be increased in the same ratio and its resistance will be increased as the square of that ratio.' This sentence was omitted later.

(8) 'And since the resistance in a denser or rarer medium is greater or less in the ratio of the density (because of the force of inertia of the particles of the medium [which is] proportional to the density.)' This sentence also was omitted.

845 THE MINT TO OXFORD

6 JUNE 1711

From the original in the Public Record Office.(1)

Reply to Letter 836

To the Rt Honble ye Earl of Oxford & Earl Mortimer
Ld High Treasurer of great Britain

May it please your Lordp

In obedience to the annexed Order of Reference of the Rt Honble the Lords Commrs of her Majties Treasury of the 8th instant,(2) upon a proposal of Sr Theodore Janssen to take any quantity of Tin not exceeding 150 Tunns, to be delivered to his Order at Leghorn & Genoa after the rate of four pounds & tenn shillings per Hundred weight to be paid one third part thereof upon the delivery of the Tin, another third part three months after & the last third part six months after, provided the quantity delivered at Genoa doth not exceed one third part of the whole: We humbly represent that the proposal is the best that We have met with & may be accepted, if your Lordp pleases, as soon as there is an opportunity of sending such a quantity of Tin to Leghorn & Genoa in her Majesties bottoms, provided the payments be made for so much Tin as shall be delivered from time to time without staying for the whole, if it should happen that any part should accidentally remain behind.

All wch is most humbly submitted to your Lordps great wisdome

C: Peyton
Is. Newton
Jn Ellis

Mint Office
6 *June* 1711

NOTES

(1) T/1, 134, no. 9. The letter was written by Newton and signed by all three officers.

(2) Newton was dozing; it was of 8th May; see Letter 836 above. See also Letter 848, note (2).

846 NEWTON TO COTES

7 JUNE 1711

From the original in Trinity College Library, Cambridge.(1)
Reply to Letter 844; for the answer see Letter 847

Sr

Yours of June 4th I received the next day & thank you for it. I am glad you received what Dr Bentley sent you & that you think the difficulty removed, except what you mention about the manner of delivering ye 37th Proposition. For clearing the sense of the first & second Paragraphs, these words may be added to the end of the second Paragraph after the word locatum.(2) *Circellus autem sustinendo vim aquæ defluentis minuet ejus velocitatem, idque in ratione qua minuit spatium per quod aqua jam transit. Nam (per Cas. 5. Prop XXXVI, & ejus Corol. 6) aqua jam transibit per spatium annulare inter circellum & latera canalis eadem velocitate qua prius transibat per canalis cavitatem totam.*

And a little after where I have these words(3) [augeatur velocitas circelli in eadem ratione et resistentia ejus augebitur in ratione duplicata] may be written these [augeatur velocitas circelli in eadem ratione & resistentia ejus augebitur in eadem ratione bis, nempe semel ob auctam quantitatem aquæ in quam circellus dato tempore agit & semel ob auctum motum quem circellus in singulas aquæ partes imprimit. Nam partes fluidi similibus motibus agitabuntur atque prius sed velocioribus et minore tempore.]

But since you are considering how to set this XXXVIIth Proposition in a cleare light I will suspend saying any thing more about it till I see your thoughts. I am

Your humble servant
IS. NEWTON

London
7th June 1711

For the Rnd Mr Roger Cotes Professor
of Astronomy at his Chamber
in Trinity College
in Cambridge

NOTES

(1) R.16.38, no. 113, printed in Edleston, *Correspondence*, pp. 40–1.

(2) 'However, the little circle by supporting the force of the down flowing water reduces its velocity, and that in the ratio of the diminution of the space through which the water now flows. For (by Case 5, Prop. 36 and its Corol. 6) the water will now flow through the annular

space between the little circle and the sides of the channel with the same velocity with which it formerly passed through the entire open space of the channel.' Cotes did not adopt this proposal, nor the next.

(3) 'let the velocity of the little circle be increased in the same ratio and its resistance will be increased in the square of that ratio...let the velocity of the little circle be increased in the same ratio and its resistance will be increased in that same ratio twice over, that is, once because of the increased quantity of water upon which the little circle acts in a given time, and once because of the increased motion which the circle impresses upon the individual particles of the water. For the particles of the fluid will be acted upon by similar motions as before but more swiftly and in a less time.'

847 COTES TO NEWTON

9 JUNE 1711

From the original in the University Library, Cambridge.[1]
Reply to Letter 846; for the answer see Letter 851

Sr.

I thank You for Yr Letter which I received last night. The most material Alterations which I had set down upon a second reading of Yr Papers are as follows. Prop. XXXVI, Paragr. 1 [acquirere potest. Ideoque per Theoremata Galilæj erit *IG* ad *IH* in duplicata ratione velocitatis aquæ per foramen effluentis ad velocitatem aquæ in circulo *AB*, hoc est, in duplicata ratione circuli *AB* ad circulum *EF*; nam hi circuli sunt reciproce ut velocitates aquarum quæ per ipsos eodem tempore & æquali quantitate adæquate transeunt. De velocitate aquæ horizontem versus &c][2] This seemes more directly to make way for Cor. 1. Supposing this alteration to be made I leave out a little above it these words [Et secetur *IG* in *O*—inter *IH* & *IG*] & a little higher alter thus [Ea vero sit uniformis velocitas glaciej descendentis ut & aquæ contingua in circulo *AB* quam aqua cadendo &c.] After the 5th Case I would add what you make ye 12th Corollary as a 6th Case.[3] It seemes to me to be properly a Case of ye Proposition. After ye 5th Corollary I would add what follows for a 6th[4] [Cor: 6 Ponderis autem pars qua sola fundum urgetur est ad pondus aquæ totius quæ fundo perpendiculariter incumbit ut circulus *AB* ad summam circulorum *AB* & *EF* sive ut circulus *AB* ad excessum dupli circuli *AB* supra fundum. Nam ponderis pars qua sola fundum urgetur est ad pondus aquæ totius in vase, ut differentia circulorum *AB* & *EF* ad summam eorundem circulorum per Cor: 4; et pondus aquæ totius in vase est ad pondus aquæ totius quæ fundo perpendiculariter incumbit ut circulus *AB* ad differentiam circulorum *AB* & *EF*. Itaque ex æquo perturbate, ponderis pars qua sola fundum urgetur est ad pondus aquæ totius quæ fundo perpendiculariter incumbit ut circulus *AB* ad summam circulorum *AB* & *EF* vel excessum dupli circuli *AB* supra fundum.] This Corollary being added may serve to give a little light to Yr *quantum sentio* in Cor: 11. I would leave out Yr remark at ye end of ye 5th Corollary vizt. [Hæc ita se habent ubi partes &c] You had given ye same caution already in ye Proposition & by placing it here You seem to restrain it to ye preceding Corollarys as if it did not equally belong to ye following ones. I would leave out Yr 6th & 7th Corollarys, for I cannot perceave yt ye 7th is a consequence of ye 6th, and what You have at ye end of ye 10th Cor. may tolerably well suffice instead of it. If these 2 Corollarys be omitted Yr 8th which I make ye 7th must begin thus[5] [Si in medio foraminis *EF* locetur circellus *PQ* centro *G* descriptus & horizonti parallelus: pondus aquæ quam circellus ille sustinet majus est &c] About ye middle of Yr 9th Corollary I would add these words [ut motus aquæ sit maxime directus] reading ye place thus [Nam ut motus aquæ sit maxime directus, columnæ illius superficies externa concurret cum basi *PQ* in angulo nonnihil acuto &c] for otherwise yt angle may be even Obtuse though what You say be granted; That *aqua sit tenuior* Within about 10 words from ye end of this Corollary you have *æqualem quam proxime*. I would omit *quam proxime*. In Yr 11th Corollary You have these words [in ratione partim simplici & partim duplicata] I cannot make any sense of these words to ye purpose & therefore would omit 'em & read thus [Et (quantum sentio) pondus quod circellus sustinet, est semper ad pondus &c.]. Your 12th Corollary as I said above I would make a 6th Case reading at ye end of it [Patebit etiam & hic Casus per Experimenta &c] I am now come to ye XXXVII Proposition which I had altered thus.[6]

Nam si vas *ABCD* fundo suo *CD* superficiem aquæ stagnantis tangat, & aqua ex hoc vase per Canalem cylindricum *EFTS* horizonti perpendicularem in aquam stagnantem effluat, locetur autem circellus *PQ* horizonti parallelus ubi vis in medio canalis, et producatur *CA* ad *K* ut sit *AK* ad *CK* in duplicata ratione quam habet excessus orificij canalis *EF* supra circellum *PQ* ad circulum *AB*: manifestum est (per Cas: 5, Cas: 6, & Cor. 1, Prop. XXXVI) quod velocitas aquæ transeuntis per Spatium annulare inter circellum et latera vasis ea erit quam aqua cadendo et casu suo describendo altitudinem *KC* vel *IG* acquirere potest.

Et, si vas sit latissimum ut lineola *HI* propemodum evanescat & altitudines *IG*, *HG* æquentur: vis aquæ defluentis in circellum (per Cor: 10)[7] erit ad pondus cylindri cujus basis est circellus ille & altitudo est $\frac{1}{2}IG$ ut EFq ad $EFq-\frac{1}{2}PQq$ quam proxime. Nam vis aquæ, uniformi motu defluentis per totum canalem, eadem erit in circellum *PQ* in quacunque canalis parte locatum.

Claudantur jam canalis orificia *EF*, *ST* et ascendat circellus in fluido undique compresso et ascensu suo cogat aquam superiorem descendere per

Spatium annulare inter circellum & latera canalis: et velocitas circelli ascendentis erit ad velocitatem aquæ descendentis ut differentia circulorum EF & PQ ad circulum PQ; et velocitas circelli ascendentis ad summam velocitatum, hoc est, ad velocitatem relativam aquæ descendentis quæ præterfluit circellum ascendentem ut differentia circulorum EF & PQ ad circulum EF, sive ut $EFq-PQq$ ad EFq. Sit illa velocitas relativa æqualis velocitati qua supra ostensum est aquam transire per idem spatium annulare dum circellus interea immotus manet, id est, velocitati quam aqua cadendo & casu suo describendo altitudinem IG acquirere potest: & vis aquæ in circellum ascendentem eadem erit ac prius per Legum Cor: 5, id est, resistentia circelli ascendentis erit ad pondus Cylindri aquæ cujus basis est circellus ille & altitudo est $\frac{1}{2}IG$ ut EFq ad $EFq-\frac{1}{2}PQq$ quamproxime. Velocitas autem circelli erit ad velocitatem quam aqua cadendo & casu suo describendo altitudinem IG acquirit ut $EFq-PQq$ ad EFq.

Augeatur amplitudo canalis in infinitum, & rationes illæ inter $EFq-PQq$ & EFq interque EFq & $EFq-\frac{1}{2}PQq$ fient rationes æqualitatis. Et propterea velocitas circelli ea nunc erit quam aqua cadendo & casu suo describendo altitudinem IG acquirere potest, resistentia vero ejus æqualis evadet ponderi Cylindri cujus basis est circellus ille & altitudo dimidium est altitudinis IG a qua cylindrus cadere debet ut velocitatem circelli ascendentis acquirat; & hac velocitate cylindrus, tempore cadendi, quadruplum longitudinis suæ describet. Resistentia autem cylindri, hac velocitate secundum longitudinem suam progredientis, eadem est cum resistentia circelli per Lemma IV; ideoque æqualis est vi qua motus ejus, interea dum quadruplum longitudinis suæ describit, generari potest quamproxime.

Si longitudo Cylindri augeatur vel minuatur; motus ejus ut & tempus quo quadruplum longitudinis suæ describit, augebitur vel minuetur in eadem ratione; adeoque vis illa qua motus auctus vel diminutus tempore pariter aucto vel diminuto generari vel tolli possit, non mutabitur, ac proinde etiamnum æqualis est resistentiæ Cylindri; nam & hæc quoque immutata manet per Lemma IV.

Si densitas Cylindri augeatur vel minuatur: motus ejus ut & vis qua motus eodem tempore generari vel tolli potest, in eadem ratione augebitur vel minuetur. Resistentia itaque Cylindri cujuscunque erit ad vim qua totus ejus motus, interea dum quadruplum longitudinis suæ describit, vel generari possit vel tolli ut densitas Medii ad densitatem Cylindri quamproxime.

Fluidum autem comprimi &c] The 3d Corollary is erroneous, I would alter it thus[8] [Iisdem positis, & quod longitudo L sit ad quadruplum longitudinis cylindri in ratione quæ componitur ex ratione $EFq-\frac{1}{2}PQq$ ad EFq semel & ratione $EFq-PQq$ ad EFq bis: resistentia cylindri &c.] In the Scholium I would make this alteration adding to ye Scheme the line *I* — *G*[9] [Sit *ABCD* rectangulum, & sint *AE* et *BE* arcus duo Parabolici axe *AB* descripti, Latere autem recto quod sit ad spatium *IG* describendum a Cylindro cadente dum velocitatem suam acquirit, ut duplum hujus spatii *IG* ad *AB*. Sint etiam *CF* & *DF* arcus alii duo Parabolici axe *CD* & latere recto quod sit prioris lateris recti quadruplum descripti, & convolutione &c.]. The Demonstration of Prop XXXVIII may thus be altered[10] Nam Globus est ad Cylindrum circumscriptum ut duo ad tria, et propterea vis illa quæ tollere possit motum omnem Cylindri interea dum Cylindrus describit longitudinem quatuor diametrorum, Globi motum omnem tollet interea dum Globus describit duas tertias partes hujus longitudinis, id est, octo tertias partes diametri propriæ. Resistentia autem Cylindri est ad hanc vim quamproxime ut densitas fluidi ad densitatem Cylindri vel Globi per Prop. XXXVII, & resistentia Globi æqualis est resistentiæ Cylindri per Lem: V, VI, VII. I will remember to alter ye 2d Corollary of this Proposition as formerly, which you had forgotten to do in Yr last Copy.

I have computed ye Table preceding ye Scholium of Prop XL & find some of Yr numbers to be amiss which I will take care to rectify; as against 0,9*G* the Space should be 0,7196609*F*, against 3*G* the space should be 4,6186570*F*, against 4*G* should be 6,6143765*F*.[11] I computed also all the Experiments & found my conclusions to agree nearly enough with yours except in ye 1st Experiment which I will alter throughout. Of ye rest the greatest difference was in ye 11th in which ye result was 46$\frac{5}{9}$ oscillations In Yr corrections You make it 46, I took care to make a right allowance for ye narrowness of ye Vessel. I desire You to send me the altitude from which ye Globes fell in ye 9th Experiment.[12] You had forgotten to mention it in Yr Copy. The six Experiments in ye Air agree also very well with my Computation, in ye 5th ye space should be 225*p*. 5*d*.

I am, Sir, Your most humble Servant
ROGER COTES

Cambridge June 9*th* 1711
For Sr Isaac Newton
at his House
in St Martin's Street
near Leicester-fields
London

NOTES

(1) Add. 3983, no. 38. Only the last paragraph of this letter was printed in Edleston, *Correspondence*, pp. 42–3, from the Trinity College MSS.

(2) This passage was printed on p. 304 (top) of the printed text; those next quoted by Cotes at the bottom of p. 303.

(3) This was done, pp. 306–7.

(4) It was printed exactly thus on p. 308.

(5) All the following changes in the Corollaries were made. It will be obvious that none of the alterations proposed here (and accepted by Newton) modify the structure of Proposition 36.

(6) Again, Cotes does not propose to change the sense of Proposition 37, whose enunciation reads: 'The resistance which arises from the cross-sectional area of a cylinder moving uniformly forwards through an infinite, compressed and inelastic medium in the direction of its length [longitudinal axis] is to the force by which its whole motion may be either generated or destroyed whilst it describes four times its own length, as the density of the medium to the density of the cylinder, approximately.' Cotes here quotes almost the whole of the proposition as printed in the form he had devised (*Principia*, 1713; pp. 310–12).

The sense of this proposition is not easy to see at first glance. Newton had already employed the principle that the resistance is the same whether the cylinder be fixed and the fluid move, or the fluid be at rest and the cylinder move; we would not expect the resistance experienced by a *fixed* cylinder to vary with its density. But here Newton is relating the resistance to the force required to *generate* the motion of the cylinder, and this clearly depends upon the mass of the cylinder, and hence upon its density.

(7) '& 11' has been added (by Cotes, apparently) in the margin, but not in the printed book.

(8) See p. 313.

(9) This became *HG* as printed (the passage is on pp. 313–14).

(10) It appears thus on p. 316.

(11) Cotes' numbers were printed.

(12) It was 182 inches.

848 NEWTON TO OXFORD

10 JUNE 1711

From the holograph copy in the Mint Papers[1]

To the Rt Honble the Earl of Oxford and Earl Mortimer,
Lord High Treasurer of great Britain.[2]

May it please your Lordp

By the weight & assay of forreign coins formerly taken in the Mint, Mexico pieces of eight unworn one with another are worth 4*s* 6*d* sterling in intrinsic value,[3] & Sevil pieces of eight (old plate) are of the same intrinsic value with those of Mexico. And according to this value eight hundred eighty eight thousand eight hundred & eighty nine pieces of eight of either Mexico or Sevill are worth two hundred thound pounds & six pence.[4]

In this recconing I have made no allowance for the wearing of the money. Exchangers reccon pieces of eight at a par wth 4*s* 3*d*$\frac{51}{95}$ without distinguishing between the several sorts of them. For the pieces of Peru are coarse & Refiners reccon them scarce worth 4*s* 3*d* a piece one wth another. If pieces of eight be taken promiscuously at a par with 4*s* 3*d*$\frac{51}{95}$, nine hundred thirty one thousand three hundred seventy two pieces of eight will be worth two hundred thousand pounds.

All wch &c.
IS. NEWTON

Mint office,
June 10 1711

NOTES

(1) Mint Papers, II, fo. 205; a draft is no. 203, dated 8 June.

(2) Robert Harley (1661–1724) was now at the zenith of his political career. He had long worked in coalition with Marlborough and Godolphin, serving as Secretary of State from 1704 to 1708, when he resigned. In August 1710 Anne dismissed Godolphin, the Treasury was placed in commission with Harley as second lord, and the Parliament that assembled in November was strongly Tory. Harley soon acquired supreme power: on 23 May 1711 he was raised to the peerage as Earl of Oxford and Earl Mortimer, and on 29 May he was appointed Lord High Treasurer, the Commission being dissolved.

(3) The Spanish silver dollar, or peso, so called because the coin was marked with the numeral 8, signifying that it was worth eight reales (hence colloquially in the United States a quarter may still be alluded to as 'two bits').

(4) Clearly Newton had been asked to supply a correct value for this coin, and he was to go into much more detail later. The reason for this interest on the part of the Treasury was that in February or early March 1711 the allied navies in the Mediterranean had captured two Genoese ships, the *San Gaetano* and the *Nostra Seigniora de Loretta*, freighted with South American silver coined into pieces of eight, and taken them as prizes into Port Mahon (Mahón, capital of the island of Minorca, captured by the British in 1708 and finally retaken by Spain only in 1782). By early June the Treasury was considering a proposal that £200000 in silver should be borrowed from this captive hoard (at a rate not exceeding 5%), to be reminted into reales in order to provide pay for the British army campaigning in Spain on behalf of Charles III; the offer price was 5*s.* 8*d.* for each piece of eight (*Cal. Treas. Books*, XXX (Part II), 1961, pp. 26, 69). On 6 July an agreement was drawn up between the four parties concerned (the Treasury, the Republic of Genoa, the naval officers who had made the capture, and the London merchants who were to handle the transaction), the value of the piece of eight being now set at 4*s.* 6*d.*, that is, $800000 borrowed would be reckoned as £180000 sterling (*ibid.*, pp. 356–7). This reduction in the sterling equivalent of the dollar was certainly effected on Newton's advice.

849 LOWNDES TO NEWTON

14 JUNE 1711

From the copy in the Public Record Office.(1)

For the answer see Letter 850

Sr. Isaac Newton and Sr. Lambert Blackwell(2) abt.
Money to be lent by the Genoes's

Gent.

My Lord Trearer. desires that you'l give Your Opinions in Writing what Her Majty. may allow per Ownes [ounce] in Sterling Money for the Eight hundred Thousand Mexico Dollars proposed to be lent by the Genouees's without regarding the Exchange I am &c: 14*th June* 1711

Wm. Lowndes

NOTES

(1) T/27, 19, 361.

(2) See Letter 848, note (4).

850 NEWTON TO OXFORD

[*c.* 15 JUNE 1711]

From the holograph draft in the Mint Papers.(1)

Reply to Letter 849

To the Rt Honble the Earl of Oxford & Earl Mortimer,
Lord High Treasurer of great Britain

May it please your Lordp.

Upon the annexed Proposal shewed me by Sr Lambert Blackwel(2) by your Lordps Order for setling the value of Mexico Dollars to be received at Port Mahon: I humbly represent that these Dollars are a penny wt worse than sterling & weigh $17\frac{1}{2}$ dwt one wth another when fresh out of the Mint & are then worth 4*s* 6*d* a piece in their intrinsic value, & if about two grains be abated for wearing they will be worth about 4*s* $5\frac{3}{4}$*d* a piece. In this valuation Sr Lambert agrees wth me very nearly: for he tells me that 1000 Dollars of this sort are found by Merchants to weigh 872 ounces Troy, within an ounce or two over or under, & at this rate a Dollar weighs 17 dwt 10gr at a Medium.

[Sevil(3) Dollars (old plate) were worth 4*s* 6*d* a piece when fresh out of the Mint, but are now much diminished by wearing, & Sevil Dollars new plate are of a lighter species being worth but 3*s* 7*d* or 3*s* 7*d*$\frac{1}{4}$ a piece. The Pillar pieces of eight are finer but more worn then the Mexico, & but few in number, the Peru pieces are coarser & most worn & diminished.(4)

After the Mexico Dollars are told out of the baggs they may be weighed by a thousand at a draught for ascertaining their value more exactly & an account may be taken of them by persons deputed on both sides, & the weights by wch they are weighed may be compared exactly wth our weights Troy by the same persons.

[If an eaven number of Dollars suppose eight or nine hundred thousand be borrowed](3) The interest at 5 per cent may be recconed in dollars & both interest & principal paid either in Dollars or in bullion [at such a rate as shall be agreed upon].(3) And I further humbly represent that Spanish Dollars are valued in London as bullion, that is a[t] [*blank*] If these Dollars should be brought to London they may be weighed at the Mint & received as bullion 1 dwt worse then standard, or if they are received by tale at Port Mahon they may be valued at 4*s* [*blank*]

That Spanish dollars are valued at London as bullion & if they should be brought to London they may be weighed at the Mint & valued as bullion 1 dwt worse than stan[dard] or at 5*s* 1$\frac{3}{4}$*d* sterling per ounce.

But if they are received by tale at Port Mahon, they may be taken at 4*s* 5*d*$\frac{3}{4}$ a piece in English mon[ey]

NOTES

(1) Mint Papers II, fo. 195; compare also another draft, *ibid.* no. 202. Presumably these drafts were composed in reply to the preceding letter (Lowndes of 14 June).

(2) A London merchant and financier who did much business with the Treasury. In 1704 he was listed by Edward Chamberlayne (*Angliæ Notitia: or the Present State of England*, p. 632) as 'Her Majesty's Envoy Extraordinary to the Great Duke of Tuscany, and the Republick of *Genoa*, and Consul at *Leghorn*'.

(3) The brackets are in the original.

(4) The first Mint in the New World was founded at Mexico under Charles V; others followed at Lima in Peru (Philip II), Potosi (Philip IV) and Bogotà (Charles II). Until 1570 the New World coins were struck with an image of the Pillars of Hercules on the reverse and the motto PLUS ULTRA; after 1570 for eighty years American silver coins were struck on the model of the Spanish dollars, having a cross on the reverse. From 1650, however, the newly-chartered Mint at Potosi reverted to the design showing the Pillars of Hercules (called 'peruleros' by the Spanish, while coins with the cross were known as 'macuquinos'). See Arthur Engel and Raymond Serrure, *Traité de Numismatique* (Part I), Paris, 1897, pp. 569–70. By Pillar Dollar, therefore, Newton could have meant either a coin struck in Mexico before 1570, or one struck at Potosi after 1650. By a 'Peru piece' here he presumably meant one struck at Lima.

851 NEWTON TO COTES

18 JUNE 1711

From the original in Trinity College Library, Cambridge.(1)
Reply to Letter 847; for the answer see Letter 854

Sr

I have read over & considered your alterations, & like them very well & return you my thanks. In ye end of Exper. 9, add, *describentes altitudinem digitorum 182*. I thank you also for correcting the numbers. I hope there will be no more occasion of stopping the press. After you have read the objection of Muys taken from Galileo's experiment of the motion of a bucket full of water you will scarce expect very much from that author.(2) I am

Sr

Your very humble servant

IS. NEWTON

St Martins street
London. June 18*th*
1711.

For the Rnd Mr Cotes Professor of Astronomy,
at his Chamber in Trinity College in the
University of Cambridge

NOTES

(1) R.16.38, no. 120, printed in Edleston, *Correspondence*, pp. 43–4.

(2) In Lemma 4, Book II, of the first edition—it was omitted from the second—Newton had correctly stated that if a vessel filled with a motionless homogeneous fluid be accelerated continuously in a straight line to the boundary of the universe (*ut interea non moveatur in orbem*) the particles of the fluid participating in this motion will remain at rest among themselves. Muys (see Letter 844, note (2), p. 355) had objected against this Galileo's argument that fluid in an *open* vessel moves relatively in the opposite sense to the acceleration. (*Dialogo...sopra i due massimi sistemi*, 1632. Fourth Day, *ad init.*)

852 NEWTON TO OXFORD

[c. 20 JUNE 1711]

From the holograph draft in the Mint Papers(1)

To the Rt Honble the Earl of Oxford & Earl Mortimer Lord High Treasurer of great Britain

May it please your Lordp

In the new Act for raising a fund of two Millions, after a recital of the resolutions of the House of Commons for encouraging the bringing plate to the Mint & adressing her Maty to direct the officers of the Mint to receive the same & of her Majts directions thereupon & of the further Resolution of the Commons that the Receipts thereof should be accepted upon any Loans or contributions upon any funds to be demanded [?] this last sessions & of the receipt of plate at the Mint to be melted down essayed & coined in pursuance of her Mats warrant: it is enacted that all the Receipts given for Plate brought into the Mint in pursuance of her Majts said Warrant before the 15th day of May be accepted upon the said fund of two millions by the Receivers thereof if tendred before ye 25th day of June; but nothing is enacted concerning the acceptance of the Receipts given for Plate imported afterwards. And what is to be done thereupon is most humbly submitted to your Lordps great Wisdome.

Is. Newton

NOTE

(1) Mint Papers II, fo. 552, undated. We have assumed that this letter was drafted soon after the passage of the Two Million Act on 12 June 1711, but before the closing date of 25 June mentioned.

See further Newton's long memorandum of 24 July (Number 858), and the Warrant of 30 July (Number 862).

853 HALLEY TO FLAMSTEED

23 JUNE 1711

From the original in the Royal Greenwich Observatory(1)

Reverend Sr

Though I am credibly informed that these sheets have been from time to time sent you from the press; yet least it should be otherwise, I have now sent you the Catalogue of the Fixt starrs, intended to be præfixt to your Book, having spared no pains to make it as compleat and correct as I could, by help of the Observations you have given us, made before the year 1706.(2) I desire you to find all the reall faults you can, not as beliving there are none, but being willing to have a work of this kind as perfect as possible: And if you signifie what's amiss, the errors shall be noted, or the sheet reprinted, if the case requires it. Pray govern your Passion, and when you have seen & considered what I have done for you, you may perhaps think I deserve at your hands a much better treatment than you for a long time have been pleasd to bestow on

Your quondam friend and not yet
profligate Enemy (as you call me)

EDM. HALLEY

London
June 23^{0}
1711

NOTES

(1) Flamsteed MSS., vol. 55, fo. 119.

(2) According to Flamsteed's autobiography (Baily, *Flamsteed*, pp. 95–6) the corrected sheets of the catalogue were left with his niece, Mrs Hodgson. Flamsteed's comment was: 'When I examined it, I found more faults in it, and greater, than I imagined the impudent editor either could, or durst have committed.' He then resolved to reprint it at his own expense.

854 COTES TO NEWTON

23 JUNE 1711

From the original in the University Library, Cambridge.(1)
Reply to Letter 851; for the answers see Letters 860, 889 and 891

Cambridge June 23d 1711

Sr.

I have received Yr Letter & delivered ye Papers to ye Printer. I hope we shall now go on without any further intermission. As for Muys,(2) I have look'd over what relates to ye Resistance of Fluids. He acknowledges yt what he offers upon yt subject at present is but crude & indigested & I am very willing to agree with him. His Objections, as far as I can understand them, do not in any wise affect Your Book, much less the new Edition of it. One Mr Green(3) of Clare-Hall has now in ye Press a Book of ye like nature with Muys, wherein I am informed he undertakes to overthrow the Principles of Yr Philosophy. I do not expect very much from him & I beleive You will not Yr self when I have told You He is a person who pretends to have solved the grand Problem of the Quadrature of ye Circle.

That ye Press may not stop, I am now looking over Yr Copy beforehand. I find nothing amiss 'till I come to Prop: 48.[4] I will choose to make my Objection against ye Corollary, wherein You have these words [Nam lineola Physica $\epsilon\gamma$, quamprimum ad locum suum primum *EG* redierit, quiescet;] This assertion cannot (I think) be reconciled wth what You assert & prove in ye Proposition vizt: [& propterea vis acceleratrix lineolæ Physicæ $\epsilon\gamma$ est ut ipsius distantia a medio vibrationis loco Ω][5] I propose to alter ye whole Proposition thus, if You approve of it.[6]

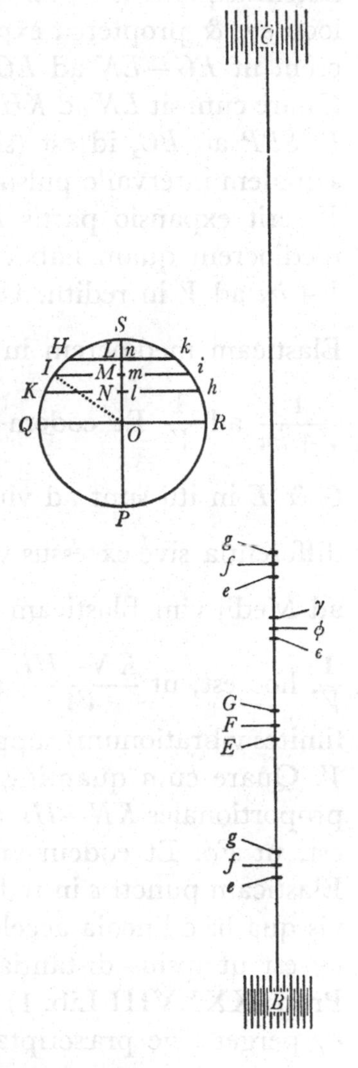

Propagentur pulsus in plagam *BC* a *B* versus *C*, & designet *BC* intervallum eorundem ab invicem. Sint *E*, *F*, *G* puncta tria Physica Medij quiescentis in recta *BC* ad æquales distantias sita; *ee*, *ff*, *gg* spatia æqualia perbrevia per quæ puncta illa motu reciproco singulis vibrationibus eunt & redeunt; ϵ, ϕ, γ, loca quævis intermedia eorundem punctorum; & *EF*, *FG* lineolæ Physicæ seu Medij partes lineares punctis illis interjectæ & successive translatæ in loca $\epsilon\phi$, $\phi\gamma$ & *ef*, *fg*. Rectæ *ee* æqualis ducatur recta *PS*, bisecetur eadem in *O*, centro *O* & intervallo *OP* describatur circulus *SIPi* & agatur diameter *QR* ad diametrum *PS* perpendicularis. Per circuli hujus circumferentiam totam cum partibus suis exponatur tempus totum vibrationis unius cum ipsius partibus proportionalibus; sic ut, completo tempore quovis *QH* vel *QHSh*, si demittatur ad *PS* perpendiculum *HL* vel *hl*, & capiatur $E\epsilon$ æqualis *OL* vel *Ol*, punctum Physicum *E* reperiatur in ϵ. Hac lege punctum quodvis *E*, eundo ab *E* per ϵ ad *e* atque inde redeundo, iisdem accelerationis & retardationis gradibus vibrationes singulas peraget cum oscillante Pendulo. Probandum est quod singula Medij puncta Physica tali motu agitari debeant. Fingamus igitur Medium tali motu a causa quacunque cieri, & videamus quid inde sequatur.

In circumferentia *PQSR* capiantur æquales arcus *HI*, *IK*, vel *hi*, *ik* eam habentes rationem ad circumferentiam totam quam habent æquales rectæ *EF*, *FG* ad pulsuum intervallum totum *BC*. Et demissis perpendiculis *IM*, *KN* vel *im*, *kn*; quoniam puncta *E*, *F*, *G* motibus similibus successive agitantur & vibrationes suas integras ex itu & reditu compositas interea peragant dum pulsus transfertur a *B* ad *C*; si *QH* vel *QHSh* sit tempus ab initio motus puncti

E, erit *QI* vel *QISi* tempus ab initio motus puncti *F*, & *QK* vel *QKSk* tempus ab initio motus puncti *G*; & propterea $E\epsilon$, $F\phi$, $G\gamma$ erunt ipsis *OL*, *OM*, *ON* in itu punctorum, vel ipsis *Ol*, *Om*, *On* in punctorum reditu, æquales respective. Unde $\epsilon\gamma$ seu $EG+G\gamma-E\epsilon$ in itu punctorum æqualis erit $EG-LN$, in reditu autem æqualis $EG+ln$. Sed $\epsilon\gamma$ latitudo est seu expansio partis Medij *EG* in loco $\epsilon\gamma$, & propterea expansio illius in itu est [ad] ejus expansionem mediocrem ut $EG-LN$ ad *EG*; in reditu autem ut $EG+ln$ seu $EG+LN$ ad *EG*. Quare cum sit *LN* ad *KH* ut *IM* ad radium *OI*, & *KH* ad *EG* ut circumferentia *PQSRP* ad *BC*, id est (si ponatur *V* pro radio circuli peripheriam habentis æqualem intervallo pulsuum *BC*) ut *OI* ad *V*, & ex æquo *LN* ad *EG* ut *IM* ad *V*: erit expansio partis *EG* punctive Physici *F* in loco $\epsilon\gamma$ ad expansionem mediocrem quam habet loco suo primo *EG* ut $V-IM$ ad *V* in itu, utque $V+im$ ad *V* in reditu. Unde vis Elastica puncti *F* in loco $\epsilon\gamma$ est ad vim ejus Elasticam mediocrem in loco *EG* ut $\frac{1}{V-IM}$ ad $\frac{1}{V}$ in itu, in reditu vero ut $\frac{1}{V+im}$ ad $\frac{1}{V.}$ Et eodem argumento vires Elasticæ punctorum Physicorum *G* & *E* in itu sunt ad vires mediocres ut $\frac{1}{V-KN}$ & $\frac{1}{V-HL}$ ad $\frac{1}{V}$, & virium differentia sive excessus vis Elasticæ puncti γ supra vim Elasticam puncti ϵ est ad Medij vim Elasticam mediocrem ut $\frac{KN-HL}{VV-V\times KN-V\times HL+KN\times HL}$ ad $\frac{1}{V}$, hoc est, ut $\frac{KN-HL}{VV}$ ad $\frac{1}{V}$ sive ut $KN-HL$ ad *V*, si modo (ob angustos limites vibrationum) supponamus *HL* & *KN* indefinite minores esse quantitate *V*. Quare cum quantitas *V* detur excessus ille est ut $KN-HL$, hoc est, (ob proportionales $KN-HL$ ad *HK*, & *OM* ad *OI*, datasque *HK* & *OI*) ut *OM*, id est, ut $F\phi$. Et eodem argumento excessus vis Elasticæ puncti γ supra vim Elasticam puncti ϵ in reditu lineolæ Physicæ $\epsilon\gamma$ est ut $F\phi$. Sed excessus ille est vis qua hæc lineola acceleratur; & propterea vis acceleratrix lineolæ Physicæ $\epsilon\gamma$ est ut ipsius distantia a medio vibrationis loco *F*. Proinde tempus (per Prop: XXXVIII Lib. 1) recte exponitur per arcum *QI*; & Medij pars linearis $\epsilon\gamma$ perget lege præscripta moveri, id est, lege oscillantis Penduli: & par est ratio partium omnium linearum ex quibus Medium totum componitur. Q.E.D.

I was going to propose an alteration of the Corollary, but I choose rather to leave it to Your self.(7) It must be made to correspond with what You have at ye end of page 372 where You cite it. I propose to alter Prop: 49th as follows.(8) p. 368, l: 28 [—ad lineolæ illius pondus ut $HK\times A$ ad $V\times EG$ sive ut $PO\times A$ ad *VV*, nam *HK* erat ad *EG* ut *PO* ad *V*.] l: 32 [—urgente vi ponderis in subduplicata ratione *VV* ad $PO \times A$ atque adeo—]. lin: ult.—in subduplicata ratione *VV* ad $PO \times A$ & subduplicata ratione $PO \times A$ conjunctim, id est, in ratione integra *V* ad *A*. Sed tempore vibrationis unius ex itu & reditu compositæ, pulsus progrediendo conficit latitudinem suam *BC*. Ergo tempus quo pulsus percurrit spatium *BC*, est ad tempus oscillationis unius ex itu & reditu compositæ, ut *V* ad *A*, id est, ut *BC* ad circumferentiam circuli cujus &c. If you approve of it these 2 following Corollaries may be added after this 49th Proposition. Cor. 1.[9] Velocitas pulsuum ea est quam acquirunt Gravia æqualiter accelerato motu cadendo & casu suo describendo dimidium altitudinis *A*. Nam tempore casus hujus cum velocitate cadendo acquisita, pulsus percurret spatium quod erit æquale toti altitudini *A*, adeoque tempore oscillationis unius ex itu & reditu compositæ percurret spatium æquale circumferentiæ circuli radio *A* descripti; est enim tempus casus ad tempus oscillationis ut radius circuli ad ejusdem circumferentiam.

Cor. 2. Unde cum altitudo illa *A* sit ut Fluidi vis Elastica directe & densitas ejusdem inverse; velocitas pulsuum erit in ratione composita ex subduplicata ratione densitatis inverse & subduplicata ratione vis Elasticæ directe.

I think ye 47th Proposition is out of its place, for ye demonstration of it proceeds upon ye supposition of ye truth of ye 48th, & therefore it ought to follow ye 48th, & moreover the 48th serves to form some Ideas which are necessary to ye understanding of ye 47th. If You agree yt these Propositions should change places, I would add ye following words at ye end of ye 47th which will then be ye 48th [Hæc Propositio ulterius patebit ex constructione sequentis][9]

I see nothing further in ye 2d Book which I could wish might be altered. In ye 3d Book under PHÆNOM: 1, The Periodical times should be 1d. 18h. 27′. 34″ 3d. 13h. 13′. 42″ 7d. 3h. 42′. 36″ 16d. 16h. 32′. 9″, & the Distances *Ex temporibus Periodicis* may be 5,667 9,017 14,384 25,299.[10] I perceive You have made use of Cassini's Tables of Jupiter's Satellits printed in 1693 in ye *Recueil d'Observations faites en plusieurs Voyages* &c.[11] but Yr numbers give ye times of ye Revolutions to Jupiter's shadow, not to ye same point of ye Ecliptick. The Revolutions to ye same point of ye Ecliptick are (by those Tables) as I have sėt 'em down above. Yr time of ye Revolution of Saturns outermost satellite differs from ye time assigned by Hugenius in his *Cosmotheoros*[12] & by Cassini in ye Philosoph: Transact:[13] but I find it is ye time which was afterwards determined by Cassini in ye *Memoires de l'Academie* 1705.[14] You have made an addition to the 3d Proposition, in which are these words[15] [Hæc ratio obtinet in Orbe Lunæ nostræ. In minore orbe motus Aphelij minor esset in triplicata ratione minoris distantiæ Lunæ a Terra, & fractio 4/243 diminui deberet in eadem ratione. Et propter hanc diminutionem vis qua Luna retinetur in orbe suo est ad vim eandem in superficie Terræ ut 1 ad $D^{2\frac{1}{243}}$ quamproxime, uti computum ineunti patebit.] I should be glad to understand this place, if it will not be too much trouble to make it out to me. I do not at present so much as understand what it is yt You assert.

Your Obliged Humble Servant
ROGER COTES

For Sr Isaac Newton
at his House
in St Martin's Street
near Leicester-Fields
London

NOTES

(1) Add. 3983, no. 13; printed in Edleston, *Correspondence*, pp. 44–50, from the Trinity College MSS.

(2) See Letter 844, note (2).

(3) Robert Greene (1678?–1730), Tutor of Clare Hall, was already known to Newton since Greene had tried to convince him that $\pi = 3\frac{1}{5}$ ('rescripsit nihil', recorded Greene). The work mentioned here was *The Principles of Natural Philosophy, In which is shown the Insufficiency of the Present Systems, to give us any Just Account of that Science: And the Necessity there is of some New Principles, In order to furnish us with a True and Real Knowledge of Nature*, Cambridge, 1712. Despite Cotes', Newton's, and Edleston's contempt for Greene, his ideas have been studied at some length by modern historians, among them most recently P. M. Heimann and J. E. McGuire, 'Newtonian Forces and Lockean Powers: Concepts of Matter in Eighteenth Century Thought', *Historical Studies in the Physical Sciences*, **3** (1971), 255–61.

(4) At Cotes' suggestion (later in this letter) the Propositions of Book II numbered 47 and 48 in the first edition of the *Principia* and in Newton's amended copy were transposed in the second edition; hence the 'Prop. 48' referred to here is Prop. 47 of the second edition. In Section 8 of Book II, beginning with Proposition 41, Newton tackled wave-motion in fluids. Proposition 44 proves that when water oscillates in a U-tube it does so with a simple harmonic motion; the next that the velocity of surface waves is as the square root of the wavelength; and this is in fact a corollary of Proposition 46, demonstrating that the velocity of propagation of such waves is $\sqrt{(g\lambda)}/\pi$, where λ is their wavelength. Next, in Proposition 47 (new numbering) Newton was to prove that, when a wave traverses a fluid, the particles oscillate with a simple harmonic motion.

(5) Cotes supposed that the statement 'the lineola comes to rest when it returns to its initial position' is inconsistent with the statement that 'the accelerative force of the lineola is proportional to its distance from the mean position of its vibration', which would be true if the 'initial position' and the 'mean position' are the same. But Newton later points out that he does *not* mean to identify these two positions and claims that there is no inconsistency. Subsequent correspondence concerning the proposition centred wholly on this point. (See Letters 860, 863, 889). In Letter 863, Cotes makes his difficulty particularly clear. In Letter 890 he finally agreed that Newton's argument needed no alteration.

(6) The form and wording of this draft version correspond pretty closely to what was

printed (*Principia* 1713; pp. 337–9), the major change proposed being in the diagram, and hence in the *sense* of the proposition. Newton rejected Cotes' suggestion, and the diagram was printed in 1713 in essentially the same form as in 1687. The 1713 diagram is given below for comparison. In the right-hand section of Newton's diagram *E*, *F*, *G* and *e*, *f*, *g* represent the extremities of the oscillations, while Cotes makes *E*, *F*, *G* the centres of oscillation, adding *e*, *f*, *g* *below* *E*, *F*, *G* to represent the second set of extremities. Cotes changes the left-hand diagram in such a way that *H*, *I*, *K* should bear the same relationship to *Q*, the *locus primus* in his argument, as they bore to *P*, the *locus primus* in Newton's discussion. This minimizes the changes of notation needed in the text itself.

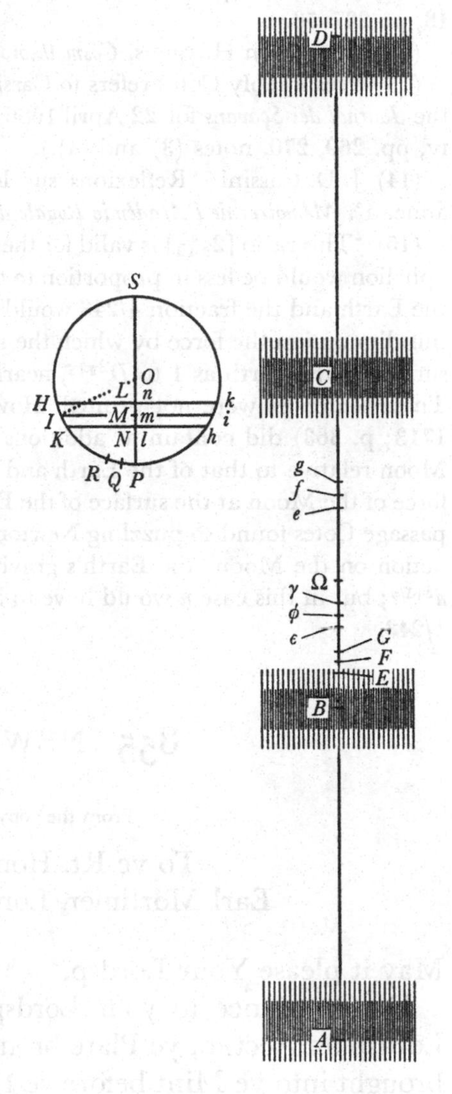

Cotes' objection is not trivial. If the medium were totally at rest, the particles would be at *E*, *F*, *G* (in Cotes' diagram), their 'mean position'. The 'initial position' that Newton chose is a position of *momentary* rest only, and it is not clear that Newton fully realized this (see Letter 860 for his own clarification). However Cotes' argument is not without ambiguity. He refers to *E*, *F*, *G* as 'puncta tria Physica Medii quiescentis' and to *EG* as the 'expansio mediocris quam habet in loco suo primo'. If he means by this the actual length of the lineola as it passes through *F*, then clearly his interpretation is wrong, and he has failed to appreciate that whilst *displacement* of the particles is zero at *F* and maximum at *f*, *pressure* is normal at *f* and maximum or minimum at *F* (during backward or forward motion respectively). If he recognized that the pressure and displacement waves were out of phase in this way, and intends by 'mediocris' the average of two values, then his argument is correct (as is Newton's also).

Cotes' argument is less complete than Newton's as it deals explicitly only with the portion of the oscillation from *E* to *f* and back (Cotes' diagram), whereas Newton treated the whole cycle.

(7) This corollary (*Principia* 1713; p. 339) was finally printed exactly as in the first edition. Had Newton accepted Cotes' rendering of Proposition 47 (new numbering) an alteration would of course have become necessary.

(8) In Proposition 49 Newton showed that the velocity of longitudinal waves might be found if the density and elastic force of the medium were known. The simplification proposed here by Cotes and the new corollaries were printed in the second edition (the mathematical relations being the same as in the first).

(9) As already noted, this was done.

(10) These more correct values are substituted (*Principia* 1713; p. 359) for those in the first edition, p. 403.

(11) *Recueil d'Observations faites en plusieurs Voyages par ordre de Sa Majesté pour perfectionner l'astronomie et la géographie. Avec divers traitez astronomiques. Par Messieurs de l'Academie Royale des Sciences* (Paris, 1693). The table of the first satellite of Jupiter was also printed in *Phil. Trans.*, **18**, pp 237–56.

(12) Christiaan Huygens, *Cosmotheoros* (The Hague, 1698).

(13) Presumably Cotes refers to Cassini's letter on the five satellites of Saturn published in the *Journal des Sçavans* for 22 April 1686 and reprinted in *Phil. Trans.* **16** (1687), 99. (See vol. IV, pp. 269, 270, notes (3) and (4).)

(14) J. D. Cassini, 'Reflexions sur les observations des satellites de Saturne et de son anneau', *Mémoires de l'Académie Royale des Sciences* (1705), p. 14.

(15) 'This ratio [$2\frac{4}{243}$] is valid for the orbit of our moon. In a smaller orbit the motion of the aphelion would be less in proportion to the third power of the lesser distance of the Moon from the Earth and the fraction 4/243 would need to be reduced in the same ratio. And because of this diminution the force by which the moon is retained in its orbit is to the same force at the surface of the Earth as 1 to $D^{2\frac{1}{243}}$, nearly, as will be evident to a beginner in computation'. These sentences were not printed. However, the new version of Book III, Prop. 3 (*Principia* 1713; p. 363) did contain as additions (i) a calculation of the Sun's attractive force on the Moon relative to that of the Earth and (ii) an indication of the way in which the centripetal force of the Moon at the surface of the Earth might be calculated. It seems possible that in the passage Cotes found so puzzling Newton meant to say that (because of the Sun's gravitational action on the Moon) the Earth's gravity acts on the Moon as though it varied inversely as $d^{2+1/n}$; but in this case n would have to be—according to Newton's own numbers—1/700, not 1/243.

855 NEWTON TO OXFORD

6 JULY 1711

From the copy in the Public Record Office[1]

To ye Rt. Honble. the Earl of Oxford & Earl Mortimer, Lord High Treasr. of Great Britain

May it please Your Lordsp.

IN Obedience to your Lordsps verbal Order, I humbly lay before your Lordsp an Acct. of ye Plate brought into ye Mint. The recpts. given for Plate brought into ye Mint before ye 15th of May Amount unto 48845£. 11*s*. 8*d*. & ye Plate was melted into 189 Ingotts & made 14640£wt. 01oz. 12dw. 6gr. of Standard Silver besides what remaines in ye Sweep.[2]

The Recpts. given for Plate brought into ye Mint from ye 15th. May to ye 19th of June[3] Inclusive Amount unto 26564£: 18*sh*: 8*d*: & the Plate was melted into 97 Ingotts & made 7940 £wt. 04 oz: 8 dwt: 01 gr of Standard Silver after ye rate of 5*sh*: 2*d* per oz will make 24615£: 2*sh*: 8*d* of New Moneys which being deducted from ye Summe of ye Recpts. leaves a Loss of 1949£: 15*sh*: 10*d*.[4]

There have been brought into ye Mint since ye 19th June 3620 oz. 10dwt. of Plate of ye Old Stad. of [*read* &] 1500 oz. of Plate of ye New, which after the rate of 5*sh.* 5*d.* Old & 5*sh.* 8*d* New will amount unto 1405£: 11*sh*: 00*d* & the silver will make about 1300 pounds of New Money in Rate. But this is not yet melted [n]or are any Recpts. given for it excepting for about 300 oz: I pray also your Lordsps directions whether this Plate shall be Coined or returned to ye Owners.(5)

The Silver to be got out of the Sweep will exceed all charges relating to ye first Melting & if your Lordsp please to allow the Melter after the rate of 5*sh.* per Ingott for battering the silver to fitt it for ye Melting Pott & for Melting it & makeing up ye Sweep (which I account a very moderate allowance) I shall be able to state the whole Acct. to your Lordsp. so soon as the sweep is made up & the Coinage finished.(6)

All wch is most humbly submitted to your
Lordsps great Wisdom

IS. NEWTON.

Mint Office
July 6*th* 1711

NOTES

(1) Mint/1, 8, 77. This is a clerical record copy. On 4 July the Treasury had resolved to instruct the Mint to make a detailed report of the plate brought in for sale, and to receive no more (*Cal. Treas. Books*, XXV (Part II), 1711, p. 79).

(2) In his letter of 15 May (Letter 840) Newton had estimated the value of all the plate then received as about £45500.

(3) On the 19th June the Two Million Act, having received the Royal Assent, was published.

(4) In the final reckoning (see below) Newton made the figures rather different, yielding a slightly smaller deficiency of £1915.

(5) Evidently, this plate was either returned to the owners or they were not allowed the recent premium on its Mint value, because no deficiency was reckoned for it.

(6) See below, Number 883.

856 CORNELIUS CROWNFIELD TO NEWTON

11 JULY 1711

From the original in the University Library, Cambridge(1)

Cambridge July ye 11*th:* 1711.

Sir,

Mr Cotes desir'd me to send you these sheets, that you may see how far we are gone, being now at a stand till he hears from you. The Last sheet Vv is only a foul Proof,[2] and will be workt off to day. Dr Bentley has deliver'd ye remainder of your Copy to Mr Cotes.

I am,
Sir,
Your most obedient Servant
CORN[ELIU]S CROWNFIELD

NOTES

(1) Add. 3983, fo. 14. The writer was Inspector of the Press; see Letter 777, note (2).

(2) This sheet ends on p. 336, early in Proposition 46. In the following pages Cotes had found the difficulties already discussed in his letter of 23 June. Newton remained silent from 18 June to 28 July because of Mint business.

857 COTES TO NEWTON

19 JULY 1711

From the original in the University Library, Cambridge.[1]
For the answer see Letter 889

Sr.

I wrote to You about a Month ago[2] concerning the 48th Proposition of Yr second Book, & the last week I desired the Printer to send You all the sheets which were printed off.[3] If You have received those sheets, You will perceive by them that ye Press is now at a stand. But having no Letter from You I fear the sheets have miscarried. The Compositor dunn's me every day, & I am forc'd to write to You again to beg Yr Resolution. I have received the last part of Yr Copy by Dr Bentley.[4] I have now read over & examined all the Calculations of ye former part which ends in the 432d page.[3] I will write to You concerning it assoon as I receive Yr Answer to my Last

I am Yr most Humble Servt
ROGER COTES

Cambridge July 19th 1711

For Sr Isaac Newton
at his House
in St Martin's Street
near Leicester Fields
London

NOTES

(1) Add. 3983, no. 15; printed in Edleston, *Correspondence*, p. 50, from the Trinity College MSS.

(2) See Letter 854.

(3) See Letter 856.

(4) According to Edleston, Bentley returned to Cambridge on 7 July, bringing the revised copy of the first edition from page 443 to the end, that is, from close to the end of Proposition 24, Book III (sig. Eee) onwards.

858 MEMORANDUM BY NEWTON

[before 24 JULY 1711]

From the original in the Public Record Office[1]

Upon the first of May the House of Commons made this Vote. Resolved that for encouraging the bringing wrought Plate into the Mint to be coined, there shall be allowed to such persons as shall so bring the same, after the rate of five shillings & five pence per ounce for the old standard & five shillings & eight pence per ounce for the new standard for all plate on wch the mark of the Goldsmiths company of London or any other City is set & for uncertain plate not so marked (being reduced to standard) after the rate of five shillings & six pence per ounce.

Upon the fift of May the Commons made these further Votes, Resolved that an humble Address be presented to her Majty that she will be pleased to give directions to the Officers of the Mint to receive in all such wrought plate as shall be brought to them & to give Receipts to such persons as shall bring the same for the amount thereof at the several rates & prices agreed by this House to be allowed for such wrought Plate as shall be brought to the Mint to be coined; & that the same be immediately coined into shillings & sixpences.

Resolved that all such Receipts to be given by the Officers of the Mint for any wrought Plate shall be accepted & taken for the full amount thereof in any payments to be made in any Loanes or any contributions upon any funds to be granted in this session of Parliament.

And upon the said Address her Maty gave directions accordingly by a Warrant dated May 10th, a copy of wch is hereunto annexed.[2]

Upon the 11th 12th & 14th of May we took in plate[3] & upon the 14th in the evening, an Order came from the Treasury to the Mint a Copy of wch is herunto annexed.[4] And upon that Order the following subscription of the Importers of plate[5] was taken for the future. We whose names are underwritten do declare that we do not expect that the value of the Plate by us this day delivered to the Officers of her Majesties Mint shall be accepted as part of the first payment of the summ for which there is a Bill now depending in the House of Commons

Upon the 17th of May the House Resolved that the Bill for regulating Hackney Coaches &c[6] should be committed to the Committee to whom the

Bill for raising 1500000 *lib.* was committed; And that it be an instruction to the said Committee to receive a clause, that the Receipts given by the Officers for Plate brought in pursuant of the Resolution of this House of the first instant be accepted as so much money in the contributions towards the said summ not exceeding two millions.

Upon the 12th of June the Bill was passed & upon the 19th was published, & from that time the Officers & people of the Mint gave notice to the Importers of Plate that the Parliament had made no provision for accepting the Receipts given for Plate since the 14th day of May.[7]

By her Majesties Warrant abovementioned the Master & Worker of her Majties Mint is authorized & required to pay the moneys produced from the said Plate into the Receipt of her Majties Exchequer & take Tallies for his discharge. But the Receipts for Plate imported since the 14th of May are not enacted to be accepted & taken in the Exchequer.

Quære. Whether her Maty may not authorise & direct the sd Master & Worker by a Warrant, a form of wch is hereunto annexed, to pay to the Importers of Plate whose Receipts are dated since the 14th of May, after the rate of five shillings per ounce imported, & so let the further account rest till the next Sessions of Parliament.[8]

[Form of Warrant drafted by Newton][9]

Now that our loving subjects may not want the use of both the Receipts & the new moneys coined out of their Plate Our will & pleasure is & we do hereby authorize & command you the said Master & Worker of our Mint to pay unto each of the importers of Plate whose Receipts are dated on or after the said 15th day of May, after the rate of five shillings per ounce of plate imported, any thing in our former Warrant to the contrary &c

[Forms of Disclaimer and Receipt drafted by Newton][9]

We whose names are underwritten do consent & agree that we will make no further demands of & from the Master & Workers of her Majts Mint for the plate of the old standard new standard & uncertain standard imported by us between the 14 day of May & the 20th day of June last past then our share of the moneys coined out of the said several sorts of plate in proportion to the gross weight of every sort imported by us. [The plate of the old standard, new standard & uncertain standard being melted assayed & coined severally] & our further share of such moneys as shall be imprested to the said Master & Worker for making good the recompence mentioned in the Receipts of our

Plate or be allowed us by Act of Parliament out of moneys in his hands. And we hereby acquit & discharge the said Master & Worker from all further demands. In witness whereof we have hereunto set our hands & seals the day & year underwritten.

Received after the rate of 5*s* per ounce the summ of the same being new moneys produced out of the plate within mentioned

[Opinion of the Attorney General on the above Memorandum][(10)]

The plate brought into ye mint on or after ye 15th day of May 1711 being melted & coyned ye notes haveing been given for ye value thereof according to her Matys directions to ye Officers of ye mint, the money wch is ye produce of ye plate belongs to her Maty & her Maty haveing directed the same to be payd into ye recei[p]t of ye Exchequer on a presumption yt the Parliamt would have made provision that those notes should be taken as money on some of ye funds given, wch not being made, I am of Opinion her Maty may by a Privy Seale direct the moneys to be applied towards dischargeing those notes as proposed

EDW. NORTHEY
July 24 1711

NOTES

(1) T1/135, no. 36; there is a holograph draft in Mint Papers, II, no. 539. The document is partially printed in Shaw, pp. 173–6.

(2) See Number 837, and *Journal of the House of Commons*, **16** (1803), 623, 629, 658.

(3) See Letter 838.

(4) See Letter 839.

(5) That is, vendors of plate to the Mint.

(6) This became the 'Act for licensing and regulating Hackney Coaches and Chairs...and for securing thereby...a yearly Fund of One Hundred eighty six thousand six hundred and seventy pounds for thirty two years to be applied to the Satisfaction of such Orders as are therein mentioned to the Contributors of any sum not exceeding Two Millions to be raised for carrying on the Warr and other her Majesties Occasions' (9 Anne c. 16); the Bill was a consolidation of two earlier Supply Bills, one for raising £1500000, the other £500000, on the security of various taxes (see *Journal of the House of Commons*, **16** (1803), 670 for 17 May).

(7) See Letter 839, note (3).

(8) This would enable Newton to pay the vendors of plate about 90% of what they were entitled to expect, leaving the provision of the remainder to the disposition of Parliament in the future.

(9) This is found in the draft only.

(10) This opinion was presented by Newton to the Treasury officers at the meeting on 27 July when the difficulty was temporarily resolved (compare *Cal. Treas. Books*, XXV (Part II), 1711, p. 85).

859 DRAFTS BY NEWTON

[before 27 JULY 1711](1)

From the Mint Papers

And whereas in pursuance of an address of the Commons of Great Britain in Parliament assembled to her Maty on or about the fifth day of May 1711, & by her Majties Warrant issued thereupon under the royal signe manual bearing date on or about ye tenth day of the same month several parcels of wrought Plate were received into her Majties Mint after the 14th day of May 1711 at such rates & prices as had been agreed unto by the said Commons: wch rates & prices for the said plate exceeded the moneys produced by the coinage of the same by the summ of one thousand nine hundred and fifteen pounds eleven shillings & six pence wch is not yet provided for: Be it further enacted that the same Deficiency or summ of one thousand nine hundred & fifteen pounds eleven shillings and six pence be supplied or made good out of the moneys wch have arisen or shall arise by the Duty on wines imported, commonly called the Coinage Duty.

[Revised version, also before 27 July][2]

A minute

Agreed by my Ld Treasurer that what moneys shall be due to Sr Isaac Newton at Christmas next for paying off the Receipts given by the Officers of the Mint for Plate, be then paid to him out of the civil list. And that the value thereof in Tin after the rate of 3*lb* 10*s* per centum be upon demand delivered to him for security thereof, to be sold after next January if the debt be not then paid.

NOTES

(1) Mint Papers, II, fo. 507. Possibly these drafts relate to the meeting at Treasury Chambers on 27 July; neither mentions the Royal Warrant of 30 July, and therefore they presumably both precede that, at any rate. The figure for the deficiency Newton gives here is that which he ultimately presented to Parliament (see Number 883). Some computations bearing on the deficiency may be found at Mint Papers II, fo. 506. He first reckoned the total paid out for plate since 10 May 1711 as £76670 2*s* 11*d*, the yield of coin minted being £71512 2*s* 4*d*, and the deficit therefore £5158 0*s* 7*d*, that is about 6·7 % of the sum paid out. Most of this loss did not concern him, as it was concealed in the certificates issued before 15 May. Then he did a similar calculation for the period directly relating to his own accounts, from 15 May onwards (probably to 19 or 20 June only), and found that from the plate for which £27824 11*s* 2*d* was paid out coin to the value of £25911 10*s* 6*d* was minted, producing a deficiency of £1913 0*s* 8*d*. This deficit is about the same as before (6·8 %). These figures were slightly amended later.

(2) Mint Papers, II, fo. 518, likewise holograph. This method of repaying the Mint from the Civil List was adopted by the Treasury on 10 August (*Cal. Treas. Books*, XXV (Part II), 1711, p. 87, and P.R.O. Mint/1, 8, 79).

860 NEWTON TO COTES

28 JULY 1711

From the original in Trinity College Library, Cambridge.[1]

Reply to Letters 854 and 857; for the answers see Letters 863 and 871

St Martins Street in Leicester Fields
London July 28th 1711.

Sr

I received your Letters & the papers sent me by the Printer But ever since I received yours of June 23 I have been so taken up with other affairs that I have had no time to think of Mathematicks. But now being obliged to keep my chamber upon some indisposition wch I hope will be over in a day or two I have taken your letter into consideration. You think that in the Corollary to the 48th Proposition these words [Nam lineola Physica $\epsilon\gamma$ quamprimum ad locum suum primum redierit, quiescet] consist not wth what I assert & prove in the Proposition, vizt [& propterea vis acceleratrix lineolæ Physicæ $\epsilon\gamma$ est [ut][2] ipsius distantia a medio vibrationis loco Ω]. But I suspect that you take the words [ad locum suum primum] in another sence then I might intend them. For when all the lineolæ physicæ $\epsilon\gamma$ are returned to their first places or places in wch they were before the vibrations began, the medium will be uniform as before & the vis acceleratrix of the lineola physica $\epsilon\gamma$ will cease, whether that lineola arrived to its first place in the beginning middle or end of the vibration.[3] For making the Corollary more intelligible, these words may be added to the end of it.[4] Partes fluidi non quiescent nisi in locis suis primis. Quamprimum in loca illa motu retardato redierint, component Medium uniforme quietum quale erat ante vibrationes excitatas.[5]

In altering the 48th Proposition you have shortned the Demonstration. If you had proposed your alteration of the Corollary I should have been better able to compare the whole wth mine.

Your emendations of Prop 49 are very well & the two Corollarys you propose may be added to it. And the 47th & 48th Propositions may change places, & at the end of the 47th these words may be added [Hæc Propositio ulterius patebit ex constructione sequentis

I will write to you about [the][2] third book in my next. I am

Sr

Your very humble Servant

IS. NEWTON

NOTES

(1) R.16.38, no. 212, printed in Edleston, *Correspondence*, p. 51.

(2) Editorial insertion. All the unannotated brackets in this letter were written by Newton.

(3) There is a faint, small 's' in the MS, which may be an afterthought. One hesitates to differ from Edleston, who printed 'vibrations', but the singular makes better sense, although the sentence is still difficult to interpret, unless we turn to Letter 889, where Newton clarifies his meaning.

(4) This was not done.

(5) 'The particles of a fluid come to rest only in their first places. As soon as they return to that place with a retarded motion, they make up a uniformly tranquil medium, such as there was before the vibrations were excited.'

861 MEMORANDUM BY NEWTON

[c. 28 JULY 1711]

From the draft in the Mint Papers[1]

By the Coinage Act & the Indenture & usage of the Mint the Master & Worker received Gold & Silver only in the mass at the just value by weight & assay to be coined. He may buy bullion of uncertain value, but not knowingly to loss, & must account for the profit. But this way of buying of bullion is not in use.

When plate or old moneys are to be coined the Importer[2] either causes the same to be melted into Ingots at his own charge before delivery or delivers it to a general Importer who causes it to be melted into ingots, & the Master of the Mint receives the ingots by weight & assay to be coined. Or if Plate or old moneys be delivered in specie to the Master he either melts the same into ingots in the presence of persons appointed to see it done, or delivers the same by weight into the custody of persons appointed to carry it to the melting pot & deliver it back to him in Ingots by weight & assay to be coined. For the Master is not to be trusted with silver of uncertain value without due checks upon him.

When the present Master & Worker was first spoken to about receiving the plate, he represented that he was ready to receive it & give receipts for the same by weight, & that some person or persons might be appointed to carry it from him by weight to the melting pot & deliver back to him by weight & assay the ingots produced & keep an account of the meltings. This was the method of coining the Vigo plate.

Some days after when the House of Commons voted an Address to her Majty to give directions to the Officers of the Mint to receive Plate the Master of the Mint was perplexed thereat & told his fellow Officers that nothing more was to be understood by that Address then that her Maty should give directions to the proper Officer or Officers, & accordingly prepared a Warrant for himself alone with blanks for the names of his fellow Officers to be inserted by the Lords Comm[ission]ers of the Treasury if they thought fit. But the Warden of the Mint(3) fell into a passion at the blanks & said he would not go into the Lords unless the blanks were first filled up, & at his desire they were filled up. Then the Master prepared a distinct Warrant for himself as Master to coin the Plate, but the Warden opposed it.

When the two Million Act(4) was published & the Master alone (after a stay of some days for the concurrence of his fellow Officers) acquainted the Lord H. Treasurer(5) wth the defect of the Act & in a second memorial, laid the state of the plate before his Lordp,(6) & in order to a third memorial was informing himself whether 5*s* per ounce would content the Importers till the Parliament met, & told the Warden that he found that it would: the Warden declared against it unless the Importers would deliver up their receipts upon payment of what the plate produced & take certificates for the remainder. Which the Importers being averse from, the Master desisted till he heard that the Officers of the Mint would be summoned to attend his Lordp & then stated the case to the Attorney General & brought the Attorneys opinion to his Lordp(7) wth the form of a Warrant for paying 5*s* per ounce to the Importers, being fully satisfied that it would have quietned them till the meeting of the Parliamt if the Warden of the Mint would have been content with an endorsement of the payments without taking back the Receipts given out for plate.

NOTES

(1) Mint Papers II, fo. 537, holograph. As usual, the paper is undated but it was certainly written after the meeting at Treasury Chambers on 27 July, and probably before the issue of the Royal Warrant on the 30th, since this is not mentioned. Presumably Newton's intention was to obtain such a Warrant.

(2) Vendor.

(3) Craven Peyton.

(4) See Memorandum Number 858, note (6).

(5) See Letter 848, note (2).

(6) See Letters 852 and 855.

(7) See the end of Number 858.

862 ROYAL WARRANT

30 JULY 1711

From the copy in the Public Record Office[1]

Warrt in relacon to plate taken in at ye
Mint on & after ye 15th May 1711

Anne R.

Whereas upon the Votes & Address of Our Commons in Parliament Assembled, We did by Warrt. under Our Royll Sign Manuall dated ye 10th day of May last[2] Authorize & Command, the Warden Master & Worker & Comptroller of Our Mint in Our Tower of London to take & receive from all persons & Bodys politick or Corporate all such wrought plate as they or any of them should bring to Our said Mint of the kinds & standards mencond in the said Votes, And to give such Rects. for the same as were deliver'd in the said Address of Our Commons, and forthwith to cause the same to be melted down & assayed, and that you ye Master & Worker of Our said Mint should imediately Coin ye same into shillings & sixpences & pay the Money produced into ye Rect. of Our Excheqr. & take Tallys for Your discharge, And Whereas in the Act of Parliament Entituled[3] [An Act for Lycencing & regulating Hackney Coaches & Chairs, & for Charging certain new Dutys on Stampt Vellom parchmt & paper & on Cards & Dice & on the Exportation of Rock salt for Ireland and for securing thereby, and by a Weekly payment out of ye Post office and by Sevll Dutys on Hides & Skinns a Yearly Fond of 186670£ for 32 years to be Applyed to the Satisfacion of such Orders as are therein mencond by the Contributors of any Summ not Exceeding Two Millions to be rais'd for Carrying on the Warr and other Her Mats. occasions] it is Enacted That all and every the Rects. given for the plate brought into the said Mint before ye 15th day of May 1711 should be accepted & taken as so much Money for the first Fourth part of the Contribucons upon the said Act, but no provision is made for accepting the Rects. dated on or after the said 15th day of May, Now That Our loving Subjects may not want the use both of the Rects. & ye New Moneys Coined out of their plate Our Will & pleasure is And We do hereby Authorize & Command You the said Master & Worker of Our Mint to pay unto Each of the Importers of Plate whose Rects. are dated on or after the said 15th day of May at the Rate of *5sh* per Ounce for ye plate Imported[4] any thing in Our former Warrt. to the Contrary in any wise notwithstanding, But You are to take Care, that an Exact Acct be kept of ye difference between the Summs so paid by You in the pursuance hereof, And the Rates at wch the said plate was deliver'd according to Your said Receipts and of the Totall Amount thereof. And for so doing this shall be Your Warrt. Given at Our Court at Windsor Castle the 30th day of July 1711 In the Tenth Year of Our Reign

By her Mats. Comand
Oxford

To ye Warden Master & Worker
& Comptroller of Our Mint

NOTES

(1) T/52, 25, 75.

(2) See above, Number 837.

(3) 9 Anne C. 16; the bill was framed by consolidating two previous supply bills on 17 May 1711, and passed on 12 June. The square brackets are in the original.

(4) This payment was less than the normal Mint rate of *5s 2d* for imported plate—no doubt because this was a temporary measure, intended to pacify the vendors of plate by paying most of what was due to them, pending a provision for the deficiency that would arise through paying them the full Parliamentary rates per ounce, but also because Peyton and Ellis were fearful of paying the vendors too much, without specific authority for so doing. See Number 864.

863 COTES TO NEWTON

30 JULY 1711

From the original in the University Library, Cambridge.[1]

Reply to Letter 860; for the answer see Letter 889

Sr.

I have read Yr Letter & find my self obliged to trouble You once more. I must beg leave to tell You I am not as yet satisfied as to the Inconsistency which I mentioned in my former Letter. You seem to say, that when the Lineola Physica $\epsilon\gamma$ is return'd to its first place (which You take to be the beginning of ye Vibration) the Medium will be uniform as at first & consequently its Vis Acceleratrix will cease. If upon ye return of the Lineola to its first place it be granted yt the Medium will be uniform,[2] I confess it must also be granted yt the Vis Acceleratrix will cease: but then if ye Vis Acceleratrix does cease in this place, it must likewise be granted that its quantity is less than in places which are nearer to ye middle of ye Vibration where it does not cease, & of consequence its quantity will not be proportionable to ye distance of ye Lineola from ye middle of the Vibration; for to be proportionable it ought not to cease in ye beginning of the Vibration, but on ye contrary it should be greater there than in any other place: and if it be greater there than in any other place, the Medium will not then be uniform. This consideration was to me the occasion of altering the Proposition. By making the middle of ye Vibration the *Locus primus* I saw this Inconsistency might be avoided. But besides this, it appeares altogether reasonable upon other accounts yt the Locus primus should be in ye middle of ye Vibration.(3) Suppose a Musical Chord to be put into motion; t'is certain it's Locus primus is ye middle of its Vibration, & consequently the Locus primus of any contiguous Lineola Physica of Air is also in the middle of its own Vibration, for the motion of the Lineola Physica follows & depends upon ye motion of ye contiguous Chord. And for ye same reason a second Lineola physica not contiguous to ye Chord but to ye first Lineola will have its Locus primus in the middle of its own vibration, since its motion depends upon the first as the first did upon ye Chord it self; & the same may be said of other Lineolæ which are yet more remote from ye Chord.

Now assoon as ye motion of ye Chord ceases in its Locus primus, that is in ye middle of its Vibration, if it should perhaps be said yt the motion of ye first Lineola would not cease of it self at ye same time, yet tis evident it will be made to cease by the Resistance of ye Chord: for being contiguous to ye Chord when it is arrived at its Locus primus or ye middle of its Vibration it can proceed no further towards ye Chord if ye Chord maintains its rest, & it cannot return back again from ye Chord as having no Vis Acceleratrix or acquired Impetus that way. And as this first Lineola ceases by ye Resistance of ye Chord so ye second ceases by ye Resistance of ye first, & so on. By this You will understand how I would alter ye Corollary: but I chose rather to refer it to Yrself as fearing I could not express my thoughts with suff[icient] clearness & brevity & exactness at ye same time. What I have represented above is not perhaps so exact as it should be, for the motions of ye Lineola must be suppos'd gradually to cease with ye motion of ye Chord; but I chose to express my self as I have done yt You might the more clearly understand me. In Altering ye Proposition(4) I altered ye 4th line of Page 366 by putting *Pl*, *Pm*, *Pn* instead of *Pn*, *Pm*, *Pl*; & in ye 2d line of Page 367 instead of [ob brevitatem pulsuum] I have put it [ob angustos limites Vibrationum] for it would be truer & more to ye purpose to say *ob magnam pulsuum distantiam* than to say *ob brevitatem pulsuum*. In Your Example taken from Mr Sauveur(5) the Latitude of ye Pulse is about 10 foot, when perhaps ye Space of Vibration is not above ye 10th of an Inch at ye utmost. If You consent to my alteration of ye Proposition the Figure must be altered; I propose to have it cut like ye Figure I sent You(6) which does better express the disproportion of ye breadth of ye Pulses & Vibrations than the former Figure. I wish You a good recovery of Yr health & am, Sr.

Yr most Humble Servant
ROGER COTES

Cambridge July 30 1711

I have wrote to You by this first opportunity yt ye Press might not stay.

For Sr Isaac Newton
at his House
in St Martin's Street
in Leicester-Fields
London

NOTES

(1) Add. 3983, no. 16, printed in Edleston, *Correspondence*, pp. 52–4, from the Trinity College MSS.

(2) Cotes is ambiguous here in his use of the word 'uniform', which can have little meaning in a vibrating medium. If, however, as seems probable, he means at the same pressure as the uniform medium at rest, then his argument is faulty, and the whole of his proposed alteration to Proposition 47 (see Letter 854, notc (6)) invalid.

(3) The argument which follows, and Newton's reply in the next letter (Letter 889), emphasize the somewhat irrelevant difficulty Cotes and Newton were labouring under. Both were concerned with relating the medium at rest to the medium in a state of essentially undamped simple harmonic motion, the analysis of whose nature constitutes the problem here, but did not recognize how complex was the concept of the *setting-up* of the waves in the medium. Nevertheless, Newton's treatment of wave-motion is a great advance on all previous attempts.

(4) Proposition 47 (new numbering), and see Letter 854.

(5) In the Scholium to Proposition 50 (*Principia* 1713; p. 344), having calculated the speed of sound in air on the theoretical principles already mentioned, Newton referred to experiments of Joseph Sauveur rather than to the less exact ones of Mersenne to which he had appealed in the first edition. Sauveur (1653–1716), professor of mathematics at the Collège Royale and member of the Académie Royale des Sciences (1696, and again at the refoundation of 1699) was the founder of the science of acoustics (his own name). In the *Histoire de l'Académie Royale des Sciences: Année 1700*, second edition, Paris 1761, p. 140 there is an account of his experiments (depending on the exact measurement of low-frequency beats created by pipes or strings of nearly equal magnitudes) to determine the wavelength of a *son fixe* ($=100$ Hz)—the velocity being unknown. He found that this tone was given by a pipe about five feet long, whence $\lambda = 10$ ft. From his computed velocity of sound Newton derived the wavelength of 10·7 Paris feet for 100 Hz.

(6) See Letter 854, note (6), where Newton's figure and Cotes' modification of it are shown.

864 DRAFT WARRANT BY NEWTON

[c. 31 JULY 1711]

From the Mint Papers(1)

Whereas by Warrant under our signe manual dated the 30th Instant we directed you the Master & Worker of our Mint in the Tower of London to pay unto each of the Importers of Plate whose Receipts are dated on or after the 15th day of May at the rate of five shillings per ounce for the plate Imported Our further Will & Pleasure is that out of such moneys as shall remain in your hands to be impested to you or advanced by you for this service you pay off the remainder of the moneys for wch the said Receipts were given, that is to say five pence per Ounce old standard eight pence per Ounce new standard & six pence per ounce uncertain standard above the aforesaid 5*s* per ounce & take back the Receipts so discharged. And for so doing this shall be your Warrant.

NOTE

(1) Mint Papers II, fo. 518. Since the Royal Warrant is mentioned as being 'dated the 30th Instant' this draft must have been written on 31 July. Allowing Newton a little forgetfulness of the date, it might have been written in the early days of August, 1711. Clearly Newton was now concerned to see that the late vendors of plate (but not after 19 June, probably) were paid not only the five shillings that could not be withheld from them on any pretext, and whose payment was now authorized by the Warrant, but extra pence allowed them by Parliament, for which the Warrant was no authority.

The further Warrant here drafted by Newton was not, it seems, ever issued; perhaps because, on 10 August, Newton's fellow Officers, Peyton and Ellis, agreed that Newton should pay to the late vendors of plate, on production of their receipts, the difference between the silver-value of their plate and the prices per ounce stated by Parliament (see P.R.O. Mint/1, 8, 79). The making up of the deficiency was ultimately provided for by resolution of the House of Commons (see Number 883).

865 NEWTON TO OXFORD

7 AUGUST 1711

From the draft in the Mint Papers(1)

To the Rt Honble the Earl of Oxford & Earl Mortimer
Ld H. Treasurer of great Britain.

May it please your Lordp

The Smith of the Mint being my servant by the ancient constitution of the Mint, and being paid by me after the rate of one penny per pound weight of gold coined & one farthing per pound weight of silver coined, and having also a Salary of 50 pounds per annum appointed by the late & present Indentures of the Mint: the last Smith[2] for the sake of that Salary was imposed upon me & behaved himself to me & others of the Mint with great insolence. Whereupon a clause was inserted into the Schedule of Salaries at the end of the Indenture of the Mint, for the ceasing of that salary upon the next voidance of the place; and the place becoming void before last Christmas, the salary is now ceased in order to a new settlement. I humbly pray therefore that such a new salary may be setled as your Lordp shall think fit & in such a manner that I may have power over the Smith as my servant for carrying on the coinage & dismiss him if he be not of good abearing according to the meaning of the Indenture of the Mint. Which may be done by appointing the Salary to me for a Smith.

All wch is most humbly submitted to your Lordps great wisdome.

I.N.

Mint Office
Aug. 7th, 1711

NOTES

(1) Mint Papers I, fo. 223.

(2) His name was Thomas Sylvester. His successor, Richard Fletcher, enjoyed a salary of £40 p.a. from Christmas 1711; he was already employed at the Mint. Compare Letter 954 below.

866 LOWNDES TO THE MINT

16 AUGUST 1711

From the original in the Mint Papers.[1]
For the answer see Letter 868

Gentlemen.

By Order of my Lord Treasurer I send you here inclosed a Spanish peece of money being a two Ryall peece;[2] My Lord directs you to Report to him the Weight, Finess, and value of the same. I am

Gentlemen
Your most humble servant
WM LOWNDES.

Treasury Chambers
Augt: 16*th*. 1711.

NOTES

(1) Mint Papers II, fo. 201.

(2) A coin of two reales, or quarter dollar. The Treasury needed information as it proposed to remint the dollars borrowed from the Genoese ships into this more convenient form. In fact, the dollars were handed over at Port Mahon on 9 September (*Cal. Treas. Papers, 1708–14* (1897), p. 310).

867 NEWTON AND PEYTON TO OXFORD

20 AUGUST 1711

From the original in the Public Record Office[1]

To the Most Honble: the Earle of Oxford and Earle Mortimer
Lord High Treasurer of Great Britain

May it please Your Lordp.

In obedience to your Lordships direction to lay a State before your Lop. of the several parcells of Tin that are now under Mortgage, the persons to whome, the places where, for what Terms, and upon what conditions, wth the Charge of keeping the said Tin: We humbly represent to your Lordp. that upon a Loan of 100000 *lib.* made July the 1st 1710; 1600 Tunns of Tin were mortgaged to Mr Moses Berenger[2] and sent into Holland at her Maties Charge, That upon a Loan of 10000 *lib.* 1st July 1710. 160 Tunns of Tinn were mortgaged to Sr. Theodore Janssen[3] and remain still in ye Mint under the key of the Store-keeper, That upon a Loan first of 40000 *lib.* 1st of September 1710. and then of 20000 *lib* more 4th October. following, 960 Tuns of Tinn were mortgaged to Sr John Lambert, Mr Edward Gibbon, and Mr. Francis Stratford and sent to Hamburgh at her Maties Charge.[4]

That these several summs were Lent at ye Interest of 6 per cent & that ye Charges of Warehouse room Commission or reward in Holland and at Hamburgh are paid out of this Interest without any further Charge to her Maty.

That Mr. Berenger is to be paid 30000 *lib.* at Christmas next, 30000 *lib.* more at Lady day next and 40000 *lib.* more at Midsummer next. Sr. Theodore Janssen is to be paid his 10000 lib. at Christmas next, and Sr John Lambert and partners 40000 *lib.* at Christmas 1712. and 20000 *lib.* at Christmas 1713. and upon these payments the said parties are to return to her Majesty the said Tin mortgaged in proportion to the Summes paid to them, yt is for every 10000 *lib.* the quantity of 160 Tunns of Tinn, and in default of payment the said several partyes are allowed to Sell and dispose of her Majties Tinn at the best Market price that can be gotten in such proportions as shall be sufficient to make good the repayment of the Principall and Interest, they ye sd severall parties being answerable to her Majty for the overplus. But we do not find that they have yet begun to sell any of the sd Tinn.

All which is most humbly submitted to your Lordships great Wisdom

CRAV. PEYTON
IS. NEWTON

Mint Office the 20th Augst
1711

NOTES

(1) T. 1/136, no. 38; it is in a clerical hand, signed by Peyton and Newton.

(2) See *Cal. Treas. Books*, XXIV (Part II), 1710, pp. 27–8. For a later agreement with this London merchant concerning this loan see *Cal. Treas. Papers, 1708–14*, p. 469.

(3) See above, Letter 845.

(4) Of these three men, Edward Gibbon (1666–1736), grandfather of the historian, made a fortune out of army contracts. At this time he was one of the Commissioners of the Customs. As a director of the South Sea Company most of his property was confiscated after the Bubble burst, but he is said to have amassed through his commercial acumen a second almost equal to the first. The other two are several times mentioned in *Cal. Treas. Papers* in connection with the export of tin. (See *Cal. Treas. Books*, XXIV (Part II), 1710, pp. 44, 89.)

868 THE MINT TO OXFORD

21 AUGUST 1711

From the holograph original in the Public Record Office.[1]
Reply to Letter 866

To the Rt Honble the Earl of Oxford & Earl Mortimer
Lord High Treasurer of great Britain

May it please your Lordp

The Spanish piece of money of two Ryalls sent by your Lordps order to the Officers of the Mint to be weighed assayed & valued is in weight 3 dwt 6 gr, in assay 3½ dwt worse then standard, & in value ten pence wanting the third part of a farthing. And twelve such pieces are worth 9*s.* 11*d*

CRAV. PEYTON
IS. NEWTON

Mint Office
21 *Aug.* 1711

NOTE

(1) T. 1/136, no. 39; draft in Mint Papers, II, fo. 201, signed by Newton alone. Ellis had ceased to be Comptroller from 11 June 1711, which accounts for the absence of his name. The letter is printed in Shaw, p. 176.

869 NEWTON TO OXFORD

28 AUGUST 1711

From the draft in the Mint Papers(1)

My Lord

I herewith send your Lordp a copy of the Report of the Officers of the Mint upon a Petition from Cornwall for a new contract for Tin at such a price as her Maty might not lose by, & for taking off a greater quantity of Tin then before.(2) A few months before the date of this Report I computed the price at 3£ 8*s* 6*d* supposing no more Tin to be received then the consumption would carry off: for preventing of wch I added the last clause of this report.

I send your Lordp also a further Report(3) upon the present state of the Tin, and remain

My Lord

Your Lordps most humble
& most obedt servant
Is. Newton

Mint Office
28 *Aug.* 1711.
Lord H. Treasurer.

NOTES

(1) Mint Papers III, fo. 483.

(2) The documents bearing on this are scanty. On 14 February 1711 a petition of the tinners of Cornwall about surplus tin was read at the Treasury, and again on an unspecified day in April a petition from the 'late Agents for tin in Cornwall' was read and referred to the Mint (*Cal. Treas. Books*, xxv (Part II), 1711, pp. 172, 625, 628). No copies of these petitions nor of the Mint report enclosed with this letter by Newton have been found. The present contracts for the Queen's purchase of tin were quite recent—that with Cornwall took effect from the first of June 1710 and that with Devon from 21 November 1710. Each was for six and a half years. Presumably the producers found that the yield of the mines was exceeding the Queen's guaranteed purchase; but as Newton well knew, even that quantity was hard to dispose of.

(3) See the next letter, of the same date.

870 NEWTON TO OXFORD

28 AUGUST 1711

From the holograph original in the Public Record Office[1]

To the Rt Honble the Earl of Oxford & Earl Mortimer,
Lord High Treasurer of great Britain.

May it please your Lordp

I humbly beg leave to lay before your Lordp a further account of the Tin. Mr Drummond[2] by a Warrant dated Octob 20th 1704, had 400 Tunns of Tin consigned to him & company at Amsterdam to be disposed of after the best rates not under 44½ Gilders per hundred weight of Holland (that is, not under 4 *lib.* 2*s.* 4*d.* per centum[3] Averdupois) for a Commission of two per cent clear of all charges & advanced 22500 *lib.* upon it at 4 per cent upon notice of its arrival at Amsterdam. And the like quantity was consigned to him June 8th 1705 on the same terms. The first parcel was sold in about 15 months, the second (by opening a trade over the Rhene[4] into France) in about 9 or 10 months. And both parcels produced by sale 61714 *lib.* clear of interest & all other charges except the duty of 3*s* per centum. Which produce is after the rate of 3 *lib.* 17*s.* 2*d* per centum. And part of this money came in sometime after the sales.

Mr Stratford[5] & Mr Free, Feb. 20th 1704/5 had 240 Tunns of Tin consigned to Hamburgh upon the like terms, & upon notice of its arrival at Hamburgh advanced 12000 *lib.* upon it at 4 per cent. Sixteen blocks were lost by an insolvent chapman, & the rest produced 18989 *lib.* 12*s.* 4*d* clear of interest & all other charges except the duty of 3*s.* per centum. Which produce is after the rate of 3 *lib.* 19*s.* 1½ per centum. And part of this money came in sometime after the sales. These Commissions were given for quickening the sale of her Maties Tin, while the Pewterers sold their own stock to the Merchants at home.

The present Commissions to sell when the time comes, are less advantageous then the former.[6] There is no lowest price set. They interrupt the course of payments. And by the great quantity of Tin lying abroad put a damp upon the Markets till the sale begins.

A merchant may have Tin here for 3 *lib.* 19*s.* 0*d.* per centum including the Duty of 3*s* per centum, & will scarce reccon the shipping it off & carrying it to Hamburgh at above 1*s* per centum. The Duty there & housing it may be 6*d* more, & the interest of the price till it arrives at Hamburgh 3*d.* Tin therefore stands the merchant in about 4 *lib.* 0*s.* 9*d* at Hamburgh. And if her Majty should sell it there to the Merchant for ready money at a set price, something must be abated of 4 *lib.* 0*s* 9*d* to incline the Merchant to buy there rather then at London. If 4 *lib.* per centum should be the price, & the charges of an Office for selling it, wch would scarce be less then 1*s* 6*d* or 2*s* per centum, be deducted; the Queen would receive but 3 *lib.* 18*s*. 0*d* or 3 *lib.* 18*s*. 6*d* per centum, & this without having any part of the money advanced.

If Mr Stratford will give after the rate of 4 *lib.* per centum & within one month after the consignement of every hundred Tuns for sale, advance the full price of 8000 *lib.*; the bargain will be manifestly more advantageous then any of the Commissions above mentioned. For besides the duty of 3*s* per centum saved to her Majty, it would revive the market at Hamburgh, bring in money by the sales, & diminish the Tin abroad, & her Majty would run no risque of selling upon trust, nor stay for any part of the money till debts can be got in. And to take 100 Tunns at a time & pay here within a month is better by four or five months interest of the money then to take only 10 or 20 Tunns at a time for ready money at Hambrugh, besides the charges of an Office for selling it there.

And as for the Objection of a monopoly, it lies as much against all the Commissions above mentioned. Merchants would not call it by this name, nor hath it the faults of a monopoly. Mr Stratford will not disable other merchants from sending Tin to that Market if they can get by it, & a high price would invite them thither. He might be enabled to sell a little lower then other merchants, (wch would promote the sale) but not to raise the price of the market.

The Tin in Mr Berangers hands, when his course of payment comes, may be sold to other Merchants by parcels to pay off his debt, and Mr Drummond may probably prove a good chapman.

The Tin sold in June July & August comes to about 11000 *lib.* & the sales in the Quarter ending at Christmas use to be less then in any other Quarter. I do not expect that the sales between this & Christmas will exceed ten or twelve thousand pounds. They may amount to eight. I hear of no Turkey Fleet(7) to go out before Christmas.

The Officers of the Mint have been of opinion that her Majesty loses something by sending Tin abroad upon Commissions & particular contracts. But the Tin is already abroad, and whether her Maty shall now be at the charge of setting up new Offices abroad for selling no more then 200 Tunns per an, in an Office, or sell it by contract or commission, is a new Question, & has made me think it my duty to state the matter thus fully to your Lordp.

All wch is most humbly submitted to your Lordps great wisdome

IS. NEWTON

Mint Office
28 *August*
1711.

NOTES

(1) T/1, 136, no. 61; there are drafts in the Mint Papers, III, fos. 482–5, 504. Compare also Newton's letter to Lord Oxford on the same subject of 22 March 1713/14, Letter 1051, vol. VI (in progress).

(2) John Drummond & Co. was a merchant house trading in Amsterdam and regularly entrusted with the sale of tin there.

(3) Newton consistently used the symbol ⊕.

(4) Rhine (L. *Rhenus*).

(5) See Letter 867.

(6) Presumably Newton refers to the loans detailed in Letter 867; since the sums advanced were unlikely to be repaid, the tin would be for the merchants to sell.

(7) That is, the Levant Company's fleet. The Company exported tin.

871 COTES TO NEWTON

4 SEPTEMBER 1711

From the original in the University Library, Cambridge.[1]
Reply to Letter 860; for the answer see Letter 889

Cambridge Sept. 4th 1711

Sr,

I received a Letter from You about a Month ago, & sent You an Answer to it the next day by the Carrier, in which I gave You my reasons why I was not satisfied as to the Inconsistency in the 48th Proposition & its Corollary which I formerly mention'd to You. I have not heard from You since that time & therefore I fear that either my Letter or Your Answer to it has miscarried. I shall be glad to know Your resolutions concerning this 48th Proposition assoon as You have leasure, that the Press may go on.[2] There were some things relating to the third Book in my former Letter, I hope You will not forget to let me know Your mind concerning them also.

I am Sr, Your most Humble Servt
ROGER COTES.

For Sr Isaac Newton
at his House
in St Martin's-Street
in Leicester-Fields
London

NOTES

(1) Add. 3983, no. 17, printed in Edleston, *Correspondence*, p. 54, from the Trinity College MSS.

(2) Despite Edleston's excuses and Newton's preoccupation with Mint affairs, no real reason for his failure to send Cotes one (recorded) word of reply is known and his silence is

hard to excuse, especially as (when he did reply) Newton merely brushed Cotes' difficulties aside. It is natural to suppose that Newton was annoyed with Cotes for pressing objections, and tired of the *Principia*. Cotes' next letter, on an entirely different matter with no allusion to the *Principia*, leads one further to imagine that Cotes received some hint that Newton did not mean to be hurried, or to be too hard pressed by his editor; compare Letter 873.

872 FLAMSTEED TO SHARP

20 SEPTEMBER 1711

From the original in the Library of the Royal Society of London[1]

The Observatory. Sept 20 ♃ 1711

...Mr. Halley has spoyled my Catalogue in printing it, and thereby put me to the trouble of reprinting it; he is doeing the same by some of my Observations and Sr Is: Newton furnishes him perfidiously with Materialls he has given him they [*sic*] places of the Moon Imparted to him in the 3 large Synopsis (that you have seen here) which he publishes the Moons Observed places from, (as I find by one of his Sheets, I have got from the press) tho they were communicated to him on this condition, precisely and expressly that he should not impart them to any one, and this reason given him for it, that they were determined by the help of a small Catalogue of the Fixed Stars I had rectified (by the help of such Observations as I had got with the Sextans) to the beginning of the year 1686, and that I intended when I had finished the Great Catalogue I was then (1694) entering upon I would correct and calculate them a new with such others as I should gaine afterwards, but Sr I: N: is to great a person to be a slave to his word, and Dr: Ha: is resolved to spoyle every thing that falls within the reach of his fingers. but of this I must give you a fuller account hereafter, at present I have other business to recommend to you...

NOTE

(1) Sharp Letters, no. 77, printed in Baily, *Flamsteed*, p. 293.

873 COTES TO JONES

30 SEPTEMBER 1711

From the original in private possession[1]

Cambridge Sept. 30*th.* 1711

Sr.

I return You my Thanks for Your Letter & the information You gave me concerning the state of Mathematicks at present in London...We have nothing of Sr Isaac Newton's that I know of in Manuscript at Cambridge, besides the first draught of his Principia as he read it in His Lectures; his Algebra Lectures which are printed; & his Optick Lectures the substance of which is for the most part containd in his printed Book,(2) but with further Improvements...I am very desirous to have the edition of ye Principia finished, but I never think the time lost when we stay for Sr Isaac's further corrections & improvements of so very valuable a Book, especially when this seems to be the last time he will concern himself with it. I am sensible his other Buisness allows him but little time for these things & therefore I cannot hasten him so much as I might otherwise do, I am very well satisfied to wait till he has leasure. I am Sr

Your Hearty Freind & Servant
ROGER COTES.

For
Mr Wm. Jones
to be left for him
at Child's Coffee-house
in St Paul's Church Yard
LONDON

NOTES

(1) Printed in Rigaud, I, pp. 258–60.

(2) That is, *Opticks*. In fact, the Latin text of the *Lectiones Opticæ* was to be separately published some years later in 1729, and there the anonymous editor points out (Preface, p. vii) that most of the experimentation on thin plates described in *Opticks* is missing from the *Lectiones*, but conversely the latter contain many geometrical demonstrations missing from *Opticks*. His copy of the *Lectiones* was one given to David Gregory; at the end he listed five pages of variants or corrections taken from the Lucasian copy in the University Library, Cambridge, to which Cotes refers (MS. Dd. 9. 67). Cotes' transcript of this MS is in Trinity College Library, MS. R. 16. 39.

874 THOMAS HARLEY TO NEWTON

3 OCTOBER 1711

From the copy in the Public Record Office.(1)
For the answer see Letter 875

Sir Isaac Newton abt. Spanish Money

Sir.

By Order of My Ld. Trearer I send you inclosed three peices of Spanish Money wch. His Lordp desires you to Essay and Certifie him the weight fineness & value thereof(2) I am &c 3*d*. 8*ber* 1711

T. HARLEY

NOTES

(1) T/27, 20, 14. Thomas Harley was the cousin of Robert Harley, the Lord Treasurer, and acted as his secretary from July 1711. In February 1714 he was sent by Oxford on a confidential mission about the succession to the Electress Sophia (see Angus McInnes, *Robert Harley, Puritan Politician* London, Gollancz, 1970, p. 159).

(2) The pieces were new coins minted by Charles III of Spain.

875 [NEWTON TO OXFORD]

[5 OCTOBER 1711]

From a holograph draft in the Mint Papers.(1) Reply to Letter 874; for the answer see Letter 876

The three Spanish pieces of two Ryals wch were sent to me by your Lordps order [to] be examined are very unequally coined. One of them weighed 3 dwt $6\frac{1}{4}$ grains & is $4\frac{1}{2}$ dwt worse than our standard. Another weighd 3dwt 13 grains & is standard scant. A third weighs 3 dwt $10\frac{2}{3}$gr & is $4\frac{1}{2}$ dwt worse. These three & that wch by your lordps order I caused to be assayed some weeks agoe & wch was $3\frac{1}{2}$ dwt worse then standard, & weighed 3dwt 6gr, are one wth another at a medium in weight 3 dwt 9gr & in Assay $3\frac{3}{16}$ dwt worse than sta[ndard] & in value $10d\frac{5}{16}$. And a Mexico piece of eight worth 4–6 is almost 5 per cent better in value then 5 of these two Ryal pieces. And 800000 such Mexico pieces of 8 will make 4189000 such two Ryal pieces

NOTE

(1) Mint Papers II, 209. This second valuation was considered by the Treasury on 5 October 1711, when Oxford ordered 'that some scheme or method be framed by comparing or adjusting the value of the said money as may be useful to the Commissioners going to Spain. Write to Sir Isaac Newton to that Purpose' (*Cal. Treas. Books*, XXV (Part II), 1961, p. 103). See Letter 876.

876 T. HARLEY TO NEWTON

6 OCTOBER 1711

From the copy in the Public Record Office.(1) Reply to Letter 875; for the answer see Letter 877

Sir Isaac Newton abt. adjusting two Ryall peices

Sir.

I have read to My Lord Trearer Your Lre of the 5th instant wherein You Report to His Lordp the Value by the weight & finess of the two Riall peices of King Charles which was sent you to Examine, and his Lordp was thereupon pleased to direct that You consider and frame some Scheme or Method for Comparing or Adjusting the Value of the said Money as may be usefull to the Gentlemen who are going as Comrs. from Her Maty. to inspect into the Affaires of Spaine &c it being part of the said Comrs. Instruccons. to Enquire & charge the pay masters with the Difference between the Value of the Species wch they receive for the Forces there, and the Value of the Species in wch they pay the said Forces, and also the profit which has been made by recoyning Money of a better into a Worse Species. I am &c *6th Oct.* 1711

T. HARLEY

NOTE

(1) T/27, 20, 22.

877 MEMORANDUM BY NEWTON

[mid OCTOBER 1711]

From a draft in the Mint Papers[1]

If the Paymasters of her Majts forces in Spain have received several summs of Mexico & Peru Dollars for paying those forces & have carried them to the Mints of K. Charles to be coined into two-ryal pieces: the Commissioners are to enquire what number of Dollars every Paymaster received, what was the weight of them before melting, what was the weight of the Ingots produced by melting, what silver was got out of the sweep, whether the weight of this silver & the Ingots equalled the weight of the Dollars wanting only the 300th part of the whole weight, what number & weight of two ryal pieces were coined out of this weight including the seigniorage[2] whether this weight of the two ryal pieces exceeded the weight of the Ingots & silver got out of the sweep by the weight of the allay put to the Ingots, & silver that is by about the hundredth part of the whole weight, whether the number of the said two ryal pieces be such as ought to be coined out of that weight, that is, about 109 pieces out of every twenty ounces, or about $109\frac{3}{4}$ pieces out of every 21 Dollars. Also how many pieces have been deteined for seigniorage & for melting the dollars into Ingots & how many the Paymaster has received out of the Mint for the Queens service & whether the number received out be in such proportion to the number of ounces coined, or to the number of Dollars melted down to be coined, as by the rules above described it ought to be. And whether the number of Dollars & two ryal pieces wch the Paymasters have charged themselves wth in their Accounts be the same wth the number of Dollars received by the Melter for the melting pot & of two ryal pieces delivered from the Mint according to the Melter &

the Mint. Or what evidence have the Paymasters by the books of [the] Melter & of the Mint by Receips or Certificates or living witnesses that they have caused all the Dollars received by them from the merchant to be melted & coined & that they have charged themselves in their accounts with all the two ryall pieces wch they have received from the Mint.

If any dispute should arise in Spain about the coarseness of the two Ryall pieces wch cannot be decided without an assay; the best way to decide the Question by ye assay is this. Let a pound weight or half a pound weight of ye two Ryall pieces be melted into a lump, and let the same weight of Mexico & pil[lar](3) Dollars be also melted into a lump. Let assays be cut off from these two lumps, by a skilful Assayer & weighed by the assay weights for the fir[e.] Let them also be weighed against one another in ye assay-scales to see th[at] they be exactly of the same weight before they be wrapt up in lead. And when ye lead is blown off from the last let the fire abate gently of it self ti[ll] the silver is ready to congeal least the silver by cooling too suddenly spring into the fire.(4) And when the assay drops are taken out of the fire let them be weighed in the Assay scales against one another exactly to find the difference of their weight. And let one or two of the Commissioners (or persons appointed by them) see all this done. Then say, As the weight of the assay drop of the two Ryal pieces is to the weight of the other assay drop so is the weight of the fine silver in a pound weight of two Ryal pieces to the wei[ght] of the fine silver in a pound weight of ye Dollars & so is the weight of all the Ingots of silver coined, to the weight of all the two Ryal pieces wch ought to be produced out of those Ingots by coinage. And wh[en] they have the whole weight & the number of two Ryal pieces in 20 pound weig[ht] they will thence know the number of two Ryal pieces in the whole weight by the Rule of three. And when they have learnt at the Mint what propo[rtion] is to be taken for seigniorage, they will know what number [the] Paymaster ought to receive out of the Mint.

NOTES

(1) Mint Papers II, fo. 190. From the reference to the Commissioners first mentioned in Thomas Harley's letter of 6 October, it seems that this was a draft reply to that letter. We assign it an arbitrary date.

(2) A deduction from the money minted to cover the costs of the Mint and its owner's profit.

(3) The paper is torn along the edge.

(4) If the metal cools too quickly, violent occlusion of the gases carries away some of the metal in 'spurting' or 'spitting'.

878 NEWTON TO [GEORGE] GREENWOOD

9 OCTOBER 1711

From a photograph of the holograph original(1)

Sr

I beg leave to acquaint you (tho I do it wth a great deal of concern) that on Saturday last, Octob 17th new stile, we had notice that the expedition against Canada under the conduct of Collonel Hill & Admiral Walker has miscarried by means of foggy & tempestuous weather when the fleet was going up the river of St. Christophers.(2) Eight transport ships with about eight hundred men on board were cast away by striking upon rocks & the rest escaped narrowly. We have not yet a distinct account of the men that are lost but Collonel Barton(3) is recconed one of them. Pray assure my Cousin your daughter that if this sad news prove true her friends here will take the best care they can of her concerns in England. But I must leave it to your own discretion to let her know it by such degrees & in such manner as may least afflict her. My Niece her sister would have written to her but for the grief she is in. I am

Sr

Your most humble Servant

London
Octob. 20*th new stile*
1711

Is. NEWTON

To Mr. Greenwood the Elder
at his house at
Roterdam

Postmark 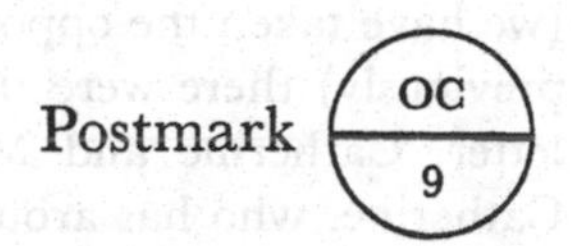

NOTES

(1) We found the photograph among materials returned to the Royal Society by the late Professor H. W. Turnbull. It is endorsed: 'Prof. W. A. E. Karunaratne, Colombo, Ceylon.' Professor Karunaratne was Professor of Pathology in the University of Colombo and it seems likely that the original letter was his property.

The historical background to its writing is as follows: attacks by the French and the English colonies upon each other, to the accompaniment of Indian wars, were the invariable consequence of a state of war between the two powers in Europe. In 'King William's War' British forces had fought their way up the St Lawrence to Quebec without capturing the city. When the Harley–St John administration took office the latter determined on a similar bold stroke, designed to rival Marlborough's victories in Europe. In May 1711 an expedition left Plymouth consisting of ten ships of the line and about thirty transports carrying 5000 troops; it was commanded by Admiral Sir Hovenden Walker (? 1656–1728) and Brigadier John Hill

(d. 1753), who had the advantage of being Mrs Masham's brother. After making contact with New England levies who were to attack Canada by land, the expedition sailed north to the mouth of the St Lawrence, which it entered with some difficulty. The disaster occurred on the second or third day in the river (22 August according to Hill's report to Dartmouth) when in bad weather, at night, a number of vessels including eight transports went ashore on the north bank of the St Lawrence near the Isle aux Oeufs. Hill's report was written on 9 September and may have been the source of Newton's information.

According to Col. King's *Journal* the wreck occurred on 23 August:

> The 23rd we had rainy and foggy weather with a very fresh gale at East. At 8 at night...the Admiral lay'd by for fear of falling foul on one side or other of the River: for it was then so excessively dark we could not see from one end to the other of our vessell. At half an hour after ten we saw land to leeward of us...we made with the utmost expedition all the sail we could to gett clear of it, but finding we could not were forced to come to an anchor near l'Isle aux Oeufs in 7 fathom water with a shoal of rocks on each quarter within a cable's length...All the night we heard nothing but ships fireing and showing lights as in the utmost distress: so that we could not but conclud that the greatest part of our Fleet was lost...

Twenty-nine officers were drowned (including one other field officer besides Robert Barton and ten captains), 79 non-commissioned officers, 597 soldiers and 35 women camp-followers. (See *Calendar of State Papers, America and West Indies*, July 1711–June 1712 (1925), pp. 88–90 and G. S. Graham (ed.), *The Walker Expedition to Quebec* (Navy Records Society, 1953); the latter includes a modern topographical account of the place of the shipwreck.)

The relationships to Newton may be elucidated as follows, with the aid of an undated document, Letter 949 (we assign it to October, 1712). Newton's mother, Hannah, took as her second husband the Reverend Barnabas Smith, by whom she had a daughter, also named Hannah, born in 1652. Some twenty years later, presumably, the younger Hannah married Robert Barton 'of Brixstoke, Gent.' (that is, of Brigstock in Northants., near Thrapston), who probably died in 1693 (vol. III, pp. 278–9). As the immediately following rough draft shows (we have taken the opportunity to print it here, though it was obviously written some years previously) there were three surviving children of this marriage: Robert, the subject of this letter, Catherine and Margaret. Robert was presumably the eldest of these three while Catherine, who has aroused so much interest, was born in 1679. (On Catherine Barton, see vol. IV, p. 350 note (1), and Augustus de Morgan, *Newton: his Friend: and his Niece*, London, 1885. She was thirty-eight years old when she married John Conduitt and thirty-nine when the first of her five children was born.)

The following document shows Newton's concern for the welfare of his half-sister's children (and incidentally refutes De Morgan's contention that Newton provided no annuity for his niece Catherine). The 1712 document further proves that Robert Barton, son of Hannah, left a widow, also (confusingly enough) named Katherine. It is thus certain that there were two women of this name in Newton's family, though it is clear that she who has aroused so much interest, the friend of Swift and Halifax, was his half-niece, not the widow, for Swift in his callous allusions to the death of Lt-Col. Barton is quite specific on this point in the *Journal to Stella*: '9 October 1711. Did I tell you that my friend Mrs Barton has a brother drowned, that went on the expedition with Jack Hill? He was a lieutenant-colonel, and a coxcomb... 14 October. I sat this evening with Mrs Barton, it is the first day of her seeing company; but I make her merry enough, and we were three hours disputing upon Whig and Tory. She grieved for her brother only for form, and he was a sad dog.'

The Greenwood connection cannot be precisely defined, but it seems extremely likely that Robert Barton married a Katherine Greenwood, who would thus be Newton's 'cousin' (relative by marriage) and Catherine's sister(-in-law). How Newton exercised his influence on behalf of the widow will be seen in Letter 949, published later. One George Greenwood was given a pass and post warrant to go to Flanders on 8 May 1697 (*C.S.P.D.*, 1692 (1927), 148). It seems reasonable to suppose that this was the father of Robert Barton's wife, whom he had married (presumably) while campaigning overseas.

(2) The St Lawrence.

(3) From Newton's later statement that Robert Barton was commissioned at age nineteen, he was born in 1677, two years before his sister Catherine, for he was commissioned ensign in Col. Emanuel Scroop How's Regiment (and in the Colonel's own Company) on 18 January 1695/6 (*C.S.P.D.*, 1696 (1913), 22). He was commissioned Lieutenant in the same Regiment on 7 July 1702 (*C.S.P.D.*, 1702–3 (1916), 377). Barton's promotion to Captain in the 37th Foot (Colonel Thomas Meredith) occurred in June 1706; he had transferred from Temple's Foot. He presumably served with the 37th at the siege of Tournai and the battle of Malplaquet (11 September 1709). It was perhaps in 1710 that Barton purchased the lieutenant-colonelcy of Kane's Regiment, which unit together with the 37th was assigned to the Quebec expedition. They moved to Ostend in April 1711. (C. T. Atkinson, *The Royal Hampshire Regiment* (1950), I, p. 26).

878*a* NEWTON TO MARTIN

[before 1711]

From the draft in the University Library, Cambridge[1]

Mr Martin

I paid lately (April 28) into the Exchequer by your hands the summ of 833 *lib* 7*s* 0*d* for an Annuity of 100*lib* for two lives at ye rate of 12 per cent.[2] I desire you would now give into the Exch. the names of the lives & bring me Orders for ye receipt of ye money as soon as they shall be signed. The first of the two lives is my own for the whole summ. And after my own life I would divide ye ann[uity] into three parcels to be paid during ye lives of 3 persons that is to say 60 pounds per annum during the life of Robert Barton the son of Hannah Barton of Brigstock in the county of North[amp]ton widdow & twenty pounds per annum during ye life of Kath[erine] Barton daughter of ye said Hannah Barton, & twenty pds per annum during ye life of Margaret Barton ye younger daughter of the said H.B.[2]

NOTES

(1) Add. 3965(18), fo. 671v, largely printed in Whiteside, *Mathematical Papers*, IV, p. 205, note (5), where it is supposed that this draft was written soon after the death of Hannah Barton's husband in 1693.

(2) £100/0·12 = £833 7*s* 0*d* to the nearest shilling, a calculation of the value of the annuity based on the assumption that its payment will continue indefinitely rather than for a fixed (or estimated) term of years (see Whiteside, *ibid.*, pp. 203–5).

879 COTES TO NEWTON

25 OCTOBER 1711

From the original in the University Library, Cambridge[1]

Sr

I hear there has this afternoon been a meeting of the Heads of our university in order to vacate Mr Whiston's Professor-ship;[2] they have agreed upon it & notice will be given that the Place is void by a Paper which will be affix'd to the School-doors[3] on Saturday next, after which the Election of a successor will follow in a very few days. I know not what Candidates will appear for it, except Mr Hussey[4] a Fellow of our College. I wish he may succeed in it, for by long acquaintance with him I know him to be of an extraordinary good understanding, temper & behaviour; & he is much the best qualified for the place of any that I know of in our University. He formerly waited upon You for Your recommendation of him to a Mathematical School at Rochester,[5] & he tells me You was pleas'd to examine him in some things for that purpose. I presume Sr to give You this trouble to beg of You that You would not engage Your self too soon to recommend any other to those of the Electors with whom You have an Interest. If there be occasion for it, I beleive Dr Bentley will in a short time write to You to beg Your recommendation of Him.

I am, Sr Your most obliged
Humble Servant
ROGER COTES.

Cambridge Octbr. 25th 1711

For
Sr Isaac Newton
at his House
in St Martin's Street
by Leicester-Fields
London

NOTES

(1) Add. 3983, no. 18.

(2) William Whiston (1667–1752), who had been acquainted with Newton since 1694, was appointed by Newton his deputy in the Lucasian professorship in Cambridge when he left the University, and after Newton's resignation of the Chair (10 December 1701) was appointed, on Newton's recommendation, to succeed him. His pronouncedly Arian views began to cause scandal from about 1707. On 30 October 1710 he was expelled from the University, Christopher Hussey being appointed to deputize for him. A year later Whiston was deprived of the Chair. Newton was never afterwards kindly towards him.

(3) The doors of what are now called the Old Schools, the University offices.

(4) Christopher Hussey (1684–1761) had been elected a Fellow of Trinity in October 1709;

he was ordained priest in 1716 and awarded the D.D. in 1731. He was never F.R.S. Cotes' efforts on his behalf were unavailing (by six votes to four), for in November 1711 Nicholas Saunderson (1682–1739) of Christ's, a blind teacher of mathematics at Cambridge (since 1707) who was granted the degree of M.A. by special dispensation from the Queen in 1711, was elected to the Lucasian professorship. He was elected F.R.S. in 1718. (See Cotes to Jones, 25 November 1711, printed in Rigaud, *Correspondence*, I, pp. 261–2.)

(5) See Bentley to Cotes, 21 May 1709, printed in Edleston, *Correspondence*, p. 1. The school was the Free Mathematical School founded at Rochester by Sir Joseph Williamson, formerly Secretary of State and P.R.S. Hussey had been unsuccessful in competition with John Colson (1680–1760; of Lichfield; M.A. Cantab. 1728—his first degree—elected Lucasian professor 1 March 1739/40) who was appointed headmaster on 1 June 1709. The school was intended to educate Rochester boys 'towards the Mathematicks and all other things which fitt them for the Sea Service'.

880 JONES TO COTES

25 OCTOBER 1711

Extract from the original in Trinity College Library, Cambridge.[1]
For the answer see Letter 881

London October. 25th. 1711

...I have nothing of news to send you; only the Germans and French have in a violent manner attack'd the Philosophy of Sr. Is: Newton, and seem resolv'd to stand by Cartes; Mr Keil, as a person concern'd, has undertaken to answere & defend some things, as Dr Friend,[2] & Dr Mead,[3] does (in their way) the rest: I wou'd have sent you ye whole Controversy, was not I sure that you know, those only are most capable of objecting against his writings, that least understand them; however, in a little time, you'l see some of these in ye *Philos. Transact.*[4]

NOTES

(1) R. 16, 38, no. 301, printed in Edleston, *Correspondence*, pp. 210–11.

(2) See Letter 830, note (6), and Edleston, *Correspondence*, p. 212.

(3) Richard Mead (1673–1754), physician, was a close acquaintance of Freind, and an ardent Newtonian. In 1704 he published a book entitled *De imperio solis ac lunæ in corpora humana et morbis inde oriunda*, in which he tried to show the influence of attraction of Sun and Moon upon the human body.

(4) See Letter 882, note (3).

881 COTES TO JONES

11 NOVEMBER 1711

From the original in private possession.(1)
Reply to Letter 880; for the answer see Letter 882

Cambridge Nov: 11. 1711

Dear Sr.

...The Controversy concerning Sr Isaac's Philosophy is a peice of News that I had not heard of unless Muys's Elementa Physices be meant.(2) I think that Philosophy needs no defence, especially when tis attack't by Cartesians. One Mr Green(3) a Fellow of Clare-Hall in our University seemes to have nearly the same design with those German & French Objectors whom you mention. His book is now in our Press & is almost finished. I am told he will add an Appendix in which he undertakes also to square the Circle. *Ex pede Herculem*, I need not recommend his Performance any further to You.

I am, Sr,
Your Obliged Freind
& Humble Servant
ROG: COTES.

for
Mr Wm. Jones
to be left for him
at Childs Coffee-House
in St Paul's Church Yard
London

NOTES

(1) Printed in Rigaud, *Correspondence*, I, pp. 260–1, and in Edleston, *Correspondence*, p. 211.
(2) See Letter 844, note (2).
(3) See Letter 854, note (3).

882 JONES TO COTES

15 NOVEMBER 1711

Extract from the original in Trinity College Library, Cambridge.(1)
Reply to Letter 881

Nov. 15*th* 1711

Dr. Sr.

I receiv'd yours of the 11th. instant...The Objections of ye writers of the Leipsic Transactions, against the Philosophy introduced in Dr. Friend's Chimical Lectures,(2) together with his answere, as also those of Wolfius, and of

Mr. Saurin of the Fr. Academy, against ye same Philosophy, with an answere by Mr. Keil, are now in the Press here, and nearly finish'd, I shall not be wanting to send them you.(3) I am concern'd to find, by Sr. Isaac, that his Book does not go forward, it is a great grieveans to be so long depriv'd of it...

NOTES

(1) R.16.38, no. 304, printed in Edleston, *Correspondence*, pp. 211–12.

(2) See Letter 830, note (6).

(3) Freind's defence was published in the *Phil. Trans.* 27 (1712), 330. Keill's answer was prepared but not printed. See Edleston, *Correspondence*, p. 213, for a long discussion of these controversies.

883 FINAL STATEMENTS OF ACCOUNT

12 DECEMBER 1711

From the *Journal of the House of Commons*, for 13 December 1711(1)

I. *The Coinage of Plate Account*(2)

	Gross Weight received.		Price allowed.			Standard Weight.				Monies produced.		
	oz.	d.wt.	£.	s.	d.	lb.	oz.	d.wt.	gr.	£.	s.	d.
Plate of the old Standard, received after the 14th Day of *May* 1711, at 5*s.* 5*d.* *per* Ounce	58,588	5	15,867	13	—	4,684	7	0	11	14,538	5	6
Plate of the new Standard, received after the same 14th Day of *May*, at 5*s.* 8*d.* *per* Ounce .	25,608	0	7,255	12	—	2,208	1	3	15	6,851	9	7
Standard of uncertain Plate, received after the same Day, at 5*s.* 6*d.* *per* Ounce . .	17,095	12	4,701	6	2	1,424	7	12	9	4,418	18	4
Silver got out of the Sweep of the said Plate . .	—	—	—	—	—	—	—	—		100	6	3
Totals . . .	101,281	17	27,824	11	2	8,317	3	16	11	25,908	19	8
DEFICIENCY . . . . £										1,915	11	6

Mint Office,
12 *December* 1711.

Cra. Peyton,
Is. Newton,
E. Phelipps.

II. *The Scottish Recoinage Account*(3)

AN ACCOUNT of what is due to the Moneyers for re-coining the Monies of *Scotland*, and their Charges incident thereunto.

	£.	s.	d.
The Silver Monies coined at *Edinburgh* by the Moneyers were, in Weight, 103,346 lb. wt.; and the Coinage thereof, at 9*d*. *per* Pound Weight, allowed to them by her Majesty's Warrant of the 12th of *July* 1707, amounted unto the Sum of £.3,875. 9*s*. 6*d*.: Whereof, they have received of Mr. *Allardes*, the late Master and Worker of her Majesty's Mint at *Edinburgh*, the Sum of £.1,429. 6*s*. 2½*d*.; and there remains due to them the Sum of	2,446	3	3½
More due to them, by the same Warrant, for the Charges of the Journies of Five Moneyers to *Edinburgh*, and Four of them back, after the Rate of Eight Pounds *per* Journey	72	—	—
More due to them, by the same Warrant, for an Allowance of 3*s*. *per diem* each, when there was not One thousand Pound Weight coined in One Week	182	2	—
TOTAL . . .	£2,700	5	3½

Mint Office,
December the 12th, 1711.

Cra. Peyton,
Is. Newton
E. Phelipps.

NOTES

(1) Vol. **17** (1803), p. 8.

(2) On 11 December 1711 the House of Commons resolved that the Mint in the Tower should lay before it an account of the deficiency produced by the coinage of plate since 14 May, and also an account of what was due to the moneyers who had worked in Edinburgh on the Scottish recoinage. These accounts having been received by the House on the 13 December, it was resolved on the following day that the sums required (totalling, as above, £4615 16*s* 9½*d*) should be granted to her Majesty to make good the deficiencies. Brief clauses on later Supply Bills implemented these resolutions.

(3) Newton does not seem personally to have been much concerned with the woes of the moneyers sent to Scotland arising from the long delay in the payment of the sums due to them since the coinage finished in 1709. Certainly such delays were no unusual experience for the moneyers. The deficiency arose from the special payments allowed the London moneyers and the differences in practice between the Tower and the Edinburgh mints.

884 LEIBNIZ TO SLOANE

18 DECEMBER 1711

From the original in the British Museum.[1]

Reply to Letter 843

Viro Celeberrimo

Domino Hans SLOANE

Godefridus Guilielmus Leibnitius S. p. d.

Quæ Dominus johannes Keilius nuper ad Te scripsit,[2] candorem meum apertius quam ante, oppugnant: quem ut ego hac ætate post tot documenta vitæ, Apologia defendam, et cum homine docto, sed novo, et parum perito rerum anteactarum cognitate, nec mandatam habente ab eo cujus interest, tanquam pro Tribunali litigem: nemo prudens æquusque probabit. Quæ ille de meo rem cognoscendi modo suspicatur, haud satis exercitatus artis inveniendi arbiter; ipsius quidem docendi causa non est cur refellam: sed norunt amici quam longe alio et ad alia proficuo itinere processerim. Frustra ad exemplum Actorum Lipsiensium[3] provocat, ut sua dicta excuset; in illis enim circa hanc rem quicquam cuiquam detractum non reperio, sed potius passim suum cuique tributum. Ego quoque et amici aliquoties ostendimus, libenter a nobis credi, illustrem fluxionum autorem per se ad similia nostris fundamenta pervenisse. Neque eo minus ego in inventoris jura venio, quæ etiam Hugenius[4] judex intelligentissimus incorruptissimusque publice agnovit: in quibus tamen mihi vendicandis non properavi, sed inventum plusquam nonum in annum pressi,[5] ut nemo me præcucurisse queri possit

itaque vestr[æ] æquitati committo, an non coercendæ sint vanæ et injustæ vociferationes, quas ipsi Newtono, viro insigni, et gestorum optime conscio, improbari arbitror, eiusque sententiæ suæ libenter daturum indicia mihi persuadeo.

Vale dabam Hanoveræ 29 decembr. 1711 [N.S.]

A Monsieur
Monsieur Hans Sloane
Secretaire dela Societé Royale
et Medecin celebre

Londres

Translation

Gottfried Wilhelm Leibniz presents a grand
salute to the very celebrated Mr. Hans Sloane

What Mr. John Keill wrote to you recently[2] attacks my sincerity more openly than [he did] before; no fair-minded or sensible person will think it right that I, at my age and with such a full testimony of my life, should state an apologetic case for it, appearing like a suitor before a court of law, against a man who is learned indeed, but an upstart with little deep knowledge of what has gone before and without any authority from the person chiefly concerned. There is no reason why I should instruct him, by rebutting his reconstruction of my way of getting to know things, who is so insufficiently experienced a judge of the manner of discovery; yet my friends know how I have proceeded by a quite different route and in pursuit of quite other objects. It is vain for him to appeal to the example of the Leipzig *Acta* [*Eruditorum*](3) in order to condone his own words, for in them I find nothing that detracts anything from anyone; rather I find that in passages here and there everyone receives his due. I, too, and my friends have on several occasions made obvious our belief that the illustrious discoverer of fluxions arrived by his own efforts at basic principles similar to our own. Nor do I lay a less claim than his to the rights of the discoverer, as Huygens(4) (who was a most clever and incorruptible judge) also acknowledged before the public—rights which I have not hastened to claim for myself but rather concealed the discovery for nine years,(5) so that no one can claim to have forestalled me.

Thus I throw myself upon your sense of justice, [to determine] whether or not such empty and unjust braying should not be suppressed, of which I believe even Newton himself would disapprove, being a distinguished person who is thoroughly acquainted with past events; and I am confident that he will freely give evidence of his opinion on this [issue].

Farewell. Hamburg, 29 December 1711 [N.S.]

NOTES

(1) MS. Sloane 4043, fos. 19–20. The letter was published in *Commercium Epistolicum* London, 1712, pp. 118–19. It was read to the Royal Society on 31 January 1711/12 and was 'Delivered to the President to Consider of the Contents thereof.' For a possible draft reply by Newton see Letter 888.

(2) See Letter 843*a*.

(3) See Letter 843*a*, note (5).

(4) We do not certainly know what published passage Leibniz may have had in mind. It is true that Huygens wrote that 'à l'egard du Probleme de la figure des verres pour assembler les rayons, lors qu'une des surfaces est donnée' those excellent geometers Newton and Leibniz had reached the same results as himself (*Traité de la Lumiere*, Leiden 1690, Preface) but this is hardly so specific an acknowledgement as Leibniz claimed. Huygens never adopted Leibniz's calculus, nor did he greatly admire it; he found fault with a good deal of the mathematical work published by Leibniz and the Bernoullis. When Newton disclosed some further glimpses of his method in the second volume of Wallis's *Opera mathematica*, Oxford, 1693, Huygens was already entering his fatal illness, though not too ill to be interested. He certainly believed that Newton and Leibniz had arrived at similar methods independently, writing to Fatio de Duillier: 'Ce peu de lumiere que vous dites avoir receu de Mr. Newton en ces matieres me fait croire qu'il sçait tout ce qu'a Mr. Leibnitz et d'avantage...Pour l'invention du calculus differentialis, il me semble, en considerant le lieu que vous citez de Mr. Newton [the 'Fluxions Scholium', *Principia*, 1687, p. 253], qu'il reconnoit la luy mesme que Mr. Leibnitz s'estoit rencontrè à avoir la mesme chose a peu pres que luy.' (*Oeuvres Complètes*, x (1905), pp. 241–2.) And in a letter to the Marquis de l'Hospital he similarly rejected Wallis's charge that Leibniz's

and Jakob Bernoulli's development of the calculus was simply an adaptation in new symbolism of the methods of Barrow and Newton (*ibid.*, p. 623).

Wallis published the mathematical correspondence between Newton and Leibniz only in the third volume of his collected works, published in 1699, though he had included long excerpts from the 1676 *epistolæ prior et posterior* in his *Algebra* of 1685, and accordingly when Newton wrote of Huygens (referring to the *Opera Mathematica*, III, p. 679, line 54): 'Hugenius literas quæ inter Newtonum et Leibnitium mediante Oldenburgo intercesserant numquam vidit' (U.L.C. Add. 3968, fo. 13r) his statement was not quite correct.

(5) Since Leibniz first described his method in print in 1684, his claim here is not unjust—indeed the beginning of this line of mathematical thinking that led to the calculus should be placed earlier still. But the methods of the calculus were not fully perfected even by 1675. See Hofmann.

885 FLAMSTEED TO SHARP

22 DECEMBER 1711

Extract from the original in the Library of the Royal Society(1)

I have had another contest with ye PR.R.S. who had formed a plot to make my instruments theirs and sent for me to a Comittee where onely himselfe & two Physitians, Dr. Sloane & another, as little skillful as himselfe were present.(2) Ye Pr[esident] ran himself into a great heat & very indecent passion. I had resolved aforehand his Kn—sh talke should not move me shewed him yt all ye Instrumts in ye Observatory were my owne the Murall Arch & Voluble Quadrant haveing been made at my own charge the rest purchased with my own Mony, except ye sextant & two clocks which were given me by Sr Jonas Moore, with Mr Towneleys Micrometer, his gift some yeares before I came to Greenwich. this netled him for he has got a letter from ye Sec[retary] of State for ye R.S. to be visitors of ye Observatory(3) & he sayd *as good have no Observatory as no Instruments*. I complained yn of my Catalogue being printed by Raymer(4) without my knowledg, & that I was *Robd of the fruits of my labours*.(5) at this he fired & cald me all the ill names Puppy &c. that he could think of.(6) All I returnd was I put him in mind of his passion desired him to govern it & keep his temper. this made him rage worse, & he told me how much I had receaved from ye Govermt in 36 yeares I had served. I asked what he had done for ye 500*lb* per Annum yt he had receaved ever since he setled in London.(7) this made him calmer but finding him goeing to burst out againe I onely told him: my Catalogue half finished was delivered into his hands on his own request sealed up. he could not deny it but said Dr Arbuthnot had procured ye Queens order for opening it. this I am persuaded was false, or it was got after it had been opened. I sayd nothing to him in return but with a little more spirit then I had hitherto shewd told them, *that God* (who was seldom spoke of with due Reverence in that Meeting) *had hitherto prospered all my labours & I doubted not would do so to an happy conclusion,* took my leave & left them. Dr. Sloane had sayd nothing all this while, the other Dr. told me I was proud & Insulted the Presidt & run into ye same passion with ye Pr[esiden]t. At my goeing out I called to Dr. Sloan, told him he had behavd himself civilly & thankt him for it. I saw Raymer(4) after, drunk a dish of Cofe wth him & told him still calmly of the Villainy of his conduct & called it *blockish* since then they let me be quiet but how long they will do so I know not nor am I solicitous;...

NOTES

(1) Sharp Letters no. 78 printed in Baily, *Flamsteed*, pp. 293–5 and in Cudworth, *Life*, p. 106; the letter is typical of Flamsteed's attitude to Newton and the Royal Society. There is a longer account of the incident in Flamsteed's diary (Baily, *Flamsteed*, p. 228).

(2) On Tuesday 16 October 1711 Newton appointed a council (or 'committee', the word first written) for 'friday come Sevennight', that is 26 October, and Hunt was instructed to desire Mr Flamsteed to meet the council that day at eleven o'clock at Crane Court 'to know if his Instruments be in order and fit to carry on the necessary coelestial observations'. The second physician present was Richard Mead.

(3) See above, Letter 814.

(4) This was Flamsteed's nickname for Halley; at the beginning of the diary entry Flamsteed notes that on entering the Royal Society's house he met Halley (so written), who invited him to drink coffee. In a letter of 5 August 1703 (Baily, *Flamsteed*, p. 751) Flamsteed says: ''tis the name of the person who made it his business to depreciate Tycho Brahe, after he had been courteously entertained by him, and assured him of his inviolable friendship'. That is, Nicolaus Raymarus (Reymer), also known as *Ursus*, who died in 1600.

(5) In the diary Flamsteed wrote: 'I told him that a frontispiece was engraved for my works, and the Prince's picture (without any notice given me of it), to present to the Queen: and that hereby I was robbed of the fruits of my labors.' The true meaning of this possibly ambiguous and certainly churlish remark (since the money for the book had come from the late Prince) is, presumably, that the money thus expended would not be available for printing further observations.

(6) In a memorably fatuous sentence, Sir David Brewster wrote of this passage (*Memoirs*, II, p. 239): 'How simple-minded must he have been in whose vocabulary the epithet given to Flamsteed was the most prominent!' But Flamsteed noted in his diary that puppy was the 'most innocent' word of abuse that Newton flung at him, and the recollection of it by Flamsteed surely reflects more on the purity of his mind, in which it rankled, than on that of Newton who knew worse words.

(7) The figure represents (at least generally) Newton's salary since he became an officer of the Mint in 1696. Newton's profits as Master were much larger.

885A LOWNDES TO THE MINT

7 JANUARY 1711/12

From the original in the Public Record Office.(1)
For the answer see Letter 928

Whitehall Treasury Chambers 7th January 1711

The Right Honoble: the Lord high Treasurer of Great Britain is pleased to referr this Petition to the Warden Master & Worker & Comptroller of her Mats: Mint, who are to consider the same & Report to his Lopp. with all convenient speed a true state of the matter therein contained together with their Opinion what is fit to be done therein

Wm Lowndes

NOTE

(1) T/1, 149, 45A, written on the petition from James Clerk and Joseph Cave to Oxford, claiming that as engravers to the Edinburgh Mint they had received no payment for making punches and other work. They valued their coin work at £120 (1*s*, £30; 6*d*, £20; other coins £70) and remaining labour at £90.

886 TAYLOUR TO NEWTON

11 JANUARY 1711/12

From the original in the University Library, Cambridge(1)

Sir

You are desired to call at the Treasury with your first Convenience to revise the draft of the Indenture(2) for the Mint at Edinburgh.

I am
Sir
Your most humble Servant
J. Taylour

Tre[asu]ry Chrs
11 *Jany* 1711

To
Sir Isaac Newton
Master and Worker of
Her Majesties Mint

NOTES

(1) Add. 3965, fo. 632. The sheet has been used by Newton for drafting some notes relating to the *Principia*, and computations. It is not easy to read.

(2) John Montgomery of Girvan had succeeded George Allardes as Master of the Edinburgh Mint in October 1709, but his indenture was to pass the Great Seal of Scotland only on 3 April 1712. The office was a sinecure since the Edinburgh Mint was now stilled for ever. Compare Letter 797, requesting a draft indenture from the Mint.

877 T. HARLEY TO NEWTON

15 JANUARY 1711/12

From the copy in the Public Record Office.(1)
For the answer see Letter 892

Sr Isaac Newton abt the printing of the Book called Historia Cœlestis

Sir

A Mem[oria]ll(2) having been presented to my Lord Treasurer for the summe of 200£ which remaines to be paid to Compleat the Charges of printing and publishing a Book Called Historia Cœlestis which was begun and Carryed on at the Expence of his late Royall Highness the Prince of Denmarke to the time of his Death and for giving a Gratuity to Doctor Hally for his great pains and service therein I send you by his Lops. Command the said Mem[oria]ll here inclosed, His Lopp. desiring you will take the trouble of Examining and Auditing the Accotts, therein mentioned, And to Report to His Lop. a State thereof together wth Yr. Opinion what sums may be fitt for His Lopp. to Order to Answer the purposes in the said Memoriall mentioned I am &c

15th Janry $17\frac{11}{12}$

T. Harley

NOTES

(1) T/27, 20, 97.

(2) Probably it was submitted by the bookseller concerned, Awnsham Churchill (see vol. IV, p. 438, note (3)).

888 NEWTON TO [?SLOANE]

[? FEBRUARY 1711/12]

From the draft in the University Library, Cambridge(1)

Sr

The papers in ye Acta Leipsica wch gave occasion to the controversy wth Mr Keil I did not see till the last summer,(2) & therefor had no hand in beginning this controversy. Mr Leibnitz thinks that one of his age & reputation should not enter into a dispute wth Mr Keil & I am of the same opinion,[3] & I think that it is as improper for me to enter into a dispute wth the author of those papers. For the controversy is between that author & Mr Keil.[4]

But Mr Leibnitz seems to say that I know how the matter stands & can put an end to the controversy if I would declare my knowledge. If he would have me declare that he is the inventor author of the differential method so far as that method differs from ye method of Fluxions: all men, even Mr Keil himself, will allow him that. If he would have me declare that he is the author of ye differential method even where the methods do agree, that is, the author of the method called by him the differential method & by me the method of fluxions: the author is the first author, & I am not yet convinced that he was the first author of that method. If he would have me approve the papers in the Acta Leipsica wch gave occasion to ye controversy wth Mr Keil; I know not what he means by that: those papers call my candor in question. After that author had asserted the invention of the Differential method to Mr Leibnitz & fortified the assertion by the credit of those that used it; he adds. *Pro differentijs igitur Leibnitianis Dn. Newtonus adhibet semperque adhibuit fluxiones quæ sint quamproxime ut fluentium augmenta æqualibus temporis particulis quam minimis genita; ijsque tum in suis Principijs Naturæ Mathematicis tum in alijs postea editis eleganter est usus &c.*[5] There is some ambiguity in the words but the most proper sense is that I always used fluxions instead of the differences of Mr Leibnitz. And is not this to make the readers beleive that I always knew the differential method of Mr Leibnitz & invented the method of fluxions by using fluxions instead of his differences. This gave occasion to Mr Keil to represent on ye contrary that Mr Leibnitz used his differences instead of my fluxions. And if this derogates from the candor of Mr Leibnitz the contrary derogates from my candor & that unjustly. For By the Letters wch passed between him & me in the years 1676 & 1677 he knows that I wrote a treatise of the methods of converging series & fluxions six years before I heard of his differential method.[6]

In the generation of lines & figures by motion, Fermat Barrow & Gregory considered the small particles by wch the quantities increased in every moment of time & thereby drew tangents. Dr Barrow called those particles moments & from him I had the language of momenta & incrementa momentanea & this language I have always used & still use as may be seen in my Analysis per Æquationes infinitas communicated by Dr Barrow to Mr Collins A.C. 1669 & in my Principia Mathematica & Quadratura Curvarum. And by putting the velocities of ye increase of quantities proportionall to the incrementa momentanea I found out the demonstration of ye method of moments & thence called it the method of fluxions. This demonstration you have in the end of ye Analysis per æquationes infinitas, and in the first Proposition of ye book of Quadratures.[7] But I do not know that Mr Leibnitz has demonstrated the differential method. So then the method of moments & the method of Fluxions is one & the same method variously named in several respects & I have always used it & Mr Leibnits by calling the moments differences has given it the name of the differential method.[8] In ye year 1664 I learnt Fermats method of drawing tangents.[9]

NOTES

(1) Add. 3968(30), fo. 438. Newton had already advised Keill and was to draw up many more statements of his case against Leibniz, few of which saw the light in any form. This draft letter, undated and unsigned, was possibly intended for Sloane, as Secretary of the Royal Society; it was obviously not meant for Keill. The allusions in the first sentences make it clear that Newton had read Leibniz's letter of 18/29 December 1711 (Letter 884), which had been laid before the Royal Society on 31 January 1711/12. Almost certainly this draft was never dispatched: its content passed into Newton's 'Extracts of ye MS Papers of Mr John Collins concerning some late improvements of Algebra' (U.L.C. Add. 3968(19)), thence into his 'Historia brevis Methodi Serierum ex Monumentis antiquis desumpta' (*ibid.* (10)) and so into the *Commercium Epistolicum*.

(2) Here the words: 'when I was told of Mr Leibnitz's letter against him' are deleted; compare Letter 830.

(3) 'therefore wish his Letters against him had not been written': deleted.

(4) '& therefore cannot be induced to set pen to paper against him cannot be induced to enter into this enter into this or write against him'; deleted.

(5) 'Accordingly, Mr. Newton employs and always has employed, instead of the differences of Mr. Leibniz, fluxions; which are approximately as the increases of the flowing quantities generated in equal least parts of time; he has used these most elegantly both in his *Mathematical Principles of Nature* and in other publications since.' As already noted, this passage occurred in the *Acta Eruditorum* review of *De quadratura* (January, 1705, pp. 30–6) written, anonymously, by Leibniz himself. (See Letter 830, note (2).)

(6) 'And I beleive he will allow the And I beleive he will allow that I found out the method before I wrote of it'; deleted.

(7) Newton means that these propositions demonstrated the basic procedures of differentiation and integration.

(8) 'But when & by what steps he found out his way of explaining it I do not know'; deleted.

(9) Newton abandons the draft when embarking on an autobiographical account of his own process of discovery, the writing of which always presented great difficulty. The allusion to Fermat here might be extended from another, later draft (Add. 3968, no. 30, fo. 441): 'I had the hint of this method from Fermat's way of drawing tangents and by applying it to abstract equations directly and invertedly, I made it general.' D. T. Whiteside has already emphasized (*Mathematical Papers*, I, p. 149, notes (4) and (5)) that Newton's repeated assertion of his early acquaintance with Fermat's work were, in historical fact, mistaken.

889 NEWTON TO COTES

2 FEBRUARY 1711/12

From the original in Trinity College Library, Cambridge.(1)

Reply to Letters 854, 857, 863 and 871; for the answer see Letter 890

London 2d Feb. $171\frac{1}{2}$

Sr

I have at length got some leasure to remove the difficulties wch have stopt the press for some time, & I hope it will stop no more. For I think I shall now have time to remove the rest of your doubts concerning the third book if you please to send them.

In reveiwing your letters I do not see but that ye XLVIIIth Proposition(2) of the second Book with its Corollary may stand. For the particles of air go from their loca prima with a motion accelerated till they come to the middle of the pulses where the motion is swiftest. Then the motion retards till the particles come to the further end of the pulses. And therefore the loca prima are in the beginning of the pulses. There the force is greatest for putting ye particle into motion if any new pulses follow. But if no new pulse follows the force ceases & the particle continues in rest. In this Proposition pag. 366. lin. 12, this emendation may be made. Quare cum sit LN ad KH ut IM ad radium OP, et KH ad EG ut circumferentia $PHShP$ ad BC; id est (si circumferentia dicatur Z et $\frac{OP \times BC}{Z}$ dicatur V,) ut OP ad $\frac{OP \times BC}{Z}$ seu OP ad V. Et ex æquo LN ad EG ut IM ad V: erit expansio partis EG, punctive physici F, in loco $\epsilon\gamma$, ad expansionem mediocrem quam pars illa habet in loco suo primo EG ut $V-IM$ ad V in itu, utque $V+im$ ad V in reditu. Unde vis elastica puncti F in loco $\epsilon\gamma$ est ad vim ejus elasticam mediocrem in loco EG ut $\frac{1}{V-IM}$ ad $\frac{1}{V}$ in itu, in reditu vero ut $\frac{1}{V+im}$ ad $\frac{1}{V}$. Et eodem argumento vires elasticæ &c See lin 27.

You stuck at a difficulty in the third Proposition of the third Book.(3) I have revised it & the next Proposition & sent you them inclosed as I think they may stand.(4) What further Observations you have made upon the third Book or so many of them as you think fit if you please to send in your next Letters, I will dispatch them out of hand. I shall be glad to have them all because I would have [the] third Book correct. I am

Your most humble Servant

IS. NEWTON

For the Rnd Mr Cotes, Professor of Astronomy, at his chamber in Trinity College in Cambridge.

NOTES

(1) R.16.38, no. 124. Printed in Edleston, *Correspondence*, pp. 56–7. See further Letter 908.

(2) As already noted, this became Proposition 47 in the second and third editions. We note, with some surprise, that Newton really did equate the *loca prima* of the particles with their positions in the quiescent medium, and imagined a *single* vibration, propagating out as a pulse from the source. The emendation that Newton goes on to suggest, including the redundant symbol Z taken over from the first edition, was not adopted by Cotes who printed the form of words he had drafted previously in his letter of 23 June (Letter 854).

(3) See Letter 854, *ad fin.*

(4) These Propositions were sent, with a new scholium to Proposition 4, on a folio sheet (MS. R.16.38, fos. 126–9). The two Propositions were published in the second edition, exactly as Newton drafted them, so we print only the scholium here. This was eventually omitted from Proposition 4 and, although parts of it were printed elsewhere, a large portion of it never appeared. (See notes below.) The draft is considerably corrected in both Newton's and Cotes' hand; we have followed Newton's final version.

889*a* DRAFT SCHOLIUM TO PROPOSITION 4, BOOK III

Scholium

Picartus[(1)] mensurando arcum gradus unius & [viginti trium][(2)] minutorum inter Amiens & Malvoisinam, invenit arcum gradus unius esse hexapedarum Parisiensium 57060. Unde ambitus Terræ est pedum Parisiensium 123249600, ut supra. Sed cum error quadringentesimæ partis digiti tam in fabrica instrumentorum quam in applicatione eorum ad observationes capiendas, sit insensibilis, et in Sectore decempedali quo *Galli* observarent Latitudines locorum respondeat minutis quatuor secundis, & in singulis observationibus incidere possit tam ad centrum Sectoris quam ad ejus circumferentiam, et errores in minoribus arcubus sint majoris momenti:* ideo *Cassinus* jussu Regio mensuram Terræ per majora locorum intervalla aggressus est, & subinde per distantiam inter Observatorium Regium *Parisiense* & villam *Colioure* in Roussilion & Latitudinum differentiam 6gr 18′, supponendo quod figura Terræ sit sphærica, invenit gradum unum esse hexapedarum 57292, prope ut *Norwoodus* noster antea invenerat. Hic enim circa annum 1635, mensurando distantiam pedum Londinensium 905751 inter *Londinum* et *Eboracum*, & observando differentiam Latitudinum 2gr 28′, collegit mensuram gradus unius esse pedum Londinensium 367196, id est, hexapedarum Parisiensium 57303.[(3)] Inter has

* Vide Historiam Academiæ Regiæ scientiarum anno 1700.

tres mensuras assumamus mensuram mediocrem hexapedarum 57230 pro gradu uno, et semidiameter Terræ erit pedum Parisiensium 19674225.[4]

Ex observationibus Astronomicis, corrigendo Theoriam Lunæ,[5] invenimus per Eclipses Lunares mediocrem ejus distantiam a Terra in Octantibus esse $60\frac{1}{4}$ semidiametrorum terrestrium quamproxime. Semidiametros intelligo ad Æquatorem ductas. Hæ semidiametri sunt omnium maximæ, et superant semidiametrum ad polum ductam milliaribus 32 circiter, ut posthac dicetur; ideoque mediocris distantia Lunæ a Terra in Octantibus, est semidiametrorum terrestrium mediocrium $60\frac{1}{2}$ circiter, id est, pedum Parisiensium 1190290612. Et hanc distantiam esse ad distantiam inter centrum Lunæ et commune gravitatis centrum Lunæ ac Terræ ut 39 ad 38 quamproxime, patebit in sequentibus per fluxum et refluxum maris; ideoque distantia inter centrum Lunæ et commune gravitatis centrum Lunæ ac Terræ, est pedum Parisiensium 1159770340.

Cum Luna revolvatur respectu fixarum diebus 27 horis 7 et minutis primis $43\frac{1}{5}$; sinus versus anguli quem Luna tempore minuti unius primi, motu medio, circum commune gravitatis centrum Lunæ ac Terræ describit, est 1275235 existente Radio 100,000000,000000. Et ut Radius est ad hunc sinum versum ita sunt pedes 1159770340 ad pedes 14, dig. 9, lin $5\frac{5}{7}$.

Et per Corollarium Propositionis III, vis qua Luna retinetur in Orbe, est ad vim eandem in superficie Terræ, in ratione quæ componitur ex ratione $177\frac{29}{40}$ ad $178\frac{29}{40}$ et ratione duplicata semidiametri Terræ ad distantiam centrorum Lunæ ac Terræ quæ fuit semidiametrorum $60\frac{1}{2}$, id est, in ratione 1 ad 3680,84502; ideoque corpus ad superficiem Terræ vi illa cadendo, tempore minuti unius secundi describet pedes Parisienses 15, dig: 1 lin: $5\frac{1}{2}$.

Longitudo Penduli[6] ad minuta secunda in vacuo oscillantis est pedum trium Parisiensium & linearum $8\frac{3}{5}$. Et altitudo quam grave in vacuo cadendo, tempore minuti unius secundi describit, est ad dimidiam longitudinem Penduli hujus, in duplicata ratione circumferentiæ ad diametrum circuli (ut indicavit *Hugenius*) ideoque est pedum Parisiensium 15, dig. 1, lin $2\frac{1}{4}$. Hic est descensus gravium in Latitudine *Lutetiæ Parisiorum* seu 48gr 50′.

Ad Æquatorem vis centrifuga corporum a diurna rotatione Terræ oriunda, est ad vim gravitatis ut 1 ad 289 circiter; et in Latitudine *Lutetiæ* minor est, idque in duplicata ratione sinus complementi 48°. 50′ ad Radium, adeoque est ad vim gravitatis ut 1 ad 669. Et hac vi descensus gravium in Latitudine *Lutetiæ* diminuitur. Descensus igitur pedum 15, dig 1, lin $2\frac{1}{4}$ augeatur parte $\frac{1}{666}$[7] seu lineis $3\frac{1}{4}$, et habebitur totus gravium descensus pedum 15, dig 1, lin $5\frac{1}{2}$ quam gravitas sola, tempore minuti unius secundi, in Latitudine 48gr 50′ efficere posset, si modo Terra quiesceret.

Vis[8] igitur qua Luna retinetur in Orbe suo, si modo descendatur ad Terram, æqualis fit vi gravitatis, idque quam accuratissime, quantum ex phænomenis colligere licuit; et propterea per argumentum superius, eadem est cum gravitate.

Translation

Scholium

By[1] measuring an arc of one degree and twenty-three minutes[2] between Amiens and Malvoisine Picard found the length of one degree to be 57 060 Paris fathoms. Whence the circuit of the Earth is 123 249 600 Paris fathoms, as above. But as an error of the fortieth part of one inch would be imperceptible both in the construction of the instruments and in their use for making observations, and as this error corresponds in the ten-foot sector used by the Frenchmen to four seconds of arc, and as the error may fall in particular observations either at the centre of the sector or at its circumference and the errors in the smallest arcs are of the greatest effect, accordingly Cassini* at the King's command tackled the measurement of the Earth with a greater distance between the places and hence, from the distance [measured] between the Royal Observatory at Paris and the town of Collioure in Roussillon, the difference in latitude being 6° 18′, found one degree to be 57 292 fathoms (supposing the Earth's shape to be spherical), close to what our countryman Norwood had found earlier. For about 1635, measuring the distance between London and York as 905 751 feet and observing the difference in latitude to be 2° 28′, he concluded that the measure of one degree is 367 196 London feet, that is 57 303 Paris fathoms.[3] Let us suppose that the mean of these three measurements is 57 230 fathoms to one degree, and the Earth's radius will be 19 674 225 Paris fathoms.[4]

* See the History of the Royal Academy of Science for the year 1700

Correcting the theory of the Moon from the astronomical observations,[5] we find by the lunar eclipses that its mean distance from the Earth in the octants is $60\frac{1}{4}$ terrestrial radii, approximately. I mean radii drawn to the equator. These radii are the largest of all, and exceed the polar radii by about 32 miles, as is explained later; and so the mean distance of the moon from the Earth in the octants is about $60\frac{1}{2}$ mean Earth-radii, that is, 1 190 290 612 Paris feet. And in the following propositions it will be made clear from the tidal ebb and flow that this distance is to the distance between the centre of the Moon and common centre of gravity of the Moon and the Earth as 39 to 38; and so the distance between the centre of the Moon and the common centre of gravity of the Moon and the Earth is 1 159 770 340 Paris feet.

As the Moon revolves with respect to the fixed stars in 27 days, 7 hours and $43\frac{1}{5}$ minutes, the versed sine of the angle that the Moon describes about the common centre of gravity of the Moon and the Earth in one minute of time, with an average motion, is 1 275 235, the radius being 10^{14}. And this radius is to this versed sine as 1 159 770 340 feet are to 14 feet, 9 inches and $5\frac{5}{7}$ lines.

And from the corollary to Proposition 3, the force by which the Moon is retained in its orbit is to the same force at the surface of the Earth in a ratio compounded of $177\frac{29}{40}$ to $178\frac{29}{40}$ and the squared ratio of the Earth's radius to the distance between the centres of the Earth and the Moon, which was $60\frac{1}{2}$ Earth-radii; that is, in the ratio of 1 to

3680·84502. And so a body falling with that force at the surface of the Earth would in one second of time describe 15 Paris feet, 1 inch, and $5\frac{1}{2}$ lines.

The length of a pendulum(6) beating seconds *in vacuo* is 3 Paris feet and $8\frac{3}{5}$ lines. And the distance fallen by a body *in vacuo* in one second of time, is to half the length of this pendulum in the squared ratio of the circumference of a circle to its diameter (as Huygens has shown) and so it is 15 Paris feet, 1 inch, $2\frac{1}{4}$ lines. This is the descent of heavy bodies at the latitude of Paris, 48° 50′.

The centrifugal force of bodies arising from the diurnal rotation of the Earth is at the equator in ratio to the force of gravity as 1 to 289, roughly; and it is less in the latitude of Paris in the squared ratio of the sine of the complement of the latitude (48° 50′) to the radius, and so it is to the force of gravity as 1 to 669. And the descent of heavy bodies at the latitude of Paris is reduced by this force. Accordingly the descent of 15 feet, 1 inch, $2\frac{1}{4}$ lines is to be increased by the fraction 1/666,(7) or $3\frac{1}{4}$ lines, to obtain the whole descent of heavy bodies which gravity by itself would cause at latitude 48° 50′ in one second of time, if the Earth were motionless: that is, 15 feet, 1 inch, $5\frac{1}{2}$ lines.

Therefore the force(8) by which the Moon is retained in its orbit, if it should be free to descend towards the Earth, becomes equal to the force of gravity and that with the utmost precision that an examination of the phenomena permits; and furthermore, from the argument above, it is identical with gravity.

NOTES

(1) The first paragraph was eventually printed, with the exception of the final sentence, as the opening to Proposition 19, Problem 3, as Newton suggested in Letter 908 (where it is referred to as Proposition 19, Problem 2). Malvoisy or Malvoisine is the name of a farm, not a village.

(2) Cotes has rectified Newton's omission as shown, but eventually 1° 22′ 55″ was used.

(3) Corrected to 57300 in Letter 908. Norwood's measure was a very rough one, as compared with those of Picard and Cassini which were effected with the greatest attainable accuracy.

(4) As Cotes pointed out (Letter 893) the mean should be 57220 fathoms. When he transferred the paragraph to Proposition 19, Problem 3, Newton simply selected Cassini's measurement as the best one.

(5) The information in this and the following two paragraphs appears, in a considerably altered form, in Corollary 7 to Proposition 37. Newton first suggested (see Letter 903) that the Scholium be transferred *in toto* to the end of Proposition 37, but on 8 April 1712 he sent further amendments to the Proposition, which superseded his previous instruction.

(6) The next two paragraphs, in considerably altered form, became the second and fourth paragraphs of Prop. 19, Prob. 3. (See Letter 908.)

(7) The fraction should read $\frac{1}{669}$, to agree with the result two lines before.

(8) Cf. Prop. 37, Corol. 7, final sentence.

890 COTES TO NEWTON

7 FEBRUARY 1711/12

From the original in the University Library, Cambridge.(1)
Reply to Letter 889; for the answer see Letter 891

Febr. 7th 171$\frac{1}{2}$

Sr.

I have received Your Letter, & as to ye buisness of Sounds I do intirely agree with You upon considering that matter over again.(2) By Your alteration of the 3d Proposition of ye IIId Book it is now very intelligible.(3) What I have observ'd concerning ye remaining part of Your Copy I will send You in the most convenient order I can. I begin with the XXXVIIth Proposition,(4) in ye 3d Section of which You have these words. [Eo autem tempore Luna distat a Sole 15$\frac{1}{2}$ gr. circiter. Et Sol in hac distantia minus auget ac minuit motum maris a vi Lunæ oriundum quam in ipsis Syzygijs & Quadraturis, in ratione Radij ad cosinum distantiæ hujus duplicatæ seu anguli 30$\frac{1}{2}$ gr. hoc est, in ratione 7 ad 6 circiter ideoque in superiore analogia pro S scribi debet $\frac{6}{7}S$] I suppose you intended to have said [in ratione duplicata Radij ad cosinum distantiæ hujus] or [in ratione diametri ad sinum versum duplicati complementi hujus distantiæ].(5) After ye same manner in the foregoing Proposition at the bottom of ye 463 Page, You have added these words [In aliis Solis positionibus vis ad mare attollendum est ut cosinus duplæ altitudinis Solis supra horizontem loci directe & cubus distantiæ Solis a Terra inverse] I suppose You intended to have said [ut sinus versus duplæ altitudinis].(6) This alteration being made in Prop. XXXVII, You will have $\frac{13}{14}S$ instead of $\frac{6}{7}S$: whence S will be to L as 1 to $5\frac{3}{28}$,(7) & in ye 4th Corollary You will have a different proportion from that of 1 to 38. In ye 3d Corollary You make use of 31′ 27″ & 32′. 12″ for the apparent diameters of ye Sun & Moon: I quæry whether it would not be more adviseable to use the Numbers of Your new Theory 32′ 15″ for ye Sun, and 31′. 16$\frac{1}{2}$″ for ye Moon.(8) Making use of these numbers & of 57′. 5″ for ye Moons Horizontal Parallax, & taking ye density of ye Sun to be to ye density of ye Earth as 100 to $398\frac{1}{17}$ as my Computation gives it: the quantity of matter in ye Moon will be to ye quantity of matter in ye Earth as 1 to $176\frac{2}{5}\times\frac{S}{L}$ or as 1 to $34\frac{2}{3}$.(9) This alteration will very much disturb Your Scholium of ye IVth Proposition as it now stands; neither will it well agree with Proposition XXXIX in which I further observe that You take ye proportion of ye semidiameters of ye Earth to be as 689 to 692:(10) but if their difference be 32 Miles, there will be another proportion, & I quæry whether here ought not to be some allowance made upon that Score. I have not examined all ye Calculations of ye Scholium to ye IVth Prop.(11) but I formerly observ'd a small difference from Your Numbers as to the descent of heavy bodies. If the length of a Pendulum which vibrates seconds be 3 feet & $8\frac{5}{9}$ lines, the descent in that time will be 15 feet 1 inch $2\frac{1}{18}$ lines: You have it $2\frac{1}{4}$ lines. And when I examined ye XIX Prop: I found ye Vis Centrifuga to be in proportion to ye Vis Gravitatis as 1 to $288\frac{7}{9}$, You have it as 1 to $290\frac{4}{5}$. In this computation I took ye measure of a degree to be 57200 Toises(12) as You had formerly stated it, the descent of heavy bodies in a second to be 15,0976 feet, the time of ye earths revolution to be 23h. 56′. 4″. If ye Vis Centrifuga be increased in ye proportion of 57230 to 57200: it will to be ye Vis Gravitatis as 1 to $288\frac{5}{8}$. I will send You some things further as I can recollect them from my loose papers of ye computations which I made about $\frac{1}{2}$ an year ago. In Your next You may be pleased to send me Your Answer to what I formerly propos'd concerning ye Periodical times of ye Satellits, for I do not yet know Your resolution as to that part of my Letter.

I am Sr, Your most obliged
& Humble Servt.
ROGER COTES

Trin: Coll. Cambridge

for
Sr Isaac Newton
at his House
in St Martin's Street
in Leicester-Fields
London
Send by Martin at ye Black bull
in bishopgat Street on Munday by Noon.

NOTES

(1) Add. 3983, no. 19, printed in Edleston, *Correspondence*, pp. 57–9 (as usual, from the draft in TCC MS. R.16.38, no. 144).

(2) This relates to Book II, Prop. 47.

(3) The modifications in the second edition are considerable (the third reprints the second); see Koyré and Cohen, *Principia*, 565–6.

(4) 'To find the force of the Moon to move the sea'; Newton's procedure is based on a comparison of the tidal forces of the sun and moon. In the Bristol Channel, for example, it was known that the height of the tide when these forces were additive was 45 feet, but when they were subtractive only 25 feet. The treatment of this question was much expanded in the copy Newton had prepared for the new edition.

(5) 'However, at that time the Moon is $15\frac{1}{2}$° [$15\frac{1}{4}$° *in Newton's MS.*] distant from the Sun, and at that distance the Sun has a less effect in increasing or diminishing the motion of the sea

caused by the force of the Moon than it has at the syzygies and quadratures, in the ratio of the radius to the cosine of double this distance, or 30½°, that is, in the ratio of 7 to 6 roughly, and so in the analogy above $\frac{6}{7}S$ should be written for S...in the doubled ratio of the cosine of this distance...in the ratio of the diameter to the versed sine of the doubled complement of this distance.' In the printed text (*Principia* 1713; p. 428) this passage disappeared; meanwhile (on 16 February) Cotes had withdrawn the amendment proposed here.

(6) In Proposition 36, 'To find the force of the Sun to move the sea'; see *Principia* 1713, p. 426 *ad calcem*, where the sentence appears in accordance with Cotes' modification, Newton having made a slip in writing 'cosine' for 'versed sine'.

(7) In his next letter Cotes withdraws this and agrees with Newton's proportion.

(8) Newton later decided on 32′ 12″ and 31′ 16½″ for the solar and lunar apparent diameters; Newton's 'new theory of the Moon' was that published in David Gregory's *Astronomiæ Physicæ et Geometricæ Elementa*, Oxford, 1702, pp. 332–36 (reprinted in Horsley, *Opera*, III, pp. 245–50 and in Baily, *Flamsteed*, pp. 735–42). An English version was also separately published in 1702, and many times reprinted.

(9) Although the figure for the Sun's relative density is very roughly correct, that for the Moon's mass is too great by a factor of more than two.

(10) 689 to 692 is the ratio of the semidiameters given in the first edition, altered to 229 to 230 in the second. As for the difference in length between the equatorial and axial radii of the Earth, now said by Newton to be 32 miles; this is about twice the true value, and nearly twice the value stated by Newton at the end of Proposition 39. For 692/689 = 1·00435, and 230/229 = 1·00437; the ratio of the radial lengths is 3963/3950 = 1·00329 whereas 3982/3950 = 1·0081. Newton gives the maximum distance as 17⅛, that is, a ratio of about 1·00430.

(11) When the passage ultimately appeared in Proposition 37, Corol. 7, the numbers were modified.

(12) The old French fathom, 1·94 metres. 57200 toises was a convenient round number. Picard (1670) had measured the degree as 57060 toises; J. D. Cassini, measuring the distance Paris–Roussillon (1700), found 57292. Both these numbers were quoted by Newton when this passage was transferred to Proposition 19 (*Principia* 1713; p. 378; see also Koyré and Cohen *Principia*, II, pp. 594–5, 807).

891 NEWTON TO COTES

12 FEBRUARY 1711/12

From the original in Trinity College Library, Cambridge.[1]
Reply to Letters 854 and 890; for the answer see Letter 893

London Feb. 12. 1711/12

Sr

In the third Book under Phænom. I, the periodical times may be 1d. 18h 27′ 34″. 3d 13h 13′ 42″. 7d 3h 42′ 36″. 16d. 16h 32′ 9″ & the distances, ex temporibus periodicis 5,667 9,017 14,384 25,299 as you have put them in yours of June 23 last. But the numbers in the Corollaries of Prop. VIII must be altered accordingly.[2] And so must one or two of ye numbers in Prop. XII & XIII.

In ye 3d section of ye XXXVIIth Proposition, I think my proportion is right. For the force of the Sun increases the force of the Moon in the Syzygies, diminishes it in the Quadratures & neither increases nor decreases it in the Octants: & therefore the distance of the Moon from the Sun must be doubled that the cosine thereof may vanish in the Octants.(3)

In the 3d Corollary of that Proposition lin 5, 6, the words should run thus [et cubus diametri Lunæ ad cubum diametri Solis inverse, id est, (cum diametri mediocres apparentes Lunæ et Solis sint 31′ 27″ & 32′ 12″) ut &c.] But instead of the Moons mean diameter 31′ 27″ may be written 31′ $16\frac{1}{2}$, & the Suns mean diameter 32′ 12″ may be every where retained, even in the Moons Theory. For 32′. 15″ is too bigg.(4)

In the Scholium to the IVth Proposition, if the length of a Pendulum wch vibrates seconds in vacuo be put 3 feet & $8\frac{2}{5}$ [lines], the descent in that time will be 15 feet 1 inch & $2\frac{1}{4}$ lines.(5)

And in the XIXth Proposition the vis centrifuga may be put into proportion to the vis gravitatis as 1 to 289, & then these corrections must be made.(6) Neare the end of the Scholium of Prop IV. for the numbers $290\frac{4}{5}$, 669 & $\frac{1}{669}$ write 289, 665, & $\frac{1}{665}$. Also pag 422 lin 9 write, ut 1 ad 289. lin. 13, ut 289 ad 288. lin 15, 289. lin 16, 288. Pag 423 lin 27, ut 1 ad 288. lin 28, pars $\frac{1}{288}$. lin 31, vis centrifuga $\frac{1}{288}$. lin ult. pars tantum $\frac{1}{288}$. Pag. 424 lin 1, ut 229 ad 228. lin 3, 19674224, seu millia[r]ium 3935. lin 5, pedum 86101 seu milliarium 17. lin 16, ut $\frac{29\times1\times5}{5\times228}$ ad 1, seu 1 ad 8. lin 29, ut 229 ad 228.

The XXXIXth Proposition must be corrected by putting the semidiameters of the earth as 228 to 229 instead of 689 to 692, or perhaps as 3919 to 3951 the difference being 32 miles.(7) I think [228 to 229] should be put for [689 to 692] & the difference of 32 miles may be allowed for in the latter part of the Proposition. But I have lost my copy of the emendation I made to that Proposition & the Lemmas preceding, & so know not how to make this correction. If you can mend the numbers so as to make ye precession of the Equinox about 50″ or 51″, it is sufficient.

I am

Your most humble Servant

Is. Newton

For the Rnd Mr Cotes Professor of Astronomy, at his chamber in Trinity College in Cambridge

NOTES

(1) R.16.38, no. 148, printed in Edleston, *Correspondence*, pp. 59–61.

(2) See *Principia* 1713; pp. 359, 370–2. Newton devoted much effort to the revision of Proposition 8 as it had stood in the first edition (see Koyré and Cohen, *Principia*, pp. 577–84);

in this Proposition and its corollaries he deduced the relative weights of bodies at the surface of the Sun, Jupiter, Saturn and the Earth, and the masses and densities of these bodies. The period and angular separation of the outermost satellite of Jupiter are employed in Corollary 1.

(3) Cotes agreed in his next letter and the paragraph remained unaltered.

(4) Again, Cotes acquiesced; see *Principia* 1713; p. 430.

(5) As Cotes points out in reply, Newton's arithmetic is astray; taking $\pi = 3{\cdot}14159$, $\pi^2 l/2 = 181{\cdot}11$ inches, hence the fractional part is 1·32 lines, not 2·25 lines. On this projected scholium see Letter 889.

(6) Proposition 19 also was greatly revised from its original form. What was at issue here was an alteration of the ratio of the force of gravity to the rotational force of the Earth at the equator from 290:1 to 289:1.

(7) The ratio finally adopted (*Principia* 1713; p. 437) was 229:230. (228/229 is some 3 parts in 100,000 less than 689/692; 229/230 is only about 1 part less. Each such part represents about seventy yards in the Earth's radius.) We have not traced the source of the ratio 3919:3951. (See Letter 890, note (10).)

892 NEWTON TO OXFORD

14 FEBRUARY 1711/12

From the copy in the Public Record Office.[1]
Reply to Letter 887

To the Right Honble. the Earle of Oxford & Earl Mortimer
Lord High Treasurer of Great Brittain

May it please your Lop.

According to Yr Lops. Order signifyed to me by Mr. Secretary Harley in his Letter of Jany. 15th last I have Examined & Audited the Accts. mentioned in the Memll. hereunto annexed abt. printing Mr. Flamsted's Observations,[2] & there is due to Mr. Churchill the Bookseller for paper & printing 98£. 11*s*. as in his Annexed Bill & to the Graver du Guernier[3] 51£. 10 for Graving the Mausoleum and Four other plates as in his annexed Bill, & to the Graver Vertue 30£ for Graving the Frontispeice,[4] and there is a plate still to be Graved which will Cost 7£. 10*s* And the rolling off and some other Graving Work Contracted for by Dr. Halley comes to 7£ ‖ 4*s* as in his annexed Bill, all these summes amount unto 194£ ‖ 15*s*. And I am Humbly of Opinion, that for reducing the Second Volume of Mr. Flamsted's observations into the same Form and Method with the First Volume, and Correcting his Catalogue of the Fix'd Stars by the Observations and Computing the places of 500 Stars more from the observations, for Compleating the Catalogue, and for Correcting the press and Supervising the whole Work Dr. Halley may deserve the Summe of 150£.[5] The performance being hard Labour for the space of one Year, And

Mr Catinaro the painter may deserve 20£. for designing what was to be Graved.[6]

All which &c.

IS NEWTON

St Martins Street
Feby 14*th* 17$\frac{11}{12}$

NOTES

(1) T/53, 21, fos. 493–4.

(2) This was not entered in the copy-book.

(3) Louis du Guernier (*c.* 1659–1716), a Protestant born in Paris, worked on the series of plates illustrating the battles of the Spanish Succession War and produced many other engravings. He had a drawing academy in Great Queen Street. The 'mausoleum' is the dedication to the defunct Prince George, shown as carved upon a tablet set in an elaborate stone memorial, with a sarcophagus below. The other plates are not identifiable.

(4) This is a portrait of Prince George supported by allegorical figures, the whole portraying his interest in navigation.

George Vertue (1684–1756) first succeeded with engravings of Kneller's portraits, from about 1709. In 1717 he was appointed engraver to the Society of Antiquaries and elected a Fellow; besides active work in his profession he collected materials for the history of art in England which were later used by Horace Walpole.

(5) Halley's labours are described in the Preface to the *Historia Cœlestis*.

(6) Little is recorded of J. B. Catenaro; he was presumably of Italian or Sicilian birth, he seems to have worked in Spain between 1692 and 1702, and his name appears on an engraved portrait of George I. The designs for the *Historia Cœlestis* are the major work attributed to him so far.

The copy-book next records Oxford's Warrant for the payment of £364 15*s* 0*d* to Newton:

After &c: Having taken into Consideration the aforegoing Report of Sr Isaac Newton touching the charges of printing Mr. Flamsted's observations Called the Historia Cœlestis, I approve of what is therein proposed, and do hereby Authorize & require you out of any Money that is or shall be Imprested to you at the Receipt of the Excheqr. for this purpose to pay unto the said Sr. Isaac Newton or his Assignes the summe of 364£..15*s* without Accot. for the purposes in the said Report mentioned, And this together with the Acquittance of the said Sr. Isaac Newton or his Assignes shall be...a sufficient Warrant. Whitehall Treasury Chambers 15th April 1712. Oxford.

893 COTES TO NEWTON

16 FEBRUARY 1711/12

From the original in the University Library, Cambridge.[1]
Reply to Letter 891; for the answer see Letter 894

Cambridge Febr. 16. 17$\frac{11}{12}$

Sr

I received Your last of the 12th of this Month. Tis very evident that the 3d section of Prop. XXXVIth ought not to be altered. I had observ'd, that in an addition which You made at the bottom of page 463,[2] *cosinus* ought to be chang'd into *sinus versus*; & thereupon (without any further consideration) I had applied the same change to ye 3d section of the following Proposition. I will observe Your directions as to the diameters of the Sun & Moon in Corol. 3d, retaining in all other places 32′. 12″ for the Sun.

In the Scholium to the IVth Proposition I think the length of the Pendulum should not be put 3 feet & $8\frac{2}{5}$ lines; for the descent will then be 15 feet 1 inch $1\frac{1}{3}$ line. I have considered how to make that Scholium appear to the best advantage as to the numbers & I propose to alter it thus.(3) To take 57220 Toises for the measure of a degree instead of 57230; for 57220 is the nearest round number to a mean amongst 57060, 57292, 57303. To take 3 feet $8\frac{10}{19}$ lines for the length of the Pendulum; for ye French sometimes make it $8\frac{1}{2}$, sometimes $8\frac{5}{9}$, & $8\frac{10}{19}$ is a mean betwixt these numbers. To take 48°. 50′ for the Latitude of Paris instead of 48° 45′ as You had put it. From these principles the following alterations may be made. *Semidiameter Terræ 19670787 ped: Distantia mediocris Lunæ a Terra 1190082614 ped. Distantia Lunæ a communi centro gravitatis 1159567675 ped. Sinus versus ped. 14, dig. 9, lin* $5\frac{5}{14}$. The latter part may be continued thus— id est, in ratione 1 ad 3680,84502; ideoque corpus ad superficiem Terræ vi illa cadendo describet pedes Parisienses 15, dig. 1, lin. $5\frac{1}{6}$

Observatum est longitudinem Penduli ad minuta secunda oscillantis in vacuo, esse pedum trium Parisiensium & linearum $8\frac{1}{2}$ seu linearum $8\frac{5}{9}$: & altitudo quam grave in vacuo cadendo tempore minuti unius secundi describit, (cum sit ad dimidiam longitudinem Penduli hujus, in duplicata ratione circumferentiæ ad diametrum circuli, ut indicavit Hugenius), erit pedum Parisiensium 15, dig.1 lin. $1\frac{9}{10}$. Hic est descensus gravium in latitudine Lutetiæ Parisiorum seu 48 gr. 50′.

Ad Æquatorem vis centrifuga corporum a diurna rotatione Terræ oriunda est ad vim gravitatis ut 1 ad 289 circiter; & in latitudine Lutetiæ minor est, idque in duplicata ratione sinus complementi Latitudinis 48°. 50′ ad Radium, adeoque est ad vim gravitatis ut 1 ad 667. Et hac vi descensus gravium in latitudine Lutetiæ diminuitur. Descensus igitur pedum 15, dig. 1, lin $1\frac{9}{10}$ augeatur parte $\frac{1}{666}$ seu lineis $3\frac{4}{15}$, & habebitur totus gravium descensus pedum 15, dig. 1. lin $5\frac{1}{6}$ quem gravitas sola, tempore minuti unius secundi in latitudine 48°. 50′ efficere posset si modo Terra quiesceret.

I have gone over the computation of ye VIIIth Proposition again, taking 32′. 12″ for the Suns diameter, for I had formerly made use of 32′. 15″. I propose these alterations. [Satellitis extimi Jovialis tempus periodicum dierum 16 & horarum $16\frac{8}{15}$]. [Pondera ad æquales distantias a centris Solis, Jovis, Saturni ac Terræ 1. $\frac{1}{1033}$. $\frac{1}{411}$ $\frac{1}{227512}$]. [Semidiametri Solis, Jovis, Saturni ac Terræ 10000, 1077. 889. 104]. [Pondera ad superficies Solis, Jovis, Saturni ac Terræ 10000. 835. 535. 410]. Densitates Solis, Jovis, Saturni ac Terræ 100. 78. 59. 396](4)

The XIIth Proposition may be altered thus. Nam cum, per Corol. 2. Prop. VIII, materia in Sole sit ad materiam in Jove ut 1033 ad 1, & distantia Jovis a Sole sit ad semidiametrum Solis in ratione paulo majore; incidet commune centrum gravitatis Jovis & Solis in punctum paulo supra superficiem Solis. Eodem argumento cum materia in Sole sit ad materiam in Saturno ut 2411 ad 1, & distantia Saturni a Sole sit ad semidiametrum Solis in ratione paulo minore; incidet &c.[5]

The XIIIth Proposition may be altered thus:[6] pag: 419. lin 18. [ut 1 ad 1033]. lin. 21. [ut 81 ad 16×1033 seu 1 ad 204 circiter]. lin. antepenult. [& $\frac{16\times81\times2411}{25}$ seu 124986] lin: ult: [ut 65 ad 124986 seu 1 ad 1923].

I observe that You have added to the XIVth Proposition a Scholium concerning the motion of ye Aphelia of the Planets, in which by supposing that of Mars to go forwards 35′ in 100 Yeares You deduce the Motion of the Earths Aphelium to be 18′. 36″. I should be glad to know whether You have found these motions to be nearly so by observations, or whether these numbers are propos'd barely as an Example; for in Your new Theory, published by Dr Gregory You make the motion of the Earths Aphelium to be 21′. 40″ in an 100 Yeares.[7] The Rule delivered in this Scholium puts me in mind of a mistake in the new Edition of Your Book which I did not observe till it was too late. In ye 16th Corollary of Prop. LXVI, Lib. 1, or in page 166, line 9th of ye new Edition You will find *ut quadratum temporis periodici corporis P directe* &c. So You had altered it in Your Copy; but I think it should be as in the former Edition, *ut tempus periodicum.* Over against Your alteration there is written in the Margin with a black-lead Pencil by another hand *quadr. temporis period* which You depended upon without considering the thing Your self.[8] I will write to You concerning the XIXth & XXth Propositions in my next.[9]

I come now to ye XXXIXth Proposition, it stands thus in Your Copy.[10] Page 470. line: 10. dele *reciproce.* lin: 26. *ut 474721 ad 4143 seu 114584 ad 10000.* Page 471 l: 20 [evaderet minor quam prius in ratione 2 ad 5. Ideoque annuus Æquinoctiorum regressus jam esset ad 20°. 11′. 46″ ut 1 ad 7330, ac proinde fieret 9″. 55‴. 9^{iv}.

Cæterum hic motus, ob inclinationem plani Æquatoris ad planum Eclipticæ minuendus est, idque in ratione &c.] You have left out all from pag: 471. l: 22 to pag: 473: l: 13. Then in page 473. lin: 27 You have [diminuendus est motus 9″. 55‴. 8^{iv}. in ratione sinus 91706 (qui sinus est complementi graduum $23\frac{1}{2}$) ad radium 100000. Qua ratione motus iste jam fiet 9″. 5‴. 46^{iv}. Hæc est annua Præcessio Æquinoctiorum a vi Solis oriunda.

Vis autem Lunæ ad Mare movendum erat ad vim Solis ut $4\frac{5}{7}$ ad 1 circiter. Et in eadem proportione est vis Lunæ ad vim Solis ad Æquinoctia movenda.

Indeque prodit annua Æquinoctiorum Præcessio a vi Lunæ oriunda 42″. 52‴. 54^{iv}, ac tota Præcessio annua a vi utraque oriunda 51″. 58‴. $40.^{iv}$

Si vis Lunæ ad Mare movendum esset ad vim Solis ut $4\frac{4}{7}$ ad 1 (nam proportionem harum virium nondum satis accurate ex phænomenis definire licuit) prodiret annua Æquinoctiorum præcessio 50″. 40‴. 43^{iv}. Quod cum Phænomenis congruit. Nam præcessio illa ex observationibus Astronomicis est 50″ vel 51″ circiter.(11)

Descripsimus jam Systema Solis, Terræ & Planetarum; superest ut de Cometis nonnulla adjiciantur.] I shall be glad to have this Proposition settled by You before we print any thing which may in any wise relate to it.

Before I conclude this Letter, I will take notice of an Objection which may seem to be against the 3d Corollary of Prop: VI. Libr. III.(12) [Itaque Vacuum *necessario* datur.] Let us suppose two Globes *A* & *B* of equal magnitudes to be perfectly fill'd with matter without any interstices of void Space; I would ask the question whether it be impossible that God should give different *vires Inertiæ* to these Globes. I think it cannot be said that they must necessarily have the same or an equal *Vis Inertiæ*. Now You do all along in Your Philosophy, & I think very rightly, estimate the quantity of matter by the *Vis Inertiæ* & particularly in this VIth Proposition in which no more is strictly proved than that the Gravitys of all Bodys are proportionable to their *Vires Inertiæ*. Tis possible then, that ye equal spaces possess'd by ye Globes *A* & *B* may be both perfectly fill'd with matter, so no void interstices may remain, & yet that the quantity of matter in each space shall not be the same. Therefore when You define or assume the quantity of Matter to be proportionable to its *Vis Inertiæ*, You must not at the same time define or assume it to be proportionable to ye space which it may perfectly fill without any void interstices; unless You hold it impossible for the 2 Globes *A* & *B* to have different *Vires Inertiæ*. Now in the 3d Corollary I think You do in effect assume both these things at once.

I am, Sr, Your
very Humble Servt
ROGER COTES

For
Sr Isaac Newton
at his House
in St Martin's Street
near Leicester-Fields
London

NOTES

(1) Add. 3983, no. 20; printed (from the draft in Trinity College, Cambridge) in Edleston, *Correspondence*, pp. 61–6.

(2) Of the first edition, to Proposition 36.

(3) This redrafting was not adopted. Ultimately, in Proposition 19, Newton adopted 57292 toises as the measure of a degree in France, 3 French feet and $8\frac{5}{9}$ lines (= 99·2 cm) for the length of the second's pendulum, and 19695539 toises for the radius of a spherical Earth. He did, however, agree to put the latitude of Paris as 48° 50′.

Semidiameter of the Earth, 19670787 feet; mean distance of the Moon from the Earth 1190082614 feet. Distance of the moon from the common centre of gravity 1159567675 feet. Versed sine 14 feet, 9 inches, $5\frac{5}{14}$ lines...that is, in the ratio of 1 to 3680.84502; and so a body falling to the surface of the Earth with that force will describe 15 Paris feet, one inch and $5\frac{1}{6}$ lines.

It is to be noted that the length of a pendulum beating seconds *in vacuo* is three Paris feet and $8\frac{1}{2}$ or $8\frac{5}{9}$ lines. Let the mean length of three feet and $8\frac{10}{18}$ lines be adopted; and the height described by a heavy body falling for one second minute in a vacuum (since this height is to half the length of this pendulum as π^2 to one, as Huygens pointed out) will be 15 Paris feet, one inch $1\frac{9}{10}$ lines. This is the fall of heavy bodies in the latitude of Paris, or 48° 50′.

The centrifugal force of bodies arising from the rotation of the Earth is, at the equator, to the force of gravity as 1 to 289, nearly; & it is less in the latitude of Paris in the square of the ratio of the sine of the complement of the latitude (48° 50′) to the radius, and so it is to the force of gravity as 1 to 667. And at the latitude of Paris the fall of heavy bodies is reduced by this force. Accordingly the fall of 15 feet, one inch $1\frac{9}{10}$ lines is to be increased by one 666th part, that is by $3\frac{4}{15}$ lines, and then one will obtain the whole fall of heavy bodies, 15 feet, one inch, $5\frac{1}{6}$ lines which gravity alone could effect in latitude 48° 50′, in one second minute of time, if the Earth were at rest.

(4) The numbers proposed in this paragraph all appear in *Principia* 1713; pp. 370–1.

(5) This paragraph too was printed: *Principia* 1713; p. 374.

(6) These are the numbers printed: *Principia* 1713, pp. 375–6.

(7) Accepting Newton's explanation of these numbers in his next letter, Cotes printed them without more ado. For 'your new Theory', see Letter 890, note (8).

(8) This mistake is entered among the corrigenda of the second edition; see also Cohen, *Introduction*, p. 215 indicating that David Gregory was responsible for it. Compare Letter 783, note (5).

(9) See Letter 896.

(10) See *Principia* 1713; pp. 437–8. The printed text is much as Cotes goes on to describe, but the numbers are almost all slightly different (and the whole is much revised from the first edition).

(11) This paragraph was not printed, because the whole presentation was tightened up later. 'If the force of the Moon for moving the sea were to be to the force of the Sun as $4\frac{4}{7}$ to 1, (for it has not proved possible to define the proportion of these forces very accurately as yet) the annual precession of the equinoxes would work out to be 50″ 40‴ 43^{iv}. Which fits the phenomena. For that precession is about 50″ or 51″ according to the astronomical observations.'

(12) Evidently this corollary was still as in the first edition, that is, it began with the sentence: 'Itaque Vacuum necessario datur.' (hence there necessarily is a vacuum). Later, two corollaries took the place of this single one; see Newton's Letter 903.

894 NEWTON TO COTES
19 FEBRUARY 1711/12

From the original in Trinity College Library, Cambridge.[1]
Reply to Letter 893; for the answer see Letter 896

Sr

In the scholium to ye IVth Proposition I should have put the length of ye Pendulum in vacuo 3 feet & $8\frac{3}{5}$ lines. It was by an accidental error that I wrote $8\frac{2}{5}$ lines. The Pendulum must be something longer in Vacuo then in Aere to vibrate seconds. You may put it either $8\frac{3}{5}$ or $8\frac{10}{19}$ as you shall think fit, the difference being inconsiderable. If you chuse $8\frac{10}{19}$, the numbers computed from thence may stand.

In the new Scholium to the XIVth Proposition, I took the motion of the Aphelium of Mars to be what Dr Halley had computed it & thence deduced the motion of the Earths Aphelium to be 18′. 36″ in an 100 years.[2] Dr Halley had formerly given me the motion of ye Aphelium of ♂ 40′ in 100 years & thence I computed the motion of the Earths Aphelium 21′. 40″: but I account the latter recconing to be more confided in, & therefore in the Theory of ye Moon you may put the motion of ye earths Aphelium 18′ 36″ in 100 years.

In ye 16th Corollary of Prop. LXVI Lib. 1 (or in pag 166 lin 9 of ye new Edition) it should be [ut tempus periodicum corporis *P* directe &c] as you well observe, & not [ut quadratum temporis periodici] as it is now printed.

In the XXXIXth Proposition these emendations may be made.[3] Pag. 470 lin 26 [ad diametrum majorem *AC* ut 228 ad 229) ut 51984 ad 457 seu 11375 ad 100.] Pag 471 lin 1 [ut 100 ad 11375 et 1000000 ad 925275 conjunctim, hoc est, ut 1000 ad 105042, ideoque motus annuli esset ad summam motuum annuli et globi ut 1000 ad 106042.] Ib. lin 7 [ut 1000 ad 106042;] Ib. lin 10 [ut 1436 ad 39343 et 1000 ad 106042 conjunctim, id est, ut 1 ad 2919. Ib. lin. 20 [evaderet minor quam prius in ratione 2 ad 5. Ideoque annuus æquinoxiorum regressus jam esset ad 20° 11′ 46″ ut 1 ad 7298, ac proinde fieret 9″, 57‴, 42^{iv}.] Pag 473 lin 27 [Cum igitur inclinatio illa sit $23\frac{1}{2}$ graduum, diminuendus est motus 9″ 57‴ 42^{iv} in ratione sinus 91706 (qui sinus est complementi graduum $23\frac{1}{2}$) ad radium 100000. Qua ratione motus ille jam fiet 9″ 8‴ 8^{iv}.[4] And a little after. Præcessio a vi Lunæ oriunda 43″. 4‴ $4^{iv}\frac{1}{2}$, ac tota Præcessio annua a vi utraque oriunda 52″ 12‴. 13^{iv}.

Si[5] vis Lunæ ad Mare movendum esset ad vim Solis ut $4\frac{1}{2}$ ad 1 (nam proportio harum virium nondum satis accurate ex phænomenis definire licuit) prodiret annua æquinoxiorum præcessio 50″ 14‴. 45^{iv}. Quæ cum phænomenis congruit. Nam præcessio illa ex observationibus Astronomicis est vel 50″ vel 51″ circiter.

Si altitudo Terræ ad Æquatorem superet altitudinem ejus ad polos miliaribus plusquam 17, materia ejus rarior erit ad circumferentiam quam ad centrum, et præcessio æquinoxiorum ob altitudinem illam augebitur & vicissim ob raritatem diminuetur.

Descripsimus jam systema Solis Terræ et Planetarum: superest ut de Cometis nonnulla adjiciantur.

For obviating the objection you make against the 3d Corollary of Prop. VI Lib. III, you may add to the end of that Corollary these words. Hoc ita se habebit si modo materia sit gravitati suæ proportionalis & insuper impenetrabilis adeoque ejusdem semper densitatis in spatijs plenis.[6]

I am Your most humble Servant

IS. NEWTON

London Feb. 19 1711/12

For the Rnd Mr Cotes Professor of Astronomy at his Chamber in Trinity College in Cambridge.

NOTES

(1) MS. R.16.38, no. 176, printed in Edleston, *Correspondence*, pp. 76–8.

(2) Halley does not appear to have published anything on the motion of the aphelia, hence it seems that this revised value for Mars must have been communicated to Newton privately, as the first value obviously was.

(3) See 1713; pp. 437–8; the numbers were again modified before printing.

(4) A passage in the first edition was later to be struck out from the second, in which the demonstration of this proposition was much shortened; see Number 912*a*.

(5) This paragraph and the next were printed but with slightly differing numbers.

(6) 'This will be so only if matter be proportional to its gravity and impenetrable besides and hence always of the same density in the spaces it fills.' This (certainly ambiguous) sentence was not printed because of the subsequent revision of these corollaries.

895 OXFORD TO NEWTON

23 FEBRUARY 1711/12

From the copy in the Public Record Office[1]

After my hearty Commendations Whereas in pursuance of an Address of the Commons of Great Brittain in Parliament Assembled made to Her Majty. on or about the Fifth day of May 1711 and by her Majts warrant issued thereupon under her Royall Sign Manual bearing Date the 10th Day of the same Month, severall Parcells of Wrought Plate were received into her Majts. Mint after the

14th Day of May 1711 at such Rates & prizes as had been agreed to by the sd Commons which Rates & prices for the said Plate Exceeded the Monies produced by the Coynage of the same by the sume of Nineteen hundred and fifteen pounds Eleaven shillings and sixpence as appears by the Anexed Certificate signed by the Warden and Comptroller of her Majts. sd Mint AND WHEREAS by an Act of this Session of Parliament Entituled (an Act for charging and continuing the Duties upon Malt Mum Cider and Perry for the Service of the Year 1712 and for applying part of the Coynage Duties to pay the Deficiency of the Value of the Plate coined and to pay for the Recoyning the old money in Scotland) It is amongst other things enacted That the said Deficiency or sum not exceeding [£1915 – 11 – 6](2) be supplyed or made good out of the moneys which have arisen or shall Arise by the Duty commonly called the Coynage Duty—These are therefore to Authorize and require you, that out of the monies that are or shall be in your hands of the Coynage Duty, you apply the said Sum of [£1915 – 11 – 6](2) to make good the above mentioned Deficiency. And for soe doing this together with the Respective Ticketts given to the severall Importers and their Acquittances thereupon shall be as well to you for payment as to the Auditors of her Majts. Imprests for allowing thereof upon your account a sufficient Warrant Whitehall Trea[su]ry Chambers 23 of February 1711

OXFORD

To my very Loving Freind Sr Isaac Newton Knt
Master and Worker of her Mats. Mint

NOTES

(1) Mint/1, 8, 80. This authorization finally concluded the business of the 'imported plate' that had dragged on for so long; compare especially Number 858 above.

(2) The amount is fully written out in words; for the justification of this sum, see Number 883 above.

896 COTES TO NEWTON

23 FEBRUARY 1711/12

From the original in Trinity College Library, Cambridge.(1)
Reply to Letter 894; for the answer see Letter 898

Febr. 23*d* 17$\frac{11}{12}$ *Cambridge*

Sr.

I received Your last. As I reviewd the XXth Proposition I perceiv'd it was by a slip of the Pen that You had put 8$\frac{2}{5}$ instead of 8$\frac{3}{5}$ lines in Your former Letter. I choose this number rather than 8$\frac{10}{19}$ for the reason which You gave & because the fraction is more simple & already in use amongst the French. I am satisfied that these exactnesses, as well here as in other places, are inconsiderable to those who can judge rightly of Your book: but ye generality of Your Readers must be gratified wth such trifles, upon which they commonly lay ye greatest stress.(2) I thank You for the information You have given me concerning the new Scholium to the XIVth Proposition. You have very easily dispatch'd the 32 Miles in Prop. XXXIXth, I think You have put that matter in the best method which the nature of the thing will bear.(3)

Your addition to ye 3d Corollary of Prop. VIth does not seem to come fully up to ye Objection. Your words are [Hoc ita se habebit si modo materia sit gravitati suæ proportionalis & insuper impenetrabilis adeoque ejusdem semper densitatis in spatiis plenis]. Now by *materia* You mean the quantity of Matter & this You had always estimated by its Vis Inertiæ, & therefore it will be supposed that You do in this place so estimate it: but if *materia* be here taken in this sense the Objection will not be obviated. Perhaps with some alteration of my words, which You may be pleased to make, the addition may stand thus [Hoc ita se habebit si modo magnitudo vel extensio materiæ in spatijs plenis, sit semper proportionalis materiæ quantitati & vi Inertiæ atque adeo vi gravitatis: nam per hanc Propositionem constitit quod vis inertiæ & quantitas materiæ sit ut ejusdem gravitas](4)

In the XIXth Proposition pag. 422 lin 9 I will put [1 ad 289] & in lin 13th [ut 289 ad 288] in line 15th [289], in line 16th [288] according to Your former directions. In the 25th & 28th lines I would omit ye fractions $\frac{2}{15}$ & write [ut 126 ad 125] & [ut 125 ad 126]: for my computation makes the former proportion to be 126,44024 ad 125,44024 & the latter to be 124,80397 ad 125,80397.(5) In Page 423 lin 11th I would put [hæ tres rationes 126 ad 125, 126 ad $125\frac{1}{2}$, & 100 ad 101]. Ib. lin 27th [ut 1 ad 289]. lin 28th [est tantum pars $\frac{1}{289}$] line 31st [vis centrifuga $\frac{1}{289}$] in ye last line [pars tantum $\frac{1}{229}$]. Page 224,(6) line 1st I would put [per polos 230 ad 229] & ye rest accordingly taking the measure of a mean degree to be 57230 Toises.

In the XXth Proposition, page 425, line 8th, You have altered thus [Unde tale confit Theorema—vel, quod perinde est, ut quadratum sinus recti Latitudinis. Et in eadem circiter ratione triplicata augentur arcus graduum Latitudinis in Meridiano. Ideoque cum Latitudo Lutetiæ &c.] I think the word *triplicata* ought to be omitted:(7) it should be [Et in eadem circiter ratione augentur arcus graduum &c]. I suppose by some inadvertency the mistake arose from this, That the degree under ye Æquator is to ye degree under the Pole as *CPcub* to *CAcub* (fig: page 422).(8) This proportion is no where mentioned in Your additional papers, but I guess You designed to have added it or something to ye same effect to make Your Rule compleat for finding the measure of a degree under any Latitude.

When I was formerly upon this place I made the following alteration in order to examine the numbers of Your Table.[9] [Unde tale confit Theorema, quod incrementum ponderis ut et mensuræ gradus unius in Meridiano pergendo ab Æquatore ad Polos sit quam proxime ut sinus versus latitudinis duplicatæ vel, quod perinde est, ut quadratum sinus recti latitudinis. Nam si M ponatur pro $\frac{AB \times PQcub - PQqq}{ABqq}$, N pro $\frac{ABqq - PQqq}{ABqq}$, & O pro $\frac{ABq - PQq}{ABq}$ (vid: fig: p. 422) erit gravitas sub Æquatore ad excessum gravitatis in alio quovis loco cujus sinus rectus latitudinis est S existente R radio, ut 1 ad $\frac{M}{RR}SS + \frac{MN}{R^4}S^4 + \frac{MNN}{R^6}S^6 +$ &c. Mensura vero gradus unius in Meridiano ad Æquatorem, erit ad excessum ejus in alio loco ut 1 ad $\frac{3O}{2RR}SS + \frac{3 \times 5OO}{2 \times 4R^4}S^4 + \frac{3 \times 5 \times 7O^3}{2 \times 4 \times 6R^6}S^6 +$ &c. Itaque cum sit AB ad PQ ut 230 ad 229, & Lutetiæ Parisiorum in latitudine 48 gr. 50′ longitudo penduli singulis minutis secundis oscillantis sit pedum trium Parisiensium & linearum $8\frac{5}{9}$; longitudines vero pendulorum æqualibus temporibus in locis diversis oscillantium sint ut gravitates: longitudo penduli sub Æquatore erit pedum trium & linearum 7,48 sub Polo erit pedum trium & linearum 9,39: mensura vero gradus unius ad Æquatorem erit Hexapedarum 56783, ad Polum erit Hexapedarum 57530, si modo inter gradus latitudinis 48 & 49 ponatur esse Hexapedarum 57200. Et simili computo confit Tabula sequens].

In making these rules I take the measure of a degree at any point of the Meridian to be proportionable to ye Radius of the curvature of ye Ellipsis at that point, or which is ye same thing to be proportionable to ye Cube of yt part of the Radius of ye curvature which is intercepted between ye point proposed in ye Ellipsis & the point where the Radius intersects ye greater Axis; and ye angle made by that intersection I take for the measure of the Latitude. Thus I had then altered ye place, but I think this exactness is not necessary; for ye following terms of these Series are inconsiderable in respect of the first, & the figure of the Earth is not exactly Elliptical & the solution of the Problem will be more simple without it, by taking ye length of ye Pendulum under the Æquator to ye length under the Poles in the proportion of 229 to 230, & the Measure of a degree at the Æquator to ye measure at ye Poles in the triplicate proportion of 229 to 230 or as 228 to 231 or 76 to 77,[10] & in both cases by making the increment from the Æquator to be as the square of ye sine of ye Latitude or as the versed sine of the doubled Latitude.

As to the Table of the lengths of Pendulums & the measures of Degrees[11] I beleive Your Readers would rather desire it were computed to ye difference of 32 Miles than to that of 17 Miles,[3] & I do not see any use of it as it now stands for which the Table made to the difference of 32 Miles may not serve. If You agree to this Proposal, I will compute it as You shall direct either by the Series or the other way. It must be placed after Your account of the Observations & thereby some small changes will be made in the context which You may be pleased to send me.

What I have further observed as to this Proposition is as follows. You have put down Goreæ Latitudo 14°. 15′. by ye Observations of Des Hayes tis 14°. 40′.[12] In Your account of Picard's experiment of an heated wire, You say [in igne posita] De la Hire says only [car M: Picard ayant exposé les corps a gelée, les mettoit ensuite aupres du feu] or near the fire.[13] By my computation the observation at Guadaloupe reduced to the Æquator gives the difference of 2,29 lines, that at Martinique 2,31 lines, exceeding Your limit of $2\frac{1}{4}$ lines;[14] the rest fall within Your limits. After [auctus in ratione differentiarum fiet milliarium 32] I would add [& diameter secundum æquatorem erit ad diametrum per polos ut 123 ad 122] for as 1,07 to 2 so is $\frac{1}{230}$ to $\frac{1}{123}$.[15] Speaking of the Shadow of the Earth in Lunar Eclipses You say [diameter ejus ab Oriente in Occidentem ducta, major erit quam diameter ejus ab Austro in Boream ducta excessu 56″ fere] I think it should be 41″: for the mean Horizontal Parallax of ye Moon in Syzygijs being 57′. 30″, the Parallax of ye Sun 10″, & the Suns mean diameter 32′. 12″; the diameter of ye Shade will be 4988″, add 70″ upon account of the Atmosphere & the diameter will be 5058″ which divided by 123 gives 41″.[16] At the end of this Paragraph You have [Et distantia mediocris centrorum Terræ & Lunæ erit $60\frac{1}{5}$ semidiametrorum Terræ] which I do'nt well understand.[17] In ye last Paragraph You have [et Pendula isochrona longiora forent in Observatorio Regio Parisiensi quam ad Æquatorem excessu semissis digiti circiter] I suppose it should be [longiora forent ad Æquatorem quam in Observatorio][18] And a little lower You have [sed & diameter umbræ Terræ—major foret—excessu 2′. 45″ seu parte duodecima diametri Lunæ] I think it should be [excessu 2′, seu parte decima sexta diametri Lunæ].[19]

In the Memoires of the Royal Academie for the Year 1708 there are one or two observations of the lengths of Pendulums, besides those which You have related in Your History from other Memoires & from the observations faites en plusieurs Voyages.[20]

Taking ye semidiameters of the Earth to be as 229 & 230 instead of 228 & 229, I have made a small alteration in Proposition XXXIXth which I will not trouble You with since I think I do understand Your thoughts as to that Proposition. The conclusion of it puts me in mind of an allowance which ought to be made in Prop. XXXVIIth on account of the Moons coming nearer to ye

Earth in Syzygijs & going further from it in Quadraturis than in her mean distance at the Octants. But this allowance would increase the number $4\frac{5}{7}$ so much as to give some disturbance to the XXXIXth Proposition & the Scholium of the IVth as they now stand, unless You think fit to ballance it some other way for there is a latitude in that XXXVIIth Proposition.(21)

I am, Sir, Your most Humble Servant

ROGER COTES

For Sr Isaac Newton at his House
in St Martin's Street in Leicester
Fields London

NOTES

(1) R.16.38, no. 178; in this case the original came to Trinity through Robert Smith, who had 'borrowed' it from Conduitt in 1738.

(2) No doubt this is to be read as an apology from Cotes to Newton, whose reply had been terse; it is perhaps hardly a fair comment on Newton's audience.

(3) As Newton emphasizes later (26 February), he believed that a difference of 17 miles between the Earth's radii 'is the least that can be & is certain upon a supposition that the Earth is uniform'; 32 miles was certainly too large a difference; hence if the difference was much greater than 17 miles it must be created by some non-uniformity in the Earth. This proviso eased any possible geodetic discrepancies. Something of the same sort had been stated in the first edition, Proposition 20, but removed from the copy for the second.

(4) 'This will be so, provided that the magnitude or extension of matter in the spaces that it fills is always proportional to the quantity of matter, the force of inertia, and the force of gravity; for it is established by this proposition that the force of inertia and the quantity of matter is as the gravity of the same.' In Newtonian mechanics there is no necessary identity of the magnitudes of gravitational and inertial mass; Cotes here makes the identity axiomatic, this wording did not appear in the final form.

(5) This was done.

(6) Cotes' (unsurprising) slip for 424; all these routine changes were made.

(7) This was done.

(8) *CA* is the semi major axis of the ellipse representing a meridian cross-section of the oblate Earth, *CP* the semi minor axis.

(9) *Summary:* From this arises the theorem, that the increase of weight and likewise of the measure of a single degree of the meridian, in passing from the equator to the poles, is approximately as the versed sine of double the latitude, or (which comes to the same thing) as the square of the sine of the latitude. For if *M* be substituted for... the gravity at the equator will be to the excess gravity in any other place whatever, the right sine of whose latitude is *S*, *R* being the radius... The measure of a single degree of the meridian at the equator will be to the excess in another place as... Thus as *AB* [the major axis] and *PQ* [the minor axis] are as 230 to 229, and Paris is at latitude 48° 50′; as the length of a pendulum beating seconds is three Paris feet and $8\frac{5}{9}$ lines, and as the lengths of pendulums keeping equal times in different places are as the gravities, the length of a [seconds] pendulum at the equator will be three feet and 7·48 lines, and at the pole three feet and 9·39 lines. The measure of a single degree at the equator will be 56783 *toises*, at the pole 57530 *toises*, provided that be-

tween 48° and 49° Lat. the degree is found to be 57200 *toises*. And by a similar computation the following Table is prepared...

(10) $77/76 \approx (230/229)^3$.

(11) This Table was an addition to the second edition.

(12) Gorée is a barren island close to the tip of Cape Verde, Senegal; its trading-post had been in the hands of the Dutch, the English, and (since 1673) the French. The post has been entirely supplanted in modern times by nearby Dakar (lat. 14° 39′). The latitude 14° 40′ was printed, as determined in 'Observations faites en l'Isle de Gorée proche le Cap Verd en Afrique' [par MM. Varin, des Hayes, et de Glos], *Mémoires de l'Académie Royale des Sciences 1666–1699* (1729), VII, pp. 447–59 (latitude, p. 450; length of seconds' pendulum = 3 feet $6\frac{5}{9}$ lines, p. 451; the measurements were made in 1682). It is curious that none of the three men who formed this scientific expedition was ever a member of the Académie; the narrative of it simply states (*ibid*., p. 432) that 'on choisit Messieurs Varin, des Hayes & de Glos, après les avoir exercez à ces sortes d'Observations; & on leur donna l'Instruction suivante...'.

(13) See Philippe de la Hire, 'Remarques sur les inégalités du mouvement des Horloges à Pendule', *Mémoires de l'Académie Royale des Sciences* (1703), pp. 285–99; the reference to the Abbé Picard is on p. 293. It is curious that in discussing the problem of the seemingly irregular variation of the length of the seconds pendulum with latitude, La Hire made no mention of Newton's treatment of this problem. The allusion here to the expansion of metal rods with rise of temperature was another addition to the second edition; as printed (*Principia* 1713; p. 386) the words are *ad ignem calefacta*. Since the phraseology criticized by Cotes is in no extant revised version of the first edition (Koyré and Cohen, *Principia*, pp. 605–10) this is further evidence of the difference between the (lost) printing copy sent to Cotes for the press, and the corrected versions still extant.

(14) See *Principia* 1713; p. 386; the limit was increased to $2\frac{1}{2}$ lines.

(15) In *Principia* 1713; p. 387 the text is *milliarium* $31\frac{1}{12}$ without Cotes' addendum.

(16) Nevertheless, 55″ was printed (*ibid*.).

(17) This sentence was deleted. Again it is not a variant listed by Koyré and Cohen *Principia*, yet it may be found in a draft of this proposition in U.L.C. Add. 3965, fo. 286v, indicating that this draft was close to Cotes' copy.

(18) This amendment was printed.

(19) The original wording was retained (with 2′ 45″ changed to 2′ 46″).

(20) See 'Extrait des Observations faites aux Indes Occidentales en 1704, 1705, & 1706 par le P. Feuillée Minime, Mathematicien du Roy', *Mémoires de l'Académie Royale des Sciences* (1708), pp. 5–16. Newton acted quickly and in his next letter sent Cotes a new paragraph reporting Feuillée's results, which was printed.

(21) There is no 'number $4\frac{5}{7}$' recorded in Proposition 37 by Koyré and Cohen, *Principia*, pp. 666–73. For its occurrence in a draft by Newton, see Letter 899, note (4).

897 T. HARLEY TO NEWTON
26 FEBRUARY 1711/12

From the copy in the Public Record Office.(1)
For the answer see Letter 900

Sr Isaac Newton abt the Counterfeiting Forreigne Gold in Ireland

Sr

His Grace ye Duke of Ormond Her Mats. Genl. and Genl. Governour of ye Kingdome of Ireland having with ye inclosed Letter to my Lord Treasurer(2) transmitted to his Lordp a Copy of a Representation(3) wch his Grace has received from ye Lords of Her Mats. most Honble. the Privy Councill of that Kingdome Proposing the making current there by Proclamation the several Pieces of Forreigne Gold therein mentioned at ye rates & prices therein named in order to prevent the Counterfeiting thereof & that Offenders therein may be punished My Lord Treasurer commands me to send You herewith ye said Letter & Representation, and is pleased to direct You to consider ye same & to Report to his Lordp as soon as conveniently You can what You think proper to be done therein I am Sr &c *26th Febry* 1711

T. HARLEY

NOTES

(1) T/27, 20, 134.

(2) Ormonde's letter, dated 4 February, is calendared in *Cal. Treas. Books*, XXVI (Part II), 1712, p. 181; as it adds it nothing material to the facts stated in the Representation with which Newton had to deal we have not printed it here. James Butler (1665–1745), second Duke of Ormonde, had been an active partisan of William III, and served again (in Spain) at the opening of the War of the Spanish Succession. He was twice Lord-Lieutenant of Ireland: in 1703–7 and from 1710 to 1713. In 1712 the Tories made him commander-in-chief and captain-general in succession to Marlborough with specific orders to remain inactive and hamper the Dutch activities against the French. After the death of Queen Anne Ormonde became a Jacobite and was attainted.

(3) See Letter 897*a*, following.

897*a* THE LORDS JUSTICES AND PRIVY COUNCIL OF IRELAND TO ORMONDE
8 JANUARY 1711/12

From a copy, in Newton's hand, in the Mint Papers(1)

Council Chamber Dublin ye 8th day of January 1711

May it please your Grace

Some of ye Judges who went the last summer circuit in this kingdome having informed this board that several Criminals had been indicted & tried before them for counterfeiting Quadruple Pistoles of Gold & other Forreign Coyne(2) wch pass in payment in this kingdom, but could not be punished pursuant to a late Act of Parliament passed in Ireland wch makes it high Treason to counterfeit forreign Coyne wch is or shall be made current in this kingdome by Proclamation of the Chief Governour & Council, because it happened by mistake that ye Quadruple Pistoles of Spanish Gold & Double Louis D'Ors & other pieces of forreign Gold which pass here in payment were omitted in the last Procla[ma]tion which declared coyn current in Ireland, tho at the same time single Pistoles & single Lewis D'ors are mentioned therein, and because we do not think proper to take upon us to Issue a Proclamation here without Directions from her Majesty, We desire your Grace will obtain her Maties Orders directing Us to Issue a Proclamation for making all such Forreign pieces of Gold as pass in payment in this kingdom current by Proclamation that such who do counterfeit them may be punished by Law vizt The Quadruple Pistole of Spanish Gold weighing 408 grains to pass at £3. 14. – The Double Pistole weighing 204 grains at £1. 17. – The Double French Louis D'or weighing 204 grains at £1. 17. – The quarter Spanish Pistole weighing 25 grains & a half at 4*s.* 7½*d.* The quarter French Pistole weighing 25 grains & a half at 4*s.* 7½*d.* The Portugal piece of Gold called a Moyder & weighing 168 grains at £1. 10. 6. The half Moyder weighing 84 grains at £– 15*s.* 3*d.* And the Quarter Moyder weighing 42 grains at 7*s.* 7*d*½. And we humbly pray your Grace will please to take the first opportunity to obtain her Majts pleasure herein because we are informed great quantities of those forreign coyns have been lately counterfeited in this kingdome to the damage of her Majts subjects, which will be prevented for the future by making them current by Proclamation according to Law.

We are
My Lord
Your Graces most humble Servants
CON. PHIPS CANC.[3]
R. INGOLDSBY[4]

ABBERCORN[5] W. KILDARE[6] CHA FIELDING P. SAVAGE
RICH. COX[7] JNO PERCIVALE[8] CHA DERING SAM DOPPING

A true copy
EDWARD SOUTHWELL[9]

NOTES

(1) Mint Papers II, fo. 236. This letter also is calendared in *Cal. Treas. Books*, XXVI (Part II), 1712, p. 181 from the record copy in P.R.O. T/14, 9, 257–9. There seem to be no significant differences.

(2) In Ireland, and indeed in the provincial towns of England, foreign gold coins were allowed to pass, and accepted in payment of taxes, at fixed rates of exchange. The details are

given later in this letter and subsequent documents. The name *pistole* (of French origin) had been used for over a century to describe the double escudo of Spain; the louis d'or was of the same weight and value. The *moidore* (=moeda d'ouro), current from 1642 to 1732, having a value of 4800 Portuguese reis, was conventionally valued at 27*s* in England.

(3) Sir Constantine Phipps (1656–1723), Lord Chancellor of Ireland, Lord Justice.

(4) Lt. Gen. Sir Richard Ingoldsby (d. 1712), commander of the forces in Ireland, Lord Justice.

(5) James Hamilton (*c*. 1661–1734), sixth Earl of Abercorn.

(6) Welbore Ellis, bishop of Kildare (1705–32).

(7) Sir Richard Cox (1650–1733), a former Lord Chancellor and Lord Justice, now Chief Justice of the Queen's Bench in Ireland.

(8) Sir John Percival (1683–1748), created Earl of Egmont in 1733, Irish parliamentarian, later one of the founders of Georgia.

(9) Edward Southwell (1761–1730), a son of Sir Robert Southwell P.R.S., Chief Secretary in Ireland from 1703 to 1707 and again from October 1710 to September 1713.

898 NEWTON TO COTES

26 FEBRUARY 1711/12

From the original in Trinity College Library, Cambridge.[1]
Reply to Letter 896; for the answer see Letter 899

Sr

I have reconsidered the third Corollary of the VIth Proposition. And for preventing the cavils of those who are ready to put two or more sorts of matter you may add these word[s] to the end of the Corollary. Vim inertiæ proportionalem esse gravitati corporis constitit per experimenta pendulorum. Vis inertiæ oritur a quantite materiæ in corpore ideoque est ut ejus massa. Corpus condensatur per contractionem pororum, & poris destitutum (ob impenitrabilitatem materiæ) non amplius condensari potest; ideoque in spatijs plenis est ut magnitudo spatij. Et concessis hisce tribus Principiis Corollarium valet.[2]

Your emendations of the XIXth Proposition may all of them stand.

In the emendation of the XXth Proposition pag 425 lin. 8 the word triplicata should be struck out as you observe. The rest may stand unto the words [Et simili computo fit Tabula sequens] correcting only the numbers as you propose & putting the numbers 229 & 230 instead of 689 & 692. The Table is computed to ye excess of 17 miles rather then to that of 32 miles, because that of 17 is the least that can be & is certain upon a supposition that the earth is uniform, that of 32 is not yet sufficiently ascertained, & I suspect that it is too big.

After the last observations of Des Hayes ending wth these words [et quod in insula S. Dominici eadem esset ped. 3, lin 7] add this Paragraph.

Denique anno 1704, P. Feuelleus invenit in Porto-belo in America longitudinem Penduli ad minuta secunda oscillantis esse pedum trium Parisiensium et linearum $5\frac{7}{12}$, id est tribus circiter lineis breviorem quam in Latitudine Lutetiæ Parisiorum; & subinde ad insulam Martinicam navigans invenit longitudinem Penduli isochroni esse pedum trium Parisiensium et linearum $5\frac{5}{6}$.(3)

Latitudo autem Paraibæ est 6gr 38′ in austrum et ea Portobeli 9gr 33′ in boream, et Latitudines insularum &c. You may here put the Latitude of Goree 14 gr 40′. I have not books by me to examin it.

Let the next Paragraph run thus. Observavit utique...ad ignem calefacta evasit pedis unius cum quarta parte lineæ...In priore casu calor major fuit quam in posteriore, in hoc vero major fuit quam calor externarum partium corporis humani. Nam metalla ad solem æstivum valde incalescunt...sed excessu quartam partem lineæ unius vix superante...differentia prodijt non minor quam $1\frac{19}{20}$ lineæ non multo major quam linearum $2\frac{2}{3}$. Et inter hos limites quantitas mediocris est $2\frac{3}{10}$. Propter calores locorum in Zona torrida negligamus tres decimas partes lineæ et manebit differentia duarum linearum circiter...jam au[c]tus in ratione differentiarum fiet milliarium plus minus 32. Est igitur excessus ille non minor quam milliarium 17, non multo major quam milliarium 32.(4)

I think the words [excessu 56″ fere] are right. For the Moons parallax 57′ 30″ must have an increase in the proportion of 32 miles to the earths semidiameter, that is an increase of 28″, wch doubled give 56″ to be added to ye diameter of ye earths shadow. For the Suns diameter & parallax remain without sensible alteration. And for ye same reason I take [excessu 2′45″] to be right.(5)

In the calculation of the Moons force (Prop. XXXVII) your scruple may be eased (I think) by relying more upon the observation of the tyde at Chepstow then on that at Plymouth, but I have mislaid my copy of the calculation. If the nearer access of the Moon to the earth in the syzygies then in the Quadratures create any difficulty be pleased to send me a copy of the calculation & I will reconsider it. The Latitude of Paris should be 48gr. 50′. I am

Your most humble servant

London Feb. 26 Is. NEWTON

$17\frac{11}{12}$

For the Rnd Mr Roger Cotes
Professor of Astronomy at his chamber
in Trinity College in Cambridge

NOTES

(1) R.16.38, no. 182.

(2) The final version still has not been reached. This draft reads: 'From pendulum experiments it is established that the force of inertia is proportional to the gravity of a body. The force

of inertia arises from the quantity of matter in a body and so is proportional to its massiness. A body is condensed by the contraction of the pores in it, and when it has no more pores (because of the impenetrability of matter) it can be condensed no more; and so in [completely] full spaces [the force of inertia] is as the size of the space. Granted these three principles the corollary is valid.'

(3) This was done (*Principia* 1713, p. 385—with variants)—see Letter 896, note (20).

(4) The middle part of this passage was yet further modified (see *Principia* 1713; p. 386).

(5) Cotes yielded to Newton, yet Newton's defence of his figures is quite illogical since he had just admitted that the *certain* increase was only 17 miles, and 32 miles was excessive. For some increase between these limits Cotes' figures are more correct.

899 COTES TO NEWTON

28 FEBRUARY 1711/12

From the original in the University Library, Cambridge.[1]
Reply to Letter 898; for the answer see Letter 903

Cambridge Febr. 28*th* $17\frac{11}{12}$

Sr

I have look'd over Your new addition to the 3d Corollary of the VIth Proposition, but I am not yet satisfied as to the difficulty, unless You will be pleased to add, That it is true upon this concession, that the Primigenial particles out of which the world may be supposed to have been fram'd (concerning which You discourse at large in the additions to Your Opticks pag: 343 & seqq:)[2] were all of them created equally dense, that is (as I would rather speak) have all the same Vis Inertiæ in respect of their real magnitude or extension *in Spatio pleno*. I call this a concession because I cannot see how it may be certainly proved either a Priori by bare abstracted reasoning; or be inferr'd from Experiments. I am not certain whether You do not Your Self allow the contrary to be possible. Your words seem to mean so, in pag: 347. lin: 5. Optic [forte etiam et diversis densitatibus diversisque viribus][3]

I do not clearly understand how You would have the alteration settled in Prop: XXth, I mean that part which begins with [Unde tale confit Theorema] & ends with [& simili computo confit Tabula sequens] You may be pleased to send me a transcript of the context, leaving void spaces for the numbers which I will take care to put in according to Your mind. You may let me know at the same time whether You choose in this place 57200 or 57230 Toises for the Measure of a degree between the Latitude of 48° & 49°. I suppose You retain $8\frac{5}{9}$ lines in the length of the Pendulum at Paris, I am satisfied that 56″ is the right increase of the Shadow of the Earth; twas my oversight in making the figure of the Shadow to be similar to that of the Earth.

As to the XXXVIIth Proposition, I take it that the Moons force must be augmented in her Syzygies & diminished in her Quadratures in the proportion of 47 to 46 nearly. Whence by my computation, if nothing else be altered in the Proposition, S will be to L nearly as 1 to $5\frac{2}{7}$. That S may be to L as 1 to $4\frac{5}{7}$ or $4\frac{7}{10}$, instead of putting $L+\frac{6}{7}S$ to $\frac{6}{7}L-\frac{6}{7}S$ as 7 to 4, it may be put $\frac{47}{46}L+\frac{6}{7}S$ to $\frac{46,6}{47,7}L-\frac{6}{7}S$ as 11 to 6. But this proportion of 11 to 6 falls without the Limits at Bristol & Plymouth. I shall therefore leave it to Your self to settle the whole Proposition, as You shall judge it may best be done. In the XXVIIIth Proposition I shall hereafter take notice, that I find the proportion to be as $69\frac{1}{24}$ to $70\frac{1}{24}$ instead of $68\frac{10}{11}$ to $69\frac{10}{11}$. I think 69 to 70 may every where be used. Your copy of ye XXXVIIth Proposition is as follows.(4)

Vis Lunæ ad mare movendum colligenda est ex ejus proportione ad vim Solis, & hæc proportio colligenda est ex proportione motuum maris qui ab his viribus oriuntur. Ante ostium fluvij Avonæ ad lapidem tertium infra Bristoliam, tempore verno & autumnali totus aquæ ascensus in conjunctione & oppositione Luminarium, observante Samuele Sturmio,(5) est pedum plus minus 45, in Quadraturis autem est pedum tantum 25. Altitudo prior ex summa virium posterior ex eorundum [*sic*] differentia oritur. Solis igitur & Lunæ in Æquatore versantium & mediocriter a Terra distantium sunto vires S & L, et erit $L+S$ ad $L-S$ ut 45 ad 25 seu 9 ad 5.

In portu Plymouthi æstus maris (ex observatione Samuelis Colepressi)(6) ad pedes plus minus sexdecim altitudine mediocri attollitur, ac tempore verno & autumnali altitudo æstus in Syzygijs superare potest altitudinem ejus in Quadraturis pedibus plus septem vel octo. Si maxima harum altitudinum differentia sit pedum octo, erit $L+S$ ad $L-S$ ut 20 ad 12 seu 5 ad 3. Donec aliquid certius ex Phænomenis constiterit, assumamus $L+S$ esse ad $L-S$ (proportione mediocri) ut 7 ad 4.

Cæterum ob aquarum reciprocos motus æstus maximi non incidunt in ipsas Luminarium Syzygias sed sunt tertij a Syzygijs ut dictum fuit, & incidunt in horam Lunarem plus minus tricesimam sextam a Syzygijs, id est in horam Solarem tricesimam septimam circiter. Oritur hic æstus ab actione Lunæ in ejus præcedente appulsu ad Meridianum loci, & hic appulsus præcedit æstum in portu Bristoliæ horis plus minus septem, ideoque incidit in horam solarem post Syzygias & Quadraturas tricesimam circiter. Eo autem tempore Luna distat a Sole $15\frac{1}{4}$ gr. circiter. Et Sol in hac distantia minus auget ac minuit motum maris a vi Lunæ oriundum quam in ipsis Syzygijs & Quadraturis, in ratione Radij ad cosinum distantiæ hujus duplicatæ seu anguli $30\frac{1}{2}$ gr. hoc est, in ratione 7 ad 6 circiter; ideoque in superiore analogia pro S scribi debet $\frac{6}{7}S$.

Sed & vis L in Quadraturis ob declinationem Lunæ diminui debet. Nam Luna in Quadraturis tempore verno & autumnali extra Æquatorem in declinatione graduum plus minus $23\frac{1}{2}$ versatur, et Luminaris ab Æquatore declinantis vis ad mare movendum diminuitur in duplicata ratione sinus complementi declinationis quamproxime, & propterea vis Lunæ in his Quadraturis est tantum $\frac{6}{7}L$. Est igitur $L+\frac{6}{7}S$ ad $\frac{6}{7}L-\frac{6}{7}S$ ut 7 ad 4. Et inde fit S ad L ut 7 ad 33 vel 1 ad $4\frac{5}{4}$.(7)

I am Sr
Your most humble Servant
ROGER COTES

For
Sir Isaac Newton
at his House
in St Martin's Street
in Leicester-Fields
London

NOTES

(1) Add. 3983, no. 21, printed in Edleston, *Correspondence*, pp. 75–8 from the draft in Trinity College.

(2) Cotes refers to the first Latin edition of *Opticks* (London, 1706); the second English edition came out only in 1717. As is well known, Newton added new *Queries* numbered 17–23 in the 1706 edition. Cotes here alludes to Query 23 (which is Query 31 in the more familiar, later English editions, the passage beginning: 'All these things being considered, it seems probable to me that God in the beginning formed matter in solid, massy, hard, impenetrable, moveable particles...').

(3) 'it may be also allowed that God is able to create particles of matter of several sizes and figures, and in several proportions to space, and *perhaps of different densities and forces*, and thereby to vary the laws of nature and to make worlds of several sorts in several parts of the universe'. The full quotation makes it clear that Newton did not contemplate such a variety in the part of the universe he was concerned with. He did not claim that his principles were valid for *all possible* worlds!

(4) Because Cotes did not copy the quotation into his draft, Edleston printed rather more from Newton's own draft (MS. 193, 194, see Edleston *Correspondence* p. 77, note) than Cotes actually copied into his letter. The text of Proposition 37, as given here, does not greatly differ from what was finally printed, but neither does it correspond closely with any of the versions considered in Koyré and Cohen, *Principia*, pp. 666–8. Newton did much juggling with the numbers before arriving at those selected for printing. Notably the ratio of L to S as $4\frac{5}{7}$ to 1 appears only here and in U.L.C. MS. Add. 3965, no. 12.

(5) See Hall & Hall, *Correspondence of Oldenburg*, IV, pp. 424–7; V, pp. 95–6. Captain Samuel Sturmy (1633–69), who had served in the West Indies trade out of Bristol, was a martyr of science; he died of a chill caught while exploring a Somerset pot-hole. His tidal observations were published in *Phil. Trans.* **3**, no. 41 (16 November 1668), 813–17.

(6) On Samuel Colepresse (d. 1669) see Hall & Hall, III, 311 note; he was a regular correspondent of Henry Oldenburg for a few years. For his tidal observations (printed in *Phil. Trans.* **3**, no. 33 (16 March 1667/8), 632–3) see Hall & Hall, IV, p. 106.

(7) Newton's draft (Edleston *Correspondence*, p. 78) continued with one further short paragraph (not as printed) and then Corollary 1 (more or less as printed).

900 NEWTON TO OXFORD

3 MARCH 1711/12

From the holograph draft in the Mint Papers.[1]
Reply to Letter 897

To the Rt Honble ye Earl of Oxford & Earl Mortimer
Lord H. Treasurer of great Britain

May it please Your Lordp

According to Your Lordp's Order signified to me by Mr Secretary Harley in his Letter of 26 Feb. last, I have considered the Letter of his Grace the Duke of Ormond, her Majts Lieutenant General & General Governour of Ireland sent to your Lordp, together with the Representation sent to his Grace from the Lords of Her Mats Most Honble Privy Council of that kingdom, mentioning a late Proclamation for making current in Ireland some pieces of forreign gold & proposing the making current there by further Proclamation several other peices of forreign gold therein named in order to prevent the counterfeiting thereof. And as to the value of the pieces I humbly represent that the Spanish Pistoles one wth another as they are brought hither by the Merchant, weigh 103 grains each at a medium, & are in fineness half a grain worse then standard, & after the rate that a Guinea is valued in England at £1. 1*s*. 6*d* are here worth 17*s*. 1*d*. & in Ireland where the silver money is raised a penny in the shilling, if they be raised in the same proportion, become worth 18*s*. 6*d*. And in proportion the Quadruple Pistole weighs 412 grains, the double Pistole 206 grains, & the quarter Pistole 25 grains & three quarters. But in the Representation the Quadruple Pistoles are said to weigh 408 the double 204 grains & the Quarter Pistole 25½ grains. Whence I gather that in the former Proclamation the weight of the Pistole was but 102 grains, wch is a grain lighter then the just weight, this grain as I conceive being abated to give a legal currency to such lighter pieces as want not above a grain of their just weight. And upon this consideration the Quadruple, Double, & Quarter Pistoles may be put in weight and value as is exprest in the Representation. And so may the double and quarter Lewis d'ors, they being of the same weight fineness & value with the double & quarter Pistoles.

The Moyders of Portugal one with another, as they are brought hither by the Merchant weigh 165¾ Grains at a medium, & are a quarter of a Grain better then standard, & in England are worth 27*s*. 8½*d*., & being raised a penny in the shilling become worth 1£ 10*s*. in Ireland. In the Representation their weight is put 168 gr wch is certainly too much, & thence it comes to pass that they are therein valued at 1£. 10*s*. 6*d*. which is 6*d*. too much. I have examined the weight of 30 parcels of Moyders conteining a thousand Moyders in each parcel & thereby found that the Moyder at a medium weighs only 165 grains & three quarters. If in favour of the lighter pieces the fraction be abated their weight & value in a new Proclamation may be put as follows. The Portugal piece of gold called a Moyder and weighing 165 grains to pass at 1£. 10*s*. The Half Moyder weighing 82½ grains at 15*s*. & the Quarter Moyder weighing 41 Grains & a quarter at 7*s*. 6*d*.(2)

Gold is over-valued in England in proportion to silver by at least 9*d*. or 10*d* in a Guinea, & this excess of value tends to increase the gold coins & diminish the silver coins of this kingdom.(3) And the same will happen in Ireland by the like overvaluing of gold in that kingdom. But its convenient that the coins should bear the same proportion to one another in both kingdoms for preventing all fraudulent practices in those that trade between them, & that the proportion be ascertained by proclamation

All wch &c

Mint Office
3d March $17\frac{11}{12}$

NOTES

(1) Mint Papers II, fo. 236v. There is a record copy in P.R.O. T/14, 9, 257–9 calendared in *Cal. Treas. Books*, XXVI (Part II), 1712, p. 182. This letter was first printed in Shaw, pp. 176–9, without stating his source. There are many rough drafts, including one on a sheet also used for *Principia* work in U.L.C. Add. 3965, fos. 281r–282v.

(2) On 17 July 1712 an Order in Council was made (this followed the 'Second Representation' submitted by Newton on 23 June) fixing the values of the coins as suggested by Newton. (See P.R.O. Mint/1, 1, 209: 'Upon reading this day at the Board a Report from Sr Isaac Newton Master of Her Matys Mint relating to the Weight and Value of Foreign Coynes in Ireland at the rates following [pistole 18*s*. 6*d* and multiples proportionately; moidore £1–10–0; ducatoon 6*s*.; piece of eight, louis d'or, rix dollar 4*s*. 9*d*; crusado 3*s*.] ordered that they do so pass in Ireland.')

(3) This is one of Newton's constant themes, richly expressed in his well-known 1717 report (see vol. VI).

901 COTES TO NEWTON
13 MARCH 1711/12

From the original in the University Library, Cambridge.(1)
Reply to Letter 898; for the answer see Letter 903

Cambridge March 13th $17\frac{11}{12}$

Sr

I received Your last of the 26th of February in due time; & by the next Post, with one or two other things, I sent You a Transcript of the XXXVIIth Proposition as it now stands in Your Copy.(2) Having received no Letter from You since that time, I fear there has been some miscarriage. About 2 sheets of the third Book are compos'd, but expecting Your answer by every Post I have not yet given leave to print them off.(3)

I am, Sr.
Your most humble Servant
ROGER COTES

For Sr Isaac Newton
at his House
in St Martins-Street
in Leicester Feilds
London

NOTES

(1) Add. 3983, no. 22, printed in Edleston, *Correspondence*, p. 79.

(2) See Letter 899.

(3) Throughout the presswork the University Press was printing off the requisite number of copies (750) of each sheet as soon as Cotes had approved the proof, and then distributing the type for reuse. Cotes presumably meant that the sheet Bbb was now set (that is, up to Proposition 14 on p. 376).

902 NICHOLAS SAUNDERSON TO JONES

16 MARCH [1711/12]

Extracts from the original in private possession(1)

Christ Coll [*Cambridge*] *March* 16

...There has been nothing publish'd here since my last to you, excepting a treatise which is not worth mentioning by one Mr Green Fellow of Clare-Hall of this University.(2) If there had been anything in it instructive or diverting I should have sent it you. But I can find nothing in it, but ill manners & elaborate nonsense from one end to the other...They are now got to the fourth Propo: of the third Book of Sr. Is. Newton's Princip: But I cannot give you so full an account of the conduct of that peece, as perhaps you would desire, or as I should have done but that I know Mr Coats maintains a correspondence with you;(3) And I d[o]ubt not but he will give you an account of it so far as he thinks Sr. Is. will be willing to have any body acquainted with it...Sr Is. Newton is much more intent upon his principia than formerly, & writes almost every post about it, so that we are in great hopes to have it out in a very little time.

NOTES

(1) Printed in Rigaud *Correspondence*, I, 264–*264. For the writer see Letter 880, note (4).

(2) See Letter 854, note (3).

(3) There seems to be none surviving from this period.

903 NEWTON TO COTES

18 MARCH 1711/12

From the original in Trinity College Library, Cambridge.[1]
Reply to Letters 896, 899 and 901

Sr

I have not yet been able fully to settle the Theory of the XIXth, XXth, XXXVIth XXXVIIth & XXXIXth Propositions, & that of the Scholium to the IVth. But I think to let the Scholium of IVth Proposition be set at the end of the XXXVIIth because it depends on a Corollary of that Proposition.[2] And therefore you may let the Press go on at present without it & set it aside till you come to the XXXVIIth Proposition.[3] But let the new Corollary to ye IIId Proposition[4] be printed at the end of that Proposition. And in the third Corollary to ye Vth Proposition strike out the word [novissimam,] & let the words in the latter part of ye Corollary run thus [Et hinc Jupiter & Saturnus prope conjunctionem se invicem attrahendo sensibiliter perturbant motus mutuos; Sol perturbat &c][5] In my copy it is prope conjunctionem novissimam. If it be so in yours, the word novissimam is better omitted.

I thank you for explaining your objection against ye third Corollary of the sixt Proposition. That Corollary & the next may be put in this manner.[6] Corol. 3. Spatia omnia non sunt æqualiter plena. Nam si spatia omnia æqualiter plena essent, gravitas specifica fluidi quo regio aeris impleretur, ob summam densitatem materiæ, nil cederet gravitati specificæ argenti vivi vel auri vel corporis cujuscunque densissimi, et propterea nec aurum neque aliud quodcunque corpus in aere descendere posset. Nam corpora in fluidis, nisi specifice graviora sint, minime descendunt. Quot si quantitas materiæ in spatio dato per rarefactionem quamcunque diminui possit, quidni diminui possit in infinitum? Corol. 4. Si omnes omnium corporum particulæ solidæ sint ejusdem densitatis neque absque poris rarefieri possint, Vacuum datur. Ejusdem densitatis esse dico quarum vires inertiæ sunt ut magnitudines. Corol. 5. Vis gravitatis diversi est generis a vi magnetica. Nam attractio magnetica non est ut materia attracta. Corpora aliqua magis trahuntur, alia minus, plurima non trahuntur; Et vis magnetica in uno et eodem corpore intendi potest & remitti, estque nonnunquam longe major pro quantitate materiæ quam vis gravitatis, et in recessu a magnete decrescit in ratione distantiæ non duplicata sed fere triplicate quantum ex crassis quibusdam observationibus animadvertere potui.

In the tenth Proposition pag. 417 lin 11 for [viginti et unius] read [triginta.] & lin. 12 for [320] read [459] & lin 17 for [800] read [850].[7]

I hope to send you the XIX & XXth Proposition emended within a Post or two.[8] I am

Your most humble Servant
Is. NEWTON.

Mar. 18*th* 1711/12

NOTES

(1) R.16.38, no. 184; printed in Edleston, *Correspondence*, pp. 79–81.

(2) This Scholium had been sent with Newton's Letter 889; only a part of it, modified, was in the end to be printed after Proposition 37. A draft of about the same date as the present letter (U.L.C., Add. 3984, no. 9) includes a completely rewritten Scholium to Proposition 4, a draft of Proposition 37 up to and including Corollary 5 and corrections to Proposition 39. Apart from their numerical values the drafts for Propositions 37 and 39 coincide with the versions sent on 8 and 22 April (Letters 909 and 912) respectively. The information in the Scholium (which bears no resemblance to the corresponding Scholium in the first edition) was largely absorbed, eventually, into the seventh Corollary to Proposition 37, which Newton sent on 22 April. The draft Scholium reads as follows:

Ex Observationibus Astronomicis corrigendo Theoriam Lunæ, invenimus per Eclipses Lunares mediocrem ejus distantiam a Terra in Octantibus esse $60\frac{1}{4}$ vel $60\frac{1}{5}$ vel (quantitate mediocri) 60,22 semidiametrorum terrestrium quamproxime. Semidiametros intelligo ad æquatorem ductas. Hæ semidiametri superant diametrum ad polum ductam milliaribus 32 circiter, ut posthac dicetur, id est parte ducentesima quadragesima sexta semidiametri, ideoque mediocris distantia Lunæ a Terra in Octantibus est semidiametrorum terrestrium mediocrium 60,46 circiter seu pedum Parisiensium 1189399723. Et hanc distantiam esse ad distantiam inter centrum Lunæ & commune gravitatis centrum Lunæ ac Terræ ut 42,3647 ad 41,3647 quamproxime, patebit in sequentibus per fluxum et refluxum maris; ideoque distantia inter centrum Lunæ et commune gravitatis centrum Lunæ ac Terræ est pedum Parisiensium 1161324470.

Cum Luna revol[va]tur respectu fixarum diebus 27 horis 7 & minutis primis $43\frac{1}{5}$, sinus versus anguli quem Luna tempore minuti unius primi motu suo medio circum commune gravitatis centrum Lunæ ac Terræ describit, est 1,275235, existente radio 100,000,000,000000. Et ut Radius est ad hunc sinum versum ita sunt pedes 1161324470 ad pedes 14,8096161.

Et per Corol. Prop III, vis qua Luna retinetur in Orbe est ad vim eandem in superficie terræ in ratione $177\frac{29}{40}$ ad $178\frac{29}{40}$ et ratione duplicata semidiametri Terræ ad distantiam centrorum Lunæ ac Terræ quæ fuit semidiametrorum 60,46, id est in ratione 1 ad 3675,959, ideoque corpus ad superficiem Terræ vi illa cadendo, tempore minuti unius secundi describet pedes Parisienses 15, dig. 1 & lin. $5\frac{1}{2}$.

(3) Obviously the compositor could not proceed beyond Proposition 18. There seem to be no cancelled leaves here, and therefore the printing of the early part of Book III must in the end have proceeded quite smoothly, though not without resetting and rearrangement of the type before imposition.

(4) This was sent on 2 February; see p. 216, note (4).

(5) It is so printed, *Principia*, p. 365.

(6) This is at last the text as printed on pp. 367–8.

(7) This was done, *ibid.*, p. 373.

(8) See Letter 908.

904 LOWNDES TO THE MINT

18 MARCH 1711/12

From the copy in the Public Record Office[1]

[To the Principal] Officers [of the] Mint abt.
Mr. Williams relating to selling her Majties Tynn

Gentlemen

By Order of My Lord Trea[su]rer I send you the inclosed paper wch his Lordp has recd from Mr Williams[2] in relation to the selling of her Mats Tynn, His Lp directs you to hear the said Mr Williams thereupon, & when you have fully considered the said paper and what he may have further to Offer on that Subject, that you will Report your Opinions upon the whole to his Lordp. I am &c 18th March 17$\frac{11}{12}$

WM LOWNDES

NOTES

(1) T/27, 20, 149.

(2) This person has been identified with John Williams, who was agent for the Queen's Printers and received payments for them (*Cal. Treas. Books* (Part II), 1712, pp. 129, 227); he was rich enough to lend the government nearly £7000 on the security of the tin stock-pile (*ibid.*, p. 268). He had proposed an enhancement of the price of tin—on the grounds that Cornwall enjoyed a virtual world monopoly—in 1705 and 1706 (see vol. IV, Letter 712). Newton always opposed his views (see Mint Papers, III, fos. 534, 545, 578–99).

Williams had been confronted with the Principal Officers of the Mint in the presence of the Lord Treasurer on 13 March, when he presented a paper (now lost) which the Treasurer referred to the Mint Officers for their consideration. See below, Letter 907.

905 OXFORD TO NEWTON

21 MARCH 1711/12

From the copy in the Public Record Office[1]

After &c. Upon consideration of the aforegoing Report[2] made to Me by the Barons of her Majts: Court of Exceqr. in Scotland relating to the Demands of the provost & Moneyers of her Majts: Mint in the Tower of London on Account of the recoynage of the moneys in Scotland upon the Union, and alsoe of the annexed Certificate in that Matter (referr'd to by the said Barons) of the Warden and Comptroller of her Majes: Mint at Edinburgh Whereby it appeares that the Sum of 2692£ 15*s* 3½*d* is still remayning due to the said provost and Moneyers for that recoynage upon the severall Allowances settled for the same by her Majts: Warrant bearing date the 12th Day of July 1707. These are in Pursuance of the Authority to me given by a Clause in an Act of Parliament Entituled [An Act for charging and continuing the Dutys upon Malt mum Cyder and Perry for the service of the year 1712 and for applying part of the Coynage Duty to pay the Deficiency of the Value of Plate coined and to pay for the recoynage of the Old Money in Scotland] to Authorize and require you out of such Money as is or shall be in your hands of the Coynage Duty to pay unto the said Provost and Monyers or to whome they shall appoint to receive the same the said sum of 2692£ ‖ 15*s* ‖ 3½*d* in full Satisfaction of what remaines due to them upon the Severall Allowances settled by her Majts said Warrant...And for so doing this being first entered with the Auditors of her Majts: Imprests...shall be as well to you for payment as to the said Auditors of Imprests For allowing thereof in your accounts a sufficient Warrant

Whitehall Treasury Chambers 21 *March* 1711/12

OXFORD

To my very loving Friend Sir Isaac Newton Knt
Master and Worker of her Majes Mint in the Tower of London.

NOTES

(1) T/53, 21, 46; there is another copy in Mint/1, 8, no. 88. We have abbreviated the purely routine phrases at the end.

(2) On 19 February 1712 T. Harley had written to the Lord Chief Barons of the Exchequer of Scotland, instructing them to report their opinion on a petition from the Moneyers of the Tower for the payment of £2700 5*s* 3½*d* which, they claimed, was still due to them for their work on the Scottish recoinage. The Barons approved an account prepared by William Drummond, Warden of the Edinburgh Mint, and William Boswell, Comptroller, showing that the proper sum due was a few pounds less than that claimed. The Baron's report and the Edinburgh Mint certificate are entered with the Warrant to Newton.

906 MARY PILKINGTON TO NEWTON

22 MARCH 1711/12

From the original in the University Library, Cambridge[1]

March ye 22
1711/2

Honrd: Sr

I have recivd nine pound & return you My Most Humble thanks; if you or my Cosin Barton[2] Please to writ to mee, you may derict to mee at Mr Middlemorr at Stanton(3) to bee left at Mr Silles at ye Han Cross in Nottingham; pray Sr doe Mee ye favour to send ye other sid of ye letter to Mrs Mary Savage(4) I am

yo[u]r Most obedient
Nece & Humble Servant

My Servics to Cos: Barton

M. Pilkington

To
Sr. Isaac Newton in
St: Martens Street Nere
lester fields
london

NOTES

(1) Add. 3966(15), fo. 210. Mary Smith (b. 1647) was Newton's half-sister by his mother's second marriage. On 22 November 1666 she had married Thomas Pilkington of Belton, Rutland, who was pricked High Sheriff of his county in 1671. Possibly the same man stood bondsman for Newton in the sum of £1000, presumably when Newton became Master of the Mint in 1699 (Brewster, *Memoirs*, II, p. 193, note). The writer of this letter was presumably the only daughter of this marriage, for (according to Newton's unrelated amanuensis Humphrey Newton: 'Mr Pilkinton, who lived at Market Overton, died in a mean condition (tho' formerly he had a plentiful estate,) whose widow with 5 or 6 children Sir Is. maintained several years together.' (Brewster, *Memoirs*, II, p. 98.) The writer and her brothers, Thomas and George, participated in the winding up of Newton's estate after his death.

(2) Catherine Barton was niece to the writer.

(3) Presumably Standon-on-the-Wolds, seven miles from Nottingham.

(4) This Newton did not do, though he may have conveyed the message otherwise. On the reverse of the other half of the sheet—that is, on the reverse of the part carrying Newton's address—is written in the same elderly hand:

March ye 22

Mrs Savage

this Comes with my thanks & to lett you know my Aunts things & my Bill was 3 pounds I am

yor: oblige
Humble Servant
M. Pilkington

Newton has used both sides of the paper for computations, and a few words relating (probably) to Book III, Proposition 37. See further Letter 955.

906a NEWTON TO — TODD

(n.d.)

From the draft in the University Library, Cambridge[1]

Mr. Todd

Whether the order I sent you to pay ye money to my Sister miscarried or whether to gain more time you are unwilling to own the receipt of it I will not affirm. But Mr. Drake is witness that such an order was sent you by ye Post enclosed in a bill of charges from himself, & letters by ye Post use not to miscarry. But be it as it will, to make sure that this come to you I have charged ye carrier to deliver it wth his own hands. About your pretenses of ye money's being ready long since & of a jugment wch you would have me beleive I had against you I do not think it material to expostulate. I shall only tell you in general that I understand your way & therefore sue you. And if you intend to be put to no further charges you must be quick in payment for I intend to loos no time. I desire you therefore to pay it to my Sister Mary Pilkington at Market Overton[2] as soon as you can & take her acquittance for your discharge. Besides the 50*lb* principall you are to pay the use[3] for an hundred pounds from the time of ye date of ye Bond Jun 20 till the date of my receipt of ye 1st 50*lb* & from that time ye use for ye other 50*lb*, excepting only 40*s* of use, wch is already payd. You are also to pay ye charges of ye suit an account of wch you will herewith receive from Mr Drake. And when I am satisfied that you have payd all this, your bond shall be delivered in here to any one you please to appoint

Sr I am

Yours

You represent that ye money was ready $\frac{1}{2}$ a year since & that I had a judgmt agt you & therefore complain that I sue you. Though I do not think it material to expostulate wth you about these things I shall only tell you in general that last midsommer not understanding yt your putting in your answer to Mr Noels creditors concerned me, though I could have been contented to have been put off to ye end of that term by this excuse, yet when no money came at ye end of yt term, I determined to write to you but once or twice more & if you paid me not before Michaelmas term to apply my self then to an Attorney, & sue out a real judgmt agt you if upon enquiry I found not that to be real wch you pretended.

NOTES

(1) Add. 3965(11), fo. 155v. We add this undated draft as further evidence of Newton's deep and long-continued concern for his family; it also displays well his passionate nature, which could manifest itself in this context as well as against Flamsteed or Leibniz.

(2) Compare Letter 906, note (1); since the mother of the writer of that letter was then still living at her marital home, it is highly likely that this draft was composed some years before 1711.

(3) Interest.

907 JOHN ANSTIS TO NEWTON

24 MARCH 1711/12

From the original in the University Library, Cambridge[1]

24 *March* 1711/12

Sr.

Mr. Peyton[2] desires that you would please (if the same be not too much trouble) to draw up a report upon Mr. Williams's Project,[3] that the same may be ready to be signed at the Mint on Wednesday next, for Mr. Williams hath already acquainted mr. Peyton that he will not then attend.[4]
With all respect I am

Your most Obedt. Servt
JOHN ANSTIS

NOTES

(1) Add. 3965(8), fo. 88r. The sheet has been used for drafts and computations relating to *Principia*, Book III, Proposition 37 in which Newton derives the ratio of S and L as 1 to 4·82366 (not used).

The writer of this letter, who was born in 1669 and died in 1744, served as M.P. for several Cornish constituencies, and was also involved directly in the Cornish tin trade. He was moreover a learned antiquary, and received (on behalf of Thomas Rymer and his publisher Awnsham Churchill) the government payments for the printing of *Fœdera*. Anstis was appointed Garter King of Arms in 1715, but shortly afterwards lost his office and suffered imprisonment on suspicion of active Jacobitism. In 1718 he recovered his position. His writings relate to heraldry and genealogy. He seems to have shared Williams' views on the price of tin.

(2) Warden of the Mint.

(3) See Letter 904. Wednesday was the regular day on which the Mint officers met to do their business—much of it previously prepared by Newton; the meaning is that Williams had been invited to meet the officers on that day, but had refused, and hence there would be no conference to modify (possibly) Newton's draft report. We have not been able to trace a copy of this report.

(4) On 14 May 1712 Williams submitted a fresh document to the Treasurer rebutting the Mint's rejection of his proposal (*Cal. Treas. Papers, 1708–14*, pp. 287–8).

908 NEWTON TO COTES
3 APRIL 1712

From the original in Trinity College Library, Cambridge.[1]
For the answer see Letter 910

London Apr. 3 1712

Sr

I have been diverted a few days wth some other intervening business, but now send you the emendations of ye XIXth XXth & XXVth[2] Propositions, as follows.

Prop. XIX. Prob. II[3]

Invenire proportionem axis Planetæ ad diametros eidem perpendiculares.

Picartus[4] mensurando arcum gradus unius et 22′. 55″ inter *Ambianum* & *Malvoisinam*, invenit arcum gradus unius esse hexapedarum Parisiensium 57060. Unde ambitus Terræ est pedum Parisiensium 123249600, ut supra. Sed cum error quadringentesimæ partis digiti tam in fabrica instrumentorum quam in applicatione eorum ad observationes capiendas sit insensibilis, et in Sectore decempedali quo *Galli* observarunt Latitudines locorum respondeat minutis quatuor secundis, et in singulis observationibus incidere possit tam ad centrum Sectoris quam ad ejus circumferentiam, et errores in minoribus arcubus sint majoris momenti:* ideo *Cassinus*[5] jussu Regio mensuram Terræ per majora locorum intervalla aggressus est, et subinde per distantiam inter Observatorium Regium *Parisiense* et villam *Colioure* in *Roussillon* & latitudinum differentiam 6 gr. 18′, supponendo quod figura Terræ sit sphærica, invenit gradum unum esse hexapedarum 57292, prope ut *Norwoodus*[6] noster antea invenerat. Hic enim circa annum 1635 mensurando distantiam pedum Londinensium 905751 inter *Londinum* et *Eboracum* & observando differentiam Latitudinum 2 gr. 28′ collegit mensuram gradus unius esse pedum Londinensium 367196, id est, hexapedarum Parisiensium 57300. Ob magnitudinem intervalli a *Cassino* mensurati, pro mensura gradus unius in medio intervalli illius id est inter Latitudines 45 gr & 46 gr usurpabo hexapedas 57292. Unde, si Terra sit sphærica, semidiameter ejus erit pedum Parisiensium 19695539.

* *Vide Historiam Academiæ Regiæ Scientiarum anno 1700*

Penduli in Latitudine *Lutetiæ Parisiorum* ad minuta secunda oscillantis longitudo est pedum trium Parisiensium & linearum $8\frac{5}{9}$. Et longitudo quod grave tempore minuti unius secundi cadendo describit est ad dimidiam longitudinem penduli hujus in duplicata ratione circumferentiæ circuli ad diametrum ejus (ut indicavit *Hugenius*) ideoque est pedum Parisiensium 15, dig. 1, lin $2\frac{1}{4}$,[7] seu linearum $2174\frac{1}{4}$.[7]

Corpus in circulo, ad distantiam pedum 19695539 a centro, singulis diebus sidereis horarum 23. 56′. 4″ uniformiter revolvens, tempore minuti unius secundi describit arcum pedum 143,6223,[8] cujus sinus versus est pedum 0,05236558, seu linearum 7,54064. Ideoque vis qua gravia descendunt in Latitudine *Lutetiæ* est ad vim centripetam corporum in Æquatore a Terræ motu diurno oriundam ut $2174\frac{1}{4}$[7] ad 7,54064.

Vis centrifuga corporum in Æquatore est ad vim centrifugam qua corpora directe tendunt a Terra in Latitudine *Lutetiæ* in duplicata ratione Radij ad sinum complementi Latitudinis illius, id est, ut 7,54064 ad 3,27.[9] Addatur hæc vis ad vim qua gravia descendunt in Latitudine *Lutetiæ*, et corpus in Latitudine *Lutetiæ* vi tota gravitatis cadendo, tempore minuti unius secundi describet lineas 2177,52,[10] seu pedes Parisienses 15, dig. 1, & lin. 5,52.[11] Et vis tota gravitatis in Latitudine illa erit ad vim centripetam corporum in Æquatore Terræ ut 2177,52[10] ad 7,54064, seu 289 ad 1.

Unde si *APBQ* figuram Terræ designet jam non amplius sphæricam sed revolutione Ellipseos circum axem minorem *PQ* genitam, sitque *ACQqca* canalis aquæ plena, a polo *Qq* ad centrum *Cc*, & inde ad Æquatorem *Aa* pergens: debebit pondus aquæ in canalis crure *ACca* esse ad pondus aquæ in crure altero *QCcq* ut 289 ad 288, eo quod vis centrifuga ex circulari motu orta partem unam e ponderis partibus 289 sustinebit ac detrahet, et pondus 288 in altero crure sustinebit reliquas. [In the rest of the XIXth Proposition proceed according to the former corrections untill you come at page 484, where read][12] ad ipsius diametrum per polos ut 230 ad 229.[13] Ideoque cum Terræ semidiameter mediocris juxta mensuram *Cassini* sit pedum Parisiensium 19695539, seu milliarium 3939 (posito quod milliare sit mensura pedum 5000) Terra altior erit ad Æquatorem quam ad Polos excessu pedum 85820, seu milliarium $17\frac{1}{6}$.

Si Planeta major sit vel minor quam Terra manente ejus densitate ac tempore periodico revolutionis diurnæ, manebit proportio vis centrifugæ ad gravitatem, & propterea manebit etiam proportio diametri inter polos ad diametrum secundum æquatorem. At si motus diurnus in ratione quacunque acceleretur vel retardetur, augebitur vel minuetur vis centrifuga in duplicata illa ratione, et propterea differentia diametrorum augebitur vel minuetur in eadem duplicata ratione quamproxime. Et si densitas Planetæ augeatur vel minuatur in ratione quavis, gravitas etiam in ipsum tendens augebitur vel minuetur in eadem ratione, et differentia diametrorum vicissim minuetur in ratione gravitatis auctæ vel augebitur in ratione gravitatis diminutæ. Unde cum Terra respectu fixarum revolvatur horis 23.56′ Jupiter autem horis 9.56′, sintque temporum quadrata ut 29 ad 5, et densitates ut 5 ad 1: differentia diametrorum Jovis erit ad ipsius diametrum minorem ut $\frac{29}{5}\times\frac{5}{1}\times\frac{1}{229}$

ad 1, seu 1 ad 8 quamproxime. Est igitur diameter Jovis ab oriente in occidentem ducta ad ejus diametrum inter polos ut 9 ad 8 quamproxime, et propterea diameter inter polos est $35''\frac{1}{2}$. Hæc ita se habent ex hypothesi quod uniformis sit Planetarum materia. Nam si materia densior sit ad centrum quam ad circumferentiam, diameter quæ ab oriente in occidentem ducitur erit adhuc major.

Jovis vero diametrum quæ polis ejus interjacet minorem esse diametro altera *Cassinus* dudum observavit, et Terræ diametrum inter polos minorem esse diametro altera patebit per ea quæ dicentur in Propositione sequente.

In the XXth Proposition(14) page 425 lin 8, read. Unde tale confit Theorema quod incrementum ponderis pergendo ab Æquatore ad Polos, sit quam proxime ut sinus versus Latitudinis duplicatæ, vel, quod perinde est, ut quadratum sinus recti Latitudinis. Et in eadem circiter ratione augentur arcus graduum Latitudinis in Meridiano. Ideoque cum Latitudo *Lutetiæ Parisiorum* sit 48gr. 50′, ea locorum sub Æquatore 00gr. 00′, et ea locorum ad Polos 90gr & duplorum sinus versi sint 11334, 00000 et 20000, existente Radio 10000, et gravitas ad Polum sit ad gravitatem ejus sub Æquatore ut 229 ad 228, & excessus gravitatis ad polum ad gravitatem sub Æquatore ut 1 ad 228: erit excessus gravitatis in Latitudine *Lutetiæ* ad gravitatem sub Æquatore, ut $1\times\frac{11334}{20000}$ ad 228 seu 5667 ad 22800[00]. Et propterea gravitates totæ in his locis erunt ad invicem ut 2285667 ad 2280000. Quare cum longitudines pendulorum æqualibus temporibus oscillantium sint ut gravitates, et in Latitudine *Lutetiæ Parisiorum* longitudo penduli singulis minutis secundis oscillantis sit pedum trium Parisiensium & $8\frac{4}{9}$ linearum, longitudo penduli sub Æquatore superabitur a longitudine synchroni penduli *Parisiensis*, excessu lineæ unius et 92 partium millesimarum lineæ. Et simili computo confit Tabula sequens(15) [*see p. 258*].

Constat autem per hanc Tabulam &c(16)

Hæc ita se habent ex hypothesi quod Terra &c

Jam vero Astronomi aliqui in longinquas regiones &c

Deinde anno 1682 D. Varini &c.

Posthac D. Couplet filius anno 1697

Annis proximis (1699 & 1700) D. Des Hayes &c

Annoque 1704 P. Feuelleus(17) invenit in Po[r]to-belo in America Longitudinem Penduli ad minuta secunda oscillantis esse pedum trium Parisiensium et linearum tantum $5\frac{7}{12}$, id est tribus fere lineis breviorem quam Lutetiæ Parisiorum, sed errante Observatione. Nam deinde ad insulam Martinicam navigans invenit longitudinem Penduli isochroni esse pedum tantum trium Parisiensium et linearum $5\frac{10}{12}$.

Latitudo(18) autem Paraibæ est 6gr. 38′ ad austrum et ea Portobeli 9gr 33′ ad boream, et Latitudines insularum Cayennæ, Goreæ, Guadaloupæ, Martanicæ [*sic*], Granadæ, Sti Christophori & Sti Dominici sunt respective 4gr 55′, 14gr 40′, 14gr 00′, 14gr 44′, 12gr 6′, 17gr 19′ & 19gr 48 ad boream. Et excessus longitudinis Penduli[19]...auxerint.

Latitudo Loci	Longitudo Penduli		Mensura gradus unius in Meridiano
Grad.	Ped.	Lin.	Hexaped.
0	3.	7,463	56907
5	3.	7,478	56913
10	3.	7,521	56930
15	3.	7,592	56957
20	3.	7,689	56995
25	3.	7,808	57041
30	3.	7,945	57095
35	3.	8,098	57154
40	3.	8,260	57218
45	3.	8,427	57283
46	3.	8,461	57296
47	3.	8,494	57309
48	3.	8,528	57322
49	3.	8,561	57335
50	3.	8,594	57348
55	3.	8,756	57412
60	3.	8,909	57471
65	3.	9,046	57525
70	3.	9,165	57571
75	3.	9,262	57602
80	3.	9,333	57626
85	3.	9,376	57653
90	3.	9,391	57659

Observavit utique D. Picartus quod virga ferrea, quæ tempore hyberno ubi gelabant frigora erat pedis unius longitudine, ad ignem calefacta evasit pedes unius cum quarta parte lineæ. Deinde D. de la Hire[19]...cum duabus tertijs partibus lineæ. In priore casu calor major fuit quam in posteriore, in hoc vero major fuit quam calor externarum partium corporis humani. Nam metalla ad solem æstivum valde incalescunt. At virga penduli[19]...quam hyberno, sed excessu quartam partem lineæ unius vix superante. Proinde[19]...differentia illa prodijt haud minor quam $1\frac{19}{20}$ lineæ, haud major quam $2\frac{1}{2}$ linearum. Et inter hos limites quantitas mediocris est $2\frac{9}{40}$ linearum. Propter calores locorum in zona torrida negligamus $\frac{9}{40}$ partes lineæ et manebit differentia duarum linearum.

Quare cum differentia illa per Tabulam præcedentem ex hypothesi quod Terra ex materia uniformiter densa const at, sit tantum $1\frac{92}{1000}$[20] lineæ: excessus altitudines Terræ ad æquatorem supra altitudinem ejus ad polos, qui erat milliarium $17\frac{1}{6}$, jam auctus in ratione differentiarum, fiet milliarium $31\frac{1}{3}$.(21) Nam tarditas Penduli sub Æquatore defectum gravitatis arguit; et quo levior est materia eo major esse debet altitudo ejus ut pondere suo materiam sub Polis in æquilibrio sustineat.

Hinc figura umbræ Terræ per eclipses Lunæ determinanda, non erit omnino circularis sed diameter ejus ab oriente in occidentem ducta, major erit quam diameter ejus ab austro in boream ducta, excessu 55″ circiter. Et parallaxis maxima Lunæ in Longitudinem paulo major erit quam ejus parallaxis maxima, in Latitudinem. Ac Terræ semidiameter maxima erit pedum Parisiensium 19764030, minima pedum 19609860 & mediocris pedum 19686945 quam proxime.(22)

Cum gradus unus mensurante *Picarto* sit hexapedarum 57060, mensurante vero *Cassino* sit hexapedarum 57292: suspicantur aliqui...(19) seu parte duodecima diametri Lunæ. Quibus omnibus experientia contrariatur. Certe *Cassinus*, definiendo gradum unum esse hexapedarum 57292, medium inter mensuras suas omnes, ex hypothesi de æqualitate graduum assumpsit. Et quamvis *Picartus* in *Galliæ* limite boreali invenit gradum paulo minorem esse, tamen Norwoodus noster in regionibus magis mensurando majus intervallum, invenit gradum paulo majorem esse quam *Cassinus* invenerat. Et *Cassinus* ipse mensuram *Picarti* ob parvitatem intervalli mensurati non satis certam & exactam esse judicavit ubi mensuram gradus unius per intervallum longe majus definire aggressus. Differentiæ vero inter mensuras *Cassini*, *Picarti* & *Norwoodi* sunt prope insensibiles & ab insensibilibus observationum erroribus facile oriri potuere, ut nutationem axis Terræ prætereams.

Pag. 424 lin penult. read 229 ad 228.(23)

The rest of the Propositions to Prop. XXXVI, may continue as they are, wth ye corrections already sent you, I will speedily send you the corrections of ye XXXVI, XXXVII, & XXXIX Propositions. I am

Your very humble Servant

ISAAC NEWTON(24)

Translation

Proposition 19. Problem 2(3)

To find the ratio between the axis of a planet and the diameters perpendicular to it.

Picard,(4) by measuring an arc of 1° 2′ 55″ between Amiens and Malvoisine...[*see p. 216 above, Letter 889a*]...57300 Paris fathoms. Because of the great size of the distance measured by Cassini, I shall adopt 57292 fathoms as the measure of a single degree in the midst of that distance, that is between latitudes 45° and 46°. Whence, if the Earth be round, its radius will be 19695539 Paris feet.

The length of a pendulum beating seconds in the latitude of Paris is 3 Paris feet and $8\frac{5}{9}$ lines. And the length that a heavy body falls in one second is to half the length of this pendulum in the squared ratio of the circumference of a circle to its diameter (as Huygens has shown) and so it is 15 Paris feet, 1 inch $2\frac{1}{4}$ lines,(7) or $2174\frac{1}{4}$ lines.(7)

A body revolving uniformly in a circle at a distance of 19695539 feet from the centre with a period of one sidereal day (23 hours, 56 minutes, 4 seconds) will in one second describe an arc of 143·6223(8) feet, whose versed sine is 0·05236558 feet or 7·54064 lines. And so the force with which heavy bodies descend in the latitude of Paris is to the centripetal force of bodies at the Equator arising from the diurnal motion of the Earth as $2174\frac{1}{4}$(7) to 7·54064.

The centrifugal force of bodies at the Equator is to the centrifugal force with which bodies are urged directly from the Earth in the latitude of Paris in the squared ratio of the radius to the sine of the complement of that latitude, that is, as 7·54064 to 3·27.(9) This force is to be added to the force with which heavy bodies descend in the latitude of Paris, and a body falling in the latitude of Paris with the whole force of gravity will in one second describe 2177·52 lines,(10) or 15 Paris feet, one inch and 5·52 lines.(11) And the whole force of gravity in that latitude will be to the centripetal force of bodies at the Earth's equator as 2177·52(10) to 7·54064 or as 289 to 1.

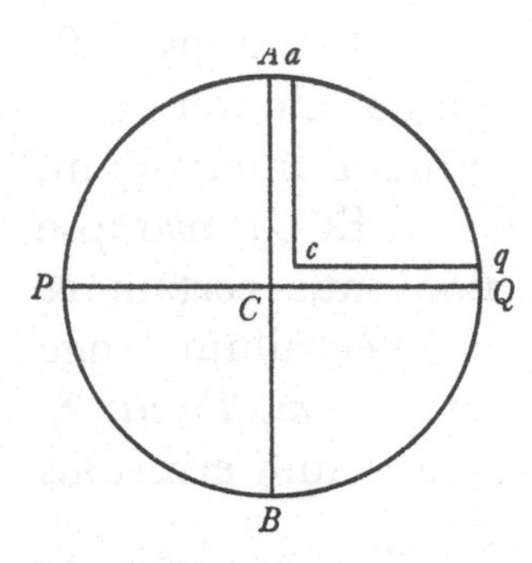

Supplied by the editors from *Principia*, Prop. 19, Book III.

Whence if $APBQ$ represents the figure of the Earth, now no longer a sphere but generated by the revolution of an ellipse about its minor axis PQ, and let $ACQqca$ be a channel full of water stretching from the pole Qq to the centre Cc and thence to the equator at Aa, the weight of the water in the arm $ACca$ of the canal should be to the weight of water in the other arm $QCcq$ as 289 to 288, because the centrifugal force arising from the circular motion will sustain and subtract one part of the 289 parts, and the weight of 288 in the other arm [$QCcq$] will balance the rest...

...to its polar diameter as 230 to 229.(13) And so as the mean radius of the Earth by Cassini's measurement is 19695539 Paris feet, or 3939 miles (assuming one mile to be 5000 feet) the Earth will be higher at the Equator than at the Poles by an excess of 85820 feet or $17\frac{1}{6}$ miles.

If a planet were greater or smaller than the Earth, but having the same density and period of diurnal revolution, the ratio between centrifugal force and gravity will remain constant, and so furthermore will the ratio between its polar and equatorial axes. But if the diurnal motion be accelerated or retarded in any proportion the centrifugal force will be increased or diminished in the square of that ratio, and furthermore the difference between the diameters will be increased or diminished roughly in that same squared ratio. And if the density of the Planet be increased or diminished in any ratio, the gravitation towards it will be increased or diminished in the same proportion, and conversely the difference between the diameters will be diminished in proportion to the increased gravity, or increased in proportion to the diminished gravity. Whence, as the

Earth's sidereal period is 23 hours 56 minutes, and Jupiter's is 9 hours 56 minutes, and the squares of these times are as 29 to 5, and the densities as 5 to 1, the difference between the diameters of Jupiter will be (in proportion to his lesser diameter) as $29/5 \times 5/1 \times 1/229$ to 1, or 1 to 8 roughly. Therefore the [equatorial] diameter of Jupiter measured from east to west is to his polar diameter as 9 to 8 approximately, and moreover the polar diameter is $35\frac{1}{2}$ seconds of arc. This is so provided that the matter of the planets is uniform. For if the matter be denser at the centre than at the circumference the equatorial diameter will be still larger.

Cassini did indeed observe a long time ago that the polar diameter of Jupiter is less than the other diameter, and the content of the following proposition makes it plain that the Earth's polar diameter is less than its other diameter...

[*From Proposition 20, the first portion in the letter being identical (except in the numerical values) with the text of the Third Edition*]

...that difference [between northern and tropical pendulums] will hardly work out to be less than $1\frac{19}{20}$ lines, or greater than $2\frac{1}{2}$ lines. And the mean between these extremes is $2\frac{9}{40}$ lines. Because of the heat of the tropical regions we may neglect 9/40 parts of a line, and the difference will remain as two lines.

Because that difference as read from the preceding Table (on the hypothesis that the Earth is composed of uniformly dense matter) is only $1\frac{92}{1000}$[20] lines, the excess of the height [reading *altitudinis*] of the Earth at the equator over its height at the poles, which was $17\frac{1}{6}$ miles, now increased in the ratio of this difference, becomes $31\frac{1}{3}$[21] miles. For the sluggishness of the pendulum at the Equator speaks for a falling off of gravity; and the lighter the matter is, the more must its height be increased so that, by its weight, it may balance the matter under the poles.

Hence the shape of the Earth to be determined by lunar eclipses will not be quite circular but its east–west diameter will be greater than its north–south diameter with an excess of about 55″. And the greatest parallax of the Moon in longitude will be a trifle greater than its greatest parallax in latitude. And the greatest semidiameter of the Earth will be 19764030 Paris feet, the least 19609860 feet and the mean 19686945, approximately.[22]

Since by Picard's measurement one degree is 57060 fathoms, and by Cassini's is 57292 fathoms, some have imagined [that the Earth is a prolate spheroid]...[19] or the twelfth part of the Moon's diameter. To all of this experience is opposed. Certainly Cassini, in defining one degree as 57292 feet, took a mean between all his measures on the hypothesis that the degrees are equal. And although Picard found the degree to be a little less at the northern frontier of France yet our countryman Norwood, measuring a greater distance in a more [northerly] region, found the degree a little greater than Cassini did. And Cassini himself concluded that because of the slightness of the distance measured Picard's measurement was not sufficiently certain and exact, whence he embarked on a measurement to define a degree over a much greater distance. Indeed the differences between the measurements of Cassini, Picard and Norwood are nearly imperceptible and could easily arise from imperceptible errors in the observations, not to say the nutation of the Earth's axis.

NOTES

(1) R.16.38, no. 185; printed in Edleston, *Correspondence*, pp. 81–8. Newton had promised the amendments he sends here in his Letters 898 and 903. Several numerical corrections have been inserted in Cotes's hand (see notes below). Cotes discusses these in his Letter 910.

(2) In what follows there is no reference to Proposition 25. This was dealt with in Letters 911 and 912.

(3) This became Proposition 19, Problem 3 in the final version, and the wording Newton suggests here was printed (see Koyré and Cohen, *Principia*, II, pp. 593–6). Newton here introduces Picard's measurements of the Earth's size, and the new measurements of G. D. Cassini, then considered to be the most accurate available. For the first and second paragraphs, cf. Letter 889.

(4) Jean Picard (1620–82) played a prominent part in the foundation of the Paris Observatory (1669). He measured the length of a degree along the meridian from a point south of Paris to Amiens, and published his results in *La Mesure de la Terre* (Paris, 1671). An abridged version appeared in English in the *Phil. Trans.*, **10** (1675), 261–72, and in 1702 a complete translation by Richard Waller was published, appended to *Memoirs for a Natural History of Animals... Royal Academy of Sciences* (London, 1702).

(5) Giovanni Domenico Cassini (1625–1712), see vol. I, p. 25, note (1). Cassini extended Picard's measurements southwards to Collioure, and published his results in the Paris *Mémoires* for 1701 (pub. 1704), pp. 169–82. Later these were shown to be inaccurate, and Newton excluded them from the third edition of the *Principia*. (See I. Todhunter, *A History of the mathematical theories of attraction and the figure of the earth*, Dover edition, 1962, p. 46 *passim*.)

(6) Richard Norwood (1590–1675), an English mathematician. In June, 1635, he measured the distance between London and York, and later published his results in *Seaman's Practice* (London, 1637).

(7) Cotes has crossed out the fraction 1/4, and substituted 1/18, but in the final version 1/4 was printed.

(8) Cotes has correctly amended this to 1436,233 as printed.

(9) Cotes has substituted 3,267, as printed.

(10) Cotes has substituted 2177,32, as printed.

(11) Cotes has substituted 5,32, as printed.

(12) Newton refers here to corrections he had sent in his Letter 891. 'Page 484' is a slip for page 424. The passage which follows is as finally printed. (See Koyré and Cohen, *Principia*, II, pp. 598–9.)

(13) In the first edition the ratio had been 692:689. As Newton had decided in Letter 898, 230:229 was finally used, with the necessary alterations to subsequent computations. A preliminary draft of this letter (U.L.C. Add. 3984, no. 8) exemplifies clearly Newton's practice of calculating and recalculating numerical values for the second edition. It reads:

Sr

I now send you the Corrections of the XIXth & XXth Propositions In the XIXth, pag 424 lin 1 read [ad ipsius diametrum per polos ut 230 ad 229. Ideoque cum Terræ semidiameter mediocris, ex mensura Cassini, ut posthac dicetur, sit pedum Parisiensium 19692788 seu milliarum 3938 (posito quod milliare sit mensura pedum 5000,) Terra altior erit ad æquatorem quam ad polos excessu pedum 85807 seu milliarum $17\frac{1}{8}$

Si Planeta vel major sit &c.

(14) The wording of this passage is the same as was finally printed (see Koyré and Cohen, *Principia*, II, pp. 601–9). The numerical values quoted are however different, for Newton has forgotten to change the ratio 229:228 into 230:229, and to make resulting alterations in subsequent values.

(15) The table differs from that finally published, again because the table Newton gives here is based on the ratio 229:228 (see note above). He has made a number of corrections which make the table difficult to read, but Cotes has written the corresponding values in a clear hand on the right hand side of the page. Cotes later recomputed the table (see Letter 910).

(16) Newton quotes here only the first few words of this and the succeeding five paragraphs. These had been supplied in an earlier draft of the proposition, now lost. The first, third, fourth, fifth and sixth paragraphs are in any case the same in all editions; the second paragraph appears only in the second edition (Koyré and Cohen, *Principia*, II, p. 605).

(17) Newton had sent Cotes this additional paragraph in Letter 898.

(18) Newton had already sent emendations for this paragraph and the one following in Letter 818. Here he introduces slight additional alterations.

(19) The omissions are Newton's.

(20) Cotes has altered $1\frac{92}{1000}$ to $1\frac{87}{1000}$, as printed.

(21) Cotes has bracketed the fractions 1/3, and inserted 7/12, the value finally printed.

(22) These numbers were not printed. See note (24) below.

(23) Newton has again forgotten to change the ratio to 230:229. This was done in the printed version.

(24) Cotes has added the following notes at the foot of this page:

Maxima	19767630		
Minima	19609820		
Mediocris	19688725	19688725	
Semd. Sph. Æqu:	19714886	19714886	
		39403611	
		19701805	Media Mediarum

Cotes commented on this calculation in his next letter. See Letter 910, note (11).

909 NEWTON TO COTES

8 APRIL 1712

From the original in Trinity College Library, Cambridge.[1]

In continuation of Letter 908; for the answer see Letters 910, 911 and 914

London Apr 8th 1712.

Sr

I sent you by Dr Bently my emendations of the 19th & 20th Propositions, & now send you those of the 36th & 37th.[2] When you have perused them I should be glad to have your thoughts upon them, & if any thing else want to be corrected before you come at ye 39th Proposition. In my next[3] I intend to send you my emendations of that Proposition. I am

Your most humble Servant
IS. NEWTON

For the Rnd Mr Cotes Professor of
Astronomy at his chamber in
Trinity College in
Cambridge.

NOTES

(1) R.16.38, no. 129; printed in Edleston, *Correspondence*, p. 88.

(2) Only one small correction to the Corollary of Proposition 36 exists (MS. R.16.38 fo. 192); it reads as follows,

'In Prop XXXVI, pag. 464 lin. 3, read 85820; & lin. 9 read, et digitorum undecim cum triente, Est enim hæc mensura ad mensuram pedum 85820 ut 1 ad 44038.'

Cotes suggested further alterations in his Letter 914 and these were adopted.

The complete redraft of Proposition 37 (MS. R.16.38 fos. 195–8) superseded the earlier one, of which Cotes had sent a copy to Newton in Letter 899. We print the new draft in full, from fos. 195–8, for subsequently it was considerably altered.

(3) See Letter 912.

909*a* DRAFT FOR PROPOSITION 37

In the emendation of Prop. XXXVII read.[1] Vis Lunæ ad mare movendum colligenda est ex ejus proportione ad vim Solis et hæc proportio colligenda est ex proportione motuum maris qui ab his viribus oriuntur. Ante ostium fluvij *Avonæ* ad lapidem tertium infra *Bristoliam*, tempore verno et autumnali totus aquæ ascensus in Conjunctione et Oppositione Luminarium, observante *Samuele Sturmio*, est pedum plus minus 45, in Quadraturis autem est pedum tantum 25. Altitudo prior ex summa virium, posterior ex earundem differentia oritur. Solis igitur & Lunæ in Æquatore versantium & mediocriter a Terra distantium sunto vires S et L, et erit $L+S$ ad $L-S$ ut 45 ad 25, seu 9 ad 5.

In portu *Plymuthi* æstus maris (ex observatione *Samuelis Colepressi*) ad pedes plus minus sexdecim altitudine mediocri attollitur, ac tempore verno et autumnali altitudo æstus in Syzygijs superare potest altitudinem ejus in Quadraturis pedibus plus septem vel octo. Si maxima harum altitudinum differentia sit pedum novem, erit $L+S$ ad $L-S$ ut $20\frac{1}{2}$ ad $11\frac{1}{2}$ seu 41 ad 23. Quæ proportio satis congruit cum priore. Ob magnitudinem æstus in Portu *Bistoliæ* [*sic*], observationibus *Sturmij* magis fidendum esse videtur, ideoque donec aliquid certius constiterit, proportionem 9 ad 5 usurpabimus.

Cæterum ob aquarum reciprocos motus, æstus maximi non incidunt in ipsas Luminarium syzygias, sed sunt tertii a syzygijs ut dictum fuit, seu proxime tertium Lunæ post syzygias appulsum ad meridianum loci, vel potius tertium post tertiam circiter vel quartam a syzygijs horam appulsum ad meridianum loci.[2] Æstas et hyems maxime vigent, non in ipsis solstitijs, sed ubi sol distat a novissimis solstitijs decima circiter vel undecima parte totius circuitus, seu gradibus plus minus 35. Et similiter maximus æstus maris oritur ab appulsu Lunæ ad meridianum loci ubi Luna distat a Sole decima vel undecima parte motus totius ab æstu ad æstum, seu gradibus plus minus septendecim cum dimidio. Et Sol in hac distantia minus auget vel diminuit motum maris a vi Lunæ oriundum quam in ipsis syzygijs et quadraturis in ratione Radij ad sinum complementi distantia hujus duplicatæ seu anguli graduum 35, hoc est, in ratione 1000000 ad 819152; ideoque in analogia superiore pro S scribi debet $0{,}819152S$.

Sed et vis Lunæ in Quadraturis, ob Declinationem Lunæ ab Æquatore, diminui debet. Nam Luna in Quadraturis vel potius in gradu $17\frac{1}{2}$ post Quadraturas, tempore Æquinoctiorum, in Declinatione graduum plus minus 22 & 21′ versatur. Et Luminaris ab Æquatore Declinantis vis ad mare movendum diminuitur in duplicata ratione sinus complementi Declinationis quamproxime. Et propterea vis Lunæ in his Quadraturis est tantum $0{,}85539968L$. Est igitur $L+0{,}81952S$ ad $0{,}85539968L-0{,}81952S$ ut 9 ad 5.[3]

Præterea diametri Orbis in quo Luna absque excentricitate moveri deberet sunt ad invicem ut 69 a 70 (per Prop. XXVIII,) ideoque distantia Lunæ a Terra in Syzygijs est ad distantiam ejus in Quadraturis ut 69 ad 70 cæteris paribus. Et distantia ejus in gradu $17\frac{1}{2}$ a syzygijs ubi æstus maximus generatur est ad distantiam ejus in gradu $17\frac{1}{2}$ a Quadraturis ubi æstus minimus generatur ut 83,8317 ad 84,8317, id est, ut 1 ad 1,0119286 vel 0,9882125 ad 1. Unde fit $1{,}0119286L+0{,}819152S$ ad $0{,}9882125\times 0{,}85539968L-0{,}819152S$ ut 9 ad 5. Et S ad L ut 1 ad $4\frac{1}{2}$.

Corol. 1. Cum igitur aqua vi Solis agitata ascendat ad altitudinem pedis unius et digitorum undecim cum triente, eadem vi Lunæ ascendet ad altitudinem pedum octo et digitorum novem. Tanta autem vis &c.

Corol. 2. Cum vis Lunæ ad mare movendum &c

Corol. 3. Quoniam vis Lunæ ad mare movendum est ad Solis vim consimilem ut $4\frac{1}{2}$ ad 1, et vires illæ (per Corol. 14 Prop. LXVI Libr. 1) sunt ut densitates corporum Lunæ & Solis & cubi diametrorum apparentium conjunctim: erit densitas Lunæ ad densitatem Solis ut $4\frac{1}{2}$ ad 1 directe et cubus diametri Lunæ ad cubum diametri Solis inverse, id est, (cum diametri mediocres apparentes Lunæ et Solis sint 31′. 16″ et 32′ 12″) ut 49112[4] ad 10000. Densitas autem Solis erat ad densitatem Terræ ut 100 ad 396 et propterea densitas Lunæ est ad densitatem Terræ ut 49112 ad 39600 seu 31 ad 25. Est igitur corpus Lunæ densius et magis terrestre quam Terra nostra.

Corol. 4. Et cum vera diameter Lunæ (ex observationibus Astronomicis) sit ad veram diametrum Terræ ut 100 ad 365, erit massa Lunæ ad massam Terræ ut 1 ad $39\frac{1}{5}$.

Corol. 5. Et gravitas acceleratrix in superficie Lunæ erit triplo minor quam gravitas acceleratrix in superficie Terræ.

Corol. 6. Et distantia centri Lunæ a centro Terræ erit ad distantiam centri Lunæ a communi gravitatis centro Lunæ ac Terræ ut $40\frac{1}{5}$ ad $39\frac{1}{5}$.

Corol. 7. Et distantia mediocris centrorum Lunæ ac Terræ æqualis erit maximis Terræ semidiametris $60\frac{1}{4}$ quamproxime. Nam Terræ semidiameter maxima fuit pedum Parisiensium 19764030. Et hujusmodi semidiametri $60\frac{1}{4}$ æquantur pedibus 1190782815. Et si hæc sit distantia centrorum Solis[5] et Lunæ, eadem (per Corollarium novissimum) erit ad distantiam centri Lunæ a communi gravitatis centro Lunæ ac Terræ ut $40\frac{1}{5}$ ad $39\frac{1}{5}$, quæ proinde est pedum 1161161352. Et cum Luna revolvatur respectu fixarum diebus 27 horis 7 & minutis primis $43\frac{1}{5}$, sinus versus anguli quem Luna tempore minuti unius primi motu suo medio circa commune gravitatis centrum Lunæ ac Terræ describit est 1,275235, existente Radio 100,000000,000000. Et ut Radius est ad hunc sinum versum ita sunt pedes 116116135 ad pedes 14,807536. Luna igitur vi illa qua retinetur in orbe, tempore minuti unius primi cadendo describeret pedes 14,807536. Et hæc vis (per Corol. Prop. III) est ad vim gravitatis nostræ in orbe Lunæ ut $177\frac{29}{40}$ ad $178\frac{29}{40}$; proindeque corpus grave in orbe Lunæ ad distantiam pedum 1190782815 a centro Terræ, vi gravitatis nostræ in Terram cadendo, tempore minuti unius primi describeret pedes 14,8908, & ad sexagesimam partem distantiæ illius, id est ad distantiam pedum 1984638 a centro Terræ, vi gravitatis in Terram cadendo tempore minuti unius secundi describeret etiam pedes 14,8908, et ad distantiam pedum 19694278 a centro Terræ cadendo eodem tempore minuti unius secundi describeret pedes 15,1217 seu pedes 15, dig. 1, et lin. $5\frac{1}{2}$. Et hac vi gravia cadunt in superficie Terræ in Latitudine urbis Lutetiæ Parisiorum, ut supra ostensum est. Et distantia pedum 19694278 paulo major est quam Terræ semidiameter mediocris, et paulo minor quam semidiameter globi cui Terra æqualis est, suntque differentiæ insensibiles; ac proinde vis qua Luna retinetur in Orbe suo ad distantiam prædictam semidiametrorum $60\frac{1}{4}$, si descendatur in Terram, congruit cum vi gravitatis quam experimur in superficie Terræ.

Corol. 8. Distantia mediocris centrorum Lunæ ac Terræ æqualis est mediocribus Terræ semidiametris $60\frac{1}{2}$ quamproxime. Nam tot semidiametri mediocres sunt pedum 1191060172.

Siquando mensuræ graduum in meridiano, longitudes pendulorum isochronorum in diversis parallelis Terræ, leges fluxus & refluxus maris, diametri apparentes Solis et Lunæ, & Lunæ parallaxis horizontalis ex phænomenis accuratius determinatæ fuerint: licebit calculum hunc omnem accuratius repetere.

Translation

The first two paragraphs are as in the second and third editions.

Moreover, because of the reciprocal motions of the waters the highest tides do not occur actually at the syzygies of the luminaries, but are a third from the syzygies as we have said, or near a third [after] the Moon's crossing the meridian of the place following the syzygies, or rather a third after about the third or fourth hour following the syzygies crossing the meridian of the place.(2) Summer and winter reach their peak not actually at the solstices but when the sun has moved round from the last solstice a tenth or eleventh part of his whole circuit, or 35 degrees more or less. And similarly the highest tide is caused in the sea by the Moon's crossing the meridian of the place when the Moon is about a tenth or eleventh part of her whole motion from tide to tide away from the Sun, that is, 17½ degrees more or less. And the Sun at this distance has a less effect in increasing or diminishing the motion of the sea caused by the Moon, than at the actual syzygies or quadratures in the ratio of the radius to the sine of the complement of double this distance [*reading* distantiæ] or of 35°, that is in the ratio of 1000000 to 819152. And so in the analogy above we should write instead of S, $0{\cdot}819152S$.

But further the force of the Moon at the quadratures ought to be lessened because of the Moon's declination from the equator. For at the quadratures or rather at 17½° after the quadratures, at the time of the equinoxes, the moon is inclined in declination through about 22° 21′. And the force of the luminary for moving the sea when she is declining away from the equator is reduced in the squared ratio of the sine of the complement of [the angle of] declination, roughly. And on this account the force of the Moon in these quadratures is only $0{\cdot}85539968L$. And therefore $L+0{\cdot}81952S$ to $0{\cdot}85539968L-0{\cdot}81952S$ is as 9 to 5.(3)

Moreover, the diameters of the orbit in which the Moon should move without eccentricity are to each other as 69 to 70 (by Proposition 28) and so the distance of the Moon from the Earth in the syzygies is to its distance at the quadratures as 69 to 70, other things being equal. And her distance at 17½° from the syzygies where the highest tide is generated is to its distance at 17½° from the quadratures where the least tide is generated as 83·8317 to 84·8317, that is, as 1 to 1·0119286 and 0·9882125 to 1. whence $1{\cdot}0119286L+0{\cdot}819152S$ to $0{\cdot}9882125\times 0{\cdot}85539968L-0{\cdot}819152S$ is as 9 to 5; and S is to L at 1 to 4½.

Corol. 1. Therefore as the water acted on by the Sun's force rises to the height of one foot eleven inches and a thirtieth, the same by the force of the Moon would rise to the height of eight feet nine inches. And a force of that amount, etc.

Corol. 2. Since the Moon's force for moving the sea, etc.

Corol. 3. Since the Moon's force for moving the sea is to the similar force of the Sun as $4\frac{1}{2}$ to 1, and those forces (by Corol. 14, Proposition 66, Book I) are as the densities of the bodies of the Moon and the Sun and the cubes of their apparent diameters conjointly, the lunar density to the solar density will be as $4\frac{1}{2}$ to 1 directly and the cube of the diameter of the Moon to the cube of the diameter of the sun inversely, that is, (since the mean apparent diameters of the Sun and the Moon are 32′ 12″ and 31′ 16″) as 49112[4] to 10000. However, the density of the Sun was to that of the Earth as 100 to 396 and for that reason the density of the Moon is to that of the Earth as 49112 to 39600 or 31 to 25. Accordingly the body of the Moon is more dense and earthy than our Earth.

Corol. 4. And as the true diameter of the Moon (from astronomical observations) is to the true diameter of the Earth as 100 to 365, the mass of the Moon to that of the Earth is as 1 to $39\frac{1}{5}$.

Corol. 5. And the accelerative gravity at the surface of the Moon will be three times less than the accelerative gravity at the surface of the Earth.

Corol. 6. And the distance of the centre of the Moon from the centre of the Earth will be to the distance of the centre of the Moon from the common centre of gravity of the Earth and the Moon as $40\frac{1}{5}$ to $39\frac{1}{5}$.

Corol. 7. And the mean distance of the centres of the Earth and Moon will be equal to $60\frac{1}{4}$ of the largest Earth-radii, approximately. For the largest Earth radius was 19764030 Paris feet. And $60\frac{1}{4}$ of such radii equal 1190782815 feet. And if this is the distance between the centres of the [Earth] and Moon, this same will be (by the last corollary) to the distance between the centre of the Moon and the common centre of gravity of the Earth and the Moon as $40\frac{1}{5}$ to $39\frac{1}{5}$, whence the latter is 1161161352 feet. And as the Moon revolves with respect to the fixed stars in 27 days, 7 hours, $43\frac{1}{5}$ seconds the versed sine of the angle which the Moon describes in one minute of time with its mean motion around the common centre of gravity of the Earth and the Moon is 1·275235, the radius being 10^{14}. And as the radius is to this versed sine so 116116135 feet are to 14·807536 feet. Accordingly the Moon, through that force by which it is retained in its orbit, would in falling for one minute describe 14·807536 feet. And this force is (by Corol. Proposition 3) to the force of our gravity in the Moon's orbit as $177\frac{29}{40}$ to $178\frac{29}{40}$; and hence a heavy body in the Moon's orbit at a distance of 1190782815 feet from the centre of the Earth, falling towards the Earth with the force of our gravity, would in one minute of time describe 14·8908 feet; and at one-sixtieth part of that distance, that is at a distance of 1984638 feet from the centre of the Earth, it would in falling towards the Earth with the force of gravity in one second of time also describe 14·8908 feet, and at a distance of 19694278 feet from the centre of the Earth, in falling for one second of time likewise, it would describe 15·1217 feet or 15 feet, one inch, and $5\frac{1}{2}$ lines. And this is the force with which heavy bodies fall at the surface of the Earth in the latitude of the city of Paris, as is shown above. And the distance of 19694278 feet is a little greater than the Earth's mean radius and a little less than the radius of a sphere equal to the Earth, and the differences are insignificant; and hence the force by which the Moon is retained in its orbit at the above-stated distance of $60\frac{1}{4}$ radii, if it should descend towards the Earth, agrees with the force of gravity that we experience on the surface of the Earth.

Corol. 8. The mean distance of the centres of the Earth and the Moon is equal to $60\frac{1}{2}$ mean Earth-radii approximately; for so many mean radii make 1191060172 feet.

When more accurate measurements have been made of degrees of latitude, and when the lengths of isochronous pendulums at different latitudes, the laws of the ebb and flow of the sea, the apparent diameters of the Sun and Moon, and the Moon's horizontal parallax are all more accurately determined from the phenomena, it will be possible to repeat this calculation with greater precision.

NOTES

(1) The first two paragraphs are given here in their final form (see Koyré and Cohen, *Principia*, II, pp. 666–7). They also differ little from the original draft which Cotes had copied in Letter 899, except in the numerical values used. The remainder of the draft (here translated) departs considerably both from the first version and from what was finally printed. Largely as a result of Cotes's objections in Letters 910 and 911 Newton was forced to revise the Propositions once more (see enclosure with Letter 912).

(2) The meaning here is not clear; the analogous passage is not rendered clearly into English in Motte–Cajori either.

(3) The number here should of course be 0·819152.

(4) Cotes has altered the number to 49151.

(5) Read: *Terræ*.

910 COTES TO NEWTON

14 APRIL 1712

From the original in the University Library, Cambridge.[1]
Reply to Letters 908 and 909; for the answer see Letter 912

Cambridge April ye 14th 1712

Sr.

I have received Your Letter by Dr Bentley & the other which You wrote since.[2] I have sent you two proof-sheets[3] for Your revisal, having made some alterations[4] in them different from Your Copy.

In Page 379 lin. 6th I have put [lin. $2\frac{1}{18}$] instead of [lin $2\frac{1}{4}$] In line 10th 1436,223 instead of 143,6223. In line 21st. 2177,32 instead of 2177,52.

In Page 382 I have put the proportion of 230 to 229 instead of 229 to 228[5] & altered ye latter part of the Page accordingly & computed ye Table anew in the next Page.[6] The latter column supposes the measure of a degree at ye Latitude of 45°. 41′ to be 57292 Toises, as I think You put it in Your Table. The two extream numbers are[7] the Cubes of 230 & 229; I found 'em to be 56908,676 & 57657,465. In ye rest ye increment from the Æquator is as the versed sine of ye doubld Latitude.

In Page 386 lin. penult. $1\frac{87}{1000}$ for $1\frac{92}{1000}$.[8] Page 387, lin:1 $31\frac{7}{12}$ for $31\frac{1}{3}$.[9] Line 11th I have put other numbers for the semidiameters of the Earth,[10] which I desire You would examine since there are different ways of coming at those numbers & I may not possibly have taken that which You like best. Line 21st. I put 95[11] Miles for 94. Line 27th, 2′. 46″[11] for 2′. 45″ Line 32d, *Norwoodus noster in regionibus magis borealibus* the word *borealibus* or something to that effect was omitted in Your Copy.[12]

In Page 389 line 26th I have put 8°. 24′[11] for 9°. 34′. In ye last Period of ye same Proposition XXIII, I have made an alteration which You will see.[13]

I think You have much improved the Method of ye whole but there seemes to be a mistake in yt Section of Prop. XXXVII which begins with, *Præterea diametri orbis in quo Luna* &c.[14] The Moons force in her Syzygies & Quadratures should be increased & diminished in the triplicate proportion of those distances to her mean distance reciprocally. Your correction is nearly according to ye duplicate proportion. I am streightned in time at present, & will explain my self more fully in my next.

Your most Humble Servant

Roger Cotes

NOTES

(1) Add. 3983, no. 23; printed (from the draft) in Edleston, *Correspondence*, pp. 91–2. The square brackets are in the original.

(2) The 'Letter by Dr. Bentley' is Letter 908; 'the other' is Letter 909.

(3) Signatures Ccc, Ddd, pp. 377–92. Cotes enclosed the sheets with his letter.

(4) Cotes had marked these alterations on Newton's Letter 908 (see p. 262, notes (7) to (10)).

(5) See Letter 908, note (14).

(6) See Letter 908, note (15).

(7) Cotes presumably meant to write 'are as the cubes', so that $(229/230)^3 =$ 56908·676/57657·465.

(8) See Letter 908, note (20).

(9) See Letter 908, note (21).

(10) Cotes had written his computation of the 'mean' semi-diameter of the earth at the end of Newton's Letter 908 (see p. 263, note (24)). This differed from Newton's value. It is not clear what either Cotes or Newton intended here, for a brief examination of the values stated shows that neither Cotes' nor Newton's values correspond to the ratio 230:229, or even 229:228. Cotes' values were eventually printed, taking the mean value of the semidiameter as 19 688 725. Newton changed the values again in his interleaved copies of the second edition, and the paragraph was omitted entirely in the third edition. (See Koyré and Cohen, *Principia*, II, p. 608.)

(11) These values were finally printed (see Koyré and Cohen, *Principia*, II, p. 609). The latest corrections to Proposition 20 were in Letter 909, but the numbers were in the original copy.

(12) It was included in the printed version.

(13) See Koyré and Cohen, *Principia*, II, p. 612–13. Cotes presumably changed both the wording and the numerical value in the sentence.

(14) Cotes, as he promises below, explained his objection in his next letter.

911 COTES TO NEWTON

15 APRIL 1712

From the original in the University Library, Cambridge.[1]
Continuation of Letter 910; for the answer see Letter 912

Trinity College April 15*th* 1712

Sr.

I hope You have received the Sheets sent You by the Carrier for Your examination, with my Letter. I come now to ye XXVth Proposition[2] which I think were better to end thus[3]...ad dies 365.6h.9′. id est. ut 1000 ad 178725 seu 1 ad $178\frac{29}{40}$. Unde ex proportione linearum *TM*, *ML*, datur etiam vis *TM*: & hæ sunt vires Solis quibus Lunæ motus perturbantur. Q.E.I. The two periods which are left out may be removed to Prop. XXXVI,[4] for I think they are of no use till we come to that Proposition. If You remove them I suppose You will at the same time alter them, by putting in lin: 14 pag. 435, instead of ye proportion of $60\frac{1}{4}$[5] to 60, the proportion of $40\frac{1}{5}$ to $39\frac{1}{5}$ if this be the proportion which may at last stand in Corol. 6th of Prop: XXXVIIth[6]

Now because the proportion of $40\frac{1}{5}$ to $39\frac{1}{5}$ is made out in the XXXVIIth Proposition, the XXXVIth & XXXVIIth ought to change places but this they cannot do because the XXXVIIth does in other respects depend upon the XXXVIth.[7] Whence it appeares that there ought to be a further alteration in the form of these Propositions, that the former of them may not depend upon ye latter. This may easily be done & I think the whole would be clearer & more Methodical if in the former Proposition the Problem were to find neither ye force of ye Sun nor ye force of ye Moon, but only their proportion to each other, & in the latter the Problem were to find the proportion of both forces to the force of Gravity. And thus ye 3d, 4th, 5th, 6th 7th & 8th Corollarys of ye XXXVIIth will belong to ye former, and ye Corollary of ye XXXVIth togather with the 1st & 2d Corollarys of ye XXXVIIth will belong to the latter. There will be this further advantage in the change, That in ye 7th Corollary of ye XXXVIIth, which will then be annex'd to ye former Proposition, a good foundation may be laid for making out the latter.

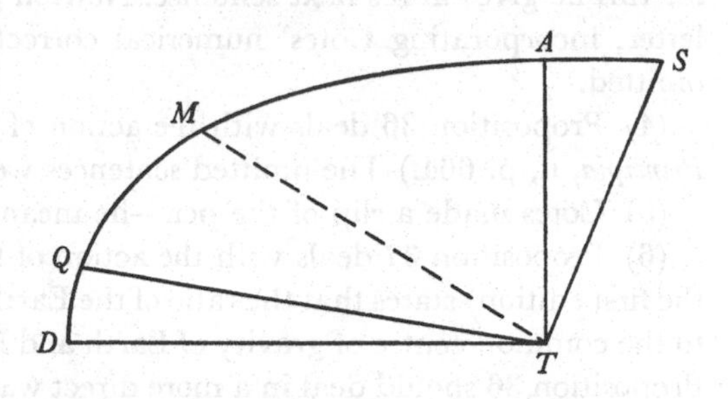

In my Letter which I yesterday wrote to You I was somewhat in haste, I just mention'd a difficulty in Prop: XXXVIIth.[8] Let *ST* be ye Moon's distance from ye Earth when she is 17°½ from her Syzygies, & *QT* be her distance at 17°½ from her Quadratures, & *MT* her mean distance in ye Octants. I think the force of ye Moon must be increased at *S* in the proportion of *MT* cub to *ST* cub, & diminished at *Q* in the proportion of *MT* cub to *QT* cub. Your last corrections increase it at *S* in the proportion of *QT* to *ST* which is nearly in the proportion of *MT* quad to *ST* quad, & diminish it at *Q* in the same proportion. I could wish when the whole is settled that the proportion of 4½ to 1 may be retained for the sake of Proposition XXXIX.(9) I think there is no Proposition in Your Book which does more deserve Your care than ye XXXIXth.

I am, Sir,
Your most Humble Servant
ROGER COTES

for
Sr Isaac Newton
at his House
in St Martin's Street
in Leicester-Fields
London

NOTES

(1) Add. 3983, no. 24; printed in Edleston, *Correspondence*, pp. 93–4, from the draft in Trinity College Library, Cambridge, R. 16. 38, no. 200.

(2) See Koyré and Cohen, *Principia*, II, pp. 618–20. The proposition deals with the Sun's perturbing effect upon the Moon's motion, by referring the system to the Earth at rest.

(3) Cotes has changed the value $178\frac{8}{11}$ of the first edition to $178\frac{29}{40}$, and Newton accepts this change in his next letter. However Cotes has also considerably condensed the end of the proposition, omitting 'Vis qua luna... $60 \times 60 \times 60 \times 178\frac{8}{11}$, seu 1 ad 638092,6.' His reason for this he gives in his next sentence. Newton gave another version of the passage in his next letter, incorporating Cotes' numerical corrections, but reinstating the sentences Cotes had omitted.

(4) Proposition 36 deals with the action of the sun upon the sea. (See Koyré and Cohen, *Principia*, II, p. 664.) The omitted sentences were not inserted there, see note (3) above.

(5) Cotes made a slip of the pen—he meant 60½.

(6) Proposition 37 deals with the action of the Moon upon the sea. Corollary 6 (absent in the first edition) states that the ratio of the Earth–Moon distance to the distance from the Moon to the common centre of gravity of Earth and Moon is 40·371 : 39·371. Cotes is suggesting that Proposition 36 should deal in a more direct way with the action of Sun upon sea, by using the basic theory of Proposition 25, but referring the system to the Moon at rest. All that would change would be the ratio in question, and the calculation could be inserted in Proposition 36. Cotes seems to have lost sight of the fact that this transfer would considerably reduce the significance of Proposition 25.

(7) This is clear from the first sentence of Proposition 37. None of the suggestions Cotes makes in the paragraph following were adopted.

(8) Cotes refers here to the paragraph 'Præterea diametri...ut 1 ad 2871400'. (See Koyré and Cohen, *Principia*, II, p. 668.) Newton accepted Cotes' correction in his next letter.

(9) Proposition 39 deals with the precession of the equinoxes, which is of course partially a result of the Moon's perturbation of the Earth's motion.

912 NEWTON TO COTES

22 APRIL 1712

From the original in Trinity College Library, Cambridge.[1]
Reply to Letters 910 and 911; for the answers see Letters 913 and 914

London Apr. 22. 1712

Sr

I have run my eye over the two proof sheets[2] & approve your corrections. The sheets may be printed off. The XXVth Proposition may end thus[3] ad dies 365.6h.9′, id est ut 1000 ad 178725 seu 1 ad $178\frac{29}{40}$. Invenimus autem in Propositione quarta quod, si Terra et Luna circa commune gravitatis centrum revolvantur, earum distantia mediocris ab invicem erit $60\frac{1}{2}$ semidiametrorum mediocrium Terræ quamproxime. Et vis qua Luna in Orbe circa Terram quiescentem ad distantiam semidiametrorum[4] 60 revolvi posset ut $60\frac{1}{2}$ ad 60 & hæc vis ad vim gravitatis apud nos ut 1 ad 60×60. Ideoque vis mediocris *ML* est ad vim gravitatis in superficie Terræ ut $1 \times 60\frac{1}{2}$ ad $60 \times 60 \times 60 \times 178\frac{29}{40}$, seu 1 ad 638092,6. Unde ex proportione linearum *TM*, *ML*, datur etiam vis *TM*. Et hæ sunt vires Solis quibus motus Lunæ perturbantur. Q.E.I. I here referr the summ of ye forces upon[5] the Sun upon the earth & Moon to the Moon alone & therefore consider the earth as resting & referr its motion to the Moon.

I am satisfied that the force of the Moon upon the Sea is in a triplicate ratio of her distance reciprocally[6] & have altered the calculations accordingly, wch I send you in the inclosed paper[7] together with the emendation of the 39th Proposition. I am

Your most humble Servant

Is. Newton

For the Rnd Mr Cotes Professor of
Astronomy at his Chamber in Trinity
College in Cambridge

NOTES

(1) R.16.38, no. 201; printed in Edleston, *Correspondence*, pp. 94–5.

(2) Cotes had sent these sheets with his Letter 910.

(3) Cotes had suggested an amendment to this passage in his last letter (see Letter 911, note (3)). Newton returns here essentially to the wording of the first edition, but with the addi-

tion of the sentence 'Invenimus...quamproxime'. It was finally printed in this form. (See Koyré and Cohen, *Principia*, II, p. 619.)

(4) Newton has inadvertently omitted part of the sentence here. This slip was corrected in the printed version.

(5) For 'upon' read 'of'.

(6) Cotes had discussed this in his last. See Letter 911, note (8).

(7) This contained emendations to Propositions 37 and 39; see below.

912 *a* THE ENCLOSED PAPER[1]

In Prop XXXVII read

Cæterum ob aquarum reciprocos motus...[*Newton's copy for the first paragraph was printed without change*].

Sed et vis Lunæ in Quadraturis...[*as printed down to*]...Et propterea vis Lunæ in his Quadraturis est tantum $0{,}8570328L$. Est igitur $L+0{,}7986355S$ ad $0{,}8570328L-0{,}7986355S$ ut 9 ad 5.[2]

Præterea diametri Orbis in quo Luna...sunt ad mediocrem ejus distantiam ut 69,100682 & 69,899318 ad $69\frac{1}{2}$. Vires autem Lunæ ad mare movendum sunt in triplicata ratione distantiarum inverse, ideoque vires in maxima et minima harum distantiarum sunt ad vim in medi[o]cri distantia ut 0,9828016 et 1,017342 ad 1. Unde fit $1{,}017342L+0{,}7986355S$ ad $0{,}9828616\times 0{,}8570328L-0{,}7986355S$ ut 9 ad 5. Et S ad L ut 1 ad 4,4815. Itaque cum vis Solis sit ad...

Corol. 1 & 2, as before. see ye Printed Copy[3]

Corol. 3. Quoniam vis Lunæ ad mare movendum est ad Solis vim consimilem ut 4,4824[4] ad 1, et vires illæ (per Corol. 14 Prop LXVI Lib. 1) sunt ut densitates corporum Lunæ et Solis & cubi diametrorum conjunctim; densitas Lunæ erit ad densitatem Solis ut 4,4824 ad 1 directe et cubus diametri Lunæ ad cubum diametri Solis inverse: id est (cum diametri mediocris apparentes Lunæ et Solis sint $31'\ 16\frac{1}{2}''$ et $32'\ 12''$) ut 4892 ad 1000. Densitas autem Solis erat ad densitatem Terræ ut 100 ad 396, et propterea densitas Lunæ est ad densitatem Terræ ut 4892 ad 3960 seu 21 ad 17. Est igitur corpus Lunæ densius et magis terrestre quam Terra nostra...[5]

Corol. 7.[6] Et mediocris distantia centri Lunæ a centro Terræ...centro Terræ et Lunæ ut 40,363 ad 39,363, quæ proinde est pedum 1161492740. Et cum Luna revolvatur...Et ut Radius est ad hunc sinum versum ita sunt pedes 1161492740 ad pedes 14,811762. Luna igitur vi illa qua retinetur in Orbe, cadendo in Terram, tempore minuti unius primi describet pedes 14,811762. Et si hæc vis augeatur...etiam pedes 14,89513. Diminuatur hæc distantia in subduplicata ratione pedum 14,89513 ad pedes 15,12028, et habebitur distantia pedum 19701651 a qua grave cadendo, eodem tempore minuti unius secundi describet pedes 15,12028, id est pedes 15, dig 1, lin 5,32. Et hac vi gravia cadunt in superficie Terræ in Latitudine urbis *Lutetiæ Parisiorum*, ut supra ostensum est. Est autem distantia pedum 19701651 paulo minor quam semidiameter...(7)

Prop. XXXVIII(8)

In the XXXIXth Proposition pag 470 lin 23 write—id est (cum Terræ diameter... [*Newton's text here is printed in the second edition on pp. 437 and 438, down to the paragraph beginning*]

Vis autem Lunæ ad mare movendum erat ad vim Solis ut 4,4824(4) ad 1 circiter...Indeque prodit annua Æquinoctiorum Præcessio a vi Lunæ oriunda 40″ 53‴ 22^{iv}, ac tota Præcessio annua a vi utraque oriunda 50″. 00‴. 42^{iv}. Et hic motus... [*the remainder of the proposition was printed as here written by Newton*].

NOTES

(1) From the original in Trinity College Library, Cambridge, MS. R.16.38, fos. 202, 203 and 208. The original draft carries many alterations by Cotes, especially in the numbers (see Letter 914). Most of this passage is nearly as printed, *Principia* 1713; pp. 427–31.

(2) The numbers printed in the book were emended by Cotes.

(3) As the later page-numbers and the texts indicate, these are the pages of the first edition, slightly modified.

(4) Newton forgot to change this to 4·4815, and Cotes did it for him.

(5) Corollaries 4, 5 and 6 as copied here by Newton only differ from the printed text in reading 39·371 for 39·363, and so 40·371 for 40·363.

(6) Newton copied the complete text of the Corollary, but we have omitted much that is identical with the printed version.

(7) The remainder of Corollary 7, and all of Corollary 8, are exactly as printed in the book.

(8) No text or emendations are given.

913 COTES TO NEWTON

24 APRIL 1712

From the original in Trinity College Library, Cambridge.(1)
Reply to Letter 912; for the answer see Letter 915

Sr.

I have received Your last, but have not yet had time to try the Calculations of the inclosed sheet. I am satisfied as to the XXVth Proposition,(2) upon reconsidering it.

In Page 441, lin: 25,(3) the first & last numbers are 368682 & 362046: they should be 368676 & 362047. The Æquation which results from hence will be(4)

$$88487{,}19-12307251{,}44x+75578{,}14xx-5082017{,}44x^3+42456{,}19x^4=0,$$

of which I find the Root to be 0,0071900057. If You approve of it I would alter the bottom of the Page thus[5] [obtinetur *x* æqualis 0,00719, & inde semidiameter *CT* fit 1,00719 & semidiameter *AT* 0,99281, qui numeri sunt ut $70\frac{1}{24}$ & $69\frac{1}{24}$ quam proxime. Est igitur distantia Lunæ a Terra in Syzygiis ad ipsius distantiam in Quadraturis (seposita scilicet Eccentricitatis consideratione) ut $69\frac{1}{24}$ ad $70\frac{1}{24}$ vel numeris rotundis ut 69 ad 70.] This will cause an alteration in the XXIXth Proposition & in the XXXIst, page 450.

I have not computed the alterations for the XXIXth,[6] not knowing whether You will chuse the whole numbers 69 & 70 or the fractions $69\frac{1}{24}$ & $70\frac{1}{24}$.

As for the other place in page 450th[7] I took the numbers 69 & 70 that I might find what alteration would arise in the conclusion of the Proposition.[8] The result of my computation is as follows. Pag: 450. lin: 18 [69 ad 70] Lin: 20, [Si capiatur angulus 16″. 21‴. 3^{iv}. 30^{v}] Page 452d. lin: 5, [erat 32″. 42‴. 7^{iv}] Lin: 8, [illud est 17‴. 43^{iv}. 11^{v}] Lin: 10, [relinquit 16″. 16‴. 37^{iv}. 42^{v}] Page 453, Lin: 22, [fit 39°. 38′. 7″, 50‴] Lin: 23, [19°. 49′. 3″. 55‴] Lin: ult: [seu 39,6355] Page 454, Lin: 3, [id est, ut 9,0827646 *ATq*] Page 455, lin: 4 [prodibit 0,1188502][9] Lin: 6, [est 1°. 29′. 58″. 3‴] Lin: 7 [subductis relinquit 18°. 19′. 5″. 52‴] Lin: 9 [relinquit 341°. 40′. 54″. 8‴] Lin: 12 [qui propterea erit 19°. 18′. 1″ 22‴]

In finding the Number 0,1188502, I supposed ye ordinate *eZ* to bisect ye base *NT* by which meanes the series for ye Area *TZeF* converged quicker than the other for the area *NeZ*, so yt on account of this Latter I would not depend upon the last figure 2, I think the other[s] are right.[10]

In Line 14th You have 19°. 20′. 31″. 1‴ from Flamsteed's Tables.[11] By Your Theory in Dr Gregory[12] tis 19°. 21′ 22″. 3‴. So in the following Proposition, page 456. Lin. 13 You have 9°. 10′. 40″; by Your Theory tis 9°. 11′. 3″[13]

There will need some other alterations in Prop. XXXIIId & its Corollary upon account of those in the preceding Proposition. You seem to depend too much upon Your Readers quickness when You say [ut rem perpendenti constabit] I hope when You review the whole You will make it easier to apprehend the agreement of the two Constructions.[14]

I do not rightly understand line 12th of page 458 [Inclinationis autem Variatio tantum augebitur per decrementum sinus *IT*, quantum diminuitur per decrementum motus Nodorum][15]

I think I had observed nothing further before we come to ye XXXVIst Proposition.

I am, Sr

Your most Humble Servant

Roger Cotes

Trinity College Apr. 24th 1712

NOTES

(1) R.16.38, no. 210; printed in Edleston, *Correspondence*, pp. 98–100.

(2) See Letters 911 and 912, *ad init.*

(3) Proposition 28, final paragraph; see Koyré and Cohen, *Principia*, II, pp. 626–7. The alteration Cotes suggests was made. The proposition concerns the major and minor axes, CT and AT, of the Moon's orbit, calculated on the assumption that the Earth lies at the centre of the ellipse.

(4) This equation is obtained by writing $CT=1+x$ and $AT=1-x$ in the equation given in the proposition itself. On a separate sheet of paper (R.16.38, fo. 209) Cotes wrote,

'Æquatio fit $88487,19-12307251,44x+75578,14xx-5082017,44x^3+42456,19x^4=0$.

'Inde $x=0,00719$, $CT=1,00719$, $AT=0,99281$ adeoque CT ad AT ut 70,041 ad 69,041, sive ut $70\frac{1}{24}$ ad $69\frac{1}{24}$ vel $70\frac{3}{73}$ ad $69\frac{3}{73}$.

'Vera Radix iterato examine est, 0071900057 ter[tio] exam[ine]:'

This solution of the equation gives the changes in the values of CT, AT and their ratio consequent upon the change Cotes had suggested in the original quartic. The values $CT=1{\cdot}00719$, $AT=0{\cdot}99281$ and $CT:AT=70\frac{1}{24}:69\frac{1}{24}$ were adopted.

(5) This wording was adopted.

(6) Newton left Cotes the choice of values; see Letter 915. Cotes computed the necessary alterations to Proposition 29 ('To find the variation of the Moon'), in Letter 916, using the values 69 and 70.

(7) Proposition 31, 'To find the hourly motion of the nodes of the Moon in an elliptic orbit,' (See Koyré and Cohen, *Principia*, II, p. 640.) The approximate values 69 and 70, and the consequent corrections Cotes suggests below, were used.

(8) Proposition 32, 'To find the mean motion of the nodes of the Moon.' (See Koyré and Cohen, *Principia*, II, p. 643.) Cotes' corrections were adopted up to the end of page 454. For corrections after that see note (9) below.

(9) Although Newton approved this, the clause containing the correction was eventually omitted. (See Koyré and Cohen, *Principia*, II, p. 645.) The four following corrections, consequent upon it, were, with negligible modification, adopted.

(10) It appears that Cotes did not know how Newton had performed this numerical quadrature, and indeed his method is still unpublished. Therefore Cotes used the method on which he had lectured at Cambridge in 1709 (which was to be published by Robert Smith in the *Opera Miscellanea* following his posthumous edition of Cotes' *Harmonia mensurarum, sive Analysis et Synthesis*, Cambridge, 1722).

(11) These were printed at the end of Flamsteed's *Doctrine of the Sphere*, London, 1680, pp. 95–104.

(12) 'Lunæ Theoria Newtoniana,' printed in David Gregory's *Astronomiæ Physicæ & Geometricæ Elementa*, Oxford, 1702, pp. 332–6.

(13) In his Letter 915, Newton explicitly requested that the values 19°. 21′. 22″. 3‴ and 9°. 11′. 3″ be used, and gave reasons. However, he must have returned to the problem later, for he has marked in the margin of the present letter the values 19°. 21′. 20″. 45‴ and 9°. 11′. 51″. He wrote to Cotes on 10 May 1712 (see Letter 919) suggesting 19°. 21′. 20″. 45‴ and 9°. 11′. 3″, and these are also the values in his annotated copy of the first edition. In his Letter 920 Cotes suggested that the first of these values be changed to 19°. 21′. 21″. 50‴, and Newton accepted this alteration in Letter 922. Hence the values finally printed were 19°. 21′. 21″. 50‴ and 9°. 11′. 3″.

(14) In Proposition 33 Newton dealt with the true motion of the Moon's nodes (see Koyré and Cohen, *Principia*, II, p. 646), but left part of the calculation to the reader. It was to this that Cotes objected; for Newtons' rather unsatisfactory reply, see Letter 915.

(15) Newton dealt with this query in his next letter.

914 COTES TO NEWTON

[26 APRIL 1712]

From the original in the University Library, Cambridge.[1]
Continuation of Letter 913; reply to Letter 912; for the answer see Letter 919

May. 26*th* 1712.[2]

Sr

I have examined Your last emendations of ye XXXVIIth Proposition.[3] I am very glad to see the whole so perfectly well settled & fairly stated, for without regard to ye Conclusion, I think the distance of $18\frac{1}{2}$ degrees ought to be taken being much better than $17\frac{1}{2}$ or $15\frac{1}{4}$.[4] And ye same may be said of ye other changes in the principles from which the conclusion is inferr'd.

In exam[in]ing Your numbers I found it necessary to alter most of 'em, I here send You others for Your approbation.

Præterea diametri Orbis in quo Luna...sunt ad mediocrem ejus distantiam ut 69,098747 & 69,897345 ad $69\frac{1}{2}$.[5] Vires autem Lunæ...ad vim in mediocri distantia ut 0,9830427 et 1,017522 ad 1.[6] Unde fit $1{,}017522L+0{,}7986355S$ ad $0{,}9830427\times 0{,}8570327L-0{,}7986355S$ ut 9 ad 5.[7] Et S ad L ut 1 ad 4,4815. Itaque cum vis Solis sit ad vim gravitatis ut 1 ad 12868200, vis Lunæ erit ad vim gravitatis ut 1 ad 2871400.[8]

Corol: 1.[9] Cum igitur aquà vi Solis agitata ascendat ad altitudinem pedis unius & undecim digitorum cum octava parte digiti, eadem vi Lunæ ascendet ad altitudinem octo pedum & digitorum octo. Tanta autem vis —

Corol: 2. Cum vis Lunæ ad mare movendum sit ad vim gravitatis ut 1 ad 2871400 —

Corol: 3. Quoniam vis Lunæ ad mare movendum est ad Solis vim consimilem ut 4,4815 ad 1...et 32'. 12") ut 4891 ad 1000. Densitas autem Solis... ad densitatem Terræ ut 4891 ad 3960 seu 21 ad 17.[10] Est —

Corol: 4...ad massam Terræ ut 1 ad 39,371[11]

Corol: 6...ut 40,371 ad 39,371[12]

Corol: 7...ut 40,371 ad 39,371, quæ proinde est pedum 1161498340...ita sunt pedes 1161498340 ad pedes 14,811833...Et hac vi Luna cadendo tempore minuti unius primi describere deberet pedes 14,89517...et habebitur distantia pedum 19701678 a qua grave cadendo, eodem tempore minuti unius secundi describet pedes 15,12028...&c.(13)

In ye XXXIXth Proposition. Vis autem Lunæ ad mare movendum erat ad vim Solis ut 4,4815 ad 1 circiter...Præcessio a vi Lunæ oriunda 40″. 52‴. 52^{iv}, ac tota Præcessio annua a vi utraque oriunda 50″. 00‴. 12^{iv}.(14)

The XXXVIth Proposition depends upon the latter part of the XXVth & must therefore stand as in the former Edition. I have altered ye Corollary of it thus.

Corol. Cum vis...ad vim gravitatis ut 1 ad 289...mensura pedum Parisiensium 85820, vis Solaris de qua egimus, cum sit ad vim gravitatis ut 1 ad 12868200 atque adeo ad vim illam centrifugam ut 289 ad 12868200 seu 1 ad 44527, efficiet ut...mensura tantum pedis unius Parisiensis & digitorum undecim cum octava parte digiti. Est enim hæc mensura ad mensuram pedum 85820 ut 1 ad 44527.(15)

I have altered the XXXVIIIth Proposition thus. Pag. 467. lin: 10...id est, ut 39,371 ad 1 & 100 ad 365 conjunctim, seu 1079 ad 100. Unde cum mare nostrum vi Lunæ attollatur ad pedes $8\frac{2}{3}$, fluidum Lunare vi Terræ attolli deberet ad pedes $93\frac{1}{2}$...excessu pedum 187...&c.(16)

I am glad to understand by Dr Bentley that You have some thoughts of adding to this Book a small Treatise of Infinite Series & ye Method of Fluxions.(17) I like the design very well, but I beg leave to make another Proposal to You. When this Book shall be finished I intended to have importun'd You, to review Your Algebra for a better Edition of it,(18) & to have added to it those things which are published by Mr Jones & what others You have by You of the like nature. These togather will make a Volume nearly of a Size with Your Principia & may be printed in the same Character. Your Treatise of ye Cubick Curves should be reprinted, for I think ye Enumeration is imperfect, there being as I reckon five cases of Æquations.(19) viz.

$$xyy+ey=ax^3+bxx+cx+d$$
$$yy+gxxy=ax^3+bxx+cx+d$$
$$xxy+ey=ax^3+bxx+cx+d$$
$$xy=ax^3+bxx+cx+d$$
$$y=ax^3+bxx+cx+d$$

I should have acquainted You with this before Mr Jones's Book(20) was published if I had known any thing of ye Printing of it, for I had observed it two or three Yeares ago. I think there are some other things of lesser moment amiss in ye Treatise.

I am, Sir, Your most Humble Servant
ROGER COTES.

Trinity College Cambridge.

For Sr Isaac Newton
at his House
in St Martin's Street
in Leicester Fields
London

NOTES

(1) Add. 3983, no 63, printed in Edleston, *Correspondence*, pp. 100–2, from the draft in Trinity College, Cambridge, but without the last two paragraphs (see also note (16) below).

(2) The letter bears the postmark 28 April, although the date 26 May is clearly written by Cotes, but on the draft in Trinity College the month May has been struck out by Cotes and replaced by April; hence Edleston gave the date of this letter as 26 April, correctly.

(3) With Letter 912, see Edleston, pp. 95–8.

(4) This is in the third paragraph of Proposition 37; it is the estimate of the distance by which the moon has passed the meridian at the time of high tide.

(5) Presumably Cotes simply means that he has recomputed these numbers from Newton's data; here Newton had given 69·100682 and 69·899318 to 69½. The numbers given here were all printed.

(6) Newton: 0·9828616 and 1·017342 to 1.

(7) Newton: $1{\cdot}017342L+0{\cdot}7986355S$ to $0{\cdot}9828616\times0{\cdot}8570323L-0{\cdot}7986355S$ as 9 to 5.

(8) Newton: S to $L=1$ to 4·4824. The remainder is added by Cotes.

(9) This corollary is not yet in its final form; the ratios have been modified by Cotes.

(10) Newton: 4,824...31° 16½'...4892 to 1000...4892 to 3960 or 21 to 17.

(11) Newton: 1 to 39·363.

(12) Newton: 40·363 to 39·363.

(13) Newton's last numbers were: 40·363 to 39·363...1161492740 feet to 14·811762... 14·89513 feet...19701651 feet.

(14) Newton: 4·4824...40″ 53‴ 22^{iv}...50″ 00‴ 42^{iv}.

(15) An emendation sent with Newton's letter of 8 April had proposed 85820 and 44038 as the two numbers.

(16) The letter as printed by Edleston ends here, although the remainder of it—written below Cotes' signature, as a postscript—appears on the draft; this portion Edleston assigned (wrongly) to a different date and printed on p. 119 of his *Correspondence*.

(17) Compare Letter 929, Cotes having by then (perhaps) forgotten this passage, which provoked no response from Newton. Bentley had returned from London early in April; he went there again about 24 May, returning to Cambridge at the end of June. Drafts of the small treatise 'Analysis per Quantitates fluentes et earum Momenta' still exist (we are informed by Dr Whiteside) one part being in private possession, and another in U.L.C. Add. 3960(6), with other sketches elsewhere. See also the preparatory allusions in Letter 829*a*.

(18) That is, *Arithmetica universalis; sive de compositione et resolutione arithmetica liber*, Cambridge, 1707. For the story of the publication of this book see Whiteside, *Mathematical Papers*, v, pp. 8–13. A second edition 'in qua multa immutantur et emendantur' appeared at London in 1722.

(19) In *Enumeratio linearum tertii ordinis* (published in *Opticks*, 1704) Newton gave four cases of the cubic; his first is the same as Cotes', his second and fourth are Cotes' fourth and fifth. Instead of Cotes' second and third, Newton's third is $y^2 = ax^3 + bx^2 + cx + d$. As D. T. Whiteside has pointed out (private communication) it is Cotes' enumeration that is imperfect, for Cotes' second case becomes Newton's third when $g = 0$ or, if $g \neq 0$, is converted by a linear transformation into the form of Case 1 when $a = 0$. Similarly, a linear transformation converts Cotes' third case to be of the form $xy^2 + ey = cx + d$, which is Case 1 when $a = b = 0$.

(20) See above, Letter 821, note (2).

915 NEWTON TO COTES

[29 APRIL 1712][1]

From the original in Trinity College Library, Cambridge.[2]
Reply to Letter 913; for the answers see Letters 916 and 917

Sr

The corrections made in your last of Apr. 24th may all stand. In ye XXIXth you may use either ye whole numbers 69 & 70 or the fractions $69\frac{1}{24}$ & $70\frac{1}{24}$.[3] In pag 455 lin 14 & pag 456 I have put the motion of the Nodes of [the] Moon from ye Equinox & should have put it from ye fixt starrs. In ye first place therefore for 19gr 20′ 31″ 1‴ write 19°. 21′. 22″. 3‴ In ye second for 9°. 10′. 40″ write 9°. 11′. 3″[4]

In pag. 458 lin 11. write. [Et in eadem ratione minuetur etiam Inclinationis Variatio.] And strike out the rest to the end of the Paragraph.[5]

In ye XXXIIIth Proposition, pag 456, instead of ye words [ut rem perpendenti constabit] may be written [ut rem perpendenti & computationes instituenti constabit][6] And the numbers in this Proposition are to be suited to ye alterations made in ye preceding Proposition as you mention.

I am

Your most humble Servant

IS. NEWTON

London Apr. 24*th*[1]
1712

NOTES

(1) The letter is dated 24 April, but this must have been a slip. The postmark is 29 April, so we assign the letter this date.

(2) R.16.38, no. 211; printed in Edleston, *Correspondence*, p. 103.

(3) See Letter 913, note (6), and Letter 916.

(4) See Letter 913, note (13).

(5) This was done (see Koyré and Cohen, *Principia*, II, p. 653); but it removed the passage

that had puzzled Cotes (see Letter 913, note (15)), instead of clarifying it. As Cotes points out in his Letter 917, the effect of the change was to consider the Moon's inclination in terms of a circular rather than an elliptic orbit. This had repercussions upon the rest of Proposition 34, and upon Proposition 35.

(6) Again, Newton has scarcely made the task easier for the reader, as Cotes had requested (see Letter 913, note (14). The required alterations to the numbers were made (see Letter 917, note (3)).

916 COTES TO NEWTON

1 MAY 1712

From the original in the University Library, Cambridge.[1]
Reply to Letter 915; for the answer see Letter 919

May. 1*st.* 1712

Sr.

I have received Your last, & taking ye whole numbers 69 & 70,[2] the alteration in Pag: 442. lin: penult: will be[3] [68,6877 ad numerum 69. Quo pacto tangens anguli *CTP* jam erit ad tangentem motus medij ut 68,6877 ad 70, & angulus *CTP* in Octantibus ubi motus medius est 45°, invenietur 44°. 27′. 28″: qui subductus de angulo motus medij 45° relinquit Variationem maximam 32′. 32″. Hæc ita se haberent si...& Variatio maxima quæ secus esset 32′. 32″, jam aucta in eadem ratione fit 35′. 10″.] You go on thus.[4]

Hæc est ejus magnitudo in mediocri distantia Solis a Terra, neglectis differentijs quæ a curvatura Orbis magni majorique Solis actione in Lunam falcatam et novam quam in gibbosam et plenam oriri possint. In alijs distantijs Solis a Terra, Variatio maxima est in ratione quæ componitur ex duplicata ratione[5] revolutionis Synodicæ Lunaris (data anni tempore) directe, et ratione anguli *CTa* directe, et triplicata ratione distantiæ Solis a Terra inverse,[6] Ideoque in Apogæo Solis Variatio maxima est 33′. 11″, & in ejus Perigæo 37′. 24″, si modo eccentricitas Solis sit ad Orbis magni semidiametrum transversam ut $16\frac{15}{16}$ ad 1000.

Hactenus Variationem investigavimus in Orbe non eccentrico, in quo utique Luna in Octantibus suis semper est in mediocri sua distantia a Terra. Si Luna propter eccentricitatem suam, magis vel minus distat a Terra quam si locaretur in hoc Orbe, Variatio paulo major esse potest vel paulo minor quam pro Regula hic allata: sed excessum vel defectum ab Astronomis per Phænomena determinandum relinquo.

I was going to diminish Your numbers 33′. 11″, & 37′. 24″ by 2″ which is nearly the diminution if those numbers are right,[7] which I confess I am forc'd to take upon trust not knowing how to state ye proportion ye of Moons Periodical revolutions nor consequently of her Synodical, in ye Apogee & Perigee of ye Sun. But I cannot fully satisfy my self about Your Rule.[8] As I take it the duplicate ratio of ye Synodical revolution & ye simple ratio of the angle *CTa* compose not the triplicate ratio of ye Synodical revolution alone, but this triplicate ratio directly & ye simple ratio of ye Periodical revolution inversly: the angle *CTa* being as the Synodical revolution directly & the Periodical revolution inversly. I have besides some scruple about introducing the ratio of ye angle *CTa*, I have not throughly considered the thing, but I quæry whether it will not be sufficient to make the compounded ratio consist only of ye duplicate ratio of ye Synodical revolution directly & the triplicate ratio of the Suns distance inversly, according to ye 16th Corol: of Prop: LXVI. Lib. 1.[9] I have transcribed the whole that You may review it & order it as You think it should stand. I am

Yr most humble Servant
ROGER COTES

I sent a Letter by the last Post which I fear has miscarried unless You wrote Yours before the receipt of it.[10]

For Sr Isaac Newton
at his House
in St Martin's Street
In Leicester Feilds
London

NOTES

(1) Add 3983, no. 25; printed (from the draft) in Edleston, *Correspondence*, pp. 103–5. The square brackets are in the original.

(2) See Letter 913, note (6).

(3) The passage was finally printed in this form (*Principia* 1713; pp. 402–3); see Koyré and Cohen, *Principia*, II, Prop. 29, p. 629.

(4) The following paragraph was transcribed by Cotes from MS. R.16.38, fos. 149, 150, which was presumably part of the original 'copy' sent by Newton. It was printed essentially in the form given here, but with two major alterations. The values 33′. 11″ and 37′. 24″ were altered to 33′. 14″ and 37′. 11″ and the words 'et ratione anguli *CTa* directe' were omitted. See also note (6) below.

(5) Newton has made an insertion mark here, and has written 'temporis' in the margin, He suggested this addition to Cotes in Letter 919, and it was adopted.

(6) In both the draft Letter (printed by Edleston from Trinity College Library, Cambridge), and Newton's manuscript version of the passage, the words 'id est, ex triplicata ratione revolutionis synodicæ Lunaris directe et triplicata ratione distantiæ Solis a Terra inverse' appear here. It is unfortunate that Cotes should have omitted them, for it is against them that he raised one of his main objections (see note (8) below). Since Cotes had Newton's original MS. of the passage, Newton was unaware of this omission and did not, later, when he accepted Cotes' objection, rule that they be left out of the final copy (see Letter 922). This however was done, presumably on Cotes' own initiative.

(7) This diminution would arise from the change to the ratio 69:70. Newton accepted the suggestion in his Letter 919, but the numbers still required further alteration; see Letter 920.

(8) Newton had suggested in the passage quoted that

$$\text{Maximum Variation of the Moon} \propto \frac{(\text{Time of Moon's synodical revolution})^2}{(\text{Sun–Earth distance})^3} \times CTa$$

where CTa is the observed angle between quadrature and syzygy. He then stated that this expression is equal to

$$\frac{(\text{Time of Moon's synodical revolution})^3}{(\text{Sun–Earth distance})^3}$$

(but this statement was omitted from Cotes' letter—see note (6) above). Cotes objects here on two counts; first, because

$$CTa \propto \frac{\text{Time of Moon's synodical revolution}}{\text{Time of Moon's periodic revolution}}$$

and substitution of this into Newton's first expression will not give his second. Second (and more important), Cotes claims that angle CTa should not in any case be included in the expression, since it has already been accounted for.

In his Letter 919, Newton argued that it was correct to include angle CTa, but Cotes replied (Letter 920) that he was still puzzled, and enlarged and clarified his reasons. In his Letter 922 Newton finally admitted that Cotes was right.

(9) Newton made no reference to this corollary in his argument; he might have done so to some advantage, although it did not directly state the result required.

(10) Presumably Letter 914, which Newton should have received before the 29th; this suggests he wrote Letter 915 on the 24th, but failed to post it before the 29th.

917 COTES TO NEWTON

3 MAY 1712

From the original in the University Library, Cambridge.(1)
Reply to Letter 915; for the answer see Letters 919 and 932

Cambridge May 3d 1712.

Sr.

I fear I give You too much trouble with my Letters, but I think this will be my last till we come to the Theory of Comets.(2) In the Corollary of the XXXIIId Proposition I put 16″. 19‴. 27^{iv}. instead of 16″. 18‴. 41$^{iv}\frac{1}{2}$.(3) I am not certain how You would compute that Motion & therefore I mention it to You, I found it by this Proportion: As 19°. 18′. 1″. 23‴. to 19°. 21′. 22″. 3‴. so 16″. 16‴. 37^{iv}. 42^{v}. to 16″. 19‴. 26^{iv}. 56^{v}.

In Your last Letter You order Pag: 458. lin: 11 thus [Et in eadem ratione minuetur etiam Inclinationis Variatio](4) This will cause some alteration in the following Corollarys & in the XXXVth Proposition, unless You design to consider the Moon's Inclination only as moving in Orbe circulari.

At the bottom of Page 461 You make use of 5°. 17′. 46″ and 5° for the extream Inclinations; In Dr Gregory's Astronomy[5] You have 5°. 17′. 20″ and 4°. 59′. 35″,[6] which I suppose You found to be more agreeable to Observations.

In the first Paragraph of the new Scholium[7] to Prop: XXXVth You have [ad 11′. 50″ circiter ascendit, & *additur* medio motui Lunæ ubi Terra pergit a Perihelio suo ad Aphelium & in opposita Orbis parte *subducitur*] As I take it the words *additur* & *subducitur* should change places. You have not mentioned how to find this Æquation in every intermediate place.

In the second Paragraph concerning the Annual Æquations of the Moon's Apogee & Node, You have forgotten to mention when they must be added & when subtracted.

In the third Paragraph You say [Per Theoriam gravitatis constitit etiam quod actio Solis in Lunam paulo major sit ubi transversa diameter Orbis Lunaris transit per Solem quam ubi eadem ad rectos est angulos cum linea Terram & Solem jungente & propterea Orbis Lunaris paulo *minor* est in priore casu quam in posteriore] I think it should be [paulo major est].

In the fourth Paragraph concerning the Æquation of the Moon arising from the position of her Nodes which You call *Semestris Secunda*, You have [additur vero medio motui Lunæ dum Nodi transeunt a Solis Sygygijs [*sic*] ad proximas Quadraturas & *subducitur* in eorum transitu a Quadraturis ad Syzygias]. As I apprehend it, the words *additur* & *subducitur* should change places.

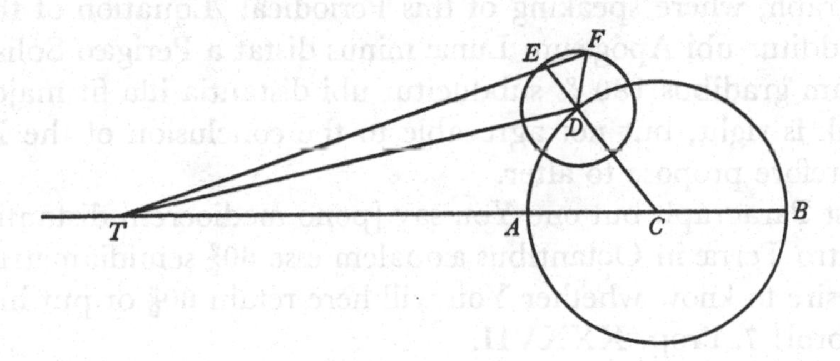

The sixth Paragraph I do not understand.[8] The Æquation which You there describe seems to be established not so much from Observations as from the Theory of gravity, but I cannot perceive how it answers Your design express'd in these words. In Perihelio Terræ propter majorem vim Solis Apogæum Lunæ velocius movetur in Epicyclo circum centrum D (I suppose it should be *centrum* C) quam in Aphelio, idque in triplicata ratione distantiæ Terræ a Sole inverse. Ob Æquationem centri Solis in argumento annuo comprehensam, Apogæum Lunæ velocius movebitur in Epicyclo in duplicata ratione distantiæ Terræ a Sole inverse. Ut idem adhuc celerius moveatur in ratione simplici distantiæ inverse, sit &c. Now the Æquation which You describe in what follows, does not in the least, as I see, depend upon the Sun's Anomaly, but intirely upon the Annual Argument of the Apogee. You will perhaps more easily perceive my difficulty if I tell You how I think the Æquation should be stated to answer what was proposd. Let *CTD* be the *Æquatio Semestris* describ'd in the preceding Paragraph; produce *CD* to *E*, so that *DE* may be to *CD* as $33\frac{7}{8}$ to 1000; make the angle *EDF* equal to the Suns Anomaly, & the line *DF* equal to *DE*, & joyn *TF*. Then will *DTF* be the second annual Æquation of the Apogee, & *TF* be the Eccentricitas Lunæ bis æquata in Apogæum Lunæ ter æqua[tum tendens.](9)

The following Paragraph concludes thus. [Ducantur rectæ duæ parallelæ *TP*, *FH* in Perigæum Solis tendentes, vel quod perinde est, capiatur angulus *GFH* æqualis distantiæ Perigæi Solis ab Apogæo Lunæ, & sit *FH* ipsi *FG* æqualis: & angulus *FTH* erit Æquatio periodica Apogæi Lunæ, & angulus *PTH* distantia Apogæi Lunæ quarto æquati a Perigæo Solis et *TH* eccentricitas tertio æquata in Apogæum quarto æquatum tendens.] Instead of which I propose the following alteration, leaving out the line *TP* in the figure.(10) [Capiatur angulus *GFH* æqualis distantiæ Apogæi Lunæ a Perigæo Solis in consequentia et sit *FH* ipsi *FG* æqualis, & angulus *FTH* erit Æquatio periodica Apogæi Lunæ, & *TH* eccentricitas tertio æquata in Apogæum quarto æquatum tendens.] This alteration will agree with what You lay down a little before in the same Paragraph, where speaking of this Periodical Æquation of the Apogee You say, [additur ubi Apogæum Lunæ minus distat a Perigæo Solis in consequentia quam gradibus 180 & subducitur ubi distantia illa fit major,] which Rule I think is right, but not agreeable to the conclusion of the Paragraph which I therefore propose to alter.

In the last Paragraph but one You say [pono mediocrem distantiam centri Lunæ a centro Terræ in Octantibus æqualem esse $60\frac{2}{9}$ semidiametris maximis Terræ] I desire to know whether You will here retain $60\frac{2}{9}$ or put instead of it $60\frac{1}{4}$ as in Corol: 7. Prop. XXXVII.

I am
Your most humble Servant
ROGER COTES.

For
Sr Isaac Newton
at his House
in St Martin's Street
in Leicester Feilds
London

NOTES

(1) Add. 3983, no. 26; printed (from the draft) in Edleston, *Correspondence*, pp. 106–9. The square brackets are in the original, except where noted.

(2) Cotes in fact wrote several more letters before broaching the subject of Cometary Theory.

(3) Newton had asked Cotes to make the necessary alterations to Proposition 33 in his Letter 915. The value $16''. 19'''. 27^{iv}$ was not in fact printed, because Newton suggested a slightly different value in his Letter 922.

(4) See Letter 915, note (5).

(5) See Letter 913, note (12).

(6) These values were adopted. See Letter 919, postscript.

(7) Compare Letter 932. We print the first draft of the 'new scholium' here for convenience (Trinity College Library, Cambridge, R.16.38, fos. 169–71; see also the extensive note in Edleston, *Correspondence*, p. 109). Newton sent no immediate answer to Cotes' queries, but some time between 20 July and 10 August sent a second draft, possibly beginning with the fifth paragraph of the Scholium ('Per eandem Gravitatis Theoriam Apogæum Lunæ...'), to part of which Cotes refers in Letter 931. This revision is now lost, but Edleston has attempted to reconstruct it (*ibid.*, p. 120). Lacking any positive reply from Newton, Cotes repeated many of his earlier queries in Letter 931, to which Newton replied promptly on 12 August.

(8) Newton, rewriting the whole of this and subsequent paragraphs, did not deal with the details of the alterations Cotes suggests here.

(9) The corner of the page is torn here.

(10) See the Figure in the draft following.

917*a* FIRST DRAFT OF THE SCHOLIUM TO PROPOSITION 35

Schol. ad pag. 462(1)

Hisce motuum Lunarium computationibus ostendere volui quod motus Lunares per Theoriam gravitatis a causis suis computari possint. Per eandem Theoriam inveni præterea quod Æquatio annua medij motus Lunæ oriatur a varia dilatatione orbis Lunæ per vim Solis juxta Corol 6 Prop. LXVI Lib. I. Hæc vis in perigæo Solis major est et orbem Lunæ dilatat, in apogæo ejus minor est & orbem illum contrahi permittit. In orbe dilatato Luna tardius revolvitur, in contracto citius; et æquatio annua per quam hæc inæqualitas compensatur, in apogæo & perigæo Solis nulla est in mediocri Solis a Terra distantia ad $11'. 50''$ circiter ascendit, et additur medio motui Lunæ ubi Terra pergit a perihelio suo ad aphelium, et in opposita Orbis parte subducitur. Assumendo radium Orbis magni 1000 et excentricitatem Terræ $16\frac{7}{8}$, hæc æquatio ubi maxima est, per Theoriam gravitatis prodijt $11'. 49''$. Sed excentricitas Terræ paulo major esse videtur, et aucta excentricitate hæc æquatio augeri debet in eadem ratione. Sit excentricitas $16\frac{15}{16}$, et æquatio maxima erit 11′. 52″.

Inveni etiam quod in perihelio Terræ propter majorem vim Solis, apogæum et nodi Lunæ velocius moventur quam in aphelio ejus, idque in triplicata ratione distantiæ Terræ a Sole inverse. Et inde oriuntur æquationes horum motuum æquationi centri Solis proportionales. Motus autem Solis est in duplicata ratione distantiæ Terræ a Sole inverse, et maxima centri æquatio quam hæc inæqualitas generat est 1gr. 56′. 26″ prædictæ Solis excentricitati $16\frac{15}{16}$ congruens. Quod si motus Solis esset in triplicata ratione distantiæ, hæc inæqualitas generaret æquationem maximam 2gr. 56′. 9″. Et propterea æquationes maximæ quas inæqualitas [*sic*] motuum apogæi et nodorum Lunæ generant, sunt ad 2gr. 56′. 9″. ut motus medius diurnus apogæi & motus medius diurnus nodorum Lunæ sunt ad motum medium diurnum Solis. Unde prodit æquatio maxima medij motus apogæi 19′. 52″, & æquatio maxima medij motus nodorum 9′. 27″.

Per Theoriam gravitatis constitit etiam quod actio Solis in Lunam paulo major sit ubi transversa diameter Orbis Lunaris transit per Solem quam ubi eadem ad rectos est angulos cum linea Terram et Solem jungente; et propterea Orbis Lunaris paulo minor est in priore casu quam in posteriore. Et hinc oritur alia æquatio motus medij Lunaris pendens a situ apogæi Lunæ ad Solem, quæ quidem maxima est cum apogæum Lunæ versatur in Octante cum Sole, et nulla cum illud ad quadraturas vel Syzygias pervenit: et motui medio additur in transitu apogæi Lunæ a Solis quadratura ad Syzygiam & subducitur in transitu apogæi a syzygia ad quadraturam. Hæc æquatio quam semestrem vocabo, in octantibus apogæi quando maxima est, ascendit ad 3′. 45″ circiter quantum ex phænomenis colligere potui. Hæc est ejus quantitas in mediocri Solis distantia a Terra. Augetur vero ac diminuitur in triplicata ratione distantiæ Solis inverse, adeoque in maxima Solis distantia est 3′. 34″ & in minima 3′. 56″ quamproxime: ubi vero apogæum Lunæ situm est extra Octantes, evadit minor, estque ad æquationem maximam ut sinus duplæ distantiæ apogæi Lunæ a proxima syzygia vel quadratura ad radium.

Per eandem gravitatis theoriam actio Solis in Lunam paulo major est ubi linea recta per nodos Lunæ ductæ transit per Solem quam ubi linea ad rectos est angulos cum recta Solem ac Terram jungente. Et inde oritur alia medij motus æquatio, quam semestrem secundam vocabo, quæque maxima est ubi nodi in Solis octantibus versantur, & evanescit ubi sunt in syzygijs et quadraturis, et in alijs nodorum positionibus proportionalis est sinui duplæ distantiæ nodi alterutrius a proxima syzygia aut quadratura; additur vero medio motui Lunæ dum nodi transeunt a Solis syzygijs ad proximas quadraturas & subducitur in eorum transitu a quadraturis ad syzygias, et in Octantibus ubi maxima est ascendit ad 47″ in mediocri Solis distantia a Terra, uti ex Theoria gravitatis colligo. In alijs Solis distantijs hæc æquatio in octantibus nodorum est reciproce ut cubus distantiæ Solis a Terra, ideoque in perigæo Solis ad 45″. in apogæo ejus ad 49″ circiter ascendit.

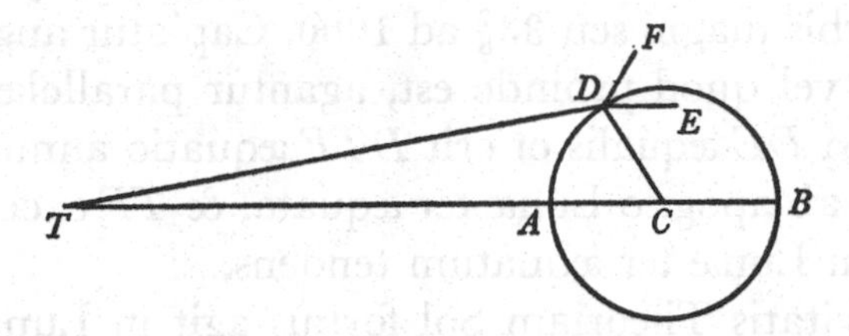

Per eandem gravitatis Theoriam apogæum Lunæ progreditur quam maxime ubi vel cum Sole conjungitur vel eidem opponitur, & regreditur ubi cum Sole quadraturam facit. Et excentricitas fit maxima in priore casu et minima in posteriore, per Corol. 7, 8, & 9 Prop. LXVI Lib. I. Et hæ æquationes per eadem Corollaria permagnæ sunt, & æquationem principalem apogæi generant, quam semestrem vocabo. Et æquatio maxima semestris est 12gr. 18′ circiter quantum ex observationibus colligere potui. *Horroxius* noster Lunam in Ellipsi circum Terram in ejus umbilico inferiore constitutam revolvi primus statuit.[2] *Halleius* superiorem Ellipseos umbilicum in Epicyclo locavit cujus centrum uniformiter revolvitur circum Terram.[3] Et ex motu in Epicyclo oriuntur inæqualitates jam dictæ in progressu & regressu apogæi et quantitate excentricitatis. Dividi intelligatur distantia mediocris Lunæ a Terra in partes 100000, et referat *T* Terram & *TC* excentricitatem mediocrem Lunæ partium 5505. Producatur *TC* ad *B* ut sit *CB* sinus æquationis maximæ semestris 12gr 18′ ad radium *TC*, et circulus *BDA* centro *C* intervallo *CB* descriptus erit epicyclus ille in quo superior Ellipseos umbilicus locatur[4] & secundum ordinem literarum *BDA* revolvitur. Capiatur angulus *BCD* æqualis duplo argumento annuo, et erit angulus *CTD* apogæi æquatio semestris, et *TD* excentricitas primo æquata in apogæum secundo æquatum tendens.

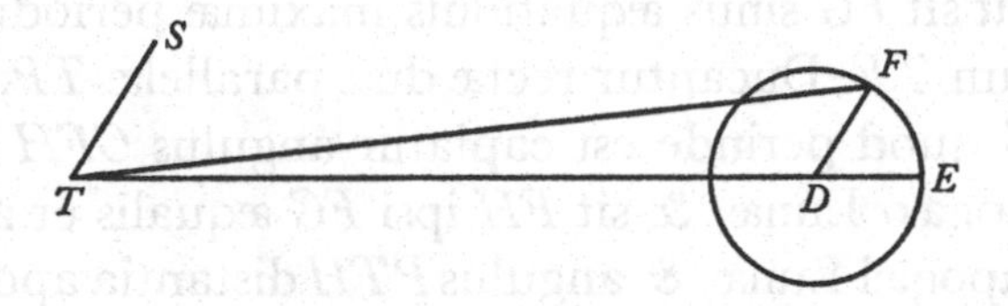

In perihelio Terræ propter majorem vim solis, Apogæum[5] Lunæ velocius movetur in epicyclo circum centrum *D*[6] quam in aphelio. idque in triplicata ratione distantiæ Terræ a Sole inverse. Ob æquationem centri Solis in Argumento annuo comprehensam Apogæum Lunæ velocius movebitur in epicyclo in duplicata ratione distantiæ Terræ a Sole inverse. Ut idem adhuc celerius moveatur in ratione simplici distantiæ inverse, sit *TD* excentricitas primo æquata, et producatur *TD* ad *E* ut sit *DE* ad *TD* ut duplum excentricitatis Solis ad radium Orbis magni seu $33\frac{7}{8}$ ad 1000. Capiatur angulus *EDF* æqualis argumento annuo, vel quod proinde est, agantur parallelæ *TS* ac *DF* solem versus, et sit *DF* ipsi *DE* æqualis et erit *DTF* æquatio annua apogæi Lunæ & *FTS* distantia Solis ab apogæo Lunæ ter æquata, & *TF* excentricitas Lunæ bis æquata in apogæum Lunæ ter æquatum tendens.

Per eandem gravitatis Theoriam Sol fortius agit in Lunam annuatim ubi apogæum Lunæ et perigæum Solis conjunguntur quam ubi opponuntur. Et inde oriuntur æquationes duæ periodicæ, una medij motus Lunæ, altera apogæi ejus: quæ quidem æquationes nullæ sunt ubi apogæum Lunæ vel conjungitur cum perigæo Solis vel eidem opponitur, et maximæ in apogæorum quadraturis. In alijs apogæorum positionibus datam habent proportionem ad invicem, suntque ut sinus distantiæ apogæorum ab invicem. Æquatio prior subducitur et posterior additur ubi apogæum Lunæ minus distat a perigæo Solis in consequentia quam gradibus 180; prior vero additur & posterior subducitur ubi distantia illa fit major. Harum æquationum quantitates maximæ per eclipses Solis et Lunæ determinandæ sunt. Et quantum sentio, Æquatio maxima apogæi ascendit ad 15′ vel 20′ circiter, sed æquatio maxima motus medij Lunæ vix ascendit ad 30″, et ob parvitatem negligi potest donec quantitas ejus ex observationibus determinetur. Producatur excentricitas Lunæ bis

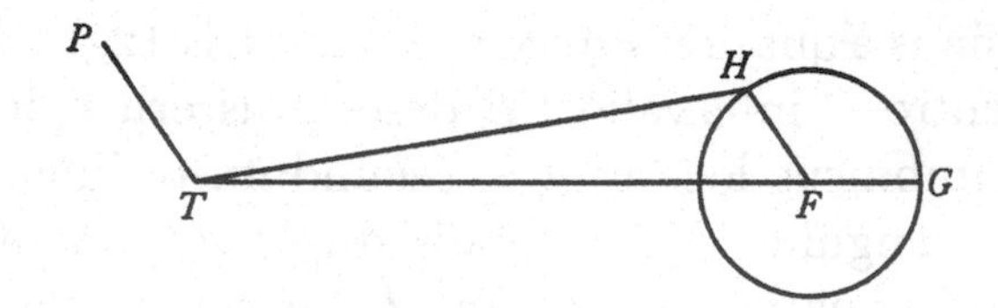

æquata *TF* ad *G*, ut sit *FG* sinus æquationis maximæ periodicæ apogæi Lunæ 15′ vel 20′ ad radium *TF*. Ducantur rectæ duæ parallelæ *TP*, *FH* in perigæum Solis tendentes, vel quod perinde est capiatur angulus *GFH* æqualis distantiæ perigæi Solis ab apogæo Lunæ, & sit *FH* ipsi *FG* æqualis et angulus *FTH* erit æquatio periodica apogæi Lunæ, & angulus *PTH* distantia apogæi Lunæ quarto æquata a perigæo Solis, et *TH* excentricitas tertio æquata in apogæum quarto æquatum tendens.

Si tres anguli *CTD*, *DTF* & *FTH* ad singulos gradus angulorum *BCDDF*, *E* et *GFH* computentur & in Tabulas referantur, et si logarithmi quoque trium distantiarum *TD*, *TF* & *TH* ad radios *TC TD* et *TF* in partes 100000 divisos simul computentur & in Tabulas referantur: aggregatum trium angulorum sub signis suis + & − erit æquatio tota apogæi, et aggregatum trium Logarithmorum erit Logarithmus excentricitatis veræ.

Habitis autem Lunæ motu medio & apogæo et excentricitate ultimum æquatis, ut et Orbis diametro transversa partium 200000; ex his eruetur verus Lunæ locus in orbe, et distantia ejus a Terra, idque per methodos notissimas. Deinde per Variationem et Reductionem ad Eclipticam dabitur ejus longitudo et latitudo vera.

Diximus(7) orbem Lunæ a viribus Solis per vices dilatari et contrahi & æquationes quasdam motuum Lunarium inde oriri. Inde etiam oritur variatio aliqua parallaxeos Lunæ, sed quam insensibilem esse judico; ideoque in computationibus motuum Lunæ, pro mediocri ejus distantia a centro Terræ semper usurpo numerum 100000 & pro Orbis diametro transversa numerum 200000, et ad parallaxim investigandam pono mediocrem distantiam centri Lunæ a centro Terræ in Octantibus æqualem esse $60\frac{2}{5}$ semidiametris maximis Terræ. Semidiametrum ejus maximam voco quæ a centro ad æquatorem ducitur, minimam quæ a centro ad polos. Et hinc fit Lunæ parallaxis horizontalis mediocris apparens in Octantibus 57′ 5″, in syzygijs 57′ 30″ in quadraturis 56′ 40″. Lunæ vero diameter mediocris apparens in Syzygijs 31. 30 in Quadraturis 31. 3 usurpari possent & polis diameter mediocris 32. 12

Et(8) cum atmosphæra Terræ ad usque altitudinem milliarium 35 vel 40 refringat Lucem Solis et refringendo spargat eandem in umbram Terræ, & spargendo lucem in confinio umbræ dilatet umbram: ad diametrum umbræ quæ per parallaxim prodit, addo minutum unum primum in eclipsibus Lunæ, vel minutum unum cum triente.

Theoria vero Lunæ primo in Syzygijs, deinde etiam in quadraturis, et ultimo in Octantibus per phænomena examinari et stabiliri debet. Et opus hocce aggressurus, motus medios Solis et Lunæ ad tempus meridianum in Observatorio regio *Grenovicensi* die ultimo mensis Decembris anni 1700 st. veteri, non incommode sequentes adhibebit, nempe motum medium Solis ♑ 20gr 43′. 50″ & apogæi ejus ♋ 7gr 44′. 30″, & motum medium Lunae ♒ 15gr 19′. 50″, & apogæi ejus ♓ 8gr 18′. 20″, & Nodi ascendentis ☊ 27 gr 24′. 20″.

Translation

Scholium for page 462 [First Edition](1)

It was my intention to show by these computations of the lunar motions that they could be calculated from their causes with the aid of the theory of gravity. Furthermore I found by the same theory that the annual equation in the mean motion of the Moon arises from the varying dilatation of the Moon's orbit [caused] by the force of the Sun, in accordance with Book I, Proposition 66, Corollary 6. This force is greater at the Sun's perigee and widens the Moon's orbit, but at the Sun's apogee it is less and allows the orbit to contract. The Moon revolves more slowly in the wide orbit, more swiftly in the contracted one, and the annual equation by which this inequality is allowed for is zero at the Sun's apogee and perigee but at the mean distances of the Sun from the Earth it amounts to about 11′ 50″, and this is to be added to the Moon's mean motion when the Earth progresses from its perihelion to its aphelion, and subtracted in the other half of the orbit. Putting the radius of the Earth's orbit at 1000 and its eccentricity at $16\frac{7}{8}$, when this equation is at its greatest it is given by the theory of gravity as 11′ 49″. But the Earth's eccentricity seems to be a little greater [than that], and when the eccentricity is increased this equation must be increased in the same proportion. Let the eccentricity be put at $16\frac{15}{16}$ and the maximum equation will be 11′ 52″.

I also found that because of the Sun's force the apogee and nodes of the Moon move more swiftly at the Earth's perihelion than they do at its aphelion, inversely as the cube of the distance of the Earth from the Sun. And this is the cause of the equations of these motions that are proportional to the equation of the Sun's centre. However, the Sun's motion is as the square of its distance from the Earth inversely, and the maximum equation of the centre generated by this inequality is 1° 56′ 26″, agreeing with the aforesaid solar eccentricity of $16\frac{15}{16}$. Because if the Sun's motion were as the cube of the distance, this inequality would generate a maximum equation of 2° 56′ 9″. And for this reason the maximal equations which the inequality of the motions of the lunar apogee and nodes generates are to 2° 56′ 9″ as the mean diurnal motion of the apogee and the mean diurnal motion of the nodes of the Moon are to the mean diurnal motion of the Sun. Whence it comes about that the greatest equation of the mean motion of the apogee is 19′ 52″ and the greatest equation of the mean motion of the nodes is 9′ 27″.

It is also established by the theory of gravity that the Sun's action on the Moon is a little greater when the transverse diameter of the lunar orbit passes through the Sun, than when it is at right angles to the line joining the Earth and the Sun, and for that reason the lunar orbit is a little smaller in the former case than in the latter. And hence arises another equation in the mean motion of the Moon depending on the position of the Moon's apogee in relation to the Sun, which is greatest when the Moon's apogee is come round to the octants of the Sun, and zero when it comes to the quadratures or the syzygies; and this is added to the mean motion in the passage of the Moon's apogee from quadrature with the Sun to syzygy and subtracted in the passage from syzygy to quadrature. This equation (which I call semiannual) amounts to about 3′ 45″ in the octants of the apogee when it is greatest, so far as I could gather from the phenomena. This is its value at the Sun's mean distance from the Earth. It is increased or diminished as the cube of the Sun's distance inversely and so is about 3′ 34″ at the Sun's greatest distance and about 3′ 56″ at his least distance. When the lunar apogee is placed away from the octants it works out less, and is to the maximum equation as the doubled sine of the distance of the lunar apogee from the nearest syzygy or quadrature, to the radius.

By the same theory of gravity the Sun's action on the Moon is a little greater when a straight line drawn through the lunar nodes passes through the Sun, then when such a line is at right angles to a line joining the Earth and the Sun. And hence arises another equation in the mean motion (which I shall call the second semiannual); this is greatest when the nodes come round to the octants of the Sun, and vanishes when they are at the syzygies and quadratures; and in other positions of the nodes it is proportional to the doubled sine of the distance of either node from the nearest syzygy or quadrature. This is to be added to the mean motion of the Moon when the nodes are passing from the solar syzygies to the nearest quadratures, and subtracted in their passage from the quadratures to the syzygies, and in the octants where it is greatest it amounts to 47″ at the Sun's mean distance from the Earth, as I deduce from the theory of gravity. At other distances of the Sun this equation in the octants of the nodes is reciprocally as the cube of the Sun's distance from the Earth, and so amounts to 45″ at the solar perigee and 49″ at his apogee, roughly.

By the same theory of gravity the lunar apogee progresses most swiftly when it is in conjunction with or opposition to the Sun, and regresses most when in quadrature with the Sun. And the eccentricity becomes greatest in the former case and least in the second, by Book I, Proposition 56, Corollaries 7, 8, 9. And, from these same corollaries, these equations are very great and generate the chief equation of the apogee, which I shall call semiannual. And the greatest semiannual equation is about 12° 18′ so far as I could ascertain from the observations. Our countryman Horrox first affirmed that the Moon revolves in an ellipse round the Earth placed at its lower focus.[2] Halley placed the upper focus of the ellipse on an epicycle, whose centre revolves uniformly about the Earth.[3] And the aforementioned inequalities in the progression and recession of the apogee arise from the motion in the epicycle. Let the mean distance between the Earth and the Moon be supposed to be divided into 100000 parts, and *T* representing the Earth let *TC* represent the mean eccentricity of the Moon, of 5505 parts. Let *TC* be produced to *B* so that *BC* may be the sine of the greatest semiannual equation 12° 18′

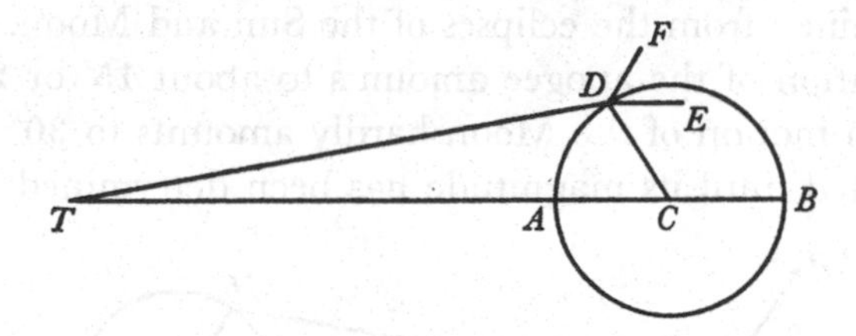

(*TC* being the radius), and the circle *BDA* (described about centre *C* with radius *CB*) will be that epicycle upon which the upper focus of the ellipse is placed and which revolves in the order of the letters *BDA*.[4] Let the angle *BCD* be taken equal to twice the annual argument, and the angle *CTD* will be the semiannual equation of the apogee, and *TD* the eccentricity first equated tending towards the apogee twice equated.

Because of the greater force of the Sun when the Earth is at perihelion, the apogee[5] of the Moon is moved more quickly in the epicycle about centre [*C*][6] than at aphelion, in the ratio of the cube of the distance from Earth to Sun inversely. Because the equation of the centre of the Sun is included in the annual argument, the Moon's apogee will be moved more quickly on the epicycle in the squared ratio of the distance between Earth and Sun inversely. So that the same may be moved more quickly still in the simple ratio of the distance inversely, let *TD* be the eccentricity equated once, and let *TD* be produced to *E* so that *DE* is to *TD* as twice the Sun's eccentricity to the radius of the

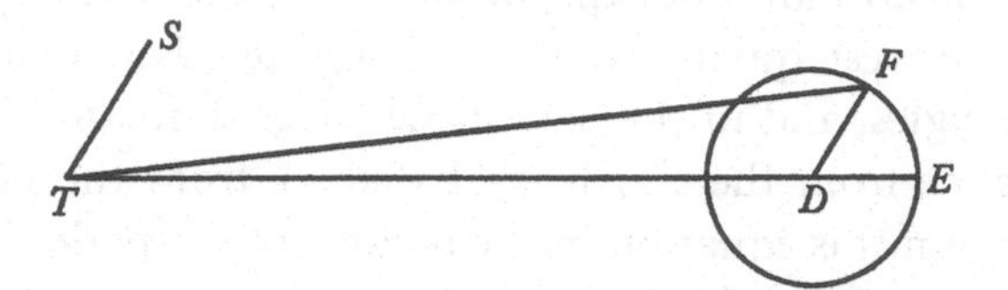

Earth's orbit, that is as $33\frac{7}{8}$ to 1000. Let the angle *EDF* be taken equal to the annual argument or (what is the same thing) let the parallels *TS* and *DF* be drawn towards the Sun, and let *DF* be equal to *DE*, and *DTF* will be the annual equation of the apogee of the Moon and *FTS* the distance of the Sun from the Moon's apogee thrice equated, and *TF* the eccentricity of the Moon twice equated tending towards the apogee of the Moon thrice equated.

By the same theory of gravity the Sun acts annually with a greater force upon the Moon when the apogee of the Moon and the perigee of the Sun are in conjunction, rather than when they are opposed. And hence arise two periodic equations, one for the Moon's mean motion, the other for its apogee; these equations are zero when the Moon's apogee is in conjunction with the Sun's perigee or when the two are in opposition, and greatest in the quadratures of the apogees. In other positions of the apogees they have a definite proportion to each other, and they are as the sine of the distance between the apogees. The first equation is subtracted and the latter added when the Moon's apogee is less removed from the solar perigee in consequence than 180°; the former is added and the latter subtracted when that distance is greater. The maximum values of these equations must be determined from the eclipses of the Sun and Moon. And so far as I can tell the greatest equation of the apogee amounts to about 15′ or 20′, but the greatest equation of the mean motion of the Moon hardly amounts to 30″ and because it is so small may be neglected until its magnitude has been determined by observation. Let

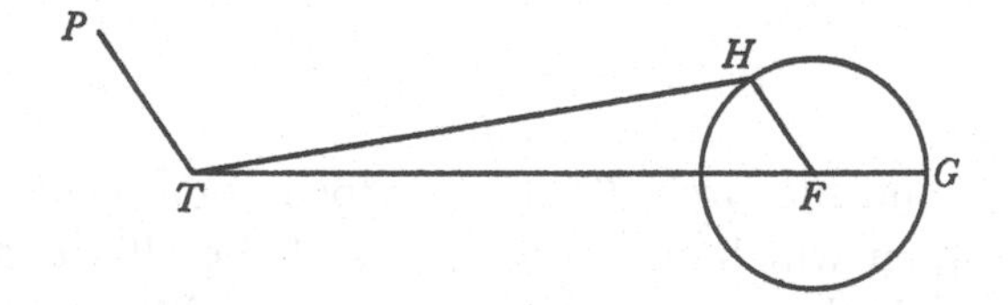

TF, the lunar eccentricity twice equated, be produced to *G*, so that *FG* may be the sine of the greatest periodic equation of the apogee of the Moon, 15′ or 20′, the radius being *TF*. Let two parallels, *TP*, *FH*, be drawn in the direction of the solar perigee, or what is the same thing let the angle *GFH* be formed equal to the distance between the perigee of the Sun and the apogee of the Moon, and let *FH* be equal to *FG*, and the angle *FTH*

will be the periodic equation of the apogee of the Moon, and the angle *PTH* the distance of the Moon's apogee four times equated from the Sun's perigee, and *TH* the eccentricity three times equated tending to the apogee four times equated.

If the three angles *CTD*, *DTF* and *FTH* are computed for every degree of the angles *BCD*, *EDF* and *GFH*, and arranged in tables; and if also the logarithms of the distances *TD*, *TF* and *TH* are likewise computed to the radii *TC*, *TD* and *TF* divided into 100000 parts and set out in tables, the aggregate of the three angles respecting their signs plus and minus will be the whole equation of the apogee, and the aggregate of the three logarithms will be the logarithm of the true eccentricity.

But when you have obtained the final equations of the mean motion of the Moon, the apogee and the eccentricity, and also the transverse diameter of the orbit of 200000 parts, from these the true place of the Moon in the orbit may be worked out and its distance from the Earth, using very well known methods. Then by means of the variation and reduction to ecliptic coordinates the true longitude and latitude will be given.

We[7] have remarked that the lunar orbit is alternately dilated and contracted by the [varying] force of the Sun, and that certain equations of the lunar motion arise from this. Thence also arises a certain variation of the Moon's parallax, but I judge it to be imperceptible; and so I always adopt in computations of the Moon's motions the number 100000 for its mean distance from the centre of the Earth, and 200000 for the transverse diameter of the orbit, and in order to investigate the parallax I suppose the mean distance in the octants between the centres of the Earth and the Moon to be equal to $60\frac{2}{9}$ of the largest Earth-radii. I call the largest Earth-radius that drawn from the centre to the equator, the least radius that from the centre to the poles. And hence it is that the Moon's mean apparent horizontal parallax in the octants is 57′ 5″, in the syzygies 57′ 30″, in the quadratures 56′ 40″. [The values] 31′ 30″ and 31′ 3″ may be adopted for the mean apparent diameter of the Moon in the Syzygies and the quadratures and 32′ 12″ for the mean polar diameter.

And[8] because the atmosphere of the Earth [extending] to a height of 35 or 40 miles refracts the light of the Sun and by refracting it scatters the light into the Earth's shadow, and by scattering the light at the bounds of the shadow broadens the shadow, I add in lunar eclipses one minute of arc, or 1′ 2″, to the diameter of the shadow as yielded by the parallax.

The theory of the Moon ought really to be checked and confirmed from the phenomena, first at the syzygies, then at the quadratures, and finally at the octants. And he who will embark on this task will not inappropriately employ the following mean motions of the Sun and Moon, [as valid] for noon, at the Royal Observatory at Greenwich, for December 31st 1700 o.s.: mean motion of the Sun, Capricorn 20° 43′ 50″ and his apogee Cancer 7° 44′ 30″; mean motion of the Moon, Aquarius 15° 19′ 50″ and her apogee Pisces 8° 18′ 20″, and the ascending node Leo 27° 24′ 20″.

NOTES

(1) MS. R.16.38, fos. 169–71, summarized by Edleston in *Correspondence*, pp. 110–12. Edleston argued convincingly that this Scholium, replacing a much shorter one in the first

edition, was no recent afterthought of Newton's, but that its loose sheets were inserted in one of Newton's interleaved copies of the first edition, and so sent to Cotes as part of the 'copy' for the second. (However, none of the textual material in MS. R.16.38 was taken into account in Koyré and Cohen, *Principia*.) The draft scholium was considerably altered by both Newton and Cotes; we have sought to print Newton's final version. Its first five and last two paragraphs were printed essentially as they are found here, though with a few important alterations (see *Principia* 1713; pp. 421–5; Koyré and Cohen, *Principia*, II, pp. 658–64). The remaining paragraphs were substantially revised in Newton's second draft (p. 287, note (7)).

(2) The 'Keplerian' theory of the Moon devised by Jeremiah Horrox (or Horrocks) was first published in his *Opera Posthuma*, London, 1672 (see Hall and Hall, *Correspondence of Henry Oldenburg*, VIII and IX), the parameters having been computed—on rather inadequate data, as he soon afterwards recognized himself—by John Flamsteed. (Compare Flamsteed to Newton, Letter 474; vol. IV, pp. 26–8.)

Horrox's theory was *not* a simple elliptical theory, for he had to account for the second inequality (evection) in the Moon's motion, discovered by Ptolemy, and the variation in the slow ($6_{+}'$ per day) revolution of the apogee of the Moon's orbit which he had discovered himself. Horrox understood that the maxima and minima of the apogee's advance (to the extent of about 25°) occurred in the solar octants, and that this was linked with the variation in the distance of the apogee from the Earth, that is with the variation in eccentricity of the Moon's orbit. Accordingly, Horrox (supposing the Earth at *T*, one focus of the lunar ellipse) did not place the empty focus at *C* (*CT* being the mean eccentricity of the Moon's orbit), but placed it at *D*, the radius *DC* being such that it subtends at *T* a maximum angle of about $12\frac{1}{2}$°. If the

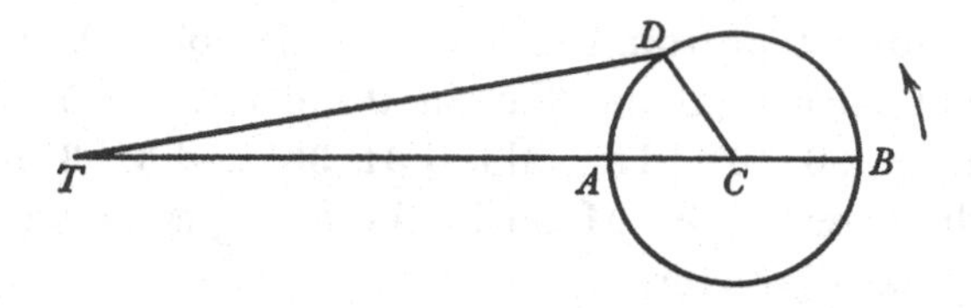

empty focus *D* completes a revolution of the circle *BDA*, centre *C*, twice in a year, there will be two maxima, two minima, and four mean motions of the lunar apogee in a year, as is required. Besides effecting the variation of the motion of the lunar apogee, this device will also cause the eccentricity *TD* to vary between the lengths *AT* and *BT*, causing the apogee and perigee of the orbit to advance towards and recede from the Earth along the line of apses (*TD* protracted each way) in the manner required by observation.

It will be evident that despite the confusingly-placed reference to Halley that follows immediately, Newton goes on in the remainder of the paragraph to describe (with his own numbers) Horrox's device for the evection.

(3) The third inequality of the Moon, known as the variation, which Halley's epicycle was designed to represent, was discovered by Tycho Brahe. This reference to Halley and Cotes' later allusion to his 'little treatise concerning the lunar theory', present an interesting puzzle. No publication of Halley's in the *Philosophical Transactions* seems to contain anything of the kind, and no modification to Horrox's theory is mentioned by biographers or bibliographers. It is not to be found with the lunar observations by Halley printed in the third edition of Thomas Streete's *Astronomia Carolina* (London, 1716). The mystery seems to be lightened by a

few pages devoted to lunar theory by Halley in the appendix to his *Catalogus stellarum australium* ... *Accedit Appendicula de rebus quibusdam astronomicis, notatu non indignis* (Thomas James impensis R. Harford, Cornhill, London, 1679). On p. 12 of this pamphlet Halley wrote:

Hujus vero genuinam ni fallor, Theoriam excogitavi, quæ cum *Horroxiana* juncta, Lunæ triplices inæqualitates, a pluribus separatim computatas, in unam solam resolvit, quæque distantiis Lunæ a Clarissimo *Cassino* constitutis, ubique intra observandi certitudinem consentit, quæque forsan defectus omnes calculi fœliciter supplere possit. Supponimus Lunæ orbitam in linea syzygiarum comprimi introrsum versus Terram, nonagesimam circiter partem mediæ distantiæ, ac simul tantundem exprimi in linea Quadraturarum, ita ut totius orbitæ compressæ Area, æqualis sit Areæ orbitæ, juxta Hypothesin *Horroxii* constitutæ, hæc orbita compressa Ellipsis est, Terra vero non occupat alterum illius focum, ut in Planetis primariis; ad ejus tamen centrum, Luna, dum in orbita sic compressa circumfertur, æquales areas æqualibus temporibus circumscribit: quod *Kepleri* præclarissimum inventum, ... nec dubito quin in Lunæ motibus non inutiliter adhibendum erit.

(If I am not mistaken, I have worked out the true theory of this which, when it is combined with that of Horrox, resolves the three lunar inequalities which many have computed separately into one single inequality, and which everywhere agrees (within the limits of observational accuracy) with the lunar distances established by the famous Cassini, and which may happily compensate for all the defects of calculation, perhaps. We suppose the Moon's orbit to be compressed inwards towards the Earth along the line of the syzygies [major axis] by about one ninetieth part of the mean distance, and likewise to be expanded by the same amount along the line of the quadratures [minor axis], so that the area of the whole compressed orbit may be equal to the area of the orbit which is defined by Horrox's hypothesis. This compressed orbit is an ellipse but the Earth is not at either of its foci, as it is [*recte*, the Sun is] in the case of the primary planets. However, the Moon while carried round in its compressed orbit, will describe equal areas in equal times, which excellent invention of Kepler's... we do not doubt to be usefully employed in the lunar motions.)

Though Halley does not specifically say so, the inequality he proposes to represent by making the ellipse less eccentric (while leaving the Earth in the same place, and its area constant) is another semi-annual inequality. Therefore it cannot be handled by simply redefining the lunar ellipse; but, as Newton says, it can be handled by introducing another epicycle still, in order to give the empty focus, or as Newton more correctly made it later in the revised version of this scholium the centre of the ellipse, the required motion to deal with this inequality; this indeed was what Tycho had done—in a different system—long before.

Halley spoke of his lunar theory to the Royal Society (minutes of 2, 9 and 23 November, and 7 December 1692; MacPike, pp. 230–1) but never, apparently, set it down in any paper in finished form; or at any rate there is no such paper in the archives of the Royal Society, and we have already printed Flamsteed's sceptical comments (vol. IV, pp. 26, 36–7). It would seem that Newton is here trying to do justice to Halley as the first to add an (arbitrary) geometrical version of the variation to Horrox's theory—Horrox himself having dealt with this in a quite different way, and non-geometrically.

In his own lunar theory Newton did not, in fact, follow Halley's model combining the evection and the variation. Instead, in Proposition 29 of Book III of the *Principia* he discussed the evaluation of the variation (33′ 14″ to 37′ 11″) without employing any geometrical model. (In practice it would of course be tabulated, like the previous inequalities.)

However, in the next paragraph of the scholium (beginning, 'Because of the greater force...') Newton does adopt, for the second of the new inequalities discovered by himself, an epicyclic model very similar to that he attributed to Halley. The appropriate figure here is a modification of that just given:

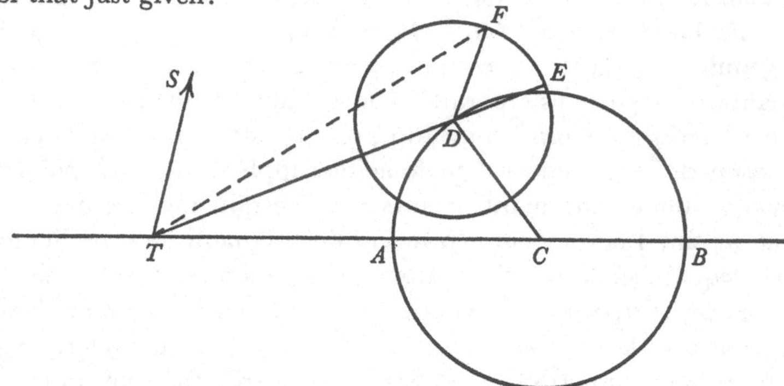

Though Newton's description does not make this very clear, the point F is now the empty focus (later, centre) of the Moon's ellipse; it rotates semiannually about the point D. The radius DF is very small; the mean Earth–Moon distance being 100000 parts, TC (the Moon's mean eccentricity) is 5505 parts, CD is 1173 parts and DF only 35 parts; the second equation of the Moon's centre (Newton's term) thus expressed amounts to only some $2\frac{1}{2}'$—a very different magnitude from the variation dealt with by Halley's epicycle. All this was made clearer in the revision of the scholium provoked by Cotes' criticisms.

The lunar theory of Halley is mentioned by David Gregory in *Astronomiæ Physicæ et Geometricæ Elementa*, Oxford, 1702, p. 331 and in the English version *The Elements of Astronomy, Physical and Geometrical*, London, 1715, II, p. 559; and by J. J. Lefrançais de Lalande, *Astronomie*, Paris, 1771, II, pp. 207–8, 218–19.

(4) In the revised draft of this scholium, under pressure from Cotes, Newton altered '*superior umbilicus*' to '*centrum*'.

(5) Presumably '*superiorem umbilicum*' should have been written.

(6) 'D' is a slip.

(7) Cotes suggested the omission of this paragraph in Letter 944 and Newton agreed.

(8) The two final paragraphs are as printed, save for changes in the numerical values proposed by Cotes in the same letter.

918 NEWTON TO THE EDITOR OF THE *MEMOIRS OF LITERATURE*

[after 5 MAY 1712]

From the draft in the University Library, Cambridge[(1)]

Sr

In your weekly paper[(2)] dated May 5 1712 I meet wth two Letters, one written by Mr Leibnitz to Mr Hartsoeker the other by Mr Hartsoeker to Mr Leibnitz in answer to ye former.[(3)] And in the Letter of Mr Leibnitz I meet wth some things reflecting upon the English I hope you will do them the justice to publish this vindication as you have printed the reflexion. He writes thus. 'It may be said in a very good sense that every thing is a continual Miracle, that is worthy of Admiration, but it seems to me that the example of a Planet wch goes round & preserves it[s] motion in its Orb without any other help but that of God, being compared wth a Planet kept in its Orb by yt matter wch constantly drives it toward ye Sun, plainly shews what difference there is between natural & rational miracles & those that are properly so called or supernatural; or rather between a reasonable explication, & a fiction invented to support an ill grounded opinion. Such is the method of those who say, after Mr de Robervals Aristarchus,(4) that all bodies attract one another by a law of nature wch God made in the beginning of things. For alledging nothing els to obtein such an effect & admitting nothing that was made by God whereby it may appear how he attains to that end, they have recourse to a miracle, that is, to a supernatural thing, wch continues for ever, when the Question is to find out a natural cause.'(5) Thus far Mr Leibnits. I know not what just occasion there was for this reflexion in a discourse forreign to this matter but its plain this was intended against some in England & I hope to make it as plain that it was undeserved. For The true state of the case is this. It has been proved by some that all bodies upon the surface of the earth gravitate towards the earth in proportion to ye quantity of matter in each of them: That the Moon tends towards the earth & all the Planets towards one another by the same law; & that by this tendency all their motions are performed. These things have been proved by mathematical demonstrations grounded upon experiments & the phænomena of nature: & Mr Leibnitz himself cannot deny that they have been proved. But he objects that because *they alledge nothing else to obteine such an effect* [he means a tendency of all bodies towards one another] *besides a law of nature wch God made in the beginning of things & admitt nothing that was made by God* (he means no vortices) *whereby it may appear how God attains to that end, they have recourse to a Miracle, & that is, to a supernatural thing wch continues for ever, when the question is to find out a natural cause.* Because they do not explain gravity by a mechanical hypothesis, he charges them wth making it a supernatural thing, a miracle & a fiction invented to support an ill grounded opinion & compares their method of philosophy to that of Mr de Robervals Aristarchus, wch is all one as to call it Romantic. They shew that there is an universal gravity & that all the phænomena of the heavens are the effect of it & with ye cause of gravity they meddle not but leave it to be found out by them that can explain it whether mechanically or otherwise. And doth this deserve to be scouted with the language of a supernatural thing, a miracle, a fiction invented to support an ill grounded opinion, & a method of philosophy after Mr Robervals Romance.

But Mr Leibnitz goes on. 'The Ancients & the Moderns who own that gravity is an occult Quality, are in the right, if they mean by it that there is a certain Mechanism unknown to them whereby all bodies tend towards the center of the earth. But if they mean that the thing is performed without any mechanism by a simple primitive quality or by a law of God who produces that effect without using any intelligible means it is an unreasonable & occult Quality, & so very occult that it is impossible that it should ever be done tho an Angel or God himself should undertake to explain it' The same ought to be said of hardness. So then gravity & hardness must go for unreasonable occult qualitys unless they can be explained mechanically.(6) And why may not the same be said of the vis inertiæ & the extension the duration & mobility of bodies, & yet no man ever attempted to explain these qualities mechanically, or took them for miracles or supernatural things or fictions or occult qualities. They are the natural real reasonable manifest qualities of all bodies seated in them by the will of God from the beginning of the creation & perfectly uncapable of being explained mechanically, & so may be the hardness of primitive particles of bodies. And therefore if any man should say that bodies attract one another by a power whose cause is unknown to us or by a power seated in the frame of nature by the will of God, or by a power seated in a substance in wch bodies move & flote without resistance & wch has therefore no vis inertiæ but acts by other laws then those that are mechanical:(7) I know not why he should be said to introduce miracles & occult qualities & fictions into ye world. For Mr Leibnitz himself will scarce say that thinking is mechanical as it must be if to explain it otherwise be to make a miracle an occult quality and a fiction.

But he goes on & tells us that God *could not create Planets that* should move round of themselves without any cause that should prevent their removing through the tangent. For a Miracle at least must keep *the Planet in.* But certainly God could create Planets that should move round of themselves without any other cause then gravity that should prevent their removing through ye tangent. For gravity without a miracle may keep the Planets in. And to understand this wthout knowing the cause of gravity, is as good a progress in philosophy as to understand the frame of a clock & the dependance of ye wheels upon one another without knowing the cause of the gravity of the weight wch moves the machine is in the philosophy of clockwork, or the understanding the frame of the bones & muscles & their connection in the body of an animal & how the bones are moved by the contracting or dilating of the muscles without knowing how the muscles are contracted or dilated by the power of ye mind, is [in] the philosophy of animal motion.(8)

NOTES

(1) Add. 3968(17), fo. 257. Sir David Brewster wrote 'May 1712' in the top right-hand corner of the recto. However, there is no reason to suppose that Newton reacted instantly against the publication; it was first mentioned to him by Cotes only in Letter 985 (dated

18 [March] 1712/13). We have placed the draft here merely for convenience. It is fairly heavily altered and rewritten, as one would expect.

(2) *Memoirs of Literature, containing an account of the state of learning at home and abroad, for the year 1712,* II, no. 18, pp. 137–43; the issue is dated 5 May 1712. See Brewster, *Memoirs,* II, pp. 282–84.

(3) Leibniz's letter is dated Hanover, 10 February 1711 (though Gerhardt, doubtless from the draft, printed 6 February); Hartsoeker's is dated Düsseldorf, 13 March 1711. They were first published, through the intervention of Leibniz's friend the Jesuit Desbosses of Cologne, in the *Mémoires pour l'Histoire des Sciences et des Beaux Arts,* for March 1712 (Trevoux, 1712), Art. XL, pp. 494–523. They were reprinted later in the *Journal des Sçavans* for December, 1712, pp. 603–25 in the Amsterdam edition. They are also included in the complete correspondence of Leibniz and Hartsoeker published by C. I. Gerhardt in Band III of his *Philosophische Schriften* Berlin, 1887. The allusion to the English school of philosophers (Leibniz mentioned neither names nor nationality) is purely incidental to the questions at issue between Leibniz and Hartsoeker.

(4) Leibniz's phrases are (Gerhardt, *ibid.*, p. 518): 'C'est ainsi que font ceux qui disent, apres l'Aristarque du feu M. de Roberval, que c'est une loye de la nature que Dieu a donnée en creant les choses, que tous les corps doivent s'attirer les uns les autres. Car n'alleguant rien que cela pour obtenir un tel effect, et n'admettant rien que Dieu ait fait qui puisse montrer comment il obtient ce but, ils recourent au miracle, c'est à dire, au surnatural, et à un surnaturel tousjours continué, quand il s'agit de trouver une cause naturelle.' It is perhaps worth commenting that Hartsoeker was fully as opposed to Newtonian mechanics as was Leibniz, remarking in his reply: 'Mais je pourrois avec raison me mocquer d'un tel [que Newton], comme je me mocquerais d'un homme qui voudroit passer pour Architecte, et qui cependant ne pourroit faire aucun batiment, quoiqu'il eust toutes sortes de bons materiaux propres pour cela.' (*ibid.*, p. 524).

Roberval's book, *Aristarchi Samii de mundi systemate partibus et motibus*, was published at Paris in 1644; he assigned a mutual attraction to all particles of matter.

(5) This line of criticism was, of course, developed by Leibniz in his much more widely read correspondence with Samuel Clarke.

(6) This riposte was not really available to Newton, since it is tantamount to making gravity an original or inherent property of matter, a view of gravity he had scornfully rejected in his (still unpublished) letters to Richard Bentley. Whereas Newton was quite content to postulate hardness as an intrinsic property of atoms or primordial particles, and his doing so seemed quite natural to contemporaries, similarly to postulate gravity as an inherent property of matter would have been (in his eyes and theirs) to have moved far towards the so-called atheistical position of attributing (like Epicurus) a *complete* sufficiency of properties to matter, thus excluding the need for the exercise of a creative power. Hence Newton's attempt in this passage to make the difficulty of explaining hardness logically equivalent to that of explaining gravity, is specious and certainly unrepresentative of his own deep personal convictions. Perhaps a realization of this contributed to his dissatisfaction with this draft.

(7) The words from 'move' to the semicolon are an interlineated addition. This is an early appearance of Newton's use of a non-Cartesian æther, here invoked as a mere conjecture. It is worth remarking on the fact that this concept is introduced as a way of avoiding Leibniz's criticism, and that this Newtonian æther does not act mechanically. It provides a *cause* of gravity, but not the quasi-Cartesian mechanical cause demanded by Leibniz.

(8) There are curious anticipations here of the General Scholium concluding the *Principia* that Newton was to write later (or, indeed, may have been writing about the same time if we should approximate the composition of the present draft to Cotes's Letter 985).

919 NEWTON TO COTES

10 MAY 1712

From the original in Trinity College Library, Cambridge.(1)
Reply to Letters 914, 916 and 917; for the answer see Letter 920

Prop. De Variatione Lunæ p. 402.(2)

Sr

I have received three letters(3) from you since my last. And the corrections wch you send me in the two first of them may all stand. In the second of them dated May 1st, you cite my words.(4) In alijs distantijs Solis a Terra Variatio maxima est in ratione quæ componitur ex duplicata ratione [temporis] revolutionis sy[n]odicæ Lunaris (dato anni tempore[)] directe, et ratione anguli *CTa* directe, et triplicata ratione distantiæ Solis a Terra inverse. Ideoque in Apogæo Solis Variatio maxima est 33′. 11″ et in ejus Perigæo 37′ 24″ si modo excentricitas Solis sit ad Orbis magni semidiametrum transversam ut $16\frac{15}{16}$ ad 1000. Here 33 11 & 37 24 may be diminished by 2″ & the word temporis may be inserted where you see it wthin the brackets. The Variatio maxima is composed of the ratios of the time, the angle *CTa*, & the sun's force, as above; because if any one of the three ratios be enlarged whilst the rest remain given, the variation will be enlarged. If the time alone be enlarged the Variation will be enlarged in a duplicate proportion, as may be gathered from the descent of falling bodies in a greater or less time. If the angle be enlarged, the Variation wch is a proportional part of ye angle will be inlarged in the same simple proportion, & the force also wch is reciprocally as the cube of ye Suns distance enlarges the Variation in proportion to it self.(5)

In pag 455 write.(6) Idem per Tabulas Astronomicas est 19. 21. 20. 45. Differentia minor est parte fere quadringentesima motus totius, et ab Orbis &c.

Pag 456 lin 13 write 9 gr. 11′. 3″(7) & lin 28 in Quadraturis autem regrediuntur motu horario 16″ 19‴. 51^{iv}.(8) I compute it thus. As *AB* to $AD+AB$ so is the mean horary motion of the Node to 16″. 19‴. 51^{iv}. I am

Sr

Your most humble Servant

Is. Newton

London
10*th May*
1712

At the bottom of pag 461 you may put the numbers 5gr. 17′. 20″ & 4gr 59′ 35″(9)

Pag 456 lin 1 instead of $38\frac{1}{3}$ write $38\frac{3}{10}$.(10)

The Lunar systeme must be altered

NOTES

(1) R.16.38, no. 218; printed in Edleston, *Correspondence*, pp. 113–14.

(2) This heading is in Cotes' hand.

(3) Newton refers to Cotes' Letters 914, 916 and 917.

(4) See Letter 916, notes (4)–(8).

(5) None of the three statements of proportionality which Newton gives here is wrong; what *is* wrong is the combination of all three to give the maximum variation. See Letter 916, note (8).

(6) This sentence, near the end of Proposition 32 (see Koyré and Cohen, *Principia*, II, p. 646), appeared with the same wording in the first edition, but with the value 19°. 20′. 31″. 1‴. The value finally printed was 19°. 21′. 21″. 50‴. (See Letter 913, note (13).)

The sentence as it eventually appeared in the second edition had one further alteration. 'quadringentesima' was correctly replaced by 'trecentesima'. Cotes suggested this change in his next letter.

(7) See Letter 913, note (3).

(8) Again, it is the numerical value in this phrase (see Proposition 33, Corollary) which is in question. The value finally printed was 16″. 19‴. 26^{iv}, as suggested by Cotes in his next letter.

(9) See Letter 917, third paragraph.

(10) This was done.

920 COTES TO NEWTON

13 MAY 1712

From the original in the University Library, Cambridge.(1)
Reply to Letter 919; for the answer see Letter 922

Sr

I have received Your last, but I am not yet clear that the ratio of the angle *CTa* ought to be introduc'd in the XXIXth Proposition,(2) though I do fully understand the reasons You give for it. As I apprehend it, the duplicate ratio of the Synodical time does it self account for the dilatation of the Angle, & therefore it ought not to be again accounted for. According to the reasoning of the 16th Corollary of Prop: LXVI, Lib: 1, the Variatio Maxima which is the angular Error of the Moon whilst she describes the half of the arch *Cpa*, is as the Square of the time imploy'd in describing that half arch directly, & the Cube of the distance from the Sun inversly: Or as the Square of the Synodical time directly & the Cube of the distance inversly. Now I think the dilatation is accounted for by taking ye angular Error which arises in the time of describing half the arch *Cpa* instead of the Error which would arise in the time of describing half ye arch *CPA*.

The thing may be considered another way, which perhaps will give more light to the understanding of my difficulty. The true Variatio maxima 35′. 10″ arises from the Arch *Cpa*, but the Variatio maxima 32′. 32″ arises from the Arch *CPA*. Now this latter by the 16th Corol of Prop: LXVI Lib I, must be altered with ye Square of the Periodical time directly & the Cube of the distance inversly, & so it will be more correct: after it is thus corrected, the corrected true Variatio maxima will be deduc'd from it, by enlarging or dilating it in the proportion of the angle *CTa* to the angle *CTA*, or in the proportion of ye Synodical to the Periodical time. Therefore the corrected true Variatio maxima will be as the Square of the Periodical time directly, the Cube of the distance inversly, the Synodical time directly & the Periodical time inversly: that is, as the Periodical & Synodical times directly & the Cube of ye distance inversly. In this Latter way I scruple not to account for the dilatation, but in the former I think it is already accounted for by taking the Square of ye Synodical time instead of ye Square of ye Periodical.

If You find the Objection to be of any moment, I desire You to send me other numbers instead of 33′. 11″ & 37′. 24″. If You choose to let ye place stand as in Your letter, yet still there must be a further alteration of those numbers besides the diminution by 2″,(3) for ye Square of ye Synodical time compounded with the ratio of the Angle *CTa*, makes not the triplicate ratio of ye Synodical time alone (upon which those numbers were computed) but that triplicate ratio directly & the ratio of the Periodical time inversly, as I observ'd in my former Letter.(4)

In pag: 455 You direct me to write:(5) Idem per Tabulas Astronomicas est 19°. 21′. 20″. 45‴. Differentia minor est parte fere quadringentesima motus totius &c. I would choose to put it thus. Idem per Tabulas Astronomicas est 19°. 21′. 21″. 50‴. Differentia minor est parte trecentisima &c. For according to Flamsteeds Tables(6) the motion of the Nodes from the Fix't Starrs in 20 Yeares or 7305 Days is 1 rev. 0 sig. 27°. 6′. 53″ & therefore in 365d. 6h. 9m, it is 19°. 21′. 21″. 50‴.(7)

The Mean horary motion of ye Nodes by the same Tables is 7″. 56‴. 56^{iv}. and as AB to $AD+AB$ or as 373 to 766 so is 7″. 56‴. 56^{iv} to 16″. 19. 26^{iv}. Therefore in pag: 456, lin, 28 I would write 16″. 19‴.

26^{iv},[7] Unless You find other reason for writing 16″. 19‴. 51^{iv}, as You put it in Your Letter.[8]

I am
Sir
Your most humble Servant
ROGER COTES

Cambridge May 13. 1712

For Sr Isaac Newton
at his House
in St Martin's Street
in Leicester Fields
London

NOTES

(1) Add. 3983, no. 27, printed (from the draft) in Edleston, *Correspondence*, pp. 114–16.
(2) See Letter 916, note (8).
(3) See Letter 916, note (7).
(4) See Letter 916, note (8).
(5) See Letter 919, note (6). The change in the values used for the calculated and observed angles necessitated this change in the fraction representing their difference.
(6) These values are derived from Flamsteed's lunar tables printed at the end of his *The Doctrine of the Sphere*, London, 1680, pp. 85–104.
(7) This value was adopted, see Newton's Letter 922.
(8) See Letter 919.

921 COTES TO NEWTON

25 MAY 1712

From the original in Trinity College Library, Cambridge.[1]
For the answer see Letter 922

Trin: College May 25th 1712

Sr,

I have not yet received an answer to my last of May 13th concerning the XXIXth Proposition; I am therefore afraid it has miscarried.[2]

I sent You by Dr Bentley a small Treatise of my own concerning Logarithms, of which the Title is, *Elementa Logometriæ*[3] togather with the Figures belonging to it. I desire the favour of You to deliver 'em to Mr Livebody[4] to be cut in Wood & to give him Your directions if he meets with any difficulty. I fear You are at this time taken up with other business, otherwise I would beg of You to peruse the Treatise. You will find I am there proposing a new sort of

Constructions in Geometry which appear to me very easy, simple & general. But I am fearfull of relying on my own Judgment alone, which possibly in this matter may be too much byass'd. What I think to be right, may to others appear whimsical & of no use & I would not willingly give them the satisfaction of laughing at my Dreams. If You think I may venture to publish it, I shall be glad to know what may want to be corrected or altered either in the Matter or Expression. I have been forc'd to use some new Terms, as *Modulus*, *Ratio Modularis*, &c.[5] If others more proper occur to You upon reading the Papers, I shall be very willing to make any alteration. I hope You will pardon this Trouble I give You. I am Sir

Your most Obliged
& Humble Servant
ROGER COTES

For Sr Isaac Newton
at his House
in St Martin's Street
in Leicester Fields
London

NOTES

(1) R.16.38, no. 221; printed in Edleston, *Correspondence*, pp. 116–17.

(2) It had not miscarried; Newton replied to it on 27 May.

(3) This was published as an article, 'Logometria Auctore Rogero Cotes', in the *Phil. Trans.* **29** (1714), 5–45 and later formed the first part of Robert Smith's posthumous edition of Cotes' works, *Harmonia Mensuraram*, Cambridge, 1722, pp. 1–41. Newton acknowledged receipt of the manuscript on 27 May, and must have sent the figures to be cut, for in his Letter of 10 August Cotes thanked him for doing so and for returning the manuscript. Any letter Newton might have sent with the manuscript is now lost (see Letter 931, note (2)) but his comments cannot have been very severe, for on 10 August Cotes asked again for criticism. In his Letter 932, Newton mentioned that he had found no need for corrections.

(4) Livebody had undertaken similar work for Newton before; see Letter 766.

(5) Implicitly founding his discussion on the standard Napierian notion that, to within a factor of proportionality, a ratio is 'measurable' by the number (*logarithmus*) of 'unit' infinitesimal ratios which combine to produce it, Cotes in his 'Logometria' (*Phil. Trans.* **29** (1714), 6–10 was led geometrically to set his measure of a ratio $y/1$ as defined by $M.dy/y = d(\mathit{Mensura}\ y)$, where dy and d (*Mensura* y) are corresponding increments of y and its *mensura*. At once *Mensura* $y = M\int_1^y 1/y.dy$, that is (in more modern Eulerian terms) the natural logarithm $M.\log_e y$; whence in all Cotesian *systemata mensurarum* (systems of measures) M is the *Mensura* of the fixed *ratio modularis* (exponential base) $e = 2{\cdot}71828\ldots$ as Cotes was readily able to compute it from the basic definition (by inverting the series expansion of *Mensura* $(1-v)$ to derive that of $e^{(\mathit{Mensura}\ y)/M}$ and then setting $y = e = \mathit{Mensura}^{-1}\ M$). It is evident, as he further goes on to say, that *Mensura* y = 'Briggian' $\log_{10} y$, on taking the *modulus* M to be $\log_{10} e = 0{\cdot}43429\ldots$ Compare J. E. Hofmann, 'Weiterbildung der logarithmischen Reihe Mercators in England. III', *Deutsche Mathematik*, **5** (1940–1), 358–75, especially 368–71 (D. T. Whiteside).

922 NEWTON TO COTES

27 MAY 1712

From the original in Trinity College Library, Cambridge.[1]

Reply to Letters 920 and 921

Sr

I have reconsidered what you write about the Variation & agree to it.[2] You may leave out the words [et ratione anguli *CTa* directe] & instead of the numbers 33′ 11″ & 37′ 24″ diminished by 2″, write 33′ 14″ & 37′ 11″.[3] For so I found them upon computing them anew.

Also in pag 455 lin 14 you may write.[4] Idem per Tabulas Astronomicas est 19gr. 21′. 21″. 50‴. Differentia minor est parte tricentissima &c And pag 456 lin. 28 you may write 16″. 19‴. 26^{iv}.

I received your papers[5] by Dr Bently & have run my eye over them, I intend to read them over again & get the cuts done for you as soon as I can find out Mr Livebody. I am

Your most humble Servant

IS. NEWTON

London
May 27 1712

NOTES

(1) R.16.38, no. 222; printed in Edleston, *Correspondence*, pp. 117–18. The square brackets appear in the original.

(2) This is with reference to Proposition 29; Newton here reverses the stand he had taken on 10 May, see Letter 916, note (8).

(3) This was done, cf. *Principia* 1713; pp. 402 (bottom) and 403 (top).

(4) This is in Propositions 32 and 33; these numbers, suggested by Cotes in his Letter 920, appear in *Principia* 1713; pp. 414, 415.

(5) See Letter 921, note (3).

923 THE MINT TO OXFORD

16 JUNE 1712

From the original in the Mint Papers.[1]

For the answer see Letter 927

To the Right Honble: the Earl of Oxford &
Earl Mortimer Lord High Treasurer of Great Britain

May it please Your Lordp.

The Office of Gravers of the Mint being constituted by a Warrant under the Broad Seal dated April the 7th. in the fourth year of Her Majties. Reign, whereby for the Advantage & safety of the Mint with respect to the Coin, John

Croker Gent. is appointed head Graver with a Salary of 200£ per annum,(2) & Samuel Bull(3) & Gabriel le Clerk Gent.(4) were made Assistants to the Cheif Graver, Each with a Salary of 80£ per annum, & upon the voidance of either of the said Assistant's places the same was to be filled up with such Probationer or Apprentice as should be presented to her Majtie: by the Warden, Master & Worker & Comptroller of the Mint, by Warrant under her Royal Signe manual: We humbly represent to Your Lordp. that the place of the said Gabriel le Clerk became void about three Years ago, & thereupon one Francis Berresford(5) hath been learning about two Years & halfe under the said Mr Croker to designe in order to Learne the Art of Graveing & is approved by the said Mr Croker as well qualified in point of Genius & Industry to make a good Workman: For which end We humbly pray Your Lordp. that out of the said 80£ per annum an Allowance may be settled upon the said Francis Beresford for binding him an Apprentice to the said Mr Croker for six years. And we humbly propose an Allowance of 35£ per annum.

All which is most humbly submitted to your Lordps. great Wisdome

C. PEYTON
IS. NEWTON
E. PHELIPPS

Mint Office
16 *June* 1712

NOTES

(1) Mint Papers I, fo. 155r; the document is in a clerical hand signed by all three officers. Exactly the same letter was written and signed over the date 12 September 1711 (*ibid.*, 165).

(2) From 1690 a seal-cutter, Henry Harris, occupied the office of chief engraver to the Mint, though incompetent. In 1697 Johann Croker (1670–1741), a German who had come to England in 1691, was appointed to do his work in return for half Harris's salary. When Harris died in 1704 Newton appointed Croker chief engraver. See also vol. IV, p. 351, note (3).

(3) See vol. IV, p. 396, note (2). Bull was second engraver, that is, enjoying the right of reversion to Croker's place.

(4) Gabriel Le Clerk, who spent his life at various German courts and never worked at the Tower Mint, was actually paid from 1704 to March 1709. He may have been an intelligence agent.

(5) He was presumably a relative of the Comptroller's Deputy (see Letter 808, note (5)).

924 TAYLOUR TO NEWTON

16 JUNE 1712

From the copy in the Mint Papers.(1)
For the answer see Letter 925

Sr

My Lord Treasurer having received from Mr Southwell the inclosed Representation from the Lords of the Privy Council in Ireland desiring some alteration to be made in a late Order of Council, & the Proclamation intended thereby relating to the currency of forreign coynes in that Kingdom(2) I am commanded (in the absence of the Secretaries) to transmit to you the said Representation with the said Order of Council & other papers wch you are desired to peruse & Report to his Lordp assoon as conveniently you can what you think proper to be done wth respect to ye alteration proposed by the Lords of the Council of Ireland. I am

Sr
Your most humble Servant
J. TAYLOUR

Trea[su]ry Chambers
16 *June* 1712
To Sr Isaac Newton Knt
Master of the Mint

NOTES

(1) Mint Papers II, fo. 241 v; this is in Newton's hand. Another copy is P.R.O. T/27, 20, 214. For the content, compare Letters 897 and 900.

(2) The original letter signed by the Irish Privy Councillors was sent to Newton and retained by him; it is now Mint Papers, II, fo. 232–3. It is dated 7 June 1712. The matters in this letter upon which Newton was to give his opinion are made sufficiently clear in his reply.

925 NEWTON TO OXFORD

23 JUNE 1712

From the draft in the Mint Papers.(1)
Reply to Letter 924

To the Rt Honble the Earl of Oxford & Earl Mortimer
Lord H. Treasurer of great Britain

May it please your Lordp

In obedience to your Lordps Order of Reference signified to me by Mr Taylour in his Letter of June 16th Instant, I have perused the Representation from the Lords of the Privy Council of Ireland touching a late Order of Council here for giving currency in that Kingdom by Proclamation to some forreign Coynes wch were omitted in a former Proclamation, a printed copy of wch they have sent, desiring a clause to be added to the said Order for making such allowance for light pieces as was made in the said Proclamation, & that the Order may comprehend also the forreign coynes mentioned in that Proclamation because the original thereof under the great seal was destroyed in the late fire wch happened there at the Council Chamber, so that the Clerk of the Council cannot now certify that the printed copy agrees with the original verbatim as the late Act of Parliament requires for the conviction of counterfeiters of those coynes. And upon comparing the said Representation with the said late Order of Council & printed Proclamation, I humbly represent that the weight of the single Pistole & Lewis d'or being in the said Proclamation put 4dwt 8gr, the weight of the double Pistole & double Lewis d'or ought in proportion to be put in a new Proclamation 8 dwt 16 gr & that of the quadruple Pistole or double doubleon 17 dwt 8 gr. And that the Moyder of Portugal (wch as the Merchants bring them hither a little worn weigh one with another 6 dwt 21¾gr, & before wearing may be a quarter of a grain heavier or above) may be put in weight 6 dwt 22gr in the same Proclamation & valued at thirty shillings. For in Ireland where an English shilling passes for thirteen pence the Moyder of this weight is worth 29*s* 11½*d* reconning gold 22 carats fine at 3£ 19*s* 8¾*d* per ounce wch is the standard value, or it is worth 30*s* 00¾*d* recconing gold 22 carats fine at 4£ per ounce as is ordinarily done: and 30*s* is a medium & the nearest round number. And a grain being allowed for wearing, this piece will be current till it weighs but 6dwt 21 gr as was stated in the late Order of Council. And after that it will be current still by abating 2*d* per grain in its value for what it wants to the weight of 6dwt 22 gr. For the latter part of the printed Proclamation concerning the allowance for light pieces & concerning the scales & weights for weighing them, I am humbly of opinion should be continued in the next Proclamation.

I humbly beg leave to represent further to your Lordp that the weights & values of the silver coynes in the printed Proclamation would answer better to one another & to the coynes themselves if two pence were taken from the value of the Crusado & eighteen or twenty grains added to the weight of the Dollars. For the Crusado is recconed in Portugal to be the tenth part of the Moyder in value & the Moyder is worth 30*s* in Ireland as above, & yet the Crusado is valued in the Proclamation at 3*s*. 2*d*. Its weight before wearing is 11 dwt 4gr. In the Proclamation its weight is put 10dwt 20gr, & a Crusado of this weight is worth but three shillings.

Rix Dollars,(2) Cross Dollars & other Dollars are in the Proclamation put of the same weight & value with pieces of eight of Mexico & Sevil, Pillar pieces

& Lewises. And yet the Dollars are much coarser & heavier than the pieces of eight & Lewises & ought to be 18 or 20 grains heavier to be of the same value. Rix Dollars before wearing weighed about 18dwt & 6 8 or 10 grains & Cross Dollars 18dwt 1gr. That they may be worth 4*s.* 9*d* (wch is their value in the Proclamation) they should weigh at least 17dwt 18gr.

I am therefore humbly of opinion that the gold coines should be of the weight & fineness expressed in the Paper hereunto annexed, & the silver ones as in the printed Proclamation, unless for the reasons above mentioned it should be thought fit to take two pence from the value of the Crusados & add eighteen grains to the weight of the Dollars.

All which is most humbly submitted to your Lordps great wisdome

Isaac Newton

Mint Office
23 *June* 1712

The weight & value of forreign coins in Ireland

The piece commonly called the Spanish quadruple Pistole of Gold or double doubleon[3] weighing seventeen penny weight eight grains to pass at three pounds & fourteen shillings.

The piece commonly called the Spanish or French double Pistole of Gold or the Doubleon or double Lewis d'Or weighing eight penny weight sixteen grains to pass at one pound & seventeen shillings.

The piece commonly called the Spanish or French Pistole of Gold weighing four penny weight eight grains to pass at eighteen shillings & six pence.

The piece commonly called the Spanish or French half pistole of Gold weighing two penny weight four grains to pass at nine shillings & three pence.

The piece commonly called the Spanish or French quarter Pistole of Gold weighing One penny weight two grains to pass at four shillings & seven pence half penny.

The piece commonly called the Moyder of Portugal weighing six penny weight twenty two grains to pass at thirty shillings

The piece commonly called the half Moyder weighing three penny weight eleven grains to pass at fifteen shillings

The piece commonly called the quarter Moyder weighing one penny weight seventeen grains & an half to pass at seven shillings & six pence.

The piece of Silver commonly called the Ducatoone[4] weighing twenty penny weight sixteen grains to pass at six shillings.

The piece commonly called the half Ducatoone weighing ten penny weight eight grains to pass at three shillings.

The piece commonly called the quarter Ducatoone weighing five penny weight four grains to pass at one shilling six pence.

The piece commonly called the piece of eight of Mexico or Sevil, the Mexico called the pillar piece & the French Lewis each weighing seventeen penny weight & the Cross Dollar Rix Dollar & all other Dollars [weighing seventeen penny weight & twenty grains](5) to pass at four shillings & nine pence.

The piece commonly called the half piece of eight of Mexico or Sevil, Pillar piece & French Lewis weighing eight penny weight & twelve grains & the half of the Rix Dollar Cross Dollar & other Dollars(5) [weighing(6) to pass at two shillings & four pence half penny.

The piece commonly called the quarter piece of eight of Mexico or Sevil, Pillar piece & French Lewis weighing four penny weight six grains & the quarter of the Rix Dollar Cross Dollar & other Dollars weighing(6) to pass at one shillings & two pence farthing

The piece commonly called the old Peru weighing seventeen penny wt to pass at four shillings & six pence

The piece commonly called the half piece of eight of Peru weighing eight penny weight twelve grains to pass at two shillings three pence

The piece commonly called the quarter piece of eight of Peru weighing four penny weight six grains to pass at one shilling & one penny half penny.

The piece commonly called the Crusado of Portugal weighing ten penny weight twenty grains, to pass at three shillings &(6)

The piece commonly called the half Crusado of Portugal weighing five penny weight ten grains to pass at One shilling &(6)

NOTES

(1) Mint Papers II, fos. 242–3. There is a similar draft on fo. 234, and a much longer version on fos. 237–40. We have not found the official version of this report, though since it was printed in Shaw (pp. 179–81) one must exist perhaps in the Mint Papers. There is no discrepancy between the draft and Shaw's text, save that the latter lacks the definitions of values. The report is not minuted in the Treasury Books.

(2) Reich-thaler or Rijksdaaler.

(3) More usually rendered 'doubloon'.

(4) The 'small ducat', the ducat being a gold coin.

(5) Square brackets in original.

(6) Lacuna in original.

926 THE ROYAL SOCIETY TO FLAMSTEED

3 JULY 1712

From the original in the Royal Greenwich Observatory(1)

Sir,

The Committee of the Royal Society appointed for that purpose, finding You have not deliverd to them a Copy of Your last Years Observations, according to the Direction of Her Majesties most gracious Letter of the 12th of December 1710.(2) do hereby demand Your compliance therewith, and now the Six Months allowed You being Elapsed, let You know, that they expect You will send Your Observations to the Societies House in Crane-Court, according to the Meaning of the said Letter. We are

Your most humble Servts.
Is Newton P.R.S
Hans Sloane
R. Mead
Edm. Halley
Abr Hill

Crane Court
July. 3. 1712

NOTES

(1) Flamsteed MSS, vol. 35, p. 129. There is a copy of this letter in the Letter-Book of the Royal Society (xvi, 1712/13, p. 1), to which the following memorandum is added:

'This is an exact Copy of a Letter which I deliverd to Mr Flamsteed (July 4. 1712) at the Royal Observatory: who returned this Answer, that he would cause a Copy to be taken of his last year's Observations, and send them accordingly to the Royal Society by Michaelmas next.
J. Thorpe'

John Thorpe was Assistant Secretary. Apparently Flamsteed ignored this letter, for a more peremptory letter was sent on 5 March 1712/13.

(2) Letter 814.

927 OXFORD TO NEWTON

10 JULY 1712

From the original in the Mint Papers.(1)
Reply to Letter 923.

After my hearty Commendations, Having considered the aforegoing Report, & approving thereof, These are to Authorize & require You, out of the Moneys that are or shall be Imprested to You, at the Rect. of Her Mats.

Excheqr. for the Use and Service of the Mint, to pay unto John Croker, the Cheif Graver of the said Mint, or his Assignes, An Allowance after the rate of Thirty five pounds per Annum; for the Support & Maintenance of Francis Berresford, And for his Care in Training him up, in the Art and Mistery of Graving; the said Francis Berresford being already bound, or being meant to be bound an Apprentice to him Which Allowance is to Commence from Xmas last, and to continue to be paid to the said John Croker for the purpose aforesaid, for & during the full end and Terme of 6 Years, in case the said Frances Berresford do so long live, & continue under his Care, Maintenance, and Tuition, As an Apprentice. And for so doing this shall be as well to You for payment, as to her Mat.'s Auditors of the Imprests, and all others concern'd in passing and Allowing thereof upon Your Account, a Sufficient Warrant; provided nevertheless, and it is hereby meant & intended, that for and during the Said Terme of Six Years, or such part thereof as the Said Allowance of Thirty five pounds per annum shall Continue to be paid as afore said, the payment of Eighty pounds per Annum (which has for some time been saved to the Crowne) And was heretofore made to Gabriel Le Clerk, as an Assistant Graver, shall not be revived.

OXFORD

Whitehall Treasury Chambers
July the 10*th* 1712

To my very loving Friend Sr Isaac Newton Knt.
Master and Worker of Her Mats Mint. and to
the Mar. & Worker thereof for the time being.

NOTE

(1) Mint Papers I, fo. 155v. The warrant is written on the verso of Letter 16 June. There is a copy in Mint/1, 8, 96–7; and yet another in British Museum MS. Add. 18,757, no. 43.

928 THE MINT TO OXFORD

16 JULY 1712

From the original in the Public Record Office.[1]
Reply to Letter 885A

To the Right Honble: the Earle of Oxford and Earle
Mortimer, Lord High Treasurer of Great Britain

May it Please your Lordship

In obedience to your Lordship's Order of Reference of the 7th of January last upon the Annexed Petition of Mr James Clerk and Mr Joseph Cave conjunct Ingravers of her Majes. Mint in Scotland, craving an Allowance for making Puncheons for the Use of that Mint, Wee humbly represent to your Lordship that we have considered the same, and finding that they are only allowed a Sallary of 50£ per Annum between them for sinking and finishing of Dyes and have no Allowance for Puncheons, wee are humbly of Opinion they be allowed for their Work of this Kind after the following Rates Vizt

For the shilling head and Reverse fifteen pounds; for the six penny Head and Reverse Ten pounds (which were the Rates allowed to the Gravers of the Mint in the Tower for the like Puncheons made by them for the late recoinage of the Monies in Scotland), also for the four penny head and Reverse Eight pounds; for the three penny head and Reverse seven pounds; for the Two penny Head and Reverse six pounds; and for the penny Head and Reverse four pounds in all Fifty pounds.

As for the Extraordinary trouble of the Officers of that Mint during the re-coinage Wee humbly certifie to your Lordsp that wee have hitherto reported no Allowance, and that wee find that Mr Caves attendance was without Order & Voluntary

All which is most humbly submitted to your Lordships great Wisdom

CRAV: PEYTON
Mint Office the 16*th July* 1712 IS. NEWTON
E. PHELIPPS

NOTE

(1) T/1, 149, no. 45, a clerical copy signed by the Principal Officers.

929 COTES TO NEWTON
20 JULY 1712

From the original in Trinity College Library, Cambridge[1]

Cambridge July 20*th* 1712

Sr.

It is now about three Weeks since Dr Bentley return'd from London. He told me, You then intended to send down Your Emendations of the Lunar Theory very soon.[2] I have not received any thing from You since that time, & am therefore apprehensive of some miscarriage. He inform'd me, You had thoughts of adding something further upon the Subject of Comets[3] & besides a small Treatise concerning the Methods of Infinite Series & Fluxions.[4] I hope You will go on with Your design: it were better that the publication of Your Book should be deferr'd a little, than to have it depriv'd of those additions. I thank You for the Picture which I have received of him:[5] 'tis much better done than the former; but I could have wish'd it had been taken from the first of Mr Thornhill's.

I am Sir
Your most Humble Servant
ROGER COTES.

For Sr Isaac Newton at his House
in St Martin's Street in Leicester
Feilds London

NOTES

(1) R.16.38, no. 223, printed in Edleston, *Correspondence*, pp. 118–19.
(2) That is, to the Scholium following Proposition 35.
(3) This was done, see Letter 951.
(4) See Letter 914, note (17).
(5) It was quite normal for a great man to present copies of an engraved portrait of himself to friends. That in question here is probably the mezzotint engraving by John Smith (*c.* 1652–1742) after the Kneller portrait of 1702, dated 1712 (see J. Challenor Smith, *British Mezzotint Portraits*, 1883, III, p. 1204). The portrait by Sir James Thornhill to which Cotes refers is perhaps that of 1710 presented to Trinity College by Richard Bentley.

930 HERCULES SCOTT TO [NEWTON]

26 JULY 1712

From the copy in the Public Record Office[1]

Edinburgh July 26: 1712

Much Honoured Sr.

I am ashamed to give you this trouble butt the great Character I heard my Brother give of You and your Civilities to him when at London Encourages Me to Apply to You for your Advice and Opinion in the particular following.

Mr. Allardes who was Master of the Mint at Edenburgh and my Brother Patrick Scott who was his Depute[2] and did manage the great Coinage here after the Union having both dyed before my Brothers Accts. were fully made up And Mr Allardes Son and my Brothers Children, being under Age We are difficulted in adjusting the Acct. of Bullion and Waste My Brother at finishing the Coinage after the Sweep was three times searched over gave in an Abreviat thereof to Mr. Allardes Freinds whereby the Waste appears to be Two Hundred thirty four pound Weight, butt what occasions matter of Debate is that my Brother when he Dyed was so violently Oppressed with Sickness that he could not apply himself to clear any of his Affairs in the mean time had lying by him One Hundred forty pound weight of Bullion in rough Ingotts. Mr Allardes

Freinds demand that Bullion to be given up to them and that my Brothers Accts. of the Coinage should be made up thereafter without any regard to that Bullion[3] On the other hand My Brothers Freinds plead the Accts. be made up according to the Mint Books which agree with the Mentioned abreviat whereby the Waste seems very probable and Equal Whereas stating the Accts. as Allardes Freinds propose and delivering up the Bullion left by my Brother diminishes the Waste so much that in our Judgment seems to bear no proportion to such a great coinage considering also there was 12838 pd. 15dwt refined by Accts. given in to the publick for which Mr. Allardes is to be allowed £501 ; 02 ; 02$\frac{3}{4}$ Sterling. I know from several Notes amongst my Brothers papers that he bought up several parcels of Bullion nott Entered in the Mint Books which I presume he designed for further supply in case the Mint had again Opened for Coinage after Allardes death, and seems to be the Bullion on hand, To prevent any Loss my Brothers poor Children may sustaine I humbly intreat You would give your Opinion what Waste may reasonably be thought to arise from so great a Coinage and whether or nott considering the Quantity refined And how much the Work was hasted it be possible there should be so small Loss as Allardes Freinds pretend to Your condescendence to this trouble will be a singular Act of Kindness and a perpetual Obligation upon

Much Honoured

Your most Humble and Most Obedient Servt

HERCULES SCOTT

Please direct for me
Mercht in Edinburgh

Much Honoured

	lb	oz	dwt	gr
The whole Quantity of Bullion brought into the Mint as per Mint Books Stand[ar]d .	104634lb	: 09oz	: 16dwt	: 23gr
Bullion Accompted for by Mr Scots Abrevial & said Books	104400	: 08	: 05	: 21
Rem[ain]s: to Waste	234	: 01	: 11	: 02
Allardes freinds demand further without Acct	108	: 10	: 08	: 22
Will remaine then to Waste	125	: 03	: 02	: 04

What I humbly desire to be Informed is whether the Waste on such a Quantity of Silver brought into the Mint whereof 12838pd. was refined can be Judged to be brought so low as 125 lb or the first Acct. 234 lb be more reasonable. There was an entire freindship & Confidence twixt Mr. Allardes & my Brother wn alive, I should be sorry if either their Representatives should sustaine any apparent Loss & could think on no better way to determine me as to my doubts in this matter then to have your Opinion I shall Intreat your Answer with Your Conveniency I am Much Hond, as Above[4]

NOTES

(1) Mint/1, 8, fo. 173.

(2) Compare vol. IV, Letter 736.

(3) Patrick Scott was responsible to Allardes, and Allardes in turn to the government. The less waste entered in Allardes' accounts, when these were presented, the better for his heirs—but it was in the Scott family's interest to keep the bullion if they could.

(4) Dr John Francis Fauquier, Newton's deputy at the Mint, replied to this appeal; Scott's answer is Letter 965.

931 COTES TO NEWTON

10 AUGUST 1712

From the original in the University Library, Cambridge.[1]
For the answer see Letter 932

Trin: Coll: August 10*th* 1712

Sr

I thank You for Your care of the wooden Cuts, which I received of the Carrier togather with the Manuscript.[2] I am glad You are not displeased with it, & I wish You had signified what Emendations might be made in it.

In my Letter of May ye 3d, I mention'd some alterations in the former part of Your Lunar Theory. You have left me uncertain as to Your resolution about them, by taking no notice of them in Your last, in which Your correction of the latter part of the Theory is set down.[3]

I observe in the beginning of it, You have chang'd [et circulus *BDA* centro *C* intervallo *CB* descriptus, erit Epicyclus ille in quo superior Ellipseos umbilicus locatur] for [Epicyclus ille in quo centrum orbis Lunaris locatur]. I quæry whether [Halleius superiorem Ellipseos umbilicum in Epicyclo locavit] should not be also chang'd into [Halleius centrum Ellipseos] I have not Dr Halley's little Treatise by me concerning the Lunar Theory.[4]

I do not yet understand the Paragraph beginning with [In Perihelio Terræ, propter majorem vim Solis &c]. As I apprehend it, the angle *EDF* in Your new Figure should be equal to the excess of ye doubled annual Argument of the Apogee above the Suns mean Anomaly as I had suppos'd in my Letter of May the 3d. Your Rule concerning that Angle is this [Et capiatur angulus *EDF* æqualis excessui argumenti annui supra distantiam Aphelii Lunæ ab Aphelio Solis] I am uncertain how You understand the words [argumenti annui] they may signify either the Annual argument of ye Moon's Apogee, or the annual argument of ye Sun, ie. the Sun's mean Anomaly.(5) I am also uncertain about the words [Aphelii Lunæ ab Aphelio Solis] I suppose it should be wrote [Apogæi Lunæ ab Apogæo Solis](6) About the end of this Paragraph You say [Et concipe centrum orbis Lunæ...interea revolvi dum punctum *D* revolvitur circum centrum *C*] I do not perceive why it should be thus.(7)

The following Paragraph is rather more obscure to me. I find I cannot form any just conceptions of it, unless You will be pleas'd to give some further light to it. The Æquation which You here call *Æquatio centri secunda* is I perceive the same with that which in Dr Gregory's Astronomy You call *Æquatio loci Lunæ Sexta*. I shall be very glad to learn from You more distinctly the reasoning by which it is established.

I am Sr Your obliged Freind
& most Humble Servant
ROGER COTES

For
Sr Isaac Newton
in St Martin's Street
in Leicester Fields
London

NOTES

(1) Add. 3983, no. 29; the letter is printed in Edleston, *Correspondence*, pp. 121–2, from the Trinity College MSS.

(2) Clearly Newton had sent to Cotes: (i) Cotes' manuscript of *Logometria* (see above, Letter 921, note (3)); (ii) wood blocks for printing the mathematical figures for it; (iii) a redraft of the Scholium to Proposition 35; (iv) a letter of explanation, perhaps very short in view of Cotes' next sentence. All of these have disappeared.

(3) Newton may have begun his copy with paragraph 5 of the Scholium, beginning: 'Per eandem Gravitatis Theoriam...' (*Principia* 1713; p. 422, foot); Cotes refers to lines 14 and 7 of p. 423, in this paragraph. His emendations were printed.

(4) Cotes' version was printed. Compare Letter 917*a*, note (3).

(5) This difficulty was removed by adding after the word *argumenti* the adjective *prædicti* (aforesaid), referring back to the previous paragraph where the sense of *argument* is clear.

(6) Newton had slipped; the printed text (*Principia* 1713; p. 424, lines 1–2) reads: 'Apogæi Lunæ a Perigæo Solis.'

(7) Cotes' difficulty, as stated here, is not at first easy to understand for Newton is stating the straightforward postulate that the centre of the moon's orbit (*F*) revolves in an epicycle (centre *D*) with centre deferent *C*. The wording as printed differs without alteration of sense. It turns out later that Cotes has expressed his difficulty badly.

932 NEWTON TO COTES

12 AUGUST 1712

From the original in Trinity College Library, Cambridge.[1]
Reply to Letters 917 and 931; for the answer see Letter 935

Sr.

Upon the receipt of yours of Aug. 10th I have looked back upon yours of May 3d wch I had forgotten. In the first paragraph of ye new Scholium to Prop XXXV, where I have [ad 11′ 50″ circiter ascendit & *additur* medio motui Lunæ ubi Terra pergit a Perihelio suo ad Aphelium et in opposita Orbis parte *subducitur*] the words additur & subducitur should change places, & after the word *ascendit* let these words be added [in alijs locis æquationi centri Solis proportionalis est,][2]

In the end of the second Paragraph add these words. Additur vero æquatio prior & subducitur posterior ubi Terra pergit a Perihelio suo ad Aphelium, & contrarium fit in opposita Orbis parte.[3]

In the third Paragraph the words [paulo minor est in priore casu] are in my copy [paulo major est in priore casu] & should be so in yours.

In the fourth Paragraph the words additur & subducitur should change places.[4]

In the beginning of the correction of the latter part of the Moons Theory you may write [Halleius centrum Ellipseos in Epicyclo locavit.][3]

In the next Paragraph beginning wth the words [In Aphelio Terræ &c][5] after the first sentence of the Paragraph the word Aphelium is written five times erroneously for the word Apogæum. Write therefore [recta *DE* versus Apogæum Lunæ...excessui Argumenti annui Apogæi Lunæ supra distantiam Apogæi Lunæ ab Apogæo Solis,[6] vel forte æqualis excessui Argumenti annui & 360gr supra distantiam Apogæi Lunæ ab Apogæo Solis...Solis ab Apogæo Lunæ...Solis ab Apogæo proprio conjunctim. The Equation described in this Paragraph I had first from observations of Lunar Eclipses, & afterwards found that it answered the Theory of gravity in the manner here described. Its quantity when greatest came to about 2′ 10″ by Eclipses.[7] By ye Theory tis 2′ 25″. I suppose you understand that the force of ye Sun for disturbing the Moons motions is reciprocally as the cube of the distance of the earth from ye Sun. The motion of the center of the Moons Orb in ye cycle *BDAB* arises from the force of the Sun, & as this force varies, the motion of the center of ye Moons Orb should vary in this cycle both as to the length of the radius *DC* & as to ye velocity of the rotation of this radius about the center *C*, supposing the suns annual motion to be always equal & uniform,[8] & that his distance from the earth only changed. But because the suns annual motion accelerates & retards in a duplicate proportion of the suns distance reciprocally & this acceleration & retardation is allowed for in the angle *BCD* so as to make the point *D* accelerate & retard in the same proportion in ye cycle *BDAB*, here is a variation of the motion of the center of the Moons Orb in the cycle *BDAB* in a duplicate proportion of the suns distance reciprocally & this without altering the length of the radius *CD*. Had this variation been in a triplicate proportion there would have been no need of any further æquation, but because it is only in a duplicate proportion, there wants a further allowance in a single proportion. And this allowance must be made wth respect to the Sun's motion & true place. If the suns true motion could be accelerated & retarded in this proportion, I would accelerate & retard the motion of the point *D* in ye epicy[c]le *BDAB* in the same proportion. But because this cannot be done, I make the allowance by the rotation of the line *DF* about ye center *D*, so that the center of the Moons orb may revolve about the center *D* in an epicycle described by the point *F*, & about ye center *C* in a curvilinear Orb with a velocity reciprocally proportional to the cube of the distance of the earth from the Sun, or directly as the force of ye Sun wch causeth this velocity; or that the velocity of the point *F* in the said curvilinear Orb be to the velocity of the point *D* in the Orb *BDAB* reciprocally as the distance of the earth from the Sun. And this will come to pass quampro[x]ime by determining ye length *DF* and the angle *EDF* as in the Theory.

The next Paragraph beginning with the words [Computatio motus hujus difficilis est] conteins only an approximation of the former Paragraph, by computing the angle at ye earth wch the line *DF* subtends at the Moon in her mean distance from the earth. For the translation of the center of the Moons Orb from *D* to *F*, creates the same translation of the whole orb of the Moon & of the Moon in its Orb from the place in wch they would otherwise be, & so makes an equation or angle at the Earth wch the line *DF* subtends at the Moon.

If the Sun did not act upon the Moon the center of the Moons orb would be in the point *C*. By the action of the Sun it is transferred from the center to the circumference of the Epicycle *BDAB*. If the earth moved uniformly in a concentric circle about the sun so that ye action of the sun upon the Moons Orb might be uniform, the center of her Orb would move uniformly in ye Epicy[c]le *BDAB*. By the inequality of the Suns action the center of the Moons orb is transferred from the center to the circumference of a secondary epicycle described with ye radius *DC* [*read DF*] about the point *D*. If the inequality of the Suns force or action on ye Moons orb arose only from the variation of the distance of the earth from ye Sun & the angular motion of the earth about the sun was uniform, the point *D* would move uniformly in the epicycle *BDAB*, the angle *BCD* wch is double to the argumentun annuum increasing uniformly & the center of the Moons orb would move uniformly about the point *D* in an Epicycle whose radius is 3*DF*. But the angular motion of the earth about the Sun not being uniform, the angular motion of the radius *CD* about the center *C* is not uniform. If the angular motion of the earth about the Sun was as the cube of the distance of the earth from the Sun reciprocally, that is as the force of the sun upon the Moons Orb, the angular velocity of the Radius *CD* about the center *C* would be in the same proportion, & the center of the Moons orb being placed in the point *D* would have a velocity in the Orb *BDAB* proportional to the force of the Sun wch causeth it, & there would be no need of a secondary Epicycle about the center *D*. But because the angular motion of the earth about the Sun is but in a duplicate proportion of the distance of the Sun reciprocally, the motion of the point *D* in the epicycle *BDA* will [be] but in a duplicate proportion & for making up this proportion a triplicate one, the center of the Moons Orb must be placed not in the point *D* but in an Epicycle about the point *D*, & the radius of the Epicycle must be but a third part of such a Radius as would make the Epicycle alone answer to a triple proportion, so that the motion of the center of the Moons orb in this Epicycle & of the point *D* about the center *C* may together compound a motion in a triplicate proportion of the distance of the earth from the Sun reciprocally.

In your papers I met wth nothing wch appeared to me to need correction. I am

Your most humble Servt
Is. NEWTON

London
Aug. 12. 1712

For the Rnd Mr Roger Cotes Professor
of Astronomy at his Chamber in
Trinity College in Cambridge

NOTES

(1) R.16.38, nos. 226–8, printed in Edleston, *Correspondence*, pp. 122–6. The square brackets appear in the original.

(2) This was later dealt with by altering the phraseology, the words *additur* and *subducitur* remaining in the original order; the addition was made (*Principia* 1713; p. 421).

(3) All this appears in the printed text.

(4) This was not in fact done, but instead 'quadratures' and 'syzygies' were transposed in the sentence to the same effect.

(5) *Recte*, 'In Perihelio Terræ...'

(6) *Recte*, '...a Perigæo Solis...' (*Principia* 1713; p. 424, line 1). The rest of this paragraph was altered later; see Number 937*a*.

(7) As Edleston points out (*Correspondence* p. 123, note) the same value was given in the *Lunæ Theoria Newtoniana* published by David Gregory. Presumably by 'observations of lunar eclipses' Newton refers to Flamsteed's observations.

(8) Newton is of course speaking metaphorically; he should have said (as he does later) 'the *earth's* annual motion...'.

933 NEWTON TO COTES

[14 AUGUST 1712]

From the original in Trinity College Library, Cambridge.(1)

For the answer see Letter 935

London. 16 *Aug*. 1712

Sr

In the Letter I wrote to you two days ago, the words [Apogæi Lunæ] were interlined after the words [excessui Argumenti annui.](2) Its better to strike out the interlined words, & at the end of the Paragraph to add this sentence.(3) [Per Argumentum annuum intelligo excessum qui relinquitur subducendo medium locum Apogæi Lunæ semel æquatum a vero loco Solis, vel a summa veri illius loci et 360gr.

Your humble Servant

Is. NEWTON

For the Rnd Mr Cotes Professor
of Astronomy at his chamber in
Trinity College in Cambridge.

NOTES

(1) MS. R. 16.38, no. 229, printed in Edleston, *Correspondence*, pp. 126–7. Newton mistook the date by two days; he had previously written on the twelfth, and the present letter is postmarked AU 14.

(2) All five Latin words had been interlineated.

(3) This instruction was superseded by Newton's new draft of 26 August (Number 937*a*).

934 CERTIFICATE BY NEWTON

14 AUGUST 1712

From the original in the Library of Trinity College, Dublin(1)

Tower of London August. 14th 1712

The weights conteined in this Box I have examined by the standard weights of her Majts Mint in the Tower & found them just. The pile of weights with a Harp graved on them is the standard for sizing & examining the penny weights

& grains & the small round weights made for forreign pieces of money current in Ireland. I presume they will agree with the weights formerly sent from hence & established by Proclamation & so may need no new authority to make them usefull.(2)

ISAAC NEWTON

NOTES

(1) MS. 2567, miscellaneous autographs no. 56, reproduced by kind permission of the Board of Trinity College, Dublin.

(2) Compare Letters 900 and 925. It is not clear why a check on the weights was needed, unless some more authoritative set had been destroyed by the fire at the Castle in Dublin.

935 COTES TO NEWTON

17 AUGUST 1712

From the original in the University Library, Cambridge.(1)
Reply to Letters 932 and 933; for the answer see Letter 937

Sr

I have received two Letters from You by the last Post & the foregoing. I thank You for the trouble You have given Your self to make the thing clearer to me, but am sorry to find You had mistaken my difficulty. I was very well satisfied as to the design of introducing a Secondary Epicycle about the point *D*: the motion which You had given the point *F* in that Epicycle was what I stuck at, and consequently Your manner also of determining the Angle *EDF*. By making the Angle *BCD* equal to the doubled Annual Argument of the Moon's Apogee, the motion of the point *D*, in the primary Epicycle *BDAB*, was not yet enough accelerated in the Earths Perihelium, nor enough retarded in the Earths Aphelium. The Secondary Epicycle was therefore added that the velocity might be in a triplicate instead of a duplicate proportion & that an increase of velocity might be made in the Earths Perihelium & a decrease be made in its Aphelium: Hence it seem'd evident to me, that the motion of the point *F* in the Secondary Epicycle ought to be such, that it might arrive at the place of its nearest distance from the point *C* in the Earths Perihelium; & there by its motion conspiring with the motion of the point *D*, might render the compound of both the swiftest: & again that it might arrive at ye place of its furthest distance from the point *C* in the Earths Aphelium & there by its motion contrary to the motion of ye point *D* might render the compound of both the slowest. Wherefore if *CD* be produced to *G* so that *DG* be equal to *DF*;(2) & on the other side between *D* & *C*, *DH* be taken equal to *DF*: tis evident that in the Earths Aphelium *DF* will coincide with *DG*, & in ye Earths Perihelium *DF* will coincide with *DH*, so revolving about the centre *D*, that the angle *GDF* may always be equal to the Suns mean Anomaly. Hence the angle *EDF* or *EDG*−*GDF* or *BCD*−*GDF* will be equal to the excess of ye doubled Annual Argument above the Suns mean Anomaly as I observ'd in my last. This is the only way according to which I can apprehend the motion of the point *F* in ye Secondary Epicycle to be regulated but I cannot perceive how it may be reconcil'd with Your way of determining the angle *EDF*, or with the time You assign for its revolution by making it equal to ye time in which the point *D* revolves about the centre *C*.

What I have here said will also affect the following Paragraph beginning with [Computatio motus hujus difficilis est &c.] But besides this there were two other difficultys contain'd in this Period [Et hæc recta [*DF*] subtendit angulum ad Terram quem translatio centri Orbis lunæ a loco *D* ad locum *F* generat, & cujus duplum propterea dici potest Æquatio centri secunda.](3) The angle at the Earth which *DF* subtends is the angle *DTF* comprehended by the lines *TD*, *TF*. I understand You thus; but I perceive by Your Letter that You do not mean the Angle *DTF*, but an angle at the Earth which is subtended by a line at the Moon equal & parallel to *DF*: so that I can now understand what follows [Et hæc Æquatio est ut sinus anguli quem recta illa *DF* cum recta a puncto *F* ad Lunam ducta continet quam proxime] which I could not do before. However I am still at a loss to understand why You take the double of that angle rather than the angle it self for the *Æquatio centri secunda.*

The following Paragraph describes the *Variatio secunda* I suppose it was deriv'd from Observations. In it the word Aphelium is twice used instead of Apogæum.

I am Sir Your most humble Servant

ROGER COTES

Cambridge Aug: 17. 1712

For
Sr Isaac Newton
at his house
in St Martin's Street
Leicester Fields
London

NOTES

(1) Add. 3983, no. 30; printed in Edleston, *Correspondence*, pp. 127–9, from the Trinity College MSS.

(2) Cotes simply draws in (mentally) the secondary lunar epicycle which (for some unknown reason) Newton had not drawn in, and which was not shown in the figure printed

(*Principia* 1713; p. 424) although there described. Hence Cotes' figure may be reconstructed thus (as it is by Edleston, p. 128, note), where *F* is the centre of the Moon's orbit and *T* is the centre of the Earth:

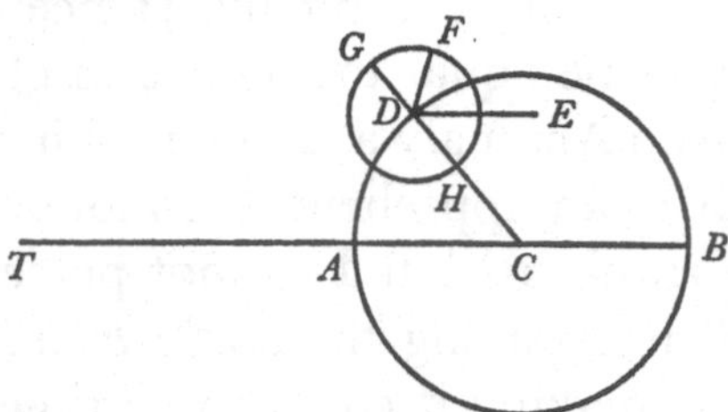

(3) 'And this straight line [*DF*] subtends an angle at the Earth which the removal of the centre of the Moon's orbit from *D* to *F* generates, and whose double may accordingly be called the second equation of the centre.' This passage was modified and extended in Newton's next draft.

936 TAYLOUR TO THE COMPTROLLERS OF THE ACCOUNTS OF THE ARMY

21 AUGUST 1712

From the copy in the Public Record Office.[1]
For the answer see Letter 948

Gentlemen

Mr. Brydges the Paymaster of Her Mats. Forces[2] having applied to My Lord Treasurer for his Directions as to the rate French Current Money shall be computed at, since her Mats. Forces in Dunkirk will receive their Subsistence in that Specie,[3] I send You by his Lordps Command Mr Brydges's Memoriall relating to that Affair here inclosed, and am to desire that You together with Sir Isaac Newton the Master [and] Worker of her Mats. Mint, (to whom You will please to signifie His Lordps Pleasure in this behalf) will consider the Subject Matter of the said Memoriall And thereupon report Your Opinion to his Lordp what is fit to be done therein Which in the absence of the Secretarys is signified by Your &c 21th Augt. 1712.

NOTES

(1) T/27, 20, 255, a record copy. The Comptrollers were Sir Philip Medows (d. 1757) and James Bruce. The former, like his father, had served on diplomatic missions and was knighted in 1700. He was appointed Comptroller in 1707. The latter was the youngest son of Robert, Earl of Ailesbury, and M.P. for Marlborough; he had been appointed in 1711.

(2) James Brydges (1673–1744), M.P. for the city of Hereford and paymaster of the forces abroad (1707–12), succeeded as ninth Baron Chandos in 1714 and was created Duke of Chandos in 1719. He accumulated great wealth from his office. His memorial, dated 19 August, is in P.R.O. T/1, 152, no. 44A and is printed in Shaw, pp. 183–4. Apart from requesting a definition of the French exchange-rate it states that 'The directions wch regulate the rate of the Holland & Brabant Money are by her Mats. Royall Signe Manual...And I am humbly of Opinion it will bee necessary that I have a like Authority for Computing the French Curr[en]t Money.'

(3) The Tory administration under Oxford and Bolingbroke had pursued its determination to make peace throughout the latter part of 1711, their greatest propaganda stroke at home being Swift's *The Conduct of the Allies*, which had appeared at the end of November. On 1 January 1712 Queen Anne created twelve new peers to maintain the Tory majority in the House of Lords. On the previous day Marlborough had been dismissed from his commands, and within a fortnight Robert Walpole was in the Tower, expelled from the House of Commons, both actions being justified by Tory charges of corruption. The Duke of Ormonde was appointed in Marlborough's place, and the victorious general was driven into exile in Holland. Now Bolingbroke eagerly led his party into a gross betrayal of England's allies during the war, and an espousal of the Jacobite cause. In May 1712 Ormonde was positively ordered to refrain from military cooperation with the Allies and further action against the French, with whom Bolingbroke had now established a secret concord; before the end, the Tories were conspiring to inform Paris of intended offensive moves by Prince Eugen and the Dutch forces. Unsupported by the English, who were marching rearwards to Ghent and Dunkirk, the Allies suffered a severe defeat at the hands of the French general Villars (13 July 1712). A few days earlier the first English troops had entered Dunkirk, which they occupied as a condition of the unilateral armistice between England and France arranged by Bolingbroke. The destruction of the fortifications and harbour of this privateer stronghold was to be the great prize of war that the Tories would present to their countrymen.

937 NEWTON TO COTES

26 AUGUST 1712

From the original in Trinity College Library, Cambridge.[1]
Reply to Letter 935; for the answer see Letter 938

London Aug 26. 1712

Sr

For removing the difficulties in the Theory of the Moon mentioned in yours of Aug. 17 I have sent you the inclosed paper conteining some alterations in the description of the latter part of that Theory. I had by mistake writ [Aphelio Solis] & changed it to [Apogæo Solis][2] & should have changed it to [Perigæo Solis,] as I have done in this paper inclosed. By considering that the angle *CDF* is the complement of ye Suns Anomaly to a circle (as I have exprest it in the paper inclosed) you may perceive that whenever the Sun is in his Apoge[e] the point *F* will fall between the points *D* & *C* & so will be in its slowest motion in the Curve line wch it describes about the center *C*. If the line *DF* kept parallel to it self the points *F* & *D* would have equal motions: but by the revolving of the point *F* about the point *D* according to the order of the signes this motion of the point *F* is subducted from the motion of the point *D*, & the difference is the motion of the point *F* in the said curve line, wch motion is therefore the slowest that it can be. And on the contrary, in the Sun's Perige[e] the line *DF* will lye in directum with the line *DC*, & the motion of the point

F[3] in the said curve line will be at the swiftest being the[3] summ of the two motions. By the inclosed paper you will understand also why I took the double of the angle subtended by a line at the Moon equal & parallel to *DF*, for the *Equatio centri secu[n]da*. The line must be doubled at the superior focus of the Moon's Orb & carried thence to the Moon. I am

Your most humble Servant

Is. Newton

For the Rnd Mr Cotes Professor of
Astronomy at his Chamber
in Trinity College in Cambridge

NOTES

(1) MS. R.16.38, fo. 233.
(2) In his Letter 932.
(3) These words are hidden under the wax of the seal.

937*a* THE ENCLOSED REVISED DRAFT[1]

...Capiatur angulus *BCD* æqualis duplo argumento annuo, seu duplæ distantiæ veri loci Solis ab Apogæo Lunæ semel æquato, et erit *CTD* æquatio secunda[2] Apogæi Lunæ et *TD* excentricitas Orbis ejus. Habitis autem Lunæ motu medio et Apogæo et excentricitate, ut et Orbis axe majore partium 200000; ex his eruetur verus Lunæ locus in Orbe et distantia ejus a Terra idque per methodus notissimas.

In perihelio Terræ propter majorem vim Solis centrum Orbis Lunæ velocius movetur in epicyclo *BDA*[3] circum centrum *C* quam in Aphelio, idque in triplicata ratione distantiæ Terræ a Sole inverse. Ob æquationem centri Solis in argumento annuo comprehensam, centrum Orbis Lunæ velocius movetur in Epicyclo illo in duplicata ratione distantiæ Terræ a Sole inverse. Ut idem adhuc velocius moveatur in ratione simplici distantiæ inverse; ab Orbis centro *D* agatur recta *DE* versus Apogæum Lunæ seu rectæ *TC* parallela, et capiatur angulus *EDF* æqualis excessui Argumenti annui prædicti supra distantiam Apogæi Lunæ a Perigæo Solis in consequentia; vel quod perinde est, capiatur angulus *CDF* æqualis complemento Anomaliæ veræ Solis ad gradus 360. Et sit *DF* ad *DC* ut dupla excentricitas Orbis magni ad distantiam mediocrem Solis a Terra et motus medius diurnus Solis ab Aphelio[4] Lunæ ad motum medium diurnus Solis ab Apogæo proprio conjunctim, id est, ut $33\frac{7}{8}$ ad 1000 et 52′. 27″. 16‴. ad 58′. 8″. 10‴ conjunctim, sive ut 3 ad 100. Et concipe centrum Orbis Lunæ locari in puncto *F*, et in Epicyclo cujus centrum est *D* et radius *DF* interea revolvi dum punctum *D* progreditur in circumferentia circuli *DABD*. Hac enim ratione velocitas qua centrum orbis Lunæ circum centrum *C* in linea quadam curva movebitur, erit reciproce ut cubus distantiæ Solis a Terra quamproxime, ut oportet.

Computatio motus hujus difficilis est, sed facilior reddetur per approximationem sequentem. Si distantia mediocris Lunæ a Terra sit partium 100000, et excentricitas *TC* sit partium 5505 ut supra: recta *CB* vel *CD* invenietur partium $1172\frac{3}{4}$, et recta *DF* partium $35\frac{1}{4}$. Et hæc recta ad distantiam *TC* subtendit angulum ad Terram quem translatio centri Orbis a loco *D* ad centrum *F* generat in motu centri hujus; et eadem recta duplicata in situ parallelo ad distantiam superioris umbilici Orbis Lunæ a Terra, subtendit eundem angulum quem utique translatio illa generat in motu umbilici, et ad distantiam Lunæ a Terra subtendit angulum quem eadem translatio generat in motu Lunæ, quique propterea æquatio centri secunda dici potest. Et hæc æquatio in mediocri Lunæ distantia a Terra est ut sinus anguli quem recta illa *DF* cum recta a puncto *F* ad Lunam ducta continet quamproxime, et ubi maxima est evadit 2′ 25″. Angulus autem quem recta *DF* et recta a puncto *F* ad Lunam ducta comprehendunt invenitur &c,

In the next Paragraph but one write *Apogæi* twice for *Aphelii*.(5)

NOTES

(1) This passage, almost word for word identical with the printed text (*Principia* 1713; pp. 423–4), was printed by Edleston, *Correspondence*, pp. 130–2. An English version may be found in Cajori, pp. 475–7, for this section survived without change into the third edition.

(2) *Principia* 1713; semestris.

(3) *Principia* 1713; last three words deleted.

(4) *Principia* 1713; Apogæo.

(5) The words 'but one' were a mistake, which misled Cotes into supposing that something was missing from his copy.

938 COTES TO NEWTON

28 AUGUST 1712

From the original in the University Library, Cambridge.(1)
Reply to Letter 937; for the answer see Letter 941

Cambridge Aug: 28th 1712

Sr.

I received Yours with the inclosed Paper, but cannot yet agree with You. In my former Letters I had supposed the point *F* to come the nearest to *C* in the Sun's Perigee & to be the furthest from *C* in the Suns Apogee: You on the contrary suppose it to be the nearest in ye Suns Apogee & the furthest in the Suns Perigee. According to Your supposition the motion of the point *F* in its curvilinear Orb will then be the swiftest when that point is at its greatest distance from the centre *C*, and slowest at its least distance from the same, for we agree that tis the swiftest in the Sun's Perigee & slowest in his Apogee: Whereas according to my supposition the swiftest motion accompanies the least distance & the slowest the greatest, as I think it ought to do.(2)

By considering that the Angle *CDF* is the complement of ye Sun's Anomaly to a circle, You say, I may perceive that whenever the Sun is in his Apogee, the point *F* will fall between the points *D* & *C*, & so will be in its slowest motion in the Curve line which it describes about the centre *C*. I do indeed perceive yt the point *F* will fall between the points *D* & *C*, but I think it will then be in its swiftest motion, not its slowest. For since the Angle *CDF* is, by supposition, the complement of the Sun's Anomaly to a circle; it follows, that as the Anomaly is continually increasing, its complement must be continually decreasing. Therefore the line *DF* does so revolve to the line *DC* as by its motion to diminish continually the Angle *CDF*. Whence it appeares that in respect of the line *DC* the line *DF* does revolve with a motion contrary to the order of the Signs. I say in respect of the moveable line *DC*, not in respect of the fixt Stars: & it is in respect of ye line *DC* that its motion must be estimated in order to compound it with the motion of the point *D* in the circle *ABD*. The motion then of the point *F* in its passage over the line *DC* or, by supposition, in the Sun's Apogee does conspire with the motion of ye point *D*, & therefore the sum of the two motions renders the motion of the point *F* in its Curvilinear Orb the swiftest in the Sun's Apogee, which ought not to be.

I think I apprehend Your meaning very well where You say, The line *DF* must be doubled at the superior focus of the Moons Orb, & carried thence to the Moon: but I cannot see any reason why the doubled line at ye superior Focus rather than the single line at the Centre should be carried to the Moon, excepting that Observations may require it.

By Your Letter I suspect that in Your Copy there is a Paragraph between that which begins with *Computatio motus hujus difficilis* &c: & that which begins with *Si computatio accuratior desideretur* &c. In my Copy they immediately follow one the other.(3)

I am Sir Your
most Humble Servant
ROGER COTES

For
Sr Isaac Newton
at his House
in St Martin's Street
in Leicester Fields
London

NOTES

(1) Add. 3983, no. 31; printed (from the draft) in Edleston, *Correspondence*, pp. 132–4.

(2) Cotes' difficulty here arises from the difference between his understanding of Newton's original phraseology, and Newton's own; this verbal difficulty Newton was able to remove in his next letter. However, there was a further, concealed point involved: should the linear or the angular velocity of the point *F* be considered? It was only in Letter 943 and after some further alteration of the phraseology that Newton convinced Cotes that it was correct to consider the linear motion of the point *F*.

(3) Compare Number 937 *a, ad fin.*

939 H. SCOTT TO NEWTON

28 AUGUST 1712

From the original in the Mint Papers(1)

Edinbr 28 *August* 1712

Much Honored
Sir

I am favoured with a letter from Mr Fauquier(2) of the 23d And am Extreamly ashamed I should have given you So much trouble through inadvertency for which I humbly beg pardon I perceive in the Accott of the Recoynage last sent I have set down 104200pd in place of 104400 And as to the 140pd of Bullion in my brothers Custodie there is 31pd 9 12dwt 2g Really belongs to Mr Allardes And the Remainder which is 108pd. 2. 7. 22g what is controverted Now I trouble you to peruse the full abstract which I have done out to a very graine according to the Mint books and Abbreviat given in by my brother at finishing the Recoynage and after the sweep was 3 tymes searched(3) Mr Allardes friends(4) make debate only because my brother had more [silver](5) lying by him then Mr Allardes can demand by the state of the [accounts](5) I humbly intreat your Opinion in the matter and refer to my last [letter](5) consider the quantity refined and other Reasons which necessarly occasion a waste I am with all duty full Respect

Much Honored
Sir
Yor most Humble & most
Obedient Servant
HERCULES SCOTT

The Much Honored
Sir Isaack Newtoun
at his logeing in St Martin-Street
near Leycester-field
London

NOTES

(1) Mint Papers, III, fo. 59. Compare Scott's previous letter of 26 July 1712 (Letter 930).

(2) We have not found this. John Francis Fauquier (d. 1726) was Newton's Deputy-Master, managing the day-to-day business for him, at a salary of £100 p.a. He is usually described as 'Dr'. Later he became a director of the Bank of England. Two sons became F.R.S., and the elder was lieutenant-governor of Virginia.

(3) The account (entered before the letter in the bound volume) shows a receipt of bullion amounting to more than 104634 lb., and a discharge of 104400 lb. Of the 234 lb. difference unaccounted for, Hercules Scott reckoned some 125 lb. (that is, 0·119%) to be proper 'waste' in any event; if this were all the waste to be allowed, however, the 108 lb. found in his brother's possession would be forfeit. Therefore Hercules argued that the *whole* discrepancy should be considered as 'waste' (amounting to 0·223%).

(4) Family or relatives.

(5) Paper torn.

940 OXFORD TO NEWTON AND HALLEY

28 AUGUST 1712

From the copy in the Public Record Office.(1)
For the answer see Letter 945

Sr Isaac Newton & Dr Hawley abt. a Mathematicall Instrument

Gentlemen

Mr Cawood(2) having presented to me the Inclosed Petition in relation to an Instrument by him invented which he alledges will be very usefull in Navigation I desire you will send to the said Mr Cawood to attend you(3) & that you will hear what he has to offer & Examine the said Instrument & let me have your opinions thereupon I am &c. 28 *Aug.* 1712

OXFORD

NOTES

(1) T/27, 20, 259. The petition mentioned is not, of course, entered and we have not found a copy of it.

(2) E. G. R. Taylor (*Mathematical Practitioners of Tudor & Stuart England*, Cambridge 1954, pp. 303, 424) writes that she was unable to trace any information about Francis Cawood, who advertised a published *Navigation completed, Longitude...found. A New Universal Chart*, price two shillings, or discover any copy of this chart, which was advertised for sale in 1710. Cawood's method relied on a 'mathematical instrument' (apparently a compass which, he claimed, always indicated true north) which he proposed to sell, and he would give instruction in the method at his house in Bartholomew Lane near the Royal Exchange.

(3) See Letter 940*a*. There is a further record of communication between Newton and Cawood in the *Sotheby Catalogue* (p. 39, Lot 173) where a sentence is quoted from a draft un-

dated letter (Newton to Cawood): 'I have not yet been able to strike such a meridian line as we were speaking of, but can do it at any time by a magnetical needle...' We have been unable to trace this letter.

940a NEWTON TO CAWOOD

[after 28 AUGUST 1712]

From the draft in the University Library, Cambridge[1]

Sr

You having presented to my Ld Treasurer a Petition in relation to your recent invented Compass admitting no variation, his Lordp has directed me & Dr Hawley to send to you to attend us & yt we should hear what you have to offer & examin your instrument. For that end Dr Hawley & I have agreed to meet at my house in St Martins street by Leicester fields at eleven of the clock to morrow morning & desire that you would then attend us wth your said instrument. I am

For Mr Francis Caywood to be left at
Mr Dowlings house at the upper end of
Angel Court neare Storeys passage
in St James's Park in
Westmr.

NOTE

(1) Add. 3972, fo. 43. For the outcome of the meeting see Letter 945.

941 NEWTON TO COTES

2 SEPTEMBER 1712

From the original in Trinity College Library, Cambridge.[1]
Reply to Letter 938; for the answer see Letter 942

Sr

The reason why the doubled line at the superior focus rather then the single one at the center should be carried to the Moon is this.[2] The angles about the superior focus are (quamproxime) proportional to the times, those about ye Center are not. And therefore if the superior focus be translated, the line drawn from it to ye Moon will keep its parallelism, & by doing so will make the same translation in the Moon.

As for your other difficulty,[3] if the line *DF* kept parallel to it self, so as being produced to cut the line *TB* in a given angle the motion of the points *D* & *F* would be always equal to one another. I do not speak of the angular motion of the lines *CD* and *CF* about the center *C* but of the local motion of the points *D* & *F* in their curvilinear Orbs wch in this case will be two equal circles. Let the circle *FMN*[4] be described wth the center *C*[5] & radius *DF* & & [*sic*] be cut by the line *CD* in the point *H* & by the line *CD* produced in the point *M*. And if the line *DF* keep parallel to it self, the increase of the angle *MDF* will be equal to the increase of the angle *BCD*. I meane that ye two angles will increase wth equal swiftness or have equal augmentations in equal times. And in this case the motions of the points *D* & *F* will be equal. But if the angle *MDF* increase but half so fast (wch is the case of the Theory), the motion of the point *F* will be accelerated neare *M* & retarded neare *N*.[4] When the line *DF* keeps parallel to it self & has no angular motion, its motion in it[s] orb will be equal to that of the point *D*. But if it has an angular motion according to the order of the letters *FMHF* (as in the Theory) that angular motion will accelerate the point *F* neare *M* & retard it neare *N*.[4] You seem to consider the angular revolution of the line *DF* or *CF* in respect of the line *DC*. I consider not the relative angular motion of the line *DF* or *CF* but the absolute linear motion of the point *F* in its linear orb described about the point *C* in the unmoved plane of the Moons orb without any relation to the angular motion of the line *CD*.

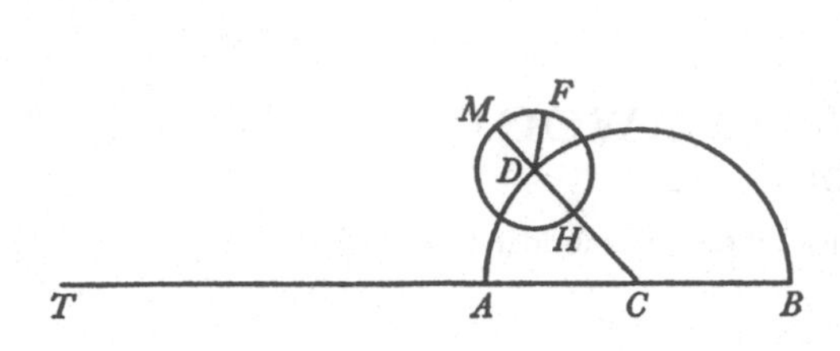

There is no Paragraph between that wch begins wth *Computatio motus hujus difficilis* &c & that wch begins wth *Si computatio accuratior desideretur* &c.[6] If the words of the paper inclosed in my last are not right, pray correct them. After these two Paragraphs there is or should be a Paragraph concerning the refraction of the Atmosphere whereby the Diameter of the earths shadow is enlarged in Lunar Eclipses. That Paragraph was (I think) in the first draught I sent you of the Moons Theory.[7] I am

Your most humble Servant
IS NEWTON

London Sept 2d
1712

For the Rnd Mr Roger Cotes Professor
of Astronomy at his Chamber in Trinity
College in Cambridge

NOTES

(1) R.16.38, no. 235; printed in Edleston, *Correspondence*, pp. 134–8.

(2) Cotes had mentioned this difficulty at the end of his previous letter.

(3) See Letter 938, note (2).

(4) For *N* read *H*. Newton has labelled his diagram poorly, and this has led to the error. In his next letter Cotes follows Newton's confusion. In related correspondence, and in the printed version, the point is referred to as *H*.

(5) Read *D*.

(6) See Letter 938, note (6).

(7) This was the eleventh paragraph of the original draft for the scholium, enclosed with Letter 917. Newton had presumably omitted it from the revised version. Cotes mentioned the matter again in his Letter 944, and Newton confirmed in his reply that he wanted it included. This was done. See Koyré and Cohen, *Principia*, II, p. 663.

942 COTES TO NEWTON

6 SEPTEMBER 1712

From the original in the University Library, Cambridge.[1]
Reply to Letter 941; for the answer see Letter 943

Cambridge Sept. 6th 1712

Sr

I received Your last, by which I do at length perceive, that You consider the absolute linear motion of the point *F* in its linear Orb described about the center *C*, & not the angular revolution of the line *CF* about the same centre, which I had before supposed You to do.[2]

I am satisfied that this linear motion of the point *F* will be accelerated near *M* & retarded near *N*:[3] & therefore if it be the linear motion which ought to be considered in Your Theory, & not the angular; You do rightly in making the angle *CDF* equal to the complement of the Suns Anomaly to a Circle, or which is the same thing, in making the angle *EDF* equal to ye excess of the Annual Argument above the distance of the Moon's Apogee from the Sun's Perigee.

But I am of opinion that You ought rather to consider the angular motion of the point *F* than the linear. And if so, since the angular revolution of the line *CF* about the centre *C*, in the unmoved plane of the Moon's Orb, is accelerated near *N*[3] & retarded near *M*; the angle *MDF* must be taken equal to the Sun's Anomaly, or which is the same thing, the angle *EDF* must be taken equal to the excess of the Annual Argument above the distance of the Moon's Apogee from the Sun's Apogee.

I will not set down other reasons for considering the angular motion rather than the linear which may admit of dispute. What I offer is as follows. I suppose these words at the end of the Paragraph answer to Observations vizt. [—subducendam si summa illa sit minor semicirculo, addendam si major. Sic habebitur—] But these words are not true by the Theory if the angle *EDF* be taken equal to the excess of the Annual Argument above the distance of the Moon's Apogee from the Sun's Perigee, as it must be taken if the linear motion be considered. And they are true by the Theory if the angle *EDF* be taken equal to the excess of the Annual Argument above the distance of the Moon's Apogee from the Sun's Apogee, as it must be taken if the angular motion be considered. Therefore the angular motion ought to be considered rather than the linear, that the Theory may answer to the Observations.(4)

Let *DL* be a line drawn from the point *D* to the Moon, then will the *Æqutioa Centri Secunda* be as ye sine of the angle *FDL*. I suppose You agree with me that the Æquation must be substracted whenever the angular distance of the

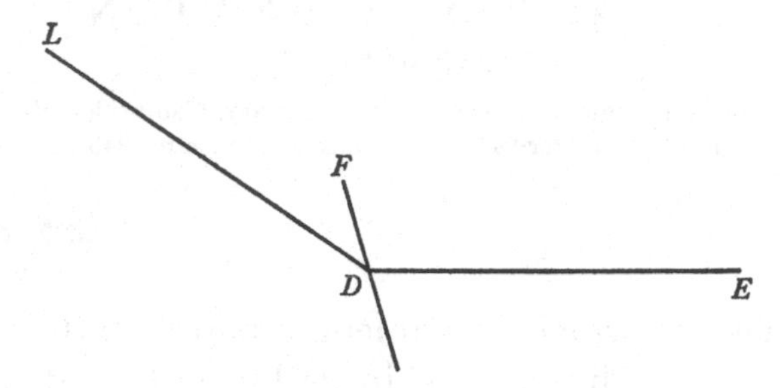

line *DL*, from the line *DF* taken according to the order of the signs is less than a semicircle & be added whenever that distance if bigger, or in other words, that it must be substracted whenever the excess of the Moon's Anomaly above the angle *EDF* is less than a semicircle, & be added whenever that excess is bigger.

If then the angle *EDF* be taken equal to the excess of the Annual argument above the distance of the Moons Apogee from the Suns Perigee: the excess of the Moons Anomaly above the angle *EDF* will be equal to the sum of the distances of the Moon from the Sun & of the Moons Apogee from the Sun's Perigee. And therefore the Æquation must be substracted when this sum is less than a semicircle & added when it is greater. Now this sum is less than a Semicircle when the sum of the distances of the Moon from the Sun & of the Moon's Apogee from the Sun's Apogee is greater than a Semicircle, & on the contrary the first sum is greater than a Semicircle when the second is less. Therefore the Æquation must be substracted when the second sum is greater than a semicircle & added when it is less. But this Rule derived from the Theory is contrary to Your Rule at the end of the Paragraph derived from Observation. From which contrariety I think it is evident, that the angle *EDF* ought not to be taken equal to the excess of the Annual Argument above the distance of the Moon's Apogee from the Sun's Perigee, & consequently the Linear motion of the point *F* ought not to be considered, but its Angular motion.

Your most Humble Servant

ROGER COTES

For Sir Isaac Newton
at his House
in St Martin's Street
in Leicester Feilds
London

NOTES

(1) Add. 3983, no. 32; printed in Edleston, *Correspondence*, pp. 136–8, from the draft in Trinity College Library, Cambridge (R.16.38, fo. 237), which is dated 7 September 1712.

(2) See Letter 938, note (2).

(3) Cotes follows Newton in confusing *H* and *N*; see Letter 941, note (4).

(4) Newton accepted this objection in his next, but dealt with it by rewording the phrase under dispute.

943 NEWTON TO COTES

13 SEPTEMBER 1712

From the original in Trinity College Library, Cambridge.(1)
Reply to Letter 942; for the answer see Letter 944

London Sept. 13*th* 1712.

Sr

If it could be supposed that the force of the sun upon the Moon for disturbing her motions could be increased wthout altering the periodical times of the sun & Moon, & that the Orb of the earth was concentric to the Sun: the line *DF* would vanish & the radius *DC* would be increased in proportion to the Sun's force without altering its angular motion about the center *C*. By the increase of the Suns force, the linear motion of the point *D* would be increased by its moving in a larger Orb, but its angular motion about the center *C* would remain the same as before. But the earths orb being excentric & the excentricity causing a variation of the Suns force upon the Moon greater then in proportion to the variation of the Suns velocity, I compensate the excess or defect of the force by a secondary epicycle described wth the radius *DF* about the center D, so that the distance CF may increase or decrease accordingly as there is an excess or defect of the suns force & by increasing or decreasing cause the linear motion of the point F in the plane of the Moons Orb to be greater or less then the linear motion of the point D in the circle BDA in proportion to the said excess or defect of the suns force.

I thank You for putting me upon examining the words(2) [*subducendam si summa illa sit minor semicirculo, addendam si major. Sic habebitur* &c.] I have compared them with my calculations of the Moons place in Eclipses & find that they must be corrected & put [—*addendam si summa illa sit minor semicirculo, subducendam si major. Sic habebitur &c.*] The Equation(3) I gathered from Observations many years ago & put it when greatest, to be 2′ 10″. The last year I gathered its quan[ti]ty from observations to be 2′ 25″ when greatest, but in describing it, committed the mistake wch I have now corrected by reviewing my old calculations. I am

Sr

Your most humble Servant

Is. Newton.

For the Rnd Mr Roger Cotes Professor of Astronomy, at his Chamber in Trinity College in Cambridge

NOTES

(1) R.16.38, no. 239; printed in Edleston, *Correspondence*, pp. 139–40.

(2) See Letter 942, note (4); the wording Newton suggests was adopted.

(3) Compare Letter 932; there 2′ 25″ was stated as a *theoretical* value for the *Æquatio centri secunda*, but now this is claimed as an observational value.

944 COTES TO NEWTON

15 SEPTEMBER 1712

From the original in the University Library, Cambridge.(1)
Reply to Letter 943; for the answer see Letter 946

Sr

I have received Your last Letter, & am now sufficiently satisfied as to the *Æquatio centri secunda*. I hope the description of the *Variatio secunda* is accurate.

The Paragraph concerning the Refraction of the Atmosphere in Eclipses was in Your first draught, but was left out in Your alteration of it.(2) There was also another Paragraph before it describing the Dimensions of the Sun's & Moon's Diameters and Parallaxes, which was also omitted in Your Paper of alterations. I am uncertain whether You would have both of them inserted or that only concerning the Effect of the Atmosphere. They stood thus.(3)

Diximus Orbem Lunæ a viribus Solis per vices dilatari & contrahi & Æquationes quasdam motuum Lunarium inde oriri. Inde etiam oritur variatio aliqua Parallaxeos Lunæ, sed quam insensibilem esse judico; ideoque in computationibus motuum Lunæ, pro mediocri ejus distantia a centro Terræ semper usurpo numerum 100000 & pro Orbis diametro transversa numerum 200000, et ad Parallaxin investigandam pono mediocrem distantiam centri Lunæ a centro Terræ in Octantibus æqualem esse $60\frac{2}{9}$ semidiametris maximis Terræ. Semidiametrum ejus maximam voco quæ a centro ad Æquatorem ducitur, minimam quæ a centro ad Polos. Et hinc fit Lunæ parallaxis horizontalis mediocris apparens in Octantibus 57′. 5″, in Syzygijs 57′. 30″, in Quadraturis 56′. 40″. Lunæ vero diameter mediocris apparens in Syzygijs 31′. 30″, in Quadraturis 31′. 3″ usurpari potest & Solis diameter mediocris 32′. 12″.

Et cum Atmosphæra Terræ ad usque altitudinem milliarium 35 vel 40 refringat lucem Solis, & refringendo spargat eandem in umbram Terræ, & spargendo lucem in confinio umbræ dilatet umbram ad diametrum umbræ quæ per Parallaxim prodit, addo minutum unum primum in Eclipsibus Lunæ, vel minutum unum cum triente.

Theoria vero Lunæ &c.

I suppose You would omit the first of these Paragraphs since the substance of it is in other parts of Your Book,[4] excepting that You have $60\frac{1}{4}$ semidiameters in Corol. 7. Prop. XXXVII. Lib. III instead of $69\frac{2}{9}$. Be pleased to send what You would have inserted.[5]

In the last Paragraph of the Scholium I suppose You have designedly altered Your first draught by putting[6] ♑ 20°. 43′. 40″ for ♑ 20°. 43′. 50″, and ♒ 15°. 20′. 00″ for ♒ 15°. 19′. 50″, and ♓ 8°. 20′. 00″ for ♓ 8°. 18′. 20″.

I am Sir

Your most Humble

Servant

ROG: COTES

Cambridge Sept. 15th 1712

For

Sr Isaac Newton

at his House

in St Martin's street

Leicester Feilds

London

NOTES

(1) Add. 3983, no. 33; printed in Edleston, *Correspondence*, pp. 140–1, from the draft in Trinity College Library, Cambridge, which omits the Latin passages.

(2) Newton had mentioned this intentional omission at the end of Letter 941.

(3) Cotes here copies the tenth and eleventh paragraphs of the draft of the Scholium to Proposition 35, which Newton had sent him on 3 May. As Cotes suggests here, the first of the two paragraphs was finally omitted and the second included. See Koyré and Cohen, *Principia*, II, p. 663 and Number 917 *a*.

(4) Cotes presumably refers here to the various corollaries of Proposition 37, but his comment is not strictly true. In particular, values for the mean horizontal parallax of the moon in its various positions were not included, though they were later inserted into the third edition as Corollary 10 (see Koyré and Cohen, *Principia*, II, p. 673).

(5) $60\frac{1}{4}$ was printed.

(6) In his next letter Newton confirmed that he wanted the values to be as in his revised draft, and these were printed.

945 NEWTON AND HALLEY TO OXFORD

18 SEPTEMBER 1712

From the original in the Public Record Office.[1]
Reply to Letter 940

Leicester Fields. Sept. 18*th*, 1712

May it please your Lordp.

According to your Lordps Order signified to us by your Letter of 28th Aug last, We have sent for Mr Cawood[2] & heard what he had to offer concerning the instrument invented by him for the improvement of navigation, which being a new sort of magnetical needle proposed to stand truly north & south without any variation, we were of opinion that such a needle would be of good use. But upon examining the needle wch he produced. & repeating the examination several days we found it weak in vertue & uncertain in its position & for the most part it had a variation like other needles, so that we are not yet satisfied that it will be of use. We have given him time to shew us something further about it: & when we meet wth satisfaction We will humbly signify it to your Lordp. We are

My Lord
Your Lordps most humble
and most obedient Servants
IS. NEWTON.
EDM: HALLEY.

Lord H. Treasurer

NOTES

(1) T/1, 152, no. 13. The letter is signed by both Newton and Halley.
(2) See Letter 940*a*.

946 NEWTON TO COTES

23 SEPTEMBER 1712

From the original in Trinity College Library, Cambridge.(1)
Reply to Letter 944; for the answer see Letter 953

Sr

I beleive it will be sufficient to insert only the last of the two Paragraphs wch you have copied in your last, vizt that wch concerns the refraction of the Atmosphere.(2) The alterations made in the last Paragraph of the Scholium were advisedly.(3) The description of the Variatio secunda is derived only from phænomena & wants to be made more accurate by them that have leasure & plenty of exact observations. The public must take it as it is. It brings the Moon nearer to the Sun in both the Quadratures. I am

Your most humble Servant

IS. NEWTON.

London Sept. 23 1712

For the Rnd Mr Cotes Professor of Astronomy in the University of Cambridge At his chamber in Trinity College.

NOTES

(1) R.16.38, no. 241; printed in Edleston, *Correspondence*, p. 141.
(2) See Letter 944, note (4).
(3) See Letter 944.

947 NEWTON TO [?MEDOWS AND BRUCE]

[? SEPTEMBER 1712]

From the holograph draft in the Mint Papers(1)

Gentlemen,

I have reexamined the paper I left in your hands last week concerning her Majts. allowance to her forces at Dunkirk per 20*s*. sterling, & send it to you under my hand.

The three Guilder piece of Holland new out of the Mint is worth in new English silver money at a medium 5*s*. 2$d\frac{3}{4}$. And thence 20*s* in new English silver money is intrinsically worth in new Dutch silver money at a medium 11

Guilders 9 stivers $5\frac{3}{4}$ deniers.(2) But 20*s*. English have of late passed in Holland for ten Guilders & 9 or 10 styvers, or at a medium for ten Guilders & $9\frac{1}{2}$ styvers. And by consequence the loss by the exchange was about one Guilder. Of this loss her Majty allowed the forces in Flanders $5\frac{1}{2}$ styvers & the forces lost about $14\frac{1}{2}$ styvers. Her Majts allowance was after the rate of $2\frac{2}{5}$ per cent & the loss of the forces after the rate of $6\frac{1}{3}$ per cent.

The French crown new species fresh out of the Mint is worth in new English silver 5*s* 1*d*. And thence 20*s* English are worth in French silver money new species 19 livres 13 sous 5,3 deniers,(2) at a medium But 20*s* English silver money pass at Dunkirk for 17*s*. 00. 00. [*read* 17 livres] And by consequence the loss by the exchange is 2 livres 13 sous 5,3 deniers And this loss being divided between her Maj't & the forces at Dunkirk in the same proportion as before in Flanders: her Majts allowance will come to $14\frac{7}{10}$ livres [*read* sous], or in the next round number to 15 sous, wch is after the rate of $3\frac{4}{5}$ per cent And the forces at Dunkirk will beare the loss of 1 livre 18 sous 5,3 deniers wch is after the rate of $9\frac{3}{4}$ per cent.

According to the proportion therefore allowed in Flanders her Majestys allowance to the forces at Dunkirk for 20*s* will amount [to] 17 livres 15 sous recconing a crown piece new species of France at 5 livres. But her Majty may vary the proportion at pleasure.

NOTES

(1) Mint Papers II, fo. 162r. Since this draft letter was not intended for Oxford, we assume it to be a part of the preliminaries preceding the submission of the rather curious report of 7 October, addressed to Oxford by the comptrollers of the accounts of the Army. It will be observed that the first part of this paper corresponds to the *second* part of Newton's memorandum on the exchanges (Number 948*a* below), but the second part of this letter puts forward a more positive recommendation than does the final report (Number 948 below).

(2) As in the English monetary system, there were twelve *deniers* (pence) to one *stuiver* or *sou* (*solidus* = shilling) and twenty *stuivers* or *sous* to the guilder or *livre*.

948 MEDOWS, BRUCE AND NEWTON TO OXFORD

7 OCTOBER 1712

From the original in the Public Record Office.(1)
Reply to Letter 936

To the Most Honble: Robert Earl of Oxford and Earl Mortimer Lord High Treasurer of Great Britain

May it please Your Lordsp.

IN Obedience to Your Lordship's Order of Reference Signified to Us by Mr. Taylour the 21st. August last, upon the Annexed Memorial of Mr. Brydges for Directions as to the Rate that French Current Money shall be

Computed at for the subsistence of Her Maties Forces in Dunkirk, and Your Lordsp. Directing Us to Consult Sir Isaac Newton thereon.

We the Comptrollers of the Accounts of the Army and Master [and] Worker of Her Majesties Mint, Do humbly Report to Your lordship, That Twenty Shillings Sterling pass at Dunkirk for Seventeen Livres, but are worth (Intrinsick Vallue) Nineteen Livres thirteen sols which Leaves a Loss of two Livres thirteen Sols or two shillings eight pence Sterling Whereas in Flanders the Loss is but about twelve Stivers, or One Shilling two pence Sterling (whereof about half was borne by the Queen) as appears by the Annexed Computation.(2)

We further Observe to Your Lordsp. That We cannot Yet learne what may be the Loss of buying Bullion of the Merchants, and sending it to Dunkirk to be coined at the next Mint for paying the Forces.

But We are Credibly Inform'd the Forces at Dunkirk are willing still to Accept of small Bills by way of Antwerp, as has hitherto been practised, and upon the old Allowance, which seems to be Cheapest for Her Majesty, and may be Continued by the present Signe Manual till further Reason occurs to alter it.(3)

We also offer to Your Lordship's Consideration whether provisions may not be sent from hence in part of their Subsistance.

All which is humbly submitted
to Your Lordship.

Comptrollers Office P. MEDOWS
Privy Garden 7th. October 1712 JA. BRUCE
IS. NEWTON

NOTES

(1) T/1, 152, no. 44. All three men signed. It is doubtful that Newton drafted this letter. It is printed in Shaw, pp. 182–3.

(2) See Number 948*a*.

(3) It is hard to see how this suggestion solves Brydges' problem of calculating the forces' pay in French money.

948*a* NEWTON'S ANNEXED MEMORANDUM

[early OCTOBER 1712]

From the holograph original in the Public Record Office.(1)

In the course of Exchange, nine pounds sterling are recconed at a par with 100 Gilders specie money of Holland, or 1£ with 11 Gilders $2\frac{2}{9}$ Styvers. But 1£ sterling lately passed in Holland only for 10 Gilders & 9 or 10 Styvers, or at a medium for 10 Gilders $9\frac{1}{2}$ Styvers. The defect is $12\frac{13}{18}$ Styvers, whereof Her Majty allowed to the Forces in Flanders $5\frac{1}{2}$ Styvers, wch is almost one half of the defect or loss by the exchange.

The par between English & French money of the new species is not yet setled by the course of exchange: but by weight & assay I find that an unworn French crown piece of the new species wch passes at Dunkirk & in France for five livres is worth 5*s* 1*d* sterling. And at this rate 20*s* sterling are worth 19 livres $13\frac{27}{61}$ sous. But 20*s* sterling pass at Dunkirk for only 17 livres. The defect or loss is 2 livres $13\frac{27}{61}$ sous, to be divided between her Maty & the forces. And as $12\frac{13}{18}$ styvers to $5\frac{1}{2}$ styvers, so are 2 livres $13\frac{27}{61}$ sous to 1 livre 3 sous, her Majts proportional part of the defect: wch added to 17 livres the current value of 20*s* sterling at Dunkirk, makes her Majts allowance for the pound sterling 18 livres 3 sous recconing a French crown new species at 5 livres. But her Maty may alter the proportion at pleasure & make the allowance in a rounder number.(2)

When 9£ sterling are recconed at a par with 100 Gilders as above, the specie money of Holland is over-valued by about $3\frac{1}{4}$ per cent. For the three Gilder piece unworn is worth only $62\frac{3}{4}$ pence sterling by the weight & assay. And thence 9£ sterling are intrinsecally worth about $103\frac{1}{4}$ Gilders. & 1£ sterling, wch lately passed at about 10 Gilders $9\frac{1}{2}$ Stivers, is worth 11 Gilders $9\frac{4}{9}$ stivers, & the loss by the exchange is about a Gilder whereof her Maty bare only $5\frac{1}{2}$ stivers, wch is about a quarter of the whole loss. And according to this proportion her Maty should beare but a quarter of the loss by the exchange at Dunkirk. But the rules of the Exchange where they are setled being generally followed, I presume it might be her Majts intention to beare about one half of the loss by the Exchange in Holland, as in the recconing first set down in this paper.

IS. NEWTON

NOTES

(1) T/1, 152, no. 44B, printed in Shaw, pp. 184–5. There are many drafts and computations relating to the Dunkirk money problem in the Mint Papers, II, fos. 155–72. None seemed sufficiently important to print here; several are similar to the letters printed here, others are lists of coins and their values. Newton must have obtained information from merchants about current values. One draft (*ibid.*, fos. 170–1) ends as this memorandum does, but begins: 'That all sorts of money are current at Dunkirk but under such values as are put upon them by the French. That the old French Ecus & Lewis dor's are lately called in...'

(2) It will be obvious that the problem to which Newton addressed himself was: given a certain exchange loss (defined as the difference between specie content of the foreign coin and current exchange value) what proportion of it should be borne by the Crown, according to the prevalent precedent? (The proportion *not* borne by the Crown simply devalued the troops allowances.) This problem does not seem to have interested his administrative colleagues at all.

949 [NEWTON TO OXFORD]
[? OCTOBER 1712]

From Newton's holograph draft in the University Library, Cambridge(1)

To the most Honble the Earl of Oxford & Earl Mortimer Ld High Treasurer of great Britain

The humble Petition of MS. Katherine Barton Widdow of Lieut. Coll. Barton [who was lost in ye late shipwreck in ye River of Canada](2) most humbly sheweth

That your Petitioners late Husband Lieutenant Col. Barton who had served her [Majesty](2) in the Army from the age of nineteen years being lost in the late shipwreck in the River of Canada together with his equipage & provisions for the expedition to Quebeck to ye value of four hundred pounds & having enjoyed his place of Lieutenant Collonel but one year for wch he paid seven hundred-pounds, (most part of the said moneys being out of your petitioners fortune) & leaving ye Petitioner wth three small children & an annuity of 100£ pr an in ye Exchange for his life being also lost to his family: your Petitioner soon after the said shipwrack petitioned her Maty for relief, & your Lordp being then moved in behalf of your Petitioner was pleased to give a very favourable answer importing that you would take care of your Petitionr in this matter. Your Petitioner therefore most humbly prays your Lordsp for a speedy relief in consideration of her great losses.

And your Petitioner shall ever pray &c

NOTES

(1) Add. 3968(14), fo. 245r. Compare Letter 878 above. The paper also carries notes, probably written later than this draft, bearing on Newton's anonymous 'review' of the *Commercium Epistolicum*.

On 23 December 1712 the Earl of Oxford directed Spencer Compton to pay Katherine Barton the sum of £30 as of her Majesty's bounty; she was one of a number of recipients of similar awards (*Cal. Treasury Papers* (Part II), 1712, p. 546). Further, on 2 July 1715, the name of Katherine Barton appears as a pensioner (£80) on the Irish Military Establishment (*ibid.* (Part II), 1715, p. 594). We have rather arbitrarily inserted this draft some space of time before the former award. The latter one can hardly have resulted from an appeal to Oxford, who had fallen from power twelve months previously. According to Brewster (*Memoirs*, II, pp. 396–7 and note) Robert Barton left three children to whom 'Sir Isaac gave away an estate in the parish of Baydon in Wiltshire' a short time before his death. This branch of the family died out. However, Robert's widow Katherine married a Colonel Gardner, and their daughter married into the (presumably unrelated) family of Cutts Barton.

(2) These words are deleted.

950 NEWTON TO HENRY INGLE

13 OCTOBER 1712

From the original in the Yale Medical Library[1]

Henry Ingle

I read over the proposal you sent me of enclosing the Pasture & stinting the Commons & approve of the same if the Cow commons be first stinted by agreement of the Freeholders under hand & seal so that after the pasture is enclosed it may not be in the power of a Jury or of the greater part of the Parish to [break the] stint.[2] I beleive that the number of Commons belonging to every farm & Cottage as well before the stint as by the stint should be expressed in the writing. And since there is an old list of the Cow-commons I beleive it will not be difficult to settle those Commons. But some having of late years transgressed in the number of sheep Commons & perhaps in that also of hors commons, I think there should be made lists of those Commons by the consent of the Parish before the stint be agreed upon. Mr Proctor or who the Parish thinks fit may draw up the form of a writing for stinting the Commons & before it is signed & sealed by the neighbours I desire that a copy may be [s]ent me to peruse. The intended enclosure of the Pasture may be also mentioned therein. And as soon as the writing is signed & sealed the enclosure may be made. I have fourteen score sheep commons of wch six score belong to ye royalty; also thirteen cow commons & an half & sevent[een] hors commons besides two hors commons wch I [lately] bought of my Cousin William Ayscough[3] wth the close next the Ling close.

I understand that the neighbours have of late years eaten the fallow Lings wth great cattel between Low Sunday & All Saints,[4] wch is contrary to an Award decreed in Chancry. I desire that they would forbear eating that piece of grownd wth any other cattel then sheep in that part of the year. Otherwise I shall cause their great cattle to be pounded.

To Pearsons cottage belong two cow commons & tenn sheep commons, to John a Manns the same & to Newtons the same To Porters none.

I beleive there will be some difficulty in setling the sheep-commons And if in the writing for inclosing the Pasture the commons for neat Beasts be stinted, I shall agree to it. The horse commons may be stinted in another writing & the Sheep Commons in a third. I return my thanks to the neighbours for ordering you to give me an account of this matter, & remain

Yr very loving friend
ISAAC NEWTON.

London 13 *Octob*
1712.

I desire you to acquaint my Tennant Tho. Percival that I have given John Newton the bearer one of the two decayd Trees on Li..n bank in the Becks, that wch is most decayed, & desire that he would let him cut it down this autumn.

NOTES

(1) Printed in C. H. Turnor, 'The Country-side in Newton's Day', *Reports and Papers of the Architectural Societies of the County of Lincoln, County of York, Archdeaconries of Northampton and Oakham, and County of Leicester*, XXXVIII (1926), p. 71. We are grateful to Dr F. Gyorgyey for supplying photocopies of this and other letters in the Yale Medical Library, and to Yale University for permission to reproduce them. It is possible that this is a draft, for Newton often added date and signature to drafts. There is no address, and it seems highly unlikely that Newton would have omitted 'Sir'. The manuscript is apparently defective and we have indicated the omission of a few illegible words; these have been supplied in square brackets from Turnor's text.

(2) Presumably the letter concerns the Manor of Woolsthorpe, of which Newton was Lord. It was worth about £30 a year to him. Henry Ingle was evidently his bailiff; he is mentioned in C. W. Foster's account of Newton's family (*Reports and Papers of...the Architectural & Archaeological Society of the County of Lincoln*, XXXIX (1928), p. 28) as the former owner of land at Colsterworth bought by Newton's distant cousin, Robert Newton. To stint a common is to restrict the number of beasts which each commoner may pasture upon it. For a similar letter see vol. II, pp. 502–4.

(3) Ayscough was the maiden name of Newton's mother.

(4) 17 April to 1 November.

951 NEWTON TO COTES

14 OCTOBER 1712

From the original in Trinity College Library, Cambridge.[1]
For the answer see Letter 953

Sr

I send you the conclusion[2] of the Theory of the Comets to be added at ye end of the book after the words [Dato autem Latere transverso datur etiam tempus periodicum Cometæ Q.E.I.]

There is an error[3] in the tenth Proposition of the second Book, Prob. III, wch will require the reprinting of about a sheet & an half. I was told of it since I wrote to you, & am correcting it. I will pay the charge of reprinting it, & send it to you as soon as I can make it ready.[4] With my service to Dr Bentley I remain

Your most humble Servant
IS. NEWTON.

London
14 *Octob.*
1712

NOTES

(1) R.16.38, no. 242; printed in Edleston, *Correspondence*, pp. 141–2.

(2) Bentley had informed Cotes in July that Newton intended this addition (see Letter 929, first paragraph). The passage (now in MS. R.16.38, fos. 252–5) is a long one appearing on pp. 476–81 of the second edition (see Koyré and Cohen, *Principia*, II, pp. 749–58: 'Cæterum Cometarum revolventium...quod sciam deprehendit').

(3) For the modifications from the first to the second editions in Proposition 10 see Koyré and Cohen, I, pp. 376–94. A. R. Hall in *Osiris*, XIII (1958), 313–17 translates the text of the first edition and gives a lengthy analysis; see also Cohen, *Introduction*, pp. 236–8 and D. T Whiteside in *Journal of the History of Astronomy*, **1** (1970), 128–9.

Newton became aware of the need for this correction about 1 October 1712; see the next document, Letter 951*a*. Proposition 10 is concerned with the motion of a body through a medium resisting as its density and as the square of the velocity of the body. Newton's method in both editions is to derive general expressions for the resistance and velocity at any point and then consider the motion along particular curved trajectories again deriving expressions for density and velocity; this enables him to assert (in the Scholium to Proposition 10) that in a medium resisting uniformly as the square of the velocity the trajectory will approximate to an hyperbola. However, particular results stated in the first edition were, as it was proved, in error by a factor of 3/2; and the fault lay not in the derivation of these results but in the opening, general analysis. Hence Newton was compelled to rework the whole, introducing new figures, so as to arrive in the version finally printed at exactly the same expressions as before, reduced by the factor of 2/3.

(4) After much prompting by Cotes, Newton at last sent the corrected version of Proposition 10 to Cotes on 6 January 1713 (Letter 961). For the printing details, see Letter 961, note (2).

951*a* NEWTON TO NIKOLAUS BERNOULLI

[? *c.* 1 OCTOBER 1712]

From the holograph draft in the University Library, Cambridge[1]

Sr

I send you inclosed the solution of ye Probleme about the density of resisting Mediums, set right.[2] I desire you to shew it to your Unkle & return my thanks to him for sending me notice of ye mistake. I am

NOTES

(1) Add. 3965(12), fo. 219v, printed in *Journal of the History of Astronomy*, **1**, 1970, 136, note 46. On the recto and the remainder of the verso of this sheet is a rough draft of the corrected proof, nearly as printed. However, Nikolaus Bernoulli's uncle Johann received no full version of the corrected proof and had to wait nearly two years before seeing the revised version in the second edition. According to his letter to de Moivre of 30 March 1714 he had then received copies of neither the *Commercium Epistolicum* nor the 1713 *Principia*. (Wollenschläger, *De Moivre*, p. 287.)

The recipient of this letter was born in Basel in 1687 and died there in 1759; Jakob [I] and Johann [I] Bernoulli were his uncles. He travelled to Holland, England and France in 1712. In 1716 he went to a chair in Padua, returning to Basel six years later. He was a correspondent of Leibniz, Euler and particularly Pierre Rémond de Monmort. He developed the accusation that Newton did not understand how to derive second differentials.

(2) That is, *Principia*, Book II, Proposition 10; compare Letter 951. Johann [I] Bernoulli (1667–1748), professor at Groningen and Basel, friend and correspondent of the Marquis de l'Hospital, of Varignon, and particularly of Leibniz, was now one of Europe's foremost mathematicians and author of the first textbook on calculus (see O. Spiess, *Der Briefwechsel von Johann Bernoulli*, Basel, 1955—unfortunately this work never progressed beyond volume I). He had for some time—like others—accumulated a list of errors in the first edition of the *Principia*, and by April 1710 he was expressing to Leibniz his interest in seeing whether Newton, in the new edition then in preparation (of which he knew from Abraham de Moivre) had corrected all of them. In particular, examining by his own methods the descent of a body in a semicircular trajectory, in a resisting medium defined as Newton had defined it in Proposition 10, Bernoulli came to a result differing by a factor of 3 to 2 from that of Newton and was able to show that Newton's expression entailed the absurd result that the vertical velocity of descent would be uniform. But he could not find the point at which Newton's argument in the 1687 *Principia* went astray. Johann Bernoulli communicated the error of Proposition 10 to Leibniz in a letter of 12 August 1710 (see Gerhardt, *Leibniz: Mathematische Schriften*, III/2, pp. 854–5.) He published a fuller exposition in the *Acta Eruditorum* for February 1713, pp. 77–97, and March 1713, pp. 115–32 and another in the *Mémoires de l'Académie Royale des Sciences* for 1711 (which appeared in 1714), pp. 47–54.

However, long before Johann Bernoulli's papers had been printed, his nephew Nikolaus, visiting London in the autumn of 1712, brought news of the error to Newton. Here the best contemporary reporter is the refugee Huguenot mathematician, Abraham de Moivre (1667–1754), who was on friendly terms with Newton. He wrote to Johann Bernoulli of his nephew's visit on 18 October 1712 (translating from the French printed in Wollenschläger, *De Moivre*, pp. 270–1):

It was a very pleasant surprise to me to see in this country your nephew [Nikolaus]: I have tried to impress on him as forcibly as I could the particular esteem I feel for his learning, his gentle and sincere conduct; and in this, Sir, I have done no more than all will do, who have the happiness of making his acquaintance; but I confess to you that his blood-relationship to yourself has increased in me the warmth of friendship and goodwill that he deserves to receive from everyone. I have had the honour of introducing him to Mr. Newton and to Mr. Halley, and I have no doubt that he will tell you that he has reason to be satisfied with the reception these gentlemen afforded him. We have together seen Mr. Newton three times, and he was so amiable as to invite us twice to dine with him; I must not omit a remarkable circumstance: your nephew having told me that he had an objection against a result in Mr. Newton's book concerning the motion of a body describing a circle in a resisting medium, and having imparted this objection to me, I thereupon, on his behalf, showed it to Mr. Newton. Mr. Newton said he would examine it and two or three days later, when I had gone to his house, he told me that the objection was valid, and that he had corrected the result; indeed, he showed me his correction, and it proved agreeable to the computation made by your nephew. Thereupon he added that he intended to see your nephew in order to thank him, and begged me to bring him to his house, which I did. Moreover, Mr. Newton affirms that this error is the simple consequence of his having considered a tangent at the wrong end, but that the foundation of his calculation and the conclusions he has drawn from it may still stand.

A curious point here is the evident supposition of de Moivre that the original author of the 'objection' was *Nikolaus*. The latter's *Addition* to his uncle's 'Lettre de 10 Janvier 1711' (*Mémoires de l'Académie Royale des Sciences*, 1711 (Paris, 1714), pp. 54–6) also claims the 'objection' for himself; on 13 November, however, Johann assumed that it was 'one of my comments that my nephew has seen here' and later (8 February 1713) referred to it again as 'Mr. Newton's error about which my nephew communicated the comment I made long ago...'. By this time Johann's own papers on the resistance of the medium were printed or in press. The division of honours seems to be that after Johann had discovered the particular case of Newton's mistake in the semicircular trajectory of Proposition 10, Nikolaus had conjectured (correctly) that the error would extend to the whole proposition, and (incorrectly) that its source was Newton's confusion of the coefficients Q, R, and S with the successive derivatives $Q = dy/dx$, $2R = d^2y/dx^2$, and $6S = d^3y/dx^3$. *Nikolaus*' error was first analysed in detail by Lagrange in 1797 (see D. T. Whiteside in *Journal for the History of Astronomy*, **1** (1970), 128–9 and notes).

Newton always insisted that he was the first to discover the mistake in the *Principia* (1687) demonstration. He devoted fifty sheets to his investigation of Proposition 10; and when he had finally found the error and its solution worked out five distinct proofs of the accuracy of the new solution (see D. T. Whiteside, *ibid.*, **1** (1970), 128–9).

952 NEWTON TO COTES

21 OCTOBER 1712

From the original in Trinity College Library, Cambridge.(1)
For the answer see Letter 953

Sr

I sent you last tuesday(2) a sheet inclosed in a Letter. It concerned the The [*sic*] Theory of Comets to be added to ye end of the book. I should be glad to hear that it came to your hands. I mentioned also an error that I was lately told of & wch wants to be set right.(3) I have heard nothing from you this month or above & should be glad of a line to know in what forwardness the Press is. I am

Your most humble Servant
IS. NEWTON

London.
Octob. 21. 1712.

For the Rnd Mr Roger Cotes
Professor of Astronomy at his Chambers
in Trinity College in Cambridge.

NOTES

(1) R.16.38, no. 256, printed in Edleston, *Correspondence*, p. 143.
(2) 14 October 1712, Letter 951.
(3) That is, the error in Proposition 10, Book II; see Letter 951, note (3).

953 COTES TO NEWTON

23 OCTOBER 1712

From the original in the University Library, Cambridge.[1]
Reply to Letters 946, 951 and 952

Sr

I received both Your last Letters, togather with the Sheet to be added at the end of the Book, which was inclosed in the former. You mentioned an Error in the Xth Proposition of the Second Book, which will require the reprinting of about a Sheet & an half. I have not review'd that Proposition to see if I might find it out, but shall stay for Your corrections.

The Sheet which is now under the Press, ends in page 492 of ye old Edition, & page 456 of the new Edition.

I have not observ'd any thing of moment to be altered in the Theory of Comets. In the new fourth Corollary of Prop. XL I have inserted after the first line [& quadratum radij illius ponatur esse partium 100000000]. Pag. 490, line. 5, I have put [in subduplicata ratione *SQ* ad *St*] instead of [in subduplicata ratione *St* ad *SQ*]. In the last page of the Book, lines 8 & 9, I design to put $2G-2C$ & $2T-2S$ for $G-C$ & $T-S$, unless You forbid it.[2] I suppose the Astronomical computations relating to the Comets are exact, having been examined both by Yourself & by Dr Halley.

I should have given You notice sooner, that I had received Your additional Sheet at the end of the Book, but that I expected Dr Bentley would have seen You before this time, for he once intended to have been at London a week ago.

I am Sir.
Your most Humble Servant
ROGER COTES.

Cambridge Octbr. 23. 1712

To
Sr Isaac Newton
at His House
in St Martin's-street
Leicester Feilds
London

NOTES

(1) Add. 3983, no. 34. There is a draft (printed in Edleston, *Correspondence*, pp. 143–4) in Trinity College Library, Cambridge (R.16.38, fo. 257). The square brackets are in the original.

(2) All these corrections were made; see pp. 454 and 476 of the second edition; the latter is not, of course, the last page of this edition.

954 NEWTON TO OXFORD

29 OCTOBER 1712

From the original in the Public Record Office.[1]
For the answer see Letter 957

To the most Honble the Earl of Oxford, & Earl Mortimer
Lord High Treasurer of great Britain

May it please your Lordp

The late Smith of the Mint Mr Tho. Silvester supporting himself against me & the other Officers of the Mint by his interest with the Officers of Ordnance whose Smith he also was:[2] to prevent the like treatment for the future by a servant, the salary of fifty pounds per annum to the Smith was in the two last Indentures of the Mint declared to cease upon the next voydance of the place, with intention to settle a smaller sallary only by Warrant. The place became voyd about two years ago, & one Richard Fletcher who formerly did the work for the coynage under Mr Silvester, hath ever since continued to do the same work, & for his service under me I have paid him out of my own allowance for coynage (according to ancient agreement) one penny per pound weight of all the gold & one farthing per pound weight of all the silver coined. But he complains very much that without a salary he is not able to carry on the service.[3] The Smith in her Mats Mint at Edinburgh being allowed a salary of thirty pounds per annum, I humbly pray your Lordp that I may be impowered by a Warrant under her Mats signe manual to pay the Smith here for the time being a salary of thirty or forty pounds per annum instead of the fifty which is taken away, & that this salary may commence from Christmas last.[4]

All which is most humbly submitted to your Lordps great wisdome.

Is. Newton

Mint Office
Octob. 29*th.* 1712.

NOTES

(1) T/1, 153, no. 21.

(2) For Newton's grievances against Sylvester see Letter 865. Sylvester had died about Christmas 1710, and Fletcher was to succeed him a year later, therefore. This suggests the possibility that Newton changed his mind about the draft letter of 7 August 1711 (Letter 865), and never sent it to the Treasury, for it is obvious that the Treasury never acted upon it. Perhaps Newton wished to satisfy himself that Fletcher was sufficiently obedient.

(3) The smith of the mint did the heavy work for the engraver in preparing puncheons and dies for striking the coin, and no doubt looked after the machinery also. There was neither work nor profit for him when Mint business was slack. The doubling of duty with the Ordnance Officers had gone on for some generations.

(4) See Number 957 of 20 November.

955 MARY PILKINGTON TO NEWTON

30 OCTOBER 1712

From the original in the Bodleian Library[1]

Oct: *ye* 30 1712

Honrd: Sr

I have reciv'd ye nine pounds you payed to Mrs Savage & return you my most Humble thanks for it.[2] I am very glad to hear of your good health & wish it may long Continue.

I have not been well this 3 weeks of a sore throt & a pain in my wright sid I have been Blooded for it & taking physick which has hinder'd mee sending my thanks sooner. I am in great hopes my Brother gorge[3] will gett to bee Steward of ye House to ye duke of Devonshire, Mr. Grovnor[4] who is ye dukes head Steward has writ to my Cosin Pilkington a bout my Brother, pray Sr give my services to my Cosin Barton & bee Please to except of dutty from

your Most obedient
Nece & Humble Servant
M. Pilkington

To
Sr Isaac Newton at his
House in St Martins Street
Near Lecester feilds
London

NOTES

(1) New College 361. The postmark is NO 3. The verso of the sheet carries a number of loose notes on ancient chronology.

(2) Compare Letter 906. It seems that Newton paid Mary Pilkington the younger £9 quarterly or half-yearly, through Mrs Savage.

(3) See Letter 906, note (1), for George.

(4) Doubtful reading; perhaps for the name usually spelled 'Grosvenor'.

956 COTES TO NEWTON

1 NOVEMBER 1712

From the original in Trinity College Library, Cambridge(1)

Sr

I here send You the Sheets as far as they are Printed off, that Your self or some freind may revise them, in order to see what Errata may be put in a Table.(2) I know not whether You have got the Copper-plate of the Comet yet done.(3) The Printer tells me there will be 750 requisite. The next week I shall be in the Countrey, when I return I suppose You will have the corrections ready which You mention'd for the Sheet to be reprinted(4)

I am Sir
Your most humble Servt.
ROGER COTES

Nov. 1*st.* 1712
For Sr Isaac Newton at his
House in St Martin's-street
Leicester feilds London

NOTES

(1) R.16.38, no. 259; printed in Edleston, *Correspondence*, p. 144.

(2) Newton sent Cotes a sheet of 'Corrigenda & Addenda' at the end of 1713 (see Letter 1029, vol. VI).

(3) See Letters 977, 979 and 980. The cuts were finally printed off at the end of March. See Letter 989.

(4) The corrections to Proposition 10, Book II were sent on 6 January 1712/13 (Letter 961).

957 ROYAL WARRANT TO NEWTON

20 NOVEMBER 1712

From the copy in the Public Record Office.(1)
Reply to Letter 954

ANNE R.

Whereas in the two last Indentures of Our Mint It was declared that the Salary of Fifty pounds per annum payable to Thomas Silvester then Smith of Our Mint in the Tower of London Should upon the next avoidance of the said place be discontinued, And Whereas Our High Treasurer of Great Britain hath laid before us a Memorial in the matter from our Trusty & welbeloved Sr Isaac Newton Knight the present Master & Worker of Our said Mint, proprosing [*sic*] that in lieu of the said former salary of fifty pounds per annum he may be impowered to pay to the Smith of our Said Mint for the time being an Allowance of Forty pounds per annum which will be an Encouragement to such Officer to performe the Duty of his Employment, and that the said Salary may commence from Christmas last; We taking the Premises into our Royal Consideration are gratiously pleased to approve thereof, Our Will & Pleasure therefore is, And We do hereby Authorize and direct the Master & Worker of Our Mint in the Tower of London now & for the time being Out of such money as is or shall be Impresteḍ to him at the Receipt of Our Exchequer out of the Coynage Duty to pay unto the Smith of our Said Mint now & for the time being the allowance of Fourty pounds per annum from Xmas aforesaid. And this shall be as well to him for payment as to the Auditor for allowing thereof from time to time upon his Accounts a sufficient Warrant, Given at Our Court at Windsor Castle the 20th day of November 1712 in the Eleventh yeare of our Reigne

By her Majties. Command
OXFORD

NOTE

(1) Mint/1, 8, 97, the Mint's record copy.

958 COTES TO NEWTON

23 NOVEMBER 1712

From the original in Trinity College Library, Cambridge.(1)
For the answer see Letter 961

Sr

I hope You have received the Sheets which I sent last, ending in Page 456 of the New Edition. We have since printed off 3 Sheets more, which take in the whole Book with the Additional Sheet, excepting about 20 lines. To fill up the following Sheet may be added a Table of the Contents of each Section, if You think fit. Dr Bentley was proposing to have subjoyned an Index to the whole, but particularly to the Third Book. If You approve of it, such an Index may soon be made.(2) If Your alterations in the Second Book are finished I desire You will be pleased to send 'em.(3)

I am Sir, Your most

Humble Servant

ROGER COTES

Cambridge Novbr. 23*d.* 1712

For Sr Isaac Newton at his
House in St Martin's Street
Leicester-Feilds London

NOTES

(1) R.16.38, no. 260, printed in Edleston, *Correspondence*, p. 145.

(2) Since signature Mmm (ending on p. 456) had been sent to Newton, the printing had proceeded through the following signatures until the last 19 lines of the work were—or would be—printed on signature Qqq1r, leaving almost all the rest of this sheet blank. Finally added to the book were the *Scholium Generale* (see Letters 961 and 977, and subsequent letters) and an *Index Rerum Alphabeticus* prepared by Cotes after discussion with Bentley (see Letters 979 and 983) covering all three books. These additions filled signature Qqq and extended over three pages of signature Rrr; the *Corrigenda* were printed on the last page of the book, Rrr2v.

(3) This refers to the correction to Proposition 10, Book II. See Letter 951, note (3).

959 T. HARLEY TO THE MINT

5 DECEMBER 1712

From the copy in the Public Record Office.(1)

For the answer see Letter 960

Officers Mint abt Halfepence &c

Gentlemen

My Lord Treasurer is pleased to Direct you to make forth and transmit to his Lordp. an Extract of the Severall Proposalls that have been Referred to You since Her Mats. Accession to the Crown for making halfe pence & Farthings And of the Reports that have been made by You thereupon I am &c

5th Decr 1712(2)

T. Harley

NOTES

(1) T/27, 20, 314, the Treasury's minute of this letter.

(2) No copper coinage was to be issued in Anne's reign, nor was any copper struck before 1718. However, this brief note set a great deal of activity in train, leading up to unsuccessful experiments in minting copper in the autumn of 1713. Progress was stopped by the fall of Oxford. But all this time the interest of speculators and copper-refiners in the Mint, as a possible market, was intense.

960 THE MINT TO OXFORD

DECEMBER 1712

From the copy in the Public Record Office.(1)

Reply to Letter 959

To the Most Honble: the Lord High Trea[su]rer of Great Britain

May it Please Your Lordship

In Obedience to Your Lordships Commands of the 5th Instant Wee humbly Lay before Your Lordship the following Extract of the Several proposals that have been Refer'd to this Office since Her Mats. Accession to the Crowne for making half pence & farthings and of the Reports that have been made thereupon Vizt.

Abel Slaney Citizen and Woollen draper of London as principal Undertaker for the Coining half pence and farthings in the Reign of the late King and Queen alledging that He was a very great Sufferer in the Changing of Tin halfpence and farthings for Copper by Tale,(2) proposed in the Year 1703 to Coin 700 Tun of Copper half pence and farthings in Seven Years of Equal value, weight and fineness of the last half pence and farthings to be melted, rolled, cut & stamped att Her Mats. Mint in the Tower, Subject to a Comptroller to be appointed by Her Majesty and at the Undertakers Expence.

In Consideration of Such Grant the said Slaney for himself & Partners proposed to give Her Majesty a Fine of 5000£ and a Rent of 1000£ per Annum by Half Yearly Payment, and to be under such Restrictions & Regulations as Her Maty should think Reasonable. Thomas Renda Esqr. Edward Ambrose & Danl: Barton, who were before partners with the said Slaney in Coining the former half pence & farthings understanding that the said Slaney designed to Intitle a new Sett of partners to the Merit of another patent upon the Termes by Him proposed did peticion that if Her Maty. thought fitt to Grant a new pattent for making Copper half pence & farthings, Strangers might nott reap the Benefitt of the Expences they had been att in performing the former pattent which they pretended was done to their Loss, butt that they might have such new pattent paying for the same what was proposed by others.

Will. Shepherd, N. Shepard and Geo. Freeman did in March $170\frac{4}{5}$ peticon to have a patent to Impower them to Coine Forty or Fifty Tunns every year for Eight or Ten years obliging themselves to make them of English Copper of equal Weight [&] Fineness with those now Currant.

The Fellow Monyers being poor and Needy and haveing no Worke in the Mint did about the same time peticon to the same Effect that out of the profitts of such Coinage they might sustain themselves until the Mint was sett to work about Gold and Silver Money.

Soon after the Union Sr. Talbot Clerk and partners did represent that having in the Year 1686 obtained Letters patents for 14 Years to putt in practice a new Invention of Furnaces for melting and refining Metals out of Oars and that by their Care and Expence great Advantage had Acrued to the Nation but that by reason of great Difficultys they mett with in the Management and the time being Expired they had nott made the hoped for Advantage they therefore did peticon that in some Recompence for their Charges and Expences they might send in two Tunns of Copper Blanks per Week into the Mint untill they had Disposed of Seven Hundred Tunns.

Mr. Chambers hearing of this proposal of Sr. Talbot Clerk represented that He and divers other persons had purchased at a very dear rate of the said Sr. Talbot Clerk and others concerned with him their Interest in the said Pattent and were afterwards incorporated by K[ing] W[illiam] and Q[ueen] M[ary] under the Name of the Governour and Comp[any] of Copper Mines in England And that having very much Improved the Copper Works and at the Charge of above £20,000 having obtained the knowledge of making Copper fine and having a greater stock in his hands then could be disposed off did propose to send 100 Tuns of Copper into the Mint at the rate of 12*d* per pound to be there Coined into half pence and Farthings at such Value as should be directed so that the Charge of Coining the same & other Incidents might be born out and that He might have 12*d* per pound to be paid to Him as fast as the Copper money should be disposed off.

William Morgan Gent. and others did in the Year 1708 peticon for a Grant for the Coining 1000 Tuns of English Copper one half into half pence and the other half into Farthings and half Farthings within the Terme of Seven Years to be of Weight and Fineness according to a Standard to be agreed to which Standard was to be at least 20lb per Centum finer and better Copper then the 700 Tuns formerly Coined: And was to be melted Assayed rolled cutt and Stamped at the Mint in the Tower Subject to a Comptroller to be appointed by Her Majesty and at the Expence of the Undertakers

By this proposal all the Copper half pence & Farthings Formerly Coined were to be taken in & Exchanged by the proposer in Tale for those of the new Stamp & so melted down

Mr. William Palmes in the Year 1710 did peticon that towards a Recompence for Losses he had sustained he might have a pattent for the Coining 700 Tuns of Copper in fourteen Years subject to such Agreement Limitations & Covenants as were made in the pattent granted for the Coining the former 700 Tuns.(3)

The Severall Reports that have been made upon those respective peticons and proposals have all been to the same Effect humbly Setting forth that all the Coinages of half pence and Farthings since the Year 1672 vizt in the Reigns of King Charles the 2d. King James the 2d. and in the Beginning of their late Majestys King William and Queen Mary were performed by Comm[issione]rs who had Money Imprested from the Exchequer to buy Copper and Tin and Coined at most at 20*d* per pound Avoirdepoise and Accounted upon Oath to the Government for the Charge and produce thereof by Tale.

That upon Calling in the Tin Farthings and half pence by reason of the Complaint made against them A pattent was Granted to Sr. Joseph Herne(4) & Others who Contracted to Change the same and to Enable them to bear that Charge they were allowed to Coine 700 Tunns at 21*d* per pound weight with a remedy of a half penny without being Accountable to the Government for the Tale the reason of which allowance ceasing Wee have all along been humbly of Opinion that the said pattent was nott to be drawn into president, especially Since the Money made thereby was light, of bad Copper, and ill Coined.

Wee have further humbly reported that its best to Coin the Copper Money as near as can be conveniently to the intrinsic value including the Charges of Coinage Sett Allowances and Incidents and reckoning the Copper att what it would sell for if the new Money should be melted down Again for which reason itt ought to be free from such mixtures as diminish the markett price And that what ever profitt arises by the Coinage Her Majesty may have itt in Her power to Gratifye whom She please therewith And therefore the former Method by Commission and upon Account seemed the more Safe Commendable & Advantageous to the Government, especially if the Method used in the Coinage of Gold and Silver be Observed as near as can be conveniently in the Coinage of Copper For thereby the Coinage may come nearer to the Intrinsic Value and will be better performed and of better Copper and by a Standing Commission any Quantity may be Coined at any time as the Uses of the Nation shall from time to time require for preventing Complaints For in the times of the peticons & proposals above mencond there was at first no want & afterwards no Considerable want of Copper Money and itt was thought Safest to Coin only what was wanted least the Coinage of too great a Quantity of [*read* at] once should occasion Complaints as it did actually in Parliament in the Coinage of the first Six Hundred Tuns of the present Copper Money.

And further upon the Peticon of Mr. Morgan there was a Verbal Report that to call in all the Copper Money then Currant would be a Loss of 70 or 80 Thousand pounds to the Government or above: and that a Thousand Tunns were too much Six or Seven Hundred Tunns being found Sufficient to Stock the Nation of England. And to an Argu[men]t of the peti[tioner]s that a new Coinage of weightier and better money would cause the Old Money to be rejected by the people and lose its Currancy it was Answered that a great Coinage Suppose of 600 or 700 Tuns might have that Effect because alone Sufficient for the Uses of the Nation but a Small Coinage nott Sufficient for that purpose was best.

This is the Tenour & Substance of the Reports which have been made upon the peticons & proposals referred to this Office during Her Majestys Reigne

All which is most humbly Submitted to Your Lo[rdshi]ps. great Wisdom

CRAVEN PEYTON
IS. NEWTON
EDWD. PHELIPPS.(5)

Mint Office the (6) *December* 1712.

NOTES

(1) Mint/1, 7, 56–8; printed in C. Wilson Peck, *English Copper, Tin and Bronze Coins in the British Museum*, 1558–1958 (London, 1964), pp. 608–10. There is another copy (lacking signature and date) in the Mint Papers, II, fos. 404–5; this is written in the same clerical hand. A related holograph draft by Newton is in Mint Papers, II, fo. 339.

(2) For the partnership of Slaney and Barton see vol. IV, p. 408 and p. 409, note (2). The coinage of tin money extended from 1684 to 1692 (see vol. IV, p. 409, note (4)).

(3) See Letter 813.
(4) See vol. IV, p. 409, note (2).
(5) Edward Phelipps or Philipps had been appointed Comptroller of the Mint on 6 June 1711 (*Cal. Treas. Books*, XXV (Part II), 1711, p. 293). He was perhaps the graduate of Trinity College, Oxford, of this same name who was called to the Bar and sat as M.P. for Ilchester (1708–14) and Somersetshire (1722–7).
(6) Blank in the copy.

961 NEWTON TO COTES

6 JANUARY 1712/13

From the original in Trinity College Library, Cambridge.[1]
Reply to Letter 958; for the answer see Letter 962

Sr

I send you enclosed the tenth Proposition of the second book corrected.[2] It will require the reprinting of a sheet & a quarter from pag 230 to pag. 240.[3] There is a wooden cut belonging to it wch I intend to send you by the next Carrier.[4] I think this Proposition as it is now done will take up much the same space as before. If not, the space about the cuts may be made a little wider or a little narrower, or the number of lines in a page may be increased or diminished by a line.[5] When this sheet & a quarter is printed off I hope your trouble of correcting will be at an end. As for making a Table to the book, I leave it to you to do what you think. I beleive a short one will be sufficient.[6] I shall send you in a few days a Scholiu[m][7] of about a quarter of a sheet to be added to the [end][8] of the book: & some are perswading me to add an Appendix concerning the attraction of the small particles of bodies.[9] It will take up about three quarters of a sheet, but I am not yet resolved about it. I am

Your humble & obedient
servant
IS. NEWTON

London.
Jan. 6. 171$\frac{2}{3}$

For the Rnd Mr Cotes Professor
of Astronomy at his Chamber
in Trinity College in
Cambridge.

NOTES

(1) R.16.38.no. 261; printed in Edleston, *Correspondence*, pp. 145–6.
(2) The mistake was first noted in Newton's letter of 14 October, Letter 951.
(3) The reprinting was, in fact, managed as follows: signature Gg4 (*Principia*; pp. 231–2) was torn away (roughly, in our copy) and the last part of Proposition 9 reprinted on the cancel,

which was attached to the stub of the old sheet. The printing of the new Proposition 10 was begun on the verso of the cancel (*ibid.*, p. 232) and continued through a completely reprinted signature Hh (*ibid.*, pp. 233–40). With signature Ii (*ibid.*, p. 241) the old text resumes at 'Reg. 1'.

(4) This was the first cut in the Proposition; the second cut was, in the second edition, the same as in the first, and indeed reappears on the unreprinted p. 241.

(5) Cotes omitted a short paragraph that had appeared in the first edition (see Letter 962, note (3)). With that done, the printer had no difficulty in fitting the new copy to the pages.

(6) In fact Cotes prepared, from about 10 March, a rather elaborate *Index Rerum Alphabeticus*, filling 6½ pages, besides the *Index Capitum* at the beginning of the book.

(7) This was the General Scholium concluding the *Principia*, which grew to considerable size. Cotes had to wait for it until Newton wrote again, on 2 March.

(8) The paper is torn by breaking of the seal.

(9) There are several draft forms of the General Scholium, of which at least two contain matter on 'the attraction of the small particles of bodies' (see Hall & Hall, *Unpublished Scientific Papers*, pp. 346–64 and Cohen *Introduction*, pp. 240–5). We print another draft that Newton wrote for the General Scholium (and later re-worked for *Opticks*) as Number 961*a*.

961*a* DE VI ELECTRICA

From a holograph draft in the University Library, Cambridge(1)

In Opticorum Libro tertio(2) ostendimus quod Radii Lucis in vicinia corporum inflectuntur licet non incidant in ipsa corpora quodque reflectuntur et refringuntur non incidendo in solidas corporum particulas sed permeando poros eorum ad parvas a particulis distantias, et quod his actionibus vices induunt facilioris reflectionis & facilioris transmissus. & in corporibus vibrationes excitent in quo calor consistit. Latet utique spiritus aliquis in corporibus omnibus quo mediante lux et corpora in se mutuo agunt. Hic spiritus non terminatur ad externas corporum partes in superficiebis mathematicis, sed paulatim rarescit & rariore sui parte e corporibus undique ad distantias parvas evagatur & paulatim cessat. Hujus actione radij ad parvas a corporibus distantias inflectuntur, idque eo magis quo sunt corporibus propiores, & permeando corpora pellucida non reflectuntur & refringuntur in unico tantum puncto sed paulatim incurvantur ut in Prop. XCIV, XCV & XCV [*sic*] Lib. 1 explicui.(3) Et præterea agendo in hoc medium excitant in eadem vibrationes perinde ut corpora mota excitant vibrationes in aere. Hæ vibrationes sunt ipsa luce velociores & radios ejus successive assequuntur & prætergrediuntur, & assequendo disponunt radios alternis vicibus ad facilem reflexionem & facilem transmissionem & luce Solis excitatæ particulas corporum agitant & his agitationibus corpora calefaciunt & ad superficies corporum reflexæ revertuntur in corpora ac diutissime perseverant.(4) Et ubi corpus calidum immergitur in corpus frigidum transeunt e corpore calido in frigidum & calorem citissime communicant. Radij vero incidendo in fundum oculi vibrationes excitant in hoc medio qui per solida nervi optici capillamenta in cerebrum delati visionem excitant. Et quemadmodum vibrationes in aere per corpora sonora excitatæ pro magnitudine sua et numero tonos omnes sonorum producunt, sic vibrationes hujus spiritus in capillamentis nervi optici latentis pro magnitudine suæ et numero colores omnes lucis efficiunt. Radii maxime refrangibiles producunt vibrationes breviores et eodem tempore plures qui sensationem excitant violaci coloris; minime refrangibiles producunt vibrationes ampliores & numero pauciores sensationem excitant rubri coloris; cæteri pro gradu refrangibilitatis vibrationes producunt intermedias qui pro magnitudine sua et numero sensationem excitant colorum intermediorum citrij flavi viridis cærulis et indici. Et cum natura simplex sit et sibi similis, & spiritus hicce medium sit deferendi impressiones lucis in cerebrum, idem erit medium deferendi impressiones sonorum odorum saporum & contactus per capillamenta nervorum aliorum in cerebrum ut & impressiones a voluntate factas in cerebro per aliorum nervorum capillamenta in musculos ad membra animalium movenda.

Est autem hic spiritus causa attractionis electricæ et lucem non tantum reflectit refringit & inflectit sed etiam emittit: id quod manifestum est per experimentum sequens.(5) Globus vitreus diametro digitorum septem vel octo, aere vacuus & circum axem celerrime rotatus, si a manu immota fricetur spiritum electricum emittit qui e regione digitorum ad internas vitri partes lucebit. Et si vitrum non sit aere omni vacuum spiritus electricus in internas vitri partes egrediens miscebitur cum aere & instar flammæ cujusdam tenuis totam vitri cavitatem implentis lucebit. Et si corpus aliquod album ad externas vitri partes in medio fere spatio inter polos locetur, & intervallo quarto circiter vel tertiæ partis digiti a vitro distet: spiritus electricus qui per frictionem excitatur & e vitro revolvente in aerem externum egreditur, impingendo in corpus illus album lucebit, et efficiet ut corpus illud album appareat lucidum instar carbonis vivi vel ligni, putrescentis in tenebris. Jam vero spiritus electricus, qui, posquam [*sic*] per frictionem e corpore egreditur, lucem per vibrationes suas emittit, medium esse potest quo flamma omnis & corpora omnia ignita & ligna putrescentia lucem emittant.

Eodem spiritu mediante particulæ corporum se mutuo trahunt ad parvas distantias; etiam absque frictione. Nam si vitra duo plana perpolita superficiebus sibi parallelis & horizonti perpendicularibus ad minimam ab invicem distantiam collocentur, & inferiore sui parte in aquam stagnantem immergantur; aqua ascendet inter vitra & altitudo ascensus erit reciproce proportionalis intervallo vitrorum. Si vitra inclinentur ad invicem et concurrant ad lineam horizonti perpendicularem, aqua ascendet inter vitra & superiore sua superficie figuram hyperbolicam efformabit cujus una asymptotos est linea concursus vitrorem et altera asymptotos est linea horizonti parallela ad superficiem aquæ stagnantis.[6] Si fistulæ vitreæ pertenues in aquam stagnantem immergantur, aqua ascendent in ijsdem ad altitudines reciproce proportionales diametris cavitatum fistularum. Et hæc experimenta a sociis Societatis Regalis[7] excogitata & coram Societate probata succedunt in vacuo Boyliano æque ac in aere aperto, ideoque pressioni atmosphæræ non debentur. Particulæ vitri ad supremam aquæ ascendentis superficiem, particulas aquæ sibi proximas & se inferiores attrahunt & attrahendo attollunt, & hæ particulæ attrahunt alias & hæ alias ad usque aquam stagnantem. Et quoniam vitrum eadem vi attrahit aquam sive distantia superficierum vitrorum sit paulo major sive eadem sit paulo minor, ideo pondus aquæ elevatæ idem est in omni casu & ut pondus idem sit. altitudo aqua est reciproce ut ejus latitudo. Et eadem de causa liquores in spongijs & substantijs spongiosis ascendere solent.

Particulas corporum ad parvas distantias se mutuo attrahere manifestum est etiam per hoc experimentum[8] Si vitra duo plana et perpolita parentur longitudine pedis unius duorumve latitudine digitorum trium vel quatuor Et eorum alterutrum horizonti parallelum statuatur & gutta olei malorum citriorum in hoc vitrum ad ejus terminum alterutrum cadat, dein vitrum alterum priori sic imponatur ut vitra intervallo quasi decimæ sextæ partis digiti ab invicem distent ad terminum illum ubi gutta olei jacet, ad alterum vero terminum sese contingant, sic, ut spatium inter vitra formam habeat cunei: gutta olei quamprimum a vitro superiore tangitur incipiet motu primum lento dein celeriore & perpetim accelerato versus concursum vitrorum moveri & motu tandem celerrimo in concursum illum ruet. Et si vitra ad terminum illum ubi concurrunt, nonnihil eleventur gutta tamen ascendet, sed motu paulo tardiori; versus occursum vitrorum scilicet attracta. Et si vitra ad terminum ubi concurrunt adhuc magis eleventur, gutta tardior evadet ac tandem quiescet, in æquilibrio inter vim gravitatis suæ et vim attractionis vitrorum suspensa: Et majori vitrorum elevatione gravitas prævalebit & gutta descendet. Et hoc experimentum succedit in vacuo æque ac in aere aperto, ideoque a vi aeris minime dependet. Oritur ascensus guttæ a vi sola vitrorum.

Per idem experimentum innotescit etiam quantitas attractionis. Si vitra erant 20 digitos longa. Gutta olei ad distantiam quatuor digitorum a concursu vitrorum stabat in æquilibrio ubi vitrum inferius inclinabatur ad horizontem in angulo graduum plus minus sex. Est autem Radius 10000 ad sinum anguli 6 gr vizt 1045 ut totum guttæ pondus, quod dicatur P, ad vim ponderis juxta planum vitri inferioris $\frac{1045}{10000}P$, cui vis attractionis versus concursum vitrorum æqualis est. Hæc autem vis est ad vim attractionis versus plana vitrorum ut sinus semissis anguli quem vitra continent ad Radium id est ut tricesima secunda pars digiti ad viginti digitos: ideoque vis attractionis versus plana vitrorum est $\frac{20\times 32\times 1045}{10000}P$ & vis attractionis versus planum vitri alter utrius est $\frac{10\times 32\times 1045}{10000}P$ seu 33,44P. Distantia vitrorum ad guttam seu crassitudo [vel] altitudo guttæ erat octogesima pars digiti ideoque pondus cylindri olei cujus diameter eadem est cum diametro guttæ & altitudo est digiti unius æquat 80P & hoc pondus est ad vim qua gutta attrahitur in vitrum alterutrum ut 80 ad 33,44. ideoque vis attractionis æquatur ponderi cylindri cujus altitudo est 33,44 pars digiti.(9) Hæc ita se habent ubi distantia guttæ a concursum vitrorum est digitorum quatuor: in alijs distantijs vis attractionis prodit per experimentum reciproce ut distantia guttæ a concursu vitrorum quamproxime. id est reciproce ut crassitudo guttæ: & propterea vis attractionis æqualis est ponderi cylindri olei cujus basis eadem est cum basi guttæ et altitudo est ad $\frac{33,44}{80}$ dig ut $\frac{1}{80}$ ad crassitudinem guttæ.(10) Sit crassitudo guttæ $\frac{1}{10000000}$ pars digiti(11) et altitudo cylindri erit $\frac{3344000}{64}$ dig seu 52550 dig.(12) id est 871 passuum. et pondus ejus æquabitur vi attractionis[.] Tanta autem vis ad cohæsionem partium corporis abunde sufficet.

Vires autem tantæ quibus particulæ corporum se mutuo attrahunt & cohærent, in fermentationibus putrefactionibus & operationibus chymicis, insignes habebunt effectus.

Translation

On the Electric Spirit

In the third book of *Opticks*(2) we showed that the rays of light are inflected in the neighbourhood of bodies although they do not fall upon the bodies themselves and that the rays are reflected and refracted not by falling upon the solid particles of bodies but by permeating their pores, at short distances from the particles, and that by these actions they create the fits of more easy reflection and more easy transmission. And within bodies they excite the vibrations of which heat consists. Certainly there is some spirit hid in all bodies, by means of which light and bodies act upon each other mutually. This spirit is not terminated in the external parts of bodies at a mathematical surface, but becomes rarer gradually, and the more rare part of it spreads out from bodies on all sides to short distances and gradually comes to an end. By the action of this spirit the rays are inflected at short distances from bodies, and to a greater extent the closer they pass to bodies; and in permeating transparent bodies they are not reflected and refracted only a single point but are gradually curved around as I have explained in Propositions 94, 95 and [96] of Book I[(3)]. And furthermore by acting on this medium they excite vibrations in it in such a way that the bodies, being set in motion, excite vibrations in the air. These vibrations are swifter than light itself and successively pursue and overtake the waves, and in pursuing them they dispose the rays into the alternate fits of being easily reflected or easily transmitted; and when excited by the light of the Sun they [the vibrations] excite the particles of bodies and by this excitation heat the bodies; and being reflected at the surfaces of bodies [internally] they turn back within the bodies and persist a very long time.[(4)] But when the hot body is immersed in a cold one they pass out of the hot body into the cold one and very quickly communicate the heat to it. When the rays fall upon the bottom of the eye they excite vibrations in this medium which, carried through the solid capillaments of the optic nerve into the brain, create vision. And just as the vibrations excited in air by resonant bodies produce all the tones of sound according to their magnitude [wavelength] and number [frequency], so the vibrations of the spirit hidden within the capillaments of the optic nerve create all the colours of light according to their magnitude and number. The most refrangible rays produce the shorter vibrations and, in the same time, more of them; these arouse [the sensation of] the colour violet; the least refrangible rays produce larger vibrations, fewer in number, which arouse [the sensation of] a red colour. The remainder produce intermediate vibrations according to their degree of refrangibility which, in proportion to their magnitude and number arouse [the sensation of] the intermediate colours orange, yellow, green, blue and indigo.

And because nature is simple and conformable to herself, and this very spirit is the vehicle for transporting the impressions of light into the brain, the same must be the vehicle for transporting the impressions of sounds, odours, taste and touch through the capillaments of other nerves into the brain, as also the impressions made by the will in the brain (through the capillaments of other [motor] nerves) into the muscles, in order to move the limbs of animals.

This spirit is, however, the cause of electrical attraction and not only reflects, refracts and inflects light but also emits it; as is manifest from the following experiment.[(5)] A glass globe, seven or eight inches in diameter, is exhausted of air and rotated very rapidly upon an axle; if the hand, kept stationary, is rubbed against it, it emits the electric spirit which will shine from the place where the fingers are into the interior of the glass. And if the glass is not wholly void of air the electric spirit issuing forth into the interior of the glass will mingle with the air and will shine like some thin flame filling the whole interior cavity of the glass. And if some white body is placed near the exterior of the glass, near the middle between its poles, separated by about the third or fourth part of an inch from the glass, the electric spirit excited by friction and issuing forth from the revolving glass into the external air and impinging upon that white body will shine, and cause that white body to glow like a live coal or a piece of putrefying wood in the dark. Now surely the electric spirit which, after it has been forced out of bodies by rubbing them emits light by its vibrations, may be the vehicle by which all flames and all bodies are ignited, and by which putrefying wood emits light.

Thanks to the same spirit the particles of bodies attract each other mutually at short distances, even without friction. For if two highly polished plane plates of glass are arranged with their surfaces parallel to each other and perpendicular to the horizon with the smallest possible interval between them, and their lower edges are dipped into still water, the water will ascend between the plates and the height of its ascent will be reciprocally proportional to the interval between them. If the plates are slanted towards one another to meet along a line perpendicular to the horizon, the water will ascend between them and with its upper surface trace a hyperbolic line, one of whose asymptotes is the line of junction of the glass plates while the other asymptote is the line parallel to the horizon made by the surface of the standing water.(6) If very slender glass tubes are dipped into still water the water ascends in them to heights reciprocally proportional to the diameters of the bores of the tubes. And these experiments devised by Fellows of the Royal Society(7) and tested before it succeed as well in the Boylian vacuum as in the open air, and so are not due to atmospheric pressure. The particles of glass at the upper surface of the rising water attract the particles of water nearest to themselves and below them and by attracting them draw them up, and these [water-] particles attract others [below them] and these yet others down to the standing water. And since the glass attracts the water with the same force whether the distance between the surfaces of the glass be a little greater or a little less, therefore the weight of water raised is in every case the same, and as the weight is the same the height of the water is reciprocally as its breadth. And for the same reason fluids commonly rise upwards in sponges and spongy substances.

That the particles of bodies attract each other mutually at short distances is also manifest from this experiment.(8) If two plane, highly polished plates of glass are prepared, a foot or two long and three of four inches wide, and one of them is set down parallel to the horizon and a drop of oil of oranges is let fall on it near one of its ends, then the second plate of glass is so laid on the first that the interval between the plates is about one-sixteenth part of an inch at the ends where the drop of oil lies, while the plates are in contact at their other ends, so that the space between the plates is wedge-shaped; as soon as the drop of oil is touched by the upper plate it begins to move (at first with a slow motion and then more quickly, being continually accelerated) towards the point of contact of the two plates. And at last dashes into the meeting-place with a very swift motion. And if the plates of glass are somewhat raised up at the ends where they meet, the drops still ascend, though with a slightly slower motion for they are attracted towards the junction of the plates. And if the plates are raised still higher at the junction end the drop moves more slowly and at last comes to rest suspended in equilibrium between the force of its gravity and the force of the attraction of the plates. And at a still higher elevation of the plates gravity predominates and the drop descends. And this experiment succeeds as well *in vacuo* as in the open air, and so is in no way dependent upon the air's force. The ascent of the drop arises from the force of the plates alone.

The quantity of the attraction may be gathered from the same experiment. If the plates were twenty inches long, the drop of oil stood in equilibrium at a distance of four inches from the junction of the plates when the lower glass was inclined to the horizon at an angle of six degrees, more or less. However, the radius 10000 is to the sine of the angle of 6° (that is, 1045) as the whole weight of the drop, which may be called P, to the force of the weight against the plane of the lower plate $\frac{1045}{10000}P$, to which the force of attraction towards the junction of the plates is equal. However, this force is to the force of attraction towards the planes of the glasses as the sine of half the angle between the plates to the radius, that is, as 1/32 inch to 20 inches; and so the force of attraction towards the planes of the glasses is $\frac{20 \times 32 \times 1045}{10000}P$ and the force of attraction towards the plane of either glass plate is $\frac{10 \times 32 \times 1045}{10000}P$, or $33{\cdot}44P$. The distance between the plates at the [site of the] drop or the thickness [or] height of the drop was 1/80 inch and so the weight of a cylinder of the oil whose diameter is the same as the diameter of the drop & whose height is one inch is equal to $80P$, and this weight is to the force by which the drop is attracted to either plate of glass as 80 to 33·44. And so the force of attraction is equal to the weight of a cylinder whose altitude is the 33·44 part of an inch.(9) This is the case when the distance of the drop from the junction of the glasses is four inches; at other distances the force appears by experiment to be reciprocally as the distance of the drop from the junction of the plates, approximately. That is to say, as the thickness of the drop; and furthermore the force of attraction is equal to the weight of a cylinder of the oil whose base is the same as the base of the drop and whose height is to $\frac{33{\cdot}44}{80}$ inches as $\frac{1}{80}$ is to the thickness of the drop.(10) Let the thickness of the drop be 10^{-7} inches (11) and the height of the cylinder will be $\frac{33{\cdot}44 10^5}{64}$ or 52550 inches(12), which is 871 paces; and the weight of this will equal the force of attraction. But so great force is abundantly sufficient to provide the cohesion of bodies.

However, these very great forces by which the particles of bodies attract each other mutually and cohere will effect remarkable results in fermentation, putrefaction and chemical reaction.

NOTES

(1) Add. 3970, fos. 427–8; we owe the reference to D. T. Whiteside. Although this paper is found among *Opticks* MSS., it clearly belongs to the *Principia* and the present period; in fact, it is obvious that the last few lines of the concluding General Scholium, as printed, summarize this draft. At the top of fo. 427r is written: 'Atque hactenus de vi gravitatis & phænomenis cælorum'; below that 'Invenimus per experimenta Optica quod' *deleted*; then 'De vi electrica'. The allusions in the draft imply that *Opticks* was already printed and connect it explicitly with the *Principia* (whose propositions are alluded to without giving the title of the book, *hujus* being understood). Moreover, the draft makes extensive use of experiments made between 1711 and 1713.

(2) In *Opticks*, Book III, Newton discussed the phenomena of inflection discovered by Grimaldi and then 'since I have not finish'd this part of my Design' concluded the volume with his Queries. That reflection is not caused by 'the impinging of Light on the solid or impervious parts of Bodies, as is commonly believed' is proved in Book II, Proposition 8. The 'fits' are first mentioned (in print) in Book II, Proposition 12. Hence Newton at first wrote 'Libro secundo' but altered the second word to 'tertio'.

(3) These are the 'optical' propositions of the *Principia* in which Newton showed that streams of particles attracted in a specific manner towards large bodies would exhibit properties analogous to those of light-rays. For an earlier description of such an æther or spirit as is described in this paragraph see vol. II, pp. 288–90 (the letter to Boyle of 28 February 1678/9)

(4) Compare *Opticks*, Book II, Proposition 12 where this notion is described as a 'Hypothesis...whether [it] be true or false I do not here consider'; and also Newton's 'Hypothesis explaining the Properties of Light', 7 December 1675 (vol. II, pp. 377–8 particularly).

(5) This celebrated experiment was first performed before the Royal Society in 1705 by Francis Hauksbee. The luminescence produced by the oscillation of mercury in a barometer had been observed much earlier. The remainder of this draft clearly illustrates the importance attached by Newton to the physical experiments of Hauksbee, to which attention has been drawn by Henry Guerlac (see *Archives internationales d'Histoire des Sciences*, **63** (1963), 113–28 and *Mélanges Alexandre Koyré*, I (1964), 228–54).

(6) This experiment on capillary attraction was first made by Brook Taylor, who surmized the hyperbolic form of the water surface between the plates, and repeated more exactly by Hauksbee (see *Philosophical Transactions*, **27**, no. 336 (for October to December 1712), 538 and 539).

(7) That is, Francis Hauksbee (elected F.R.S. in 1705) and Brook Taylor, Newton was in the Chair when these various experiments were performed before the Society.

(8) This experiment also was made by Hauksbee; it is described in *Philosophical Transactions*, **27**, no. 332 (for October to December 1711) and no. 334 (for April to June 1712), 395–6 and 473–4). This experiment was first discussed in print by Newton (together with Hauksbee's other studies on capillarity) in the second English edition of *Opticks* published in 1717, in his additions to Query 31. These passages do not, of course, occur in the preceding Latin edition of 1706; they are obviously re-worked from the present draft though more cautious and showing some discrepancies. In *Opticks* Newton writes: 'Now by some Experiments of this Kind, (made by Mr. *Hauksbee*) it has been found that the Attraction [of the drop] is almost reciprocally in a duplicate Proportion of the distance of the middle of the Drop from the Concourse of the Glasses...' (Dover reprint, 1952, p. 393.)

(9) The words 'ideoque...digiti' are interlineated; for 33·44 read $\frac{33\cdot 44}{80^2}$.

(10) That is, $h = \frac{33\cdot 44}{80^2 . b}$.

(11) This value is of the same order of magnitude as that of the least separation of two glass surfaces pressed together, according to *Opticks*, Book II, Part II, Dover reprint, 1952, p. 233.

(12) *Read*: 52250 inches, equal to 4354 feet.

962 COTES TO NEWTON

13 JANUARY 1712/13

From the original in Trinity College Library, Cambridge.[1]

Reply to Letter 961; For the answer see Letter 977

Cambridge Jan. 13*th* 1713

Sr.

I have considered Your alteration of Prop: X Lib. II. and am well satisfied with it. I observe that You have increased the Resistance in the proportion of 3 to 2, which is the only change in Your Conclusions, arising from hence (as I apprehend it) that in the new Figure *LH* is to *NI* as *Roo* to $Roo+3So^3$, whereas in ye former Figure *kl* was to *FG* as *Roo* to $Roo+2So^3$.(2) Some things in Your Paper I have altered, they are not worth Your notice, being only faults in transcribing.(3) I have this day received the Wooden Cut. I shall expect the Scholium at ye end of the Book & the Appendix at Your leasure.

I am Sir

Your obliged Freind
& Humble Servant
ROGER COTES.

For Sr. Isaac Newton
at his House in St. Martin's Street
Leicester Feilds London

NOTES

(1) R.16.38, no. 266, printed in Edleston, *Correspondence*, pp. 146–7.

(2) The relation of the lines *LH, NI* is correctly given by Cotes (see *Principia* 1713; p. 234, ll. 5–6), but strictly the relation of *kl* to *FG* in the first edition was as Ro^2-So^3 to Ro^2+So^3 (p. 264, ll. 26–7). However, as the ratio approaches equality one may write it as Cotes does.

(3) Edleston points out that Cotes also omitted a short paragraph, which may be rendered in English as follows: 'One might imagine the projectiles to continue in the chords of the arcs *GH, HI, IK* and to be acted on by the forces of gravity and resistance only at the points *G, H, I* and *K*, just as in the first proposition of Book I the body was acted upon by an intermittent centripetal force; then the chords may be diminished to infinity in order to render the forces continuous. And the solution of the problem would by this method work out very easily.' This passage would have opened the scholium on p. 240. (Compare Whiteside, *Journal for the History of Astronomy*, 1 (1970), 136–7, note 48.)

963 NEWTON TO THE CHANCELLOR OF THE EXCHEQUER

16 JANUARY 1712/13

From the original in the Public Record Office(1)

Letter from the Master and Worker of the Mint to
Mr. Chancellour concerning the Value of Spanish Money

Mint Office Janry. 16*th* 171$\frac{2}{3}$

Sr

Our Assay Master being out of Town and his Clerk Sick in bed I got the Two Spanish Peices Assayed at Goldsmith's Hall(2) The Peice of King Charles weighed 3d.wt. 12gr. and in fineness was four Peny Weight worse then Standard, The Value thereof in English Money is 10*d.* & 3/5ths of a Peny The Peice of King Philip weighed 4dwts $\frac{1}{4}$ grain and was One Ounce Two Peny

Weight worse then Standard and in Value 11*d* half Peny English They seem to be Quarter Peices of Eight of the New Species and in the nearest Round Numbers, Five of them may be reckon'd worth a Mexico or Pillar Piece of Eight if I had had Two or Three More of each sort I could have made a better Judgement of them, but by the best Judgement I can make of these Two only it seems to Me that they may be dim[in]ished in Value of One Fift Part or in the Proportion of five to four. I am

Sr.

Your most Obedient and
most Humble Servant
IS. NEWTON

NOTES

(1) T/1, 158, no. 15. This is a clerical copy not actually signed by Newton. The Chancellor and Under-Treasurer of the Exchequer at this time was Robert Benson (d. 1731); he was M.P. for the City of York and had been one of the Treasury Commissioners after the fall of Godolphin in 1710; upon being created Baron Bingley (21 July 1713) he resigned his office (1 November 1713) and was sent on an embassy to Madrid.

(2) The occasion for this assay is not known; the pieces were presumably of two reals (compare Letter 875).

964 NEWTON TO HERCULES SCOTT

[28 JANUARY 1712/13]

From a draft in the Mint Papers.[1]

For the answer see Letter 986

Sr.

I have been long indebted to you for your Letters & was in good hopes that the Question you wrote to me about would have been decided without me.[2] But understanding that it is still depending, I here send you my thoughts about it.

I imploy a Melter to melt all the gold & silver coyned & allow him thirteen pence per pound weight Troy for melting the gold: whereof I reccon at least 3*d* for potts & fire, & the other 10*d* for wast & charges of making up the sweep. Whence the wast doth not amount to five grains in the pound weight of gold. And the wast in silver cannot be much more.

Because I do not make up the sweep my self I cannot speak of this matter by by own experience. But consulting my Melter about it, he told me that the wast in melting was about 6 grains per pound weight of silver, but in refining it was about double to that in melting. And afterwards he told me that in one parcel he had found the wast in melting amount to 14 grains per pound weight. But this I suspect was by some accident, [or falshood in his servants][(3)] For the Goldsmiths reccon the wast so little that they have perswaded the Crown to make no allowance for it in making the money in our Mint, whereas in your Mint the Master is allowed to put twelve grains of Copper into every pound weight of silver when the silver is molten & they are pouring it off into the moulds, & this is done to make amends for the wast of the copper wch fumes away in the melting.

In the year 1707 when the money current in Scotland was to be recoined, we wrote to the Officers of your Mint that we were not allowed to put any copper into the pot for making recompence for the wast of the copper wch fumed away in the melting, & that they were to conform themselves to ye practise of our mint. But they replied that by the flaming coales wch they used in melting, a greater wast was caused then in our Mint, so that unless they were still allowed to putt the 12 grains of copper into the pot, they could not coin the money standard. Whereupon this allowance was connived at.[(4)]

How much the copper fumes away in your meltings by the flaming of the coals I do not know: but I reccon that when the allowance of 12 grains per pound weight was instituted, it was deemed a sufficient recompence for the wast made by fuming away: whereas in our Mint as I said we have no allowance made for recompencing that wast. And thence I gather that the wast in your Mint after making up the sweep ought to be less by some grains then the wast in our Mint: & that the wast upon the whole coinage (if the sweep be well made up) must be under 234 pounds weight & may be so little as not to exceed 125 pounds weight.

For whilst 12 grains of copper are added to every pound weight of silver in every melting & a pound weight of silver makes but about half a pound weight of money: there are about 24 grains of copper added to every pound weight of money coined. And this addition diminishes the whole wast, & should make it 24 grains per pound weight less in your mint then in ours, supposing the wast by the fuming away of the metal to be alike in both mints* And the wast by the flaming coals in your Mint should be 24 grains per pound weight more then in ours to make the whole wast which remains after making up the sweep, equal in both Mints. I am

Your very humble servant
IS. NEWTON

* I mean that yr 24 grains are more then enough to make good all your wast.

NOTES

(1) Mint Papers I, fo. 181. The date is given in Scott's reply, dated 21 March (*ibid.*, 188). There is a very rough draft of this letter in U.L.C. Add. 3965(12), fo. 362r.

(2) Newton means that he hoped Fauquier would have settled the business for him; it can hardly be said that Newton's own letter is decisive on the point at issue.

(3) This phrase has been deleted.

(4) See vol. IV, p. 500, Letter 729.

965 H. SCOTT TO FAUQUIER

29 JANUARY 1712/13

From the copy in the Public Record Office[1]

To Dr. fauquier *Edinburgh Janry* 29: 17$\frac{12}{13}$

Sr.

I was Honoured with your Missive of the 23 August past desiring I would explaine to Sr Isaac Newton what I had writ anent the Recoinage occasioned by a Difference arisen betwixt Mr Allardes Representatives and those of my Brother patrick Scott who was his Depute I writ Sr Isaac upon the 28 sd month and on the 27 Novr.[2] with an exact abrevial of the whole Bullion Imported into the Mint and the Waste arising thereupon I never had any return from Sr Isaac, Sr it would be a singular Kindness if you would Communicate to me Sr. Isaacs thoughts of the Matter for I am loth to importune him further Seing I hear he has been Valetudinary which [I] am heartily Sorry for: I beg leave also to intreat your own Opinion and what other experienced people belonging to your Mint Judge of it wch I request you will nott deny me My good Freind Mr. Thos. Scott will call for your Answer and transmitt itt to my hands: here under is an Extract of the whole Recoynage from the Mint Books[3] that you may have it under Consideration hoping you'll pardon this trouble and Freedom I am

Sr

Your Most Humble and most
Obedient Servt.

HERCULES SCOTT

NOTES

(1) Mint/1, 8, 174, a record copy.

(2) For the former see Letter 939; we have not found the latter.

(3) The summary account shows 104634 lb. of bullion taken into the Edinburgh Mint, of which over 265 lb. was unaccounted for. If the 140 lb. of bullion found in the possession of the late Patrick Scott were credited to the Mint, the deficiency (the 'waste') would become

125 lb. Hercules Scott thought this too low an allowance; and further, adding 31 lb. from his brother's 140 lb. that was in any case due to Allardes' heirs to the Mint's share, he reckoned the maximum waste at 234 lb. which he thought not excessive. This is peculiar accounting, but suited his book.

966 NEWTON TO OXFORD

29 JANUARY 1712/13

From the holograph original at Longleat House(1)

Mint Office
Jan. 29. 171$\frac{2}{3}$

May it please your Lordp

Her Maties Assaymaster of the Mint Mr Daniel Brattel(2) died yesterday about noon, & the place requires a man well qualified for skill & experience to carry on the assays of the gold & silver with a steady hand. Of this sort very few persons are to be met with, & I do not know one better qualified then Mr Charles Brattel the brother of the deceased.(3) In his brothers absence he has frequently acted for him in this service to the satisfaction of the Officers of the Mint, so that we know his ability by experience. We are now in the middle of a coinage of gold, & for carrying on the service without interruption it would be convenient that a new Assaymaster were speedily appointed.(4) It is a Patent place with a salary of 200*lb* per annum & 20*lb* per annum for a Clerk. I am

My Lord
Your Lordps most humble
& most obedt servant

Ld. H. Treasurer   Is. Newton

NOTES

(1) Portland Papers, IX, fo. 189.

(2) According to Craig, *Newton*, p. 86, 'he had obtained the post in succession to his father, Sir John, one of the founder members of the Royal Society, who had been appointed in 1665'. However, there was no F.R.S. of this name. Daniel apparently succeeded his father (upon the latter's death) in 1698, when he petitioned for an increase of salary after working as assistant Assaymaster for twenty years without pay (Mint Papers, I, fo. 104).

(3) He was ultimately appointed. His petition (dated 31 January 1712/13) to succeed Daniel whom he had assisted for fifteen years (i.e. since 1698) is in Mint Papers, I, fo. 96, referred by the Treasury to the Mint. In the same volume of MSS. as the present letter (Portland Papers, fo. 190) is an undated representation of Importers of Bullion in his support.

(4) This was not to be, as other candidates pressed forward vigorously. The most active of them was Catesby Oadham, who alleged that Charles Brattel 'having a place in the Custom house and being Ignorant of the preparing of Aqua Fortis and of the government of his fire, Can't be able to do the Business of an Assay Master...' (*ibid.*, fo. 192). In the end a practical trial was instituted.

967 LOWNDES TO THE MINT
31 JANUARY 1712/13
From the holograph original in the Mint Papers(1)

Whitehall Trea[su]ry Chambers 31*th Janry* 17$\frac{12}{13}$

The Most honoble the Lord High Treasurer is pleased to Referr this Petition to the Warden, Master & Worker & Comptroller of Her Majesties Mint in ye Tower of London who are to Examine the Petitioners Qualifications for the Place desired and to Report their Opinion to his Lordp thereupon.

WM LOWNDES

NOTE

(1) Mint Papers I, fo. 96. The instruction is written on Charles Brattell's beautifully calligraphic petition to Oxford to succeed his brother Daniel as Assaymaster. Compare Letter 966.

968 LOWNDES TO THE MINT
3 FEBRUARY 1712/13
From the copy in the Public Record Office(1)

Officers of ye Mint abt. a Comparison of the Coynes of France & Holland wth yt of England

Gentlemen

My Lord Treasurer Commands me to send you the inclosed paper Containing a Comparison of the Silver & Gold Coynes of France & Holland with the Gold & Silver Species of England, His Lordp directs You to peruse & consider the same, And let him have Your Opinion thereupon in a Report as soon as conveniently You can I am &c 3*th Febry* 1712/3

WM LOWNDES

NOTE

(1) T/27, 20, 342. We have found no reply to this letter, though there are tables (not holograph) of the values of French and Dutch coins in the Mint Papers, II, fos. 136, 156, 159 and 163.

969 LOWNDES TO THE MINT

4 FEBRUARY 1712/13

From the copy in the Public Record Office(1)

Officers Mint abt. Mr Kemp & Mr Cumberlege & proprietors of ye Copper Works at Great Marlow

Gentlemen

My Lord Treasurer Commands me to send You the inclosed Petitions of Mr John Kemp, & Mr. John Cumberlege each of them praying to succeed Mr. Brattle, late Assay Master of the Mint, his Lordp directs You to consider the same, with the other Applications that have been made for the said Office, And that you propose to his Lordp the fittest person in Your Judgement for filling up the said Vacancy.(2) I am &c the *4th Febry* 1712/3

WM LOWNDES

This incloses also a proposition of the proprietors of the Copper Works at Great Marlow for supplying Blank pieces of Copper for Coyning farthings & halfepence, which you'll consider with all other papers now under your Consideration relating to the said Coynage, and make your Report to his Lordp thereupon.(3)

NOTES

(1) T/27, 20, 343. Again, we have found no reply; but for the Assaymaster appointment, see Letter 992, and for the previous report on copper see Letter 960.

(2) Compare Letters 966 and 967.

(3) At Longleat House, in the Portland Papers, IX, fo. 245, there is an undated proposal from John Pery and Partners of Temple Mills, Great Marlow, to supply copper for coinage, which is perhaps that mentioned here.

970 BOLINGBROKE TO THE ROYAL SOCIETY OF LONDON

7 FEBRUARY 1712/13

From the copy in the Letter Book of the Royal Society(1)

Whitehall 7th. February 1712/3

Gentlemen

Her Majesty having been graciously pleased to direct that Instructions should be given hence forward to her Ministers that go abroad to contribute all they can in their several stations towards promoting the design for which the Royal Society was first instituted by corresponding as occasion may require with the President and fellows of the said Society, and by procuring as satisfactory Answers as possible to such Enquiries as may be sent from time to time; This to inform you thereof and to acquaint you at the same time that as her Majesty intends shortly to dispatch a minister to the Court of Mosco,[(2)] if you please to prepare a draught of Instructions whereby he may be usefull to you in those parts, I will lay them before the Queen for her Commands. I am also directed to let you know that her Majtys. intention is that I should write to such of her Ministers in my Department abroad, if you desire any particular to be recommended to them. I am

Gentlemen

Your most humble Servant

BOLINGBROKE

[P.S.]

I have in the best manner I could returned your humble thanks to her Majesty for the Orders which she has been pleased to give: Her Majesty received your Complimt. very graciously and was pleased to express her Intention of Countenancing and Encouraging your Studies.

B:

NOTES

(1) *Letter Book* xv (1713–23), p. 1; the letter is printed in Weld, *History*, I, p. 420. Henry St John had been created Viscount Bolingbroke in 1712.

(2) Charles Whitworth (later Baron Whitworth) who had been in Moscow and St Petersburg as ambassador extraordinary since 1705, with one interruption, left Russia in June 1712; his successor (as Minister-resident) was George Mackenzie, who received his instructions in May 1714 and arrived in St Petersburg in September (see D. B. Horn, *British Diplomatic Representatives, 1689–1789*, Royal Historical Society, Camden Third Series, XLVI, 1932, pp. 110–11).

In the Letter of Instructions to Mackenzie (dated 20 May 1714) there is the following paragraph (P.R.O., F.O. 90, LXXII, unfoliated):

'You are to contribute all that lyes in Your Power towards promoting the Design for which the Royal Society, for the emprovement of Natural knowledge was established at London by Our late Royal Uncle King Charles the 2d of blessed Memory, by corresponding as occasion may require wth. the President and Fellows of the said Society, & by procuring as satisfactory answers as possible to such Enquirys as may be sent to you by them from time to time.'

971 LOWNDES TO THE MINT
12 FEBRUARY 1712/13

From the copy in the Public Record Office[1]

Officers of the Mint abt Weight of 100*lb* in Money according to the Indenture of ye Mint

Gentlemen

The Tellers of the Exchequer, having in pursuance of my Lord Treasurer's Commands laid Accots. before him, of what the money received there in 100 *lb* do generally Weigh, with the difference between that weight & what 100 *lb* in Money should weigh according to the Indenture of the Mint, This by his Lordps Order incloses to You the said Accots. for Your perusall and consideration.[2] I am &c the 12*th Febry* 1712/3

Wm Lowndes

NOTES

(1) T/27, 20, 348. Compare the Treasury Minute of 27 February, Number 975 below.

(2) In effect, the Exchequer was complaining because the taxes were paid in light money, and thought it the Mint's business to stop the abuse.

972 TREASURY REFERENCE TO THE MINT
12 FEBRUARY 1712/13

From the original minute in the Public Record Office[1]

The three aforegoing proposals[2] were referred by order of my Lord Treasurer the 12th of Febry 1712 to the Warden Master & Worker and Comptroller of her Mats. Mint in the Tower of London to consider and Report their Opinion thereupon

signed

Wm Lowndes

NOTES

(1) INDEX 4623, p. 110, first printed in *Cal. Treas. Books*, XXVII (Part II), 1713, pp. 118–19, where further details may be found. Compare also Letter 960.

(2) These are three proposals for supplying copper to the Mint for coining into halfpence and farthings, at various rates, from (*a*) John Pery and partners of the Copper and Brass Works called Temple Mills near Great Marlow, Bucks.; (*b*) Charles Hore; (*c*) Charles Parry, who claims that his father lost £7000 through bringing British copper to perfection. For further action on them see Number 996.

973 DERHAM TO NEWTON

20 FEBRUARY 1712/13

From the original in King's College Library, Cambridge[1]

Upminster ♀ *Feb*: 20. 171$\frac{2}{3}$.

Much Hond Sr

As I was perusing the *Commerc. Epist.*[2] wch ye R.S. honoured me with, it came into my mind yt in some of Mr Collins's L[ette]rs to Mr Townley of Lanc.[3] (now in my hands) there was something relating to that subject, & looking over Mr Towneleys papers, I found a long L[ette]r of Mr Collins's giving a sort of Historical account of the matter, That in Sept. [1668][4] Mercator published his *Logar.*,[5] one of wch he sent to Dr Wallis, &c.... another to Dr Barrow, who thereupon sent him up some papers of Mr Newton (now his successor)[6] by wch, & some other Communications &c. it appears ye method was invented some years before by Mr Newton, & generally applied, &c. Then follows an account of your Method, & of Mr Gregories performance in yt kind,[7] with what Mr Gregory had written to him about it in Feb: 1671 & Jan: 1672,[8] &c.

There is a great deal more, too long to speak of. But if you think the papers may be of use to you, at your request I will bring them wn I next come to London, to be looked over, or transcribed: but I am engaged not to part with any of them out of my power. I have also divers of Mr Sluse's[9] L[ette]rs & other papers of his from Rome & Leige to Mr Towneley, but they being in French, I cannot as yet give any account whether there be anything relating to your matter in them. Not meeting you when I was last in town I shall take this occasion to acquaint you yt I have tryed Mr Huygens's Glass of 122 feet[10] at ♄, ♀, ☽, & ♂ & some Fixt *, & hope shortly to have a view of ♃ also. I believe it by far the best long Glass I ever looked through, representing those Celestials very clean & well. But I can hardly think Mr Huygens could see tollerably through it with the Eye-glass accompanying it, wch is but 6 inches Focus: I therefore make use of Eye-glasses of a longer Focus.[11] I am not yet so well accommodated for strict Observations with this Glass, as to tell you any thing of ♄ Sat. [*sic*] &c. For I am forced to raise a long Ladder, & send my man up with ye glass, neither have I a good Eye-glass to my mind, only some Spectacle-glasses. But would you, or other of my friends that have interest enough, procure me a small Prebend, to enable me to be at charges without injuring my wife & children, I promise you I would stick at no charge to get an Apparatus for this noble Glass, to make it as serviceable to the R.S. as in me lies: & to accomplish some other matters also for their Service. Be pleased to excuse my presumption thus upon your friendship & favour, wch I desire may be no otherwise troublesome to you, than if any thing happens in your way, & you have no other friend capable of it, you would, for the service of ye R S. as well as my self, think of me, & at the same time pardon

Most Hond Sr

Your affectionately humble Servant

WM. DERHAM

If you have any Commands direct
them to Upminster near Rumford,
Essex, by ye General [Pos]t.(12)

To Sr Isaac Newton
in Martin-Street
near Leicester-fields
London
wth great care

NOTES

(1) Keynes MS. 95A; printed in Brewster, *Memoirs*, II, pp. 519–20. William Derham (1657–1735), B.A. Oxford 1679, ordained priest in 1682, had been vicar of Upminster since 1689. He was elected F.R.S. in 1702.

(2) *Commercium Epistolicum D. Johannis Collins et aliorum de analysi promota: jussu Societatis Regiæ in lucem editum*, London, 1712; see the Introduction, p. xxvii. The volume was distributed as a gift; few copies were sold. Cotes had received a copy from Jones by 13 February (Rigaud, *Correspondence*, I, p. 262).

(3) Richard Towneley (1629–1707), see Charles Webster in *Transactions of the Historic Society of Lancashire and Cheshire*, **118** (1966), 51–76. Rigaud, *Correspondence*, I, pp. 188–95 prints only two letters from Towneley to Collins (of 15 April and 13 May 1672) from which, however, it is clear that they exchanged letters frequently. The second letter refers to Newton.

(4) The date has been almost completely removed in breaking the seal.

(5) *Logarithmotechnia*.

(6) Newton succeeded Barrow as Lucasian Professor of Mathematics at Cambridge on 29 October 1669. Barrow first mentioned Newton (not by name) to Collins on 20 July 1669; see also Collins to Gregory of 25 November 1669 (vol. I, pp. 13–15).

(7) James Gregory had published several important mathematical works by this time, but what is significant is that his mathematical advances were made known to Collins through their mutual correspondence (see H. W. Turnbull, *James Gregory Tercentenary Memorial Volume*, London, 1939).

(8) Gregory's letters to Collins of 15 February 1670/1 and 17 January 1671/2 are printed in Turnbull, *ibid.*, pp. 168–72, 210–12.

(9) For René François de Sluse, see vol. I, p. 28, note (1). The correspondence between Sluse and Towneley is mentioned by the latter in the second of the letters noted in (3) above, and in Hall & Hall, *Oldenburg* IX, Letter 2076. Towneley was in fact Sluse's first correspondent in England. All the letters in Derham's possession seem to have been lost or destroyed since.

(10) See vol. III, p. 192, note (2). There is a note of Derham's observations with this objective in the Journal Book of the Royal Society, under date 12 November 1713, and he referred to them later in *Astro-theology* (London, 1715). Later the same objective was employed by James Pound (1669–1724) who, through Newton's influence, acquired the old maypole from the Strand to use as its support.

(11) The shorter the focal length of the eyepiece (or eye lens) the greater the magnification and the smaller the field of view.

(12) The paper is torn away with the seal.

974 LOWNDES TO THE MINT

26 FEBRUARY 1712/13

From the copy in the Public Record Office.(1) For the answer see Letter 981

[To the Principal] Officers [of the] Mint abt raysing the Price of Tyn in Cornwall.

Gentlemen

By Order of My Lord Trea[su]rer I Send You inclosed Mr Anstis's Memo-[ria]ll(2) to his Lordp touching the raysing the Price of One Shilling & Six Pence more on every Hundred of Tyn hereafter to be sold in Cornwall which You are to consider and Report Your Opinion thereon to his Lordp with all Convenient Speed. I am &c 26 Feby 17$\frac{12}{13}$

WM LOWNDES

NOTES

(1) T/27, 20, 363.

(2) This is in P.R.O. T/1, 159, no. 21A, and dated two days before. Anstis (see Letter 907, Note (1)) argued that while a difference of three shillings a hundredweight in the price of tin as between London and Cornwall was reasonable in time of war because of high rates of freight and insurance, the difference should be reduced in peace time, thereby bringing in an increased revenue to the Crown of £400 or more yearly.

975 NEWTON AT THE TREASURY

27 FEBRUARY 1712/13

From the original minute in the Public Record Office.(1)

Whitehall Treasury Chambers 27th Feby 1712

Present

Lord Treasurer Mr Chancellor [of the Exchequer]

Lord Hallifax Mr Attorney & Mr Solicitor come in

The Officers of the Mint Receivers of Customs & Excise and Tellers Clerks are called in

Mr Peyton produced a Crown & severall half Crownes diminished by Washing

Mr Attorney & Sollicitor say diminishing is within ye law agst Treason

Mr Pauncefoot [of the Customs] sais the Tallow chandlers at first brought great quantities of Counterfeit money.

Mr Ferne [of the Excise] says, there is not much bad money comes to his hands

Mr Lilly [Receiver General of the Post Office] says he always tells and weighs the money. Much brass was brought at first but not at present, he finds the weight wanting to be 2 or 3 ounces in a bag.

They all say they cull ye money they suspect.

Mr Attorney saies by old laws in force the money in ye Exchequer is to be taken by weight.

Mr Peyton says half ye silver brought to ye Mint to be coyned is new bullion & they suspect it to be money melted.

The Officers of ye Mint say they have mett with new money clipt and edged again

Sr Isaac Newton saies 6*d* bags are too light abt 14 or 15 oz the 1*s* bags abt 7 oz & $\frac{1}{2}$ the Crownes & $\frac{1}{2}$ Crownes about 4 oz in every 100 lb by wearing & unlawfull diminishing together.

My Lord orders that the reasonable wearing only of ye 6*d* 1*s* & Crowns or half Crownes separately be adjusted & a Medium taken from thence for ye ordinary wear of a bag in wch those peices are promiscuously putt.

The Officers of ye Mint are to consider how to prevent the Counterfeiting of the Copper halfpence & farthings.

NOTE

(1) T/29, 20, 45; the many contractions in this minute have been expanded. The minute was first printed in *Cal. Treas. Books*, XXVII (Part II), 1713, p. 14. The official copy sent to the Mint (covering only a part of what is printed above) is in the Mint Papers, II, fo. 83. Some notes by Newton relative to the matter are in Mint Papers, II, p. 88v, while both sides of the same sheet also contain the matter printed in the next document (Number 976). Compare Lowndes' Letter 971.

976 NEWTON TO [?THE EDITORS OF THE *ACTA ERUDITORUM*]

[FEBRUARY 1712/13]

From drafts in the Mint Papers[1]

I

In the Acta Leipsica of Febr. 1712[2] I meet wth an account of the Analysis per quantitatum series fluxiones ac differentias published by Mr Jones, wch being of a piece wth the Account of Sr Isaac Newtons book de Quadratura Curvarum published in ye Acta Leipsica of January 1705, I beg leave to write to you about it.

The author of this Acct wishes that Mr Jones had published integrum commercium epistolicum Collinsianum or at least more extracts thereof, it being of consequence to know the times when men of note fell into their meditations: & yet has omitted to tell his Readers yt the said Analysis was written in the year 1669, wch was some years before Mr Leibnitz understood the differential method, & continues to assert the first invention of the differential method to Mr Leibnitz

II

In the Acta Eruditorum mensis February 1712 meeting wth an Account of the Analysis per Quantitatum series fluxiones ac differentias lately published by Mr Jones, wherein the Demonstration of the number of the Lines of the third Order is desired, I take the liberty to acquaint you that ye Demonstration is the book it self. The Reader is there taught how to find the plagæ crurorum infinitorum with the Asymptotes of the crura Hyperbolica: & all the variety of the crura infinita of the lines of the third sort are there reduced to four cases wch give the four equations there set down, & therefore those 4. equations are the lines of that Order.

And whereas Mr Jones had in the Preface to yt Book preferr'd his method of rationes primæ & ultimæ to ye method of infinitely littles & the Author of ye said Account replies that ye difference is only in the mode of speaking both methods proceeding alike by quantities infinitely little: I beg leave to acquaint you that the difference between the methods is real. In the method of rationes primæ & ultimæ quantities are never considered as infinitely little. The whole computation is done in finite quantities by the Geometry of Euclid untill you come to an Equation & then by reducing the equation & making one of the quantities vanish you obtain the equation desired.

NOTES

(1) The first draft is from Mint Papers II, fo. 88v, the second from the recto of the same sheet. Each page also contains a draft passage from the General Scholium of the *Principia*, the second passage being in fact the last words of the book in its second edition. Presumably the drafting of the General Scholium was completed before Letter 977, and so these other drafts may have been written about the same time. Certainly this draft – whether intended for the *Acta Eruditorum*, the *Philosophical Transactions*, or an individual – was never sent as a letter; for its development, ultimately into Newton's *Recensio of the Commercium Epistolicum* (1715), see Whiteside, *Mathematical Papers* II, pp. 263–73.

(2) See *Acta Eruditorum*, February 1712 (published in the same year), pp. 74–7, reprinted in Whiteside, *loc. cit.* pp. 259–62.

977 NEWTON TO COTES

2 MARCH 1712/13

From the original in Trinity College Library, Cambridge.[1]
Reply to Letter 962; for the answer see Letter 980

Sr

The inclosed is the Scholium[2] wch I promised to send you, to be added to the end of the book. I intended to have said much more about the attraction of the small particles of bodies, but upon second thoughts I have chose rather to add but one short Paragraph about that part of Philosophy. This Scholium finishes the book.[3] The cut for the Comet of 1680 is going to be rolled off.[4] I am

Your most humble & obedient
Servant

ISAAC NEWTON

London
2d March[5]
171$\frac{2}{3}$.

For the Revnd Mr Roger Cotes
Professor of Astronomy, at his
Chamber in Trinity College
in
Cambridge

NOTES

(1) R.16.38, no. 268; printed in Edleston, *Correspondence*, p. 147.

(2) Trinity College Library, Cambridge, R.16.38, fos. 269, 270, 272. See Letters 961 and 962. Newton wrote several drafts of the *Scholium Generale* before deciding on its final form. See Cohen, *Introduction*, pp. 240–5, and Hall & Hall, *Unpublished Scientific Papers*, pp. 348–64.

(3) See Letter 961, note (9).

(4) The figures of the comet are identical in the two editions, but it seems that the plate has been reworked and 'Pag. 465' (correctly) added. No doubt Newton had the necessary 750 pulls run off in London.

(5) The letter is postmarked 3 March (a Tuesday), and Newton's next letter makes it clear that it was indeed posted on this date.

978 NEWTON AND THE COMMITTEE TO FLAMSTEED

5 MARCH 1712/13

From the original in the Royal Greenwich Observatory(1)

Sir

The Committee of the Royal Society appointed for that purpose, do hereby give You Notice, that they expect that You will in due time send them a Copy of Your Observations made the last Year and ending with December 1712. to the Societies House in Crane-Court, according to the directions of Her Majesties most Gracious Letter of the 12th of December 1710.(2)

We are
Your humble Servts
Is. Newton
F Robartes
Hans Sloane
Abr Hill
Edm: Halley

Crane-Court
March 5. $17\frac{12}{13}$.

To the Reverend
Mr John Flamsteed
at the Royal Observatory
at Greenwich
These

NOTES

(1) Flamsteed MSS, vol. 35, fo. 121 The letter is signed by all Committee members present.

(2) The Royal Warrant, Letter 814: compare Letter 926. It had been agreed at a meeting of the Royal Society on 26 February that the Greenwich Committee should meet on 5 March in order to draft and despatch this letter to Flamsteed.

Flamsteed's observations for 1712 were delivered, sealed, to the Society on 25 June 1713, with a request that the seal should be broken only at a full meeting of the Society in Flamsteed's presence.

979 NEWTON AND BENTLEY TO COTES

5 MARCH 1712/13

From the original in Trinity College Library, Cambridge.(1)
Continuation of Letter 977; for the answer see Letter 980

Sr

I sent you by last tuesdays Post the last sheet of ye Principia,(2) & told you that the cut for ye Comet of 1680 was going to be rolled off. But we want the page where it is to be inserted in the book. I think ye page is 462 or 463.(3) Pray send me wch it is, that it may be graved upon the Plate for directing the Book-binder where to insert it. I am

Your most humble Servant
IS. NEWTON

London
5 *March* 171$\frac{2}{3}$

I(4) have Sr Isaac's Leave to remind you of what You and I were talking of, An alphabetical Index,(5) & a Preface in your own Name;(6) If you please to draw them up ready for ye press, to be printed after my Return to Cambridg, You will oblige

Yours
R. BENTLEY.

For the Rnd Mr Roger Cotes
Professor of Astronomy, at his
Chamber in Trinity College
in
Cambridge

NOTES

(1) R.16.38, no. 273; printed in Edleston, *Correspondence*, p. 148.

(2) See Letter 977; the sheet contained the *Scholium Generale*.

(3) The large, folded plate should face p. 465.

(4) Bentley's postscript is in his own hand.

(5) Bentley and Cotes had discussed this some time before 29 November 1712; see Letters 958 and 961. Cotes wrote to Bentley on 10 March 1713 (Letter 983) that he was ready to start compiling the Index. It was completed by the end of April, when Cotes sent a copy of it to William Jones; see Letter 993, and Cotes' next letter to Jones, Letter 995.

(6) Cotes's concern over the Preface resulted in considerable further correspondence. See Letter 980, note (3).

980 COTES TO NEWTON
8 MARCH 1712/13

From a draft in Trinity College Library, Cambridge.(1)
Reply to Letters 977 and 979; for the answer see Letter 984

Sr

I received both Your Letters with the last sheet of the Book inclosed in the former of them. The Paragraph beginning with Cæterum Trajectoriam quam Cometa descripsit &c, which is in the 497th page of the former Edition, falls in the 465th page of the new Edition. This is the place to which I suppose You would refer the Cut for the Comet.(2)

I intend in a day or two to set about the Alphabetical Index.(3) I will write to Dr Bentley concerning the Preface by ye next Post. I am Sr

Yours &c.

March 8. $17\frac{12}{13}$

NOTES

(1) R.16.38, no. 274; printed in Edleston, *Correspondence*, p. 148. The original letter is lost.

(2) See Letter 979, note (3).

(3) The following passage is here deleted in the draft: 'As for the Preface, I believe it may be of some advantage be very proper that some things should be said in my name which cannot so conveniently be mention'd by Your self. I am not insensible of the little arts which have been used by a certain foreing [*sic*] Mathematician [i.e. Leibniz] to lessen the credit of Your Book abroad.' Cotes must have decided to leave to Bentley the task of tackling Newton about the projected preface. Whilst clearly willing to print the preface in his own name, Cotes would have preferred Newton or Bentley to write it (see Cotes' letter to Bentley, Letter 983). Bentley, after discussion with Newton, sent Cotes on 12 March (Letter 984) a very brief outline of what the preface ought to contain—a sketch of the contents of the work, a comparison with the first edition, and a note on the calculus controversy—but still put upon Cotes the task of actually writing it. Cotes endeavoured to obtain further advice by sending Newton, in Letter 985, details of what he hoped to include about the Newtonian experimental philosophy and about the calculus. Newton proffered a little help over the first of these, and in particular over the rôle of hypotheses in the *Principia* (see Letter 988) but none over the second, and indeed insisted that he did not want to see the preface again before publication (see Letter 989).

The preface as finally printed contained no outline of the work nor details of changes from the first edition. (See Koyré and Cohen, *Principia*, I, pp. 19–35; there is an English translation in Cajori, *Principia*, pp. xx–xxxii). The former was made superfluous by the inclusion of an *Index Capitum Totius Operis*, and the latter because Newton dealt with it himself (albeit briefly) in the *Auctoris Præfatio*. There was no mention of the calculus controversy, despite the plans to include something about it. The preface was thus completely taken up with a discussion of the Newtonian philosophy as contrasted to that of Descartes, with particular reference to the nature of the gravitational force. Samuel Clarke corrected the preface for Cotes, and offered some criticisms; see Letter 1001.

981 THE MINT TO OXFORD

9 MARCH 1712/13

From the original in the Public Record Office.(1) Reply to Letter 974

To the most Honble the Earle of Oxford & Earle
Mortimer Lord High Treasurer of great Britain

May it please your Lordp

In obedience to your Lordps Order of Reference of 26th of last Febr. upon the Memorial of Mr Anstis for raising the price of the Tynn in Cornwal from 3£. 13*s* to 3£. 14*s*. 6*d* per [centum] merchants weight: We have considered the same & are humbly of opinion that the price be raised to 3£ 15*s*, so that for the future it may be sold in Cornwall but one shilling per [centum] cheaper then in London; the difference of the freight from London & from Cornwall to places abroad being but small.

All wch is most humbly submitted to your Lordps great wisdome

C: PEYTON
Is. NEWTON
E PHELIPPS

Mint Office
9 *Mar.* 171$\frac{2}{3}$

NOTE

(1) T/1, 159, no. 21. The letter is in Newton's hand throughout, signed by himself and his colleagues. It is endorsed: 'Agreed'.

982 T. HARLEY TO NEWTON AND PHELIPPS

9 MARCH 1712/13

From the original in the Public Record Office.(1) For the answer see Letter 990

Whitehall Trea[su]ry Chambers 9*th March* 17$\frac{12}{13}$

The Rt honoble. the Lord high Trea[su]rer of Great Britain is pleased to Referr the Mem[oria]ll with the Bills annext, to the Master Worker & Comptroller of her Mats. Mint, who are to consider the Services alledged to be done, examine the reasonableness of the said Bills and to Report their Opinion to his Lordp, what is fit to be done therein

T. HARLEY

NOTE

(1) T/1, 159, no. 19A. The instruction is written on the foot of the Memorial to Oxford from Craven Peyton, Warden of the Mint, dated 16 April 1712 [!] in which he states that there is due to the late Robert Weddell, Peyton's deputed clerk, the sum of £467 15*s* 2*d* for salary,

allowances and charges arising from Weddell's prosecution of currency offenders in the period from 1 June 1710 to Christmas 1711. Peyton seeks directions to pay the sum owing to the widow, Mrs Martha Weddell, from the £400 per annum allowed for the purpose out of the Coinage Duty.

Weddell, who had served Newton when Warden in the same way, is several times mentioned in vol. IV (where he is wrongly given the name Thomas in the Index).

983 COTES TO BENTLEY

10 MARCH 1712/13

From the original in the University Library, Cambridge.(1)
Reply to Letter 979, for the answer see Letter 984

Sr,

I received what You wrote to me in Sr Isaac's Letter. I will set about the Index in a day or two. As to the Preface(2) I should be glad to know from Sr Isaac with what view he thinks proper to have it written. You know the Book has been received abroad with some disadvantage, & the cause of it may easily be guess'd at.(3) The *Commercium Epistolicum*(4) lately published by order of the Royal Society gives such indubitable proof of Mr Leibnitz's want of candour that I shall not scruple in the least to speak out the full truth of the matter if it be thought convenient. There are some peices of his looking this way, which deserve a censure, as his *Tentamen de Motuum Cœlestium causis*.(5) If Sr Isaac is willing that something of this nature may be done, I should be very glad if, whilst I am making the Index, he would be pleasd to consider of it & put down a few Notes of what he thinks most material to be insisted on. This I say upon Supposition that I write the Preface my self. But I think it will be much more adviseable that You or He or Both of You should write it whilst You are in Town. You may depend upon it that I will own it, & defend it as well as I can, if hereafter there be occasion.

I am, Sir,
Your most obliged
& humble Servant
ROGER COTES.

Cambridge March. 10*th*
$171\frac{2}{3}$

For
The Revrd Dr Bentley
at Cotton House
Westminster

NOTES

(1) Add. 3983, no. 36; printed in Edleston, *Correspondence*, p. 149 from the draft in Trinity College Library, Cambridge.

(2) See Letter 980, note (3).

(3) It was nearly twenty-six years since the *Principia* had been first received abroad, and then by no means 'with disadvantage'.

Presumably Cotes meant that Newton's reputation and authority had not grown as fast abroad, in that interval, as they had in England, and this was true; the prestige of Leibniz had grown faster. Cotes was of course aware of the critical reception accorded to the writings of such Newtonians as Freind (see Letter 880); yet he was mistaken if he supposed that the continental attitude to Newtonian mathematical physics had become more hostile, rather than less, with the passage of years. There is little doubt that Cotes was being carried in an anti-Leibnizian stream without himself having a thorough first-hand knowledge of the facts and issues involved.

(4) *Commercium Epistolicum.* The report was dated 24 April 1712; the first printed copies were shown to the Society on 8 January 1712/13. Cotes received a personal copy (with three others for local distribution) on 6 February. This first edition was for the most part privately distributed, few copies were ever put on sale. (See also the variorum edition of J. B. Biot and F. Lefort, Paris, 1856.)

(5) 'Tentamen de motuum cœlestium causis', *Acta Eruditorum* for February, 1689, pp. 82–96. For an analysis of Leibniz's paper and Newton's objections, see Aiton, 'The Celestial Mechanics of Leibniz in the light of Newtonian criticsm', *Annals of Science*, **18** (1962) 31–41. For the criticism of Leibniz's paper by Newton and Keill see vol. VI.

984 BENTLEY TO COTES

12 MARCH 1712/13

From the original in Trinity College Library, Cambridge.(1)
Reply to Letter 983; for the answer see Letter 985

At Sr Isaac Newton's
March 12.

Dear Sir,

I communicated your Letter to Sr. Isaac, who happened to make me a visit this morning, & we appointed to meet this Evening at his House & there to write you an Answer. For ye Close of your Letter, wch proposes a Preface(2) to be drawn up here, and to be fatherd by you, we will impute it to your Modesty; but you must not press it further, but go about it your self. For ye subject of ye Preface, you know it must be to give an account, first of ye work it self, 2dly of ye improvements of ye New Edition; & then you have Sr Isaac's consent to add what you think proper about ye controversy of ye first Invention. You your self are full Master of it, & want no hints to be given you: However when it is drawn up, You shall have His & my Judgment, to suggest any thing yt may improve it. Tis both our opinions, to spare the *Name* of M. Leibnitz, and abstain from all words or Epithets of reproch: for else, yt will be ye reply (not that its untrue) but yt its rude & uncivil. Sr. Isaac presents his service to you.

I am
Yours
R. BENTLEY

For Mr. Roger Cotes
Professor of Astronomy
at Trinity College
in Cambridge

NOTES

(1) R.4.42, no. 12; printed in Edleston, *Correspondence*, p. 150.
(2) See Letter 980, note (3).

985 COTES TO NEWTON

18 [MARCH] 1712/13

From the original in the University Library, Cambridge.[1]
Reply to Letter 984; for the answer see Letter 988

Cambridge Febr. 18*th*[2] 171$\frac{2}{3}$

Sr.

I have received Dr. Bentley's Letter in answer to that which I wrote to him concerning the Preface. I am very well satisfied with the directions there given, & have accordingly been considering of the matter. I think it will be proper besides the account of the Book & its Improvements to add something more particularly concerning the manner of Philosophizing made use of & wherein it differs from that of De Cartes & others. I mean in first demonstrating the Principles it imploys[3]. This I would not only assert but make evident by a short deduction of the Principle of Gravity from the Phænomena of Nature, in a popular way, that it may be understood by ordinary Readers & may serve at the same time as a Specimen to them of the Method of the whole Book. That you may the better understand what I aim at, I think to proceed in some such manner.[4]

Tis one of the primary Laws of Nature, That all Bodys preserve in their State, &c. Hence it follows yt Bodys which are moved in Curve-lines & continually hindred from going on along ye Tangents to those Curve-lines must incessantly be acted upon by some force sufficient for that purpose. The Planets (tis matter of fact) revolve in Curve-lines. Therefore &c.

Again, tis Mathematically demonstrated that *Corpus omne, quod movetur &c*

pr. 2. Lib. 1 and *Corpus omne, quod radio &c. pr. 3 Lib. 1.* Now tis confess'd by all Astronomers, that ye Primary Planets about ye Sun & ye Secondary about their respective Primarys do describe Areas proportional to the times. Therefore ye Force by which they are continually diverted from the Tangents of their Orbits is directed & tends towards their Central Bodies. Which Force (from what cause whatever it proceeds) may therefore not improperly be call'd Centripetal in respect of ye revolving Body & Attractive in respect of the Central.

Furthermore, tis Mathematically demonstrated that [Cor. 6, Prop. 4, Lib. 1] and [Cor. 1, Prop. 4 Lib 1.] But tis agreed on by Astronomers, that &c. or &c. Therefore the Centripetal forces of the Primary Planets revolving about ye Sun & of ye Secondary Planets revolving about their Primary ones, are in a duplicate proportion &c.

In this manner I would proceed to the 4th Prop. of Lib III & then to ye 5th. But in the first Corollary of the 5th I meet with a difficulty[5], it lyes in these words *Et cum Attractio omnis mutua sit* I am persuaded they are then true when the Attraction may properly be so call'd, otherwise they may be false. You will understand my meaning by an Example. Suppose two Globes *A* & *B* placed at a distance from each other upon a Table, & that whilst *A* remains at rest *B* is moved towards it by an invisible Hand. A by-stander who observes this motion but not the cause of it, will say that *B* does certainly tend to the centre of *A*, & thereupon he may call the force of the invisible Hand the Centripetal force of *B*, or the Attraction of *A* since ye effect appears the same as if it did truly proceed from a proper & real Attraction of *A*. But then I think he cannot by virtue of the Axiom [Attractio omnis mutua est] conclude contrary to his Sense & Observation, that the Globe *A* does also move towards the Globe *B* & will meet it at the common centre of Gravity of both Bodies. This is what stops me in the train of reasoning by which as I said I would make out in a popular way the 7th Prop. Lib. III.[6] I shall be glad to have Your resolution of the difficulty, for such I take it to be. If it appeares so to You also; I think it should be obviated in the last sheet of Your Book which is not yet printed off, or by an Addendum to be printed with ye Errata Table. For 'till this Objection be cleared I would not undertake to answer any one who should assert You do *Hypothesim fingere* I think You seem tacitly to make this Supposition that the Attractive force resides in the Central Body[7].

After this Specimen I think it will be proper to add some things by which Your Book may be cleared from some prejudices which have been industriously laid against it. As that it deserts Mechanical causes, is built upon Miracles & recurrs to Occult qualitys. That you may not think it unnecessary to answer such Objections You may be pleased to consult a Weekly paper call'd *Memoirs of Literature* sold by Ann Baldwin. In ye 18th Number of the Second Volume of those Papers, which was published May 5th, 1712(8) You will find a very extraordinary Letter of Mr. Leibnitz to Mr. Hartsoeker which will confirm what I have said. I do not propose to mention Mr. Leibnitz's name, t'were better to neglect him; but the Object[ions] I think may very well be answered & even retorted upon the maintainers of Vortices.

After I have spoke of Your Book it will come in my way to mention the improvements of Geometry upon which it is built, & there I must mention the time when these Improvem[ents] were first made & by whom they were made.(9) I intend to say nothing of Mr. Leibnitz but desire You will give me leave to appeal to ye Commercium Epistolicum to vouch what I shall say of Your Self, & to insert into my Preface the very words of the Judgment of the Society (page 120 Comm: Epist) that Foreigners may more generally be acquainted with the true state of the Case. I am Sr, Your most Humble Servt.

ROGER COTES

For Sr Isaac Newton
at his House
in St. Martin's Street
Leicester Feilds
London

Postmark: M[R] 20

NOTES

(1) Add 3983, no. 35; printed in Edleston, *Correspondence*, pp. 151–4, from the draft in Trinity College Library, Cambridge.

(2) Read 18 March. But Newton in his next refers to 'yours of Feb 18th'.

(3) This, in fact, was the discussion which eventually occupied the whole of the preface; see Letter 980, note (3).

(4) The preface, in fact, began, not in this way, but with a general description of the effects of gravity upon terrestrial bodies, and a discussion of the law of equal action and reaction. Cotes here gives the essence of what was to be included in paragraphs 10 to 20. (See Koyré and Cohen, *Principia*, I, pp. 22–7.)

(5) The first corollary to Proposition 5 stated that all the planets and their satellites gravitate towards one another. Cotes seems naively to have misunderstood the implications of the third law here. Newton in his reply (Letter 988) answers Cotes' query very cursorily, but in the longer draft (see Letter 988, note (1)) he devotes considerable space to the question. He raises two main objections—first that the result Cotes suggests is contrary to experience, and second that it results logically in a contravention of the first law of motion, as he had explained in Book I. (See Koyré and Cohen, *Principia*, I, p. 69; it is clearly from here that Cotes derives his own example.)

(6) The 7th Proposition of Book III states that there is a 'power of gravity pertaining to all bodies proportional to the several quantities of matter which they contain'.

(7) Despite Newton's clear dismissal of Cotes' problem, he was nevertheless worried about the role of hypotheses, for in his next letter he specifies relevant additions to the *Scholium Generale*.

(8) See Letter 918.

(9) This was not done; see Letter 980, note (3).

986 H. SCOTT TO NEWTON

21 MARCH 1712/13

From the original in the Mint Papers.(1)
Reply to Letter 964

Edinburgh 21 *March* 1713

Much Honoured
Sir

I am Honoured with yours of the 28th January for which do return my hearty thanks, I have considered the Severall points you touch at in relation to the Wast in melting from which I observe you Conclude the waste by the practice used in the Mint here may be reckoned about equall to what the Melter in the Towr does undergo; and that in the late Coynage here the waste may be only 125 pounds, I notice also that your Melters loss is reckoned from 6 to 10 grains per pound, and that it falls out Somtymes to be more But when I consider the different wayes of Melting here and with you I still think that in all Events the Melter here must suffer a greater loss, and therefor beg leave to lay before you the Circumstances upon which my Jugeing in this matter is founded. The allowance of putting copper into the Pot whilst the standart Silver is melting is not given by connivance as a favour to the Melter but out of necessity because of the Violent heat of our Coall, and that without such a practice it would be next to impossible to bring out the Silver at due Standart, And seeing the Melter in the Towr who has not that difficulty suffers also a loss from 6 to 10 gr: pr pd and somtymes higher it seems rationall to me that the melting here will occasion a greater loss, for the Melter here undergoes the common loss with those that Melt with a moderat equall heat, and likewise the loss in the pots refyneing and as the loss in refyneing, even by the practice of the Towr, is reckoned double what Occurs in common melting, So I conclude that in the melting here the waste may be reckoned double what [it] is in the Towr for in Effect the melting with our common coall is almost Equivalent to refyneing. I shall suppose 300 weight Standart Silver put into a pott, immediately upon melting $\frac{1}{3}$ thereof is taken out and cast into the Moulds, but before that can be done, the remaining Silver is Overfined and needs an Addition of Copper to reduce it to standart, and before the 2d part can be cast into the

Moulds the remaining 3d part is again Overfyned and must be reduced by Addition of Alloy, so in every pott the Silver is twice refyned which in my Opinion will Occasion a Supervenient waste above that of Ordinary melting. The practice of proportioning alloy to a pot of standart Silver whilst it is melting and casting into the Moulds has been adjusted by long Experience and is understood but by very few and is so nice that if the workpeople do not dispatch the pott in due tyme, the Moulds prove above Standart and must be remelted which often happened, and frequently the fire and heat proved unexpectedly violent that the alloy could not be proportioned, and were necessitat to take of the pot and clean it and begin anew, and seing the allowance of 125 pd weight for wast mentioned in your letter is upon supposition of Ordinary Equall melting for the whole Coynage, there must necessarly be an additional Wast for the forsd reasons, there being also a large quantity necessarly refined upwards of 12000 weight for which Mr Allardes got a Considerable Sume allowed him by the public. As to what the London Goldsmiths have suggested about the Waste being so inconsiderable as not to deserve an allowance from the Government I'm perswaded has been malicious for our Goldsmiths of Knowledge and reputation are of another Opinion. Mr Allardes and my brother did treat with the best of them and frankly offered 4*d* per pd weight for bearing the waste which was more then allowed to the Melter in the Towr but none of them would undertake it but at a greater allowance, so that I still think the Account I made up from the Mint books agreeing to the abbreviat thereof given in by my Brother ought not to be quarelled by Mr Allardes friends by which although the work was so much hasted the Waste appears to be considerably less then of any Coynage ever we had formerly, I would therefor humbly beg you againe to consider the Account and the reason presently Suggested and give me your advice in the matter. I told you in my former that it appears from my Brothers papers that he bought in Bullion which he did not carry into the Books and its known he designed to have provided as much as would pay of the Bank how soon the Mint should be opened, And if Mr Allardes friends shall disallow so great a part of the waste as he now Contraverts my brothers Children will suffer a loss far beyond any benefit he had by serving Mr Allardes in that matter For his Sallary was Sixteen pound 13/4*d* per annum when there was Coynage and the late Recoynage continued only about two years besides this he had no other benefite except a complement at Mr Allardes death for the Service in the first Melting from the Bank for which Mr Allardes had a full allowance given him out of the Equivalent that melting being no part of his duty as Master. Sir I am doing in this for a widow and four infants which emboldens me so far to trouble you, had Mr Allardes himself been alive I should have had no difficulty for he knew very well what Service my brother did him in that matter and would never have done any thing wherby probably there might arise any loss to my brothers family. I am with all dutyfull respect

Much Honoured Sir

Your most humble & most obedient servant

HERCULES SCOTT

NOTE

(1) Mint Papers I, fo. 188.

987 TREASURY REFERENCE TO THE MINT

21 MARCH 1712/13

From the minute in the Public Record Office[1]

John Roos Her Mats. Chief Engraver[2] Craves payment of 565£. 12*s*. 7*d* for Silver & Engraving of several publique Seales pursuant to Her Mats. Directions under Her Royal Signe Manual[.] 21 March 1712 Ref to ye Principal Officers of ye Mint who are to peruse Her Mats. W[arran]ts directing ye several publique Seales to be prepared & the respective Certificates of their being delivered pursuant thereto & Examine into ye Reasonableness of the prices set down for ye same & thereupon make Report with their opinion.

W. LOWNDES.

NOTES

(1) INDEX 4623, p. 114.

(2) Craig, *Newton*, p. 54 gives his Christian name as William. The engraving and cutting of seals was distinguished from the coin and medal work, of which Croker was in charge. A warrant for the payment of the whole sum due to Roos giving details of the seals cut (including one for Virginia and another for Maryland), dated 26 May 1713, is in *Cal. Treas. Books*, XXVII (Part II), 1713, pp. 226–7; an authorization for the payment upon the recommendation of the Mint Officers, dated 10 June 1713, is in *ibid*. p. 36. We have not found the Mint report.

988 NEWTON TO COTES

28 MARCH 1713

From the original in Trinity College Library, Cambridge.[1]
Reply to Letter 985

Sr

I had yours of Feb 18th,[2] & the Difficulty you mention[3] wch lies in these words [Et cum Attractio omnis mutua sit] is removed by considering that as in Geometry the word Hypothesis is not taken in so large a sense as to include the

Axiomes & Postulates, so in experimental Philosophy it is not to be taken in so large a sense as to include the first Principles or Axiomes wch I call the laws of motion. These Principles are deduced from Phænomena & made general by Induction: wch is the highest evidence that a Proposition can have in this philosophy. And the word Hypothesis is here used by me to signify only such a Proposition as is not a Phænomenon nor deduced from any Phænomena but assumed or supposed wthout any experimental proof. Now the mutual & mutually equal attraction of bodies is a branch of the third Law of motion & how this branch is deduced from Phænomena you may see in the end of the Corollaries of ye Laws of Motion, pag. 22. If a body attracts another body contiguous to it & is not mutually attracted by the other: the attracted body will drive the other before it & both will go away together wth an accelerated motion in infinitum, as it were by a self moving principle, contrary to ye first law of motion, whereas there is no such phænomenon in all nature.

At the end of the last Paragraph but two now ready to be printed off I desire you to add after the words [nihil aliud est quam Fatum et Natura.] these words:(4) [Et hæc de Deo: de quo utique ex phænomenis disserere, ad Philosophiam experimentalem pertinet.]

And for preventing exceptions against the use of the word Hypothesis I desire you to conclude the next Paragraph in this manner(4) [Quicquid enim ex phænomenis non deducitur Hypothesis vocanda est, et ejusmodi Hypotheses seu Metaphysicæ seu Physicæ use Qualitatum occultarum seu Mechanicæ in Philosophia experimentali locum non habent. In hac Philosophia Propositiones deducuntur ex phænomenis & redduntur generales per Inductionem. Sic impenetrabilitas mobilitas & impetus corporum & leges motuum & gravitatis innotuere. Et satis est quod Gravitas corporum revera existat & agat secundum leges a nobis expositas & ad corporum cælestium et maris nostri motus omnes sufficiat.

I have not time to finish this Letter but intend to write to you again on Tuesday. I am

Your most humble Servant

IS. NEWTON

London. 28 *March*
1713.

For The Reverend Mr Roger Cotes
Professor of Astronomy, at
his Chamber in Trinity
College in
Cambridge.

NOTES

(1) MS. R.16.38, no. 382, printed in Edleston, *Correspondence*, pp. 154–6. The draft of this letter (U.L.C. Add. 3984, no. 14, fo. 1) differs considerably from the version Cotes received. It reads:

Sr

I like your designe of adding something more particularly concerning the manner of Philosophizing made use of in the Principia & wherein it differs from the method of others, vizt by deducing things mathematically from principles derived from Phænomena by Induction. These Principles are the 3 laws of motion. And these Laws in being deduced from Phænomena by Induction & backt with reason & the three general Rules of philosophizing are distinguished from Hypotheses & considered as Axioms. Upon these are frounded [*sic*] all the Propositions in the first & second Book. And these Propositions are in the third Book applied to the motions of ye heavenly bodies.

And first because ye Planets move in Curve lines, it follows from the first Axiom or Law of Nature that they are incessantly acted upon by some force wch continually diverts them from a rectilinear course.

Again from Prop. 2 & 3 Lib 1, it follows that this force is directed towards the central bodies about wch the Planets move

And by Prop. 6 Corol 4 Lib. 1 & Corol 1 Prop 45 Lib. 1, that this force in receding from the central Body decreases in a duplicate proportion of ye distance. &c

And when you come at the difficulty you mention in the first Corollary of the 5t Proposition of the third Book, wch lies in these words *Et cum Attractio omnis mutua sit*: the Objection you mention may be proposed & answered in this manner. 1 That it is but an Hypothesis not founded upon any one Observation. 2 That it is attended wth the absurd consequence described pag. 22, namely that a body attracted by another body without mutually attracting it would go to the other body & drive it away before it with an accelerated motion in infinitum, contrary to ye first law of Motion. And such an absurd Hypothesis wch would disturb all nature, is not to be admitted in opposition to the first & third Laws of motion wch are grownded upon Phænomena. For that all attraction is mutual & mutually equal follows from both those laws. One may suppose that bodies may by an unknown power be perpetually accelerated & so reject the first law of motion. One may suppose that God can create a penetrable body & so reject the impenetrability of matter. But to admitt of such Hypotheses in opposition to rational Propositions founded upon Phænomena by Induction is to destroy all arguments taken from Phænomena by Induction & all Principles founded upon such arguments. And therefore as I regard not Hypotheses in explaining the Phenomena of nature so I regard them not in opposition to arguments founded upon Phænomena by Induction or to Principles setled upon such arguments. In arguing for any Principle or Proposition from Phænomena by Induction, Hypotheses are not to be considered. The Argument holds good till some Phænomenon can be produced against it. This Argument holds good by the third Rule of philosophizing. And if we break that Rule, we cannot affirm any one general law of nature: we cannot so much as affirm that all matter is impenetrable. Experimental Philosophy reduces Phænomena to general Rules & looks upon the Rules to be general when they hold generally in Phænomena. It is not enough to object that a contrary phænomenon may happen but to make a legitimate objection, a contrary phenomenon must be actually produced. Hypothetical Philosophy consists in imaginary explications of things & imaginary arguments for or against such explications, or against the arguments of Experimental Philosophers founded upon Induction. The first sort of Philosophy is followed by me, the latter too much by Cartes, Leibnitz & some others. And according to the first sort of Philosophy the three Laws of motion are proposed as general Principles of Philosophy tho founded upon Phænomena by no better Argument then that of Induction without exception of any one Phænomenon. For the impenetrability of matter is grownded upon no better an Argument. And the mutual equality of Attraction (wch is a branch of the third Law of motion) is backt by this further argument [,] that is if the attraction between two bodies was not mutual and mutually equall they would not stay in rerum natura. The body wch is most strongly attracted would go to the other & press upon it, & by the excess of its pressure both would go away together with a motion accelerated in infinitum. If a great mountain upon either pole of the earth gravitated towards the rest of ye earth more then the rest of the earth gravitated towards the mountain, the the [*sic*] weight of the mountain would drive the earth from the plane of ye Ecliptic & cause it, so soon as it could extricate it self from the systeme of ye Sun & Planets, to go away in infinitum wth a motion perpetually accelerated. Thus the Objection wch you mention is not only an Hypothesis & on that account to be excluded [from] experimental Philosophy, but also introduces a principle of self motion into bodies wch would disturbe the whole frame of nature, & in the general opinion of mankind is as remote from the nature of matter as impenetrability [*read* penetrability] is recconed to be. Experimental philosophy argues only from phænomena, draws general conclusions from the consent of phænomena, & looks upon the conclusion as general when ye consent is general without exception, tho the generality cannot be demonstrated a priori. In Mathematicks all Propositions not demonstrated mathematically are Hypotheses, but some are admitted as Principles under the name of Axioms or Postulates wthout being called Hypotheses. So in experimental Philosophy its proper to distinguish Propositions into Principles, Propositions & Hypotheses, calling those Propositions wch are deduced from Phænomena by proper Arguments & made general by Induction (the best way of arguing in Philosophy for a general Proposition) & those Hypotheses wch are not deduced from Phænomena by proper arguments. But if any man will take the word Hypothesis in a larger sense, he may extend it, if he pleases to the impenetrability of matter the laws of motion & the Axioms of Geometers. For it is not worth the while to dispute about the signification of a word.

What has been said, doth not hinder the body *B* from being moved by an invisible hand towards the resting body *A*: [*ends*]

(2) This is Letter 985, wrongly dated by Cotes 18 February 1712/13.

(3) See Letter 985, note (3).

(4) The additional words were printed (see Koyré and Cohen, *Principia*, II, pp. 763–4). It is curious that these famous sentences were written by Newton as an afterthought to an addendum, merely in response to Cotes' doubts which reacted upon Newton's earlier sensitive response to Leibniz's criticism of Newtonian gravitation (Letter 918). However, Newton had already inserted a declaration against hypotheses into the new Query 20 of the 1706 *Optice* (p. 314); in the second English edition of *Opticks* (1717) he was to reword this passage more forcibly against redundant mechanical hypotheses, and to extend the penultimate paragraph of Query 31 into a little essay on the non-hypothetical scientific method (see Alexandre Koyré, 'Les Queries de l'*Optique*', *Archives Int. d'Hist. des Sciences*, pp. 50–1, 1960).

989 NEWTON TO COTES

31 MARCH 1713

From the original in Trinity College Library, Cambridge.[1]
Continuation of Letter 988

London, 31 *Mar*. 1713.

Sr

On saturday[2] last I wrote to you, representing that Experimental philosophy proceeds only upon Phenomena & deduces general Propositions from them only by Induction. And such is the proof of mutual attraction. And the arguments for ye impenetrability, mobility & force of all bodies & for the laws of motion are no better. And he that in experimental Philosophy would except against any of these must draw his objection from some experiment or phænomenon & not from a mere Hypothesis, if the Induction be of any force.

In the same Letter, I sent you also an addition to the last Paragraph but two & an emendation to the last Paragraph but one in the paper now to be printed off in the end of the Book.

I heare that Mr Bernoulli has sent a Paper of 40 pages to be published in the Acta Leipsica relating to what I have written upon the curve Lines described by Projectiles in resisting Mediums.[3] And therein he partly makes Observations upon what I have written & partly improves it. To prevent being blamed by him or others for any disingenuity in not acknowledging my oversights or slips in the first edition I beleive it will not be amiss to print next after the old Præfatio ad Lectorem, the following Account of this new Edition.[4]

In hac secunda Principiorum Editione, multa sparsim emendantur & nonnulla adjiciuntur. In Libri primi Sect. II, Inventio virium quibus corpora in Orbibus datis revolvi possint, facilior redditur et amplior. In Libri secundi Sect. VII Theoria resistentiæ fluidorum accuratius investigatur & novis experimentis confirmatur. In Libro tertio Theoria Lunæ & Præcessio Æquinoctiorum ex Principijs suis plenius deducuntur, et Theoria Cometarum pluribus et accuratius computatis Orbium exemplis confirmatur.

28 *Mar*. 1713. I. N.

If you write any further Preface, I must not see it. for I find that I shall be examined about it.[5] The cuts for ye Comet of 1680 & 1681 are printed off[6] & will be sent to Dr Bently this week by the Carrier. I am

Your most humble Servant
ISAAC NEWTON

For the Rnd Mr Cotes Professor of
Astronomy in the University of
Cambridge. At his Chamber
in Trinity College in Cambridge

NOTES

(1) R.16.38, no. 284; printed in Edleston, *Correspondence*, pp. 156–7. In Newton's draft of this letter (U.L.C. Add. 3984, no. 15) the first paragraph differs considerably from the version received by Cotes:

Sr

On Satturday last I wrote to you representing that Experimental philosophy proce[e]ds only upon Phenomena & makes Propositions general by Induction from them. In this Philosophy neither Explications nor Objections are to be heard unless taken from Phænomena. Nor are Propositions here made general by arguments a priore by [*read* but] only by Induction without exception. And upon such an Induction the mutuall and mutually equal Attraction is founded. One may suppose that there may be bodies penetrable or immoveable or destitute of force, or with attraction mutually unequal, but such suppositions without any instance in Phænomena are mere hypotheses & have no place in experi[ment]al Philosophy: & to introduce them into it would be to overthrow the Argument from Induction upon wch all the general Propositions in this Philosophy are built.

. . . .

(2) 28 March 1712/13 (Letter 988).

(3) *Acta Eruditorum* for February 1713 (pp. 77–95) and March 1713 (pp. 115–32); the paper concerns Newton's mistake in Book II, Proposition 10. (See Letter 951, note (3).) Newton learned of this from De Moivre, who in turn had been informed by Johann Bernoulli's letter of 18 February 1713 N.S.; see Wollenschläger, *De Moivre*, p. 282.

(4) The passage appears as the *Auctoris Præfatio in Editionem Secundam*. (See Koyré and Cohen, *Principia*, I, p. 18.) A rough draft on the same sheet contains after *facilior redditur* the additional sentences: 'Cognita vi centripeta qua corpus in circulo quovis circa centrum ejus revolvi possit, statim habetur vis quo corpus in Ellipsi quavis circa centrum ejus revolvi possit per Scholium Prop. X. Et ex hac statim habebitur vis qua, corpus in hac Ellipsi circa punctum aliud revolvi possit per Cor. 3. VII'

(5) See Letter 980, note (3).

(6) See Letter 956, note (3).

990 NEWTON TO OXFORD

[31 MARCH 1713]

From the holograph draft in the Mint Papers.(1)
Reply to Letter 982

To the most Honble the Earl of Oxford & Earl Mortimer
Lord High Treasurer of great Britain

May it please your Lordp

In obedience to your Lordps Order of Reference upon the Bills of Mr Robt Weddel for prosecuting coyners during the space of nineteen months ending last Christmas was a twelve-month, & upon the Memorial of the Warden of the Mint concerning the same: We have considered the services alledged to be done & examined the reasonableness of the said Bills, & humbly represent to your Lordp that the services so far as We can find, were done; & allowing to Mr Weddel for travelling charges 6*d* per mile, 2*s* per post stage & 15*s* per day abroad (as was setled by the late Lord Treasurer & allowed also by the late Lords Commissioners of the Treasury), & to the men who assisted him in apprehending & prosecuting criminals, 6*s* 6*d* per day for each man & horse during their journeys: the travelling charges amount unto 186£. 12*s*. 6*d*. wch wth Mr Weddels salary, a Bill of Mr Fords, the fees of receiving the money allowed upon the last account, & the charges of a law suit wth a coyner upon the Privy Seal, make up the summ of 337£, 1*s*. 6*d*. wch we humbly think reasonable to be allowed.

That there are other charges wch admitt of no vouchers or strickt examination, whereof the fees of Council & other Court Charges upon the trials of persons (vizt for Indictmts, swearing of witnesses, attending wth Records, &c) amount unto about 62£. 10*s*. And these charges being necessary for carrying on the prosecutions, We are humbly of opinion that the Prosecutors account of them, where nothing appears false or unreasonable, should be accepted.

And the maintainance of witnesses during their attendance on courts of justice, & pocket expenses in apprehending & examining people accused & in attending on Judges & Justices of the Peace & paying for their Warrants & for stationery ware & Post Letters, amount further to about 68£.

But considering that about one half of the money set down in Mr Weddels Account hath been advanced by him without interest for carrying on the prosecutions, & that the charges of receiving the money due upon this Account are not set down therein; both wch may amount to above 40£; & that Mr Weddel was a very good prosecutor & if prosecutors be discouraged they may be induced to pay themselves by taking money for favouring or protecting coyners: We are humbly of opinion that the Bills of Mr Weddel amounting in the whole to 467£. 15*s*. 2 be allowed.

All which &c

NOTE

(1) Mint Papers I, fo. 468. An earlier draft occurs at *ibid.*, fo. 473. The clerical transcript sent to the Treasury, and signed by Newton and the Comptroller, Edward Phelipps, is in P.R.O. T/1, 159, no. 19. There is no material difference between the two versions. Newton prepared two detailed accounts of all the expenses involved in Weddell's pursuit of criminals (Mint Papers I, fos. 471, 469); these were not submitted to the Treasury.

991 T. HARLEY TO THE MINT

2 APRIL 1713

From the copy in the Public Record Office.(1)
For the answer see Letter 992

[To the Principal] Officers [of the] Mint abt Mr. Oadham
& abt [applicants] for ye place of Assaymaster

Gentlemen

My Lord Treasurer Commands me to transmit to you the inclosed petition of Mr [Catesby] Oadham praying a Tryall as to his qualifications for the place of Assay Master of the Mint, his Lordp is pleased to direct you to cause a Tryall to be made before you by him & the rest of the Petitioners for the said office who have been Referred to you (if they desire it) And that you will then make your Report to his Lordp wch of them you consider to be the most expert & fittest person for that Imploymt. I am &c 2d April 1713

T. HARLEY

NOTE

(1) T/27, 20, 395; compare Letter 966. The petition is not copied in this minute. Compare Longleat House, Portland Papers, IX, fo. 192.

992 THE MINT TO OXFORD

10 APRIL 1713

From the holograph original at Longleat House.(1)
Reply to Letter 991

To the most Honble the Earl of Oxford
& Earl Mortimer Lord High Treasurer of
Great Britain

May it please your Lordp

According to your Lordps Order upon the Petition of Mr Catesby Oadham for a tryal as to his Qualifications for ye place of Assaymaster of the Mint, signified to us by Mr Secretary Harly in his Letter of ye 2d Instant, that we should cause a tryal to be made before us by him & the rest of the Petitioners for the said Office whose Petitions have been referred to us (if they desired it,) and that we should then make our Report to your Lordp which of them We conceived to be the most expert & fittest person for that imployment; We have caused such a trial to be made before us by Mr Charles Brattel & the said Mr Oadham, the other Petitioners upon notice having waved a trial;(2) & we humbly conceive Mr Brattel to be the more expert & fitter person for that imployment.[3]

All which &c

Mint Office
10 *Apr.* 1713

CRAV: PEYTON
IS NEWTON
E: PHELIPPS

NOTES

(1) Portland Papers, IX, fo. 193. There is a draft in the Mint Papers I, fo. 193.

(2) This trial was held on 8 April.

(3) Catesby Oadham carried on his campaign by submitting a certificate signed by importers of bullion to the Mint affirming that he was the better candidate; Newton replied to this on 26 August 1713.

993 JONES TO COTES

29 APRIL 1713

Extract from the original in Trinity College Library, Cambridge.[1]
For the answer see Letter 995

London Aprill 29th. 1713

...I am mightily pleas'd to see the End of the Principia, and return you many thanks for the very Instructive Index, that you have taken the pains to add, and hope 'twill not be long before we shall see the beginning of that Noble Book...

NOTE

(1) R.16.38, no. 315; printed in Edleston, *Correspondence*, pp. 223–4.

994 T. HARLEY TO THE MINT

2 MAY 1713

From the copy in the Public Record Office[1]

[To the Principal] Officers of the Mint abt Coyning Farthings

Gentlemen

My Lord Trea[su]rer being much Importuned about the Business of Coyning Farthings is desirous of hearing that matter and the pretences of the Severall Petitioners that have been Referr'd to You for Liberty to Coyne the same his Lordp therefore, if Friday next in the forenoone[2] be time sufficient for the severall Petitioners to Attend, is pleased to Direct you to give them notice and also to attend Your Selves at that time Accordingly, in case it be not sufficient be pleased to let me know & his Lordp will appoint such other time as you shall think fit for doing thereof. You'l please to bring all Papers with You which may concerne this hearing particularly the Votes which have been made by the house of Commons about Farthings or the Coynage of them. In the meane time (as My Lord is informed) Farthings are wanted & would be of use & Service at Port Mahon & Gibralter. his Lordp directs You to consider thereof and let him know at your Attendance Your Opinions how & in what manner the same may best be supplyed, I am &c *2d May* 1713

T. HARLEY

NOTES

(1) T/27, 20, 414. For the previous Mint Report see Letter 960.

(2) The second being a Saturday, 'next Friday' was the eighth. See Number 996 for the proceedings at the Treasury.

995 COTES TO JONES

3 MAY 1713

Extract from the draft in Trinity College Library, Cambridge.[1]

Reply to Letter 993

...I am glad you can approve of that Index to the Principia. It was not design'd to be of any use to such Readers as Your self, but to those of ordinary capacity. I hope the whole Book may be finished in a fortnight or three weeks. I have lately been out of Order, or it might have been done by this time....

NOTE

(1) R.16.38, no. 316v; printed in Edleston, *Correspondence*, p. 224.

996 NEWTON AT THE TREASURY

8 MAY 1713

From the original minute in the Public Record Office[1]

Whitehall Treasury Chambers 8th May 1713

Present

Lord Treasurer. Mr Chancellor [of the Exchequer]

Officers of the Mint, & severall persons concerned, for themselves or others in proposalls for Coyning of Copper Farthings & half pence are called in. An Abstract of the severall proposals are [*sic*] read with opinion of the Officers of the Mint thereupon, also a particular Report of the sd Officers upon a petition of Cha. Hore is read[2]—then severall Votes of the House of Commons concerning Farthings & halfe pence in the years 1694, 1695, 1698, 1699, 1700 & 1708. Lord Treasurer asks the Officers of the Mint Whether [there is] any occasion for a further Coynage of them for Great Britain. Sr Isaac Newton says about 2 years ago he made it his Businesse to inform himself from most parts & he did not find there was any want of them then, nor doth he yet thinke there is.[3] Lord Tr If there should be a want of them, would it be best to Coyn them according to the Value of the old ones Or to make them as near as may be to the Intrinsick Value. Officers of the Mint say they are of Opinion they should be Coyned as near the Intrinsick Value as may be to prevent Counterfeiting. Objected by one of the proposers, that they would then be too heavy & burthensome. Lord Tr That need not be for the Value may be made up in workmanship—Another objects That then the old ones would not go. Lord Tr are none of the Farthings now in Use of greater Value than others. Officers of the Mint answer Yes. Lord Trer cannot there be a Standard for Copper. Mr Hore says that will be very difficult. Sr Is: N. We account Copper at a proper Standard that will bear hammering when it is red hot. Lord Treasurer asks what that Copper may be Worth Sir Is: N. about 11*d* $\frac{1}{2}$ per pound or 12*d* at most. Lord Tr may not Farthings be made in Fineness & Workmanship so as not to allow of Counterfeiting Officers [of the] Mint Yes. They persist that they ought to be Coyned in the Mint & no where else because of the danger of Trusting proper Tools in any other hands & that they should be made as near as may be of intrinsick value. Lord Trer cannot Copper Money be made in Value in proportion to the Silver Money, and any body be allowed to import Copper into the Mint to be Coyned as they do Silver. Sir Is: N. said, If Rules are set for the Importers he sees no Objection at present, but will consider of it. Lord Tr: there must be a Regulation as to the Quantity to be received in to avoid the Coyning of too great a Quantity. One of the proposers saies tho' he never heard any thing of this kind offerd before, he thinks it would be a very great Encouragemt to the Copper Manufacture & great Quantitys would be sent to the plantations. Ld Tr directs the Officers of the Mint to Consult with the proposers & prepare such a Regulation as they shall thinke most proper to be laid before the Q. in Councill for setling the Standard & Value of the Copper—The Charges of the Workmanship—and such Rules as they shall thinke proper to be setled for the Importers of Copper into the Mint and for the Coyning & delivering out of the Same, and to attend his Lordp therewith assoon as may be.

NOTES

(1) T/29, 20, fos. 81–2. For the summons to the Mint see Letter 994.

(2) See Letter 996*a*.

(3) Compare Letter 815.

996a NEWTON TO OXFORD

[? FEBRUARY 1712/13]

From the draft in the University Library, Cambridge(1)

To the most Honble the Earl of Oxford & Earl
Mortimer, Lord High Treasurer of great Britain

May it please your Lordp

In obedience to your Lordps Order of Reference of ye 12th of Feb last upon the Petition & Proposal of Charles Hore for the sole making & vending in 30 years, the quantity of 700 Tunns of half pence & farthings of such fine copper as when wrought into vessels would be worth 2*s* 6*d* per pound weight, & to cut a pound weight of such copper into no more then 28 pence, & that an affidavit shall be made by his workman of the fineness of the Copper suitable to a sample given in: We have considered the same, & are humbly of opinion that it may be dangerous to have any sort of coynage or coining tools out of the Mint; that the cheaper the copper is the less temptation there will be to counterfeit the copper money, & that very good copper for this purpose may be had for about 11 pence halfpenny per pound weight; That an affidavit of ye goodness of the copper by the servant of the Petitioner be not relied upon, but the copper be assayed in the Mint whenever there shall be a coinage. That a Patentee who coins ye money without account may make great profit by coining it light & of bad metal & therefore it should be coined upon account; That whatever a Patentee gets by the coinage increases the temptation to counterfeit the money when coined & Therefore it should be coined to ye just value of the copper workmanship & incident charges as nearly as may be; And that the coinage of 700 Tunns in 30 years, wch is after ye rate of 23⅓ Tunns per annum, would in a few years create clamours, the people having twice complained in Parliament against too much copper money & not yet begun to move for a greater quantity.

All wch is most humbly submitted

NOTE

(1) Add. 3984, no. 15v, intended as a partial reply to Letter 972. It was probably never sent as a letter, since a 'report' concerning Charles Hore was read on 8 May. We print the draft here as a further example of Newton's consistent opinions on this topic.

997 NEWTON'S RECOMMENDATION OF MACHIN

12 MAY 1713

From a copy in the British Museum[1]

To the Right Honoble Sr. Rich. Howse Knt. Lord Majr. and the Rest of the Committee for Gresham Affairs on the Citty's side only:[2]

The humble Peticon of John Machin.

Showeth,

That your Peticoner being informed that your Astronomy Lecturer at Gresham College being now voyd by the resignation of Mr Torriano,[3] That your Peticoner hath been for abt. four Years past admitted a Fellow of the Royall Society,[4] and hath made a very great Progress in the Mathematicks, and doubts not but hee is every way qualifyed to supply the said place, as will appeare by a Certificate hereunto annext.

And your Peticoner doth hereby oblige himselfe constantly to reside in the said College, and to Read the Lectures himselfe, and to comply with such orders, as from time to time shall be made by this Comittee Therefore most humbly prays Your honours and Wors[hi]ps to make choyce of him to supply the sd. Lecture. And hee shall ever pray.

JOHN MACHIN

[John Keill's recommendation follows, then:]

I do hereby certify that Mr. John Machin, who is well known to me, is studious, sober, & learned in the Latin tongue & in the Mathematicks, & not only understands Geometry and Astronomy in a Vulgar manner, but is a great Master therein. And that when his late Royal Highness Prince George of Denmark appointed the Honble Fra Robartes Esqr. Sr. Chr Wren, my self, Dr. Gregory & Dr. Arbuthnot, to take care that Mr. Flamsted's Observations should be well printed, we appointed Mr. Machin as the fittest person we could think of, to examine the Manuscript before it went into the press,[5] so that the Edition might be more correct, and he performed this task to our satisfaction.

Leicester Feilds
May the 12. 1713 ISAAC NEWTON

[Then follow recommendations by Edmond Halley and James Pound[6]]

NOTES

(1) MS Add. 6194, p. 210. John Machin (d. 1751), a competent mathematician, had served on the *Commercium Epistolicum* committee (Introduction, p. xxv) and was to serve as Secretary of the Royal Society from 1718 to 1747. He was appointed to the Gresham Professorship on 16 May.

(2) Gresham College was administered (under the terms of the will of Sir Thomas Gresham) by a joint Committee consisting of the Lord Mayor and representatives of the City of London, and the Master and representatives of the Mercers' Company. The will also provided (see Ward, *Lives*, pp. 20–1) that the four chairs of divinity, astronomy, music and geometry should be at the disposition of the City, the other three chairs being at the disposition of the Company.

(3) The grandfather of Alessandro Torriano (1667–1716), having first fled to Geneva because of his Protestant sympathies, found refuge in London in 1620. His father, George, was a London merchant. Alessandro was educated at St John's College, Oxford, and elected Gresham Professor of Astronomy on 31 July 1691. On 30 November following he was elected F.R.S. He took Orders, held a country living, and served as chaplain to the Earl of Manchester.

(4) Machin was elected F.R.S. on 30 November 1710.

(5) Cf. vol. IV, p. 538. As D. T. Whiteside has pointed out to us, the fact that Newton had a high opinion of Machin is confirmed by a comment of Conduitt's, 'Sr I. told me that Machin understood his principia better than any body that Halley was the best astronomer that Machin the best Geometer.' (King's College Library, Cambridge; Keynes MS 130. 6^A, fo. 12v.) Newton went so far as to include in the third edition of the *Principia* a short paper by Machin on lunar theory (see Koyré and Cohen, *Principia*, II, pp. 648–51). An extended version of this was included as an appendix to Motte's English translation of 1729.

(6) James Pound (1669–1724) had been elected F.R.S. on 20 December 1699; after his return from Madras he created an observatory at Wanstead, Essex, where he was assisted by his nephew, James Bradley, and co-operated with William Derham. He was known for his careful observations of the satellites of Jupiter and Saturn.

998 NEWTON TO OXFORD

22 MAY 1713

From the holograph original at Longleat House[1]

To the most Honble the Earl of Oxford & Earl
Mortimer, Lord H. Treasurer of Great Britain.

May it please your Lordp

Since the last trial of the Pix[2] there has been coined above a million of money & therefore I humbly pray your Lordp that there may be a new tryall of the Pix this summer.

All wch is most humbly submitted to your Lordps great wisdome

Is. Newton

Mint Office,
22th of May 1713.

Sr Isaac desires some day towards ye end of July—(3)

NOTES

(1) Portland Papers, XI, fo. 201.

(2) On 21 August 1710; see Letter 816.

(3) This sentence is added in another hand. The trial was eventually appointed for 7 August 1713 (see Number 1000 (II)).

999 JAMES NEWTON TO NEWTON

[n.d.]

From the original in the Cambridge University Library(1)

Honoured Sr

I hope your goodness will pardon this my bouldness and to lett your good honour know that I am poor James Newton that is latley come out of sickness and humbley begs for godallmity sake as your goodness seaved my life heither too for to signe this note and Sr John Newton will give me some relief and If Ever I trouble againe I will suffer death

and in so Doing your poor petitioner in duty bound will Ever pray

James Newton

For
Sr Isaac Newton att his
house in St. Martins Street
near Lecester feilds
London
These

NOTE

(1) Add. 4005, fo. 69, printed in Hall and Hall, *Unpublished Scientific Papers*, p. 318. We place this begging letter—typical of many others—arbitrarily at this point. Newton used the paper for a draft relating precisely (as he notes) to the second edition of the *Principia*, p. 472, line 13 (also printed in Hall and Hall, *loc. cit.*). Obviously the letter was written after 1710, and re-used by Newton after the spring of 1713, but conceivably several years later as he was thrifty in these matters. There is no real reason to believe that the writer was a genuine relative of Newton's.

1000 ORDERS OF COUNCIL

24 JUNE 1713

From copies in the Public Record Office[1]

(I) [Directing Newton to prepare Peace Medals[2]]

At the Court at Kensington 24th June 1713
Present
The Queen's most Excellent Maty.

Her Maty having been pleased to Declare That she would have Meddals Stampt to perpetuate the Memory of the happy Conclusion of the Peace And there having been this day presented to her Maty a Draft of such a Medal, her Maty was pleased to Approve thereof, And to Order, as it is hereby Ordered, That the Master of Her Maties Mint do forthwith cause to be made & prepared such Number of Medals of Gold, and of such Value according to the abovesaid Draft as shall be directed under Her Mats. Royall Signet & Sign Manuall, And to be distributed in such manner as her Maty shall think fit. And the Rt Honble the Lord High Treasurer of Great Britain is to give the necessary Directions herein.

J. Povey

[Here follows a Treasury order to Newton]

Let the Master & Worker of Her Mats Mint take Care that her Mats. pleasure signifyed in the above written Order of Council be duely Complyed with so far as appertains unto him Whitehall Trea[su]ry Chambers 3d July 1713.

(II) [Giving directions for a trial of the Pyx]

At the Court at Kensington 24th June 1713
Present
The Queen's most Excellent Maty in Councill

It is this day Ordered by Her Maty in Councill, That the Lords of Her Mats. most honble. Privy Councill do meet at the House inhabited by the Usher within the Receipt of her Maty's Exchequer Westminster, on Fryday the 7th of August next at Nine of the Clock in the Morning for the Tryall of her Mats. Coynes in the Pix of the Mint within the Tower of London, And the

Rt honble. the Lord High Chancellour of Great Britain is directed to require the Warden & Company of the Goldsmiths of London to Summon a Jury of Working Goldsmiths to give Attendance on their Lordps at the place aforesaid, And the Rt honble the Lord High Treasurer of Great Britain is to direct the Warden Master & Worker and Comptroller of her Mats Mint within the Tower of London, with the rest of the officers therein concerned to be then present

JOHN POVEY

[Here follows the Order from Oxford]

Let the Warden Master & Worker, and Comptroller of Her Majesties' Mint within the Tower of London and the rest of the officers concerned take Notice of Her Mats pleasure Signifyed in the above written Order of Councill White-hall Trea[su]ry Chambers 29th June 1713

OXFORD

NOTES

(1) T/54, 22, 86–7.
(2) The Treaty of Utrecht had been signed on 31 March (O.S.).

1001 COTES TO SAMUEL CLARKE

25 JUNE 1713

From the draft in Trinity College Library, Cambridge[1]

Cambridge June 25th 1713

Sr

I received Your very kind Letter,[2] I return You my thanks for Your corrections of the Preface, & particularly for Your advice in relation to that place where I seem'd to assert Gravity to be Essential to Bodies. I am fully of Your mind that it would have furnish'd matter for Cavilling, & therefore I struck it out immediately upon Dr Cannon's mentioning Your Objection to me, & so it never was printed. [The impression of the whole Book was][3] finished about a week ago.

My design in that passage was not to assert Gravity to be essential to Matter, but rather to assert that we are ignorant of the Essential propertys of Matter & hat in respect of our Knowledge Gravity might possibly lay as fair a claim to that Title as the other Propertys which I mention'd. For I understand by Essential propertys such propertys without which no others belonging to the same substance can exist: and I would not undertake to prove that it were impossible for any of the other Properties of Bodies to exist without even Extension.(4)

Be pleased to present my humble Service to Sr Isaac when You see him next, & let him know that the Book is finished

I am Sr
Your much Obliged Freind
& Humble Servant
R C

To Dr Clark

NOTES

(1) R. 16.38 no. 316; printed in Edleston, *Correspondence*, pp. 158–9. Samuel Clarke (1675–1729) was ably suited to comment upon Cotes' preface. He published several editions of his Latin translation of Rohault's *Traité de Physique*, with annotations of his own which relied increasingly in the later editions on Newton's work.

In the 1690's his interests turned towards theology; in 1704 and 1705 he gave the Boyle lectures (later published as *The Being and Attributes of God*) in which he used Newtonian concepts of space to support his theological views. In 1705 Newton commissioned him to translate the *Opticks* into Latin.

(2) We have not found the letter of Clarke to Cotes referred to here. Clearly Cotes had sent Clarke a draft of the Preface, and Clarke had returned it with suggested corrections.

(3) The words in square brackets are deleted in the manuscript.

(4) Cotes has deleted an additional sentence, 'Properties are wont to define matter to be an Extended Substance.'

1002 BENTLEY TO NEWTON

[30 JUNE 1713]

From the original in Trinity College Library, Cambridge(1)

Dear Sir

At last Your book is happily brought forth; and I thank you anew yt you did me the honour to be its conveyer to ye world.(2) You will receive by the Carrier, according to your Order, 6 Copies; but pray be so free as to command what more you shall want. We have no Binders here, yt either work well or quick; so you must accept of them in Quires. I gave Roger [Cotes] a dozen, who presents one to Dr[s] Clark(3) & Whiston. This I tell you, yt you may not give double. And on yt account I tell you, That I have sent one to the Treasurer,(4) Ld Trevor,(5) & Bp of Ely.(6) We thought it was properest for You to present Dr Halley: so you will not forget him. I have sent (though at great abatement) 200 already to France & Holland: the Edition in England to ye last buyer is 15*s* in quires: & we shall take care to keep it up so, for ye honour of ye Book.(7) I can think of nothing more at present. but shall expect your Commands, if you have any thing to order me. I am with all respect & esteem

Your affectionate humble Servt
RI: BENTLEY

Tuesday
Trin: Coll.

For Sr. Isaac Newton
at his House in
Martin Street near
Leicester Fields London

NOTES

(1) R.4.47, fo. 24; printed in Brewster, *Memoirs*, II, p. 254. Since 1 July (the date of the postmark) was a Wednesday, the letter was presumably written on 30 June.

(2) Cotes' Preface is dated 12 May; Newton presented a copy to the Queen on 27 July. Bentley's name is nowhere publicly associated with the second edition of the *Principia*. Bentley was (as we would say) its publisher, and so pocketed the receipts. Conduitt's story of Newton's reply to his question, why did he let Bentley undertake the book, is well known ('Why, he was covetous, and I let him do it to get money.' Brewster, *Memoirs*, I, p. 314 note; More, *Newton*, p. 557). See the Appendix.

(3) Samuel Clarke.

(4) Presumably the Earl of Oxford; Bentley missed no opportunities.

(5) Thomas Trevor (1658–1730), a barrister who became in turn solicitor-general, attorney-general, and Chief Justice of the Common Pleas. He served as one of the commissioners for the Union with Scotland (1706) and was given a barony in 1712. Newton was apparently acquainted with him; see Edleston, *Correspondence*, p. 228.

(6) John Moore (1646–1714) occupied in turn the sees of Norwich (1691–1707) and Ely (1707 to 31 July 1714). In 1715 his splendid library was given by George I to the University of Cambridge.

(7) These statements confirm that Bentley was in effect the publisher of the book. On 31 October 1713 Flamsteed wrote to Sharp: 'Dr Bentley puts the price 18*s*; and so much mine cost me. I am told he sent Sir I. Newton half-a-dozen, and made him pay 18*s*. a piece for them. Perhaps this was a contrivance: possibly it is not true.' (Baily, *Flamsteed*, p. 305.) This price has often been quoted as that of a *bound* volume; but if Flamsteed had bought the book bound he would (in the context) have said as much. The price to the trade was 13*s*; see the Appendix.

1003 MEMORANDUM CONCERNING A COPPER COINAGE

[1713]

From the holograph drafts in the Mint Papers(1)

Considerations about the Coynage of Copper Moneys

1. That six or at most seven hundred Tunns of Copper moneys are sufficient for all England, & am of opinion that there is above three quarters of that quantity now in the nation.(2)
2. That whenever there shall be a coinage of such moneys, the coynage do not exceed 20 or 30 or at most 40 Tunns per annum. For this money should have time to spread eavenly without making a clamour any where before the nation be sufficiently stockt.
3. That it be in the power of the Queen or Lord Treasurer to diminish or stop the coinage at pleasure for preventing clamours if there should be occasion.
4. That the moneys be coined in the Mint, it being unsafe to have coining Tools & coinage abroad.
5. That ye moneys be well coined & a pound weight cut into no more then twenty pence, least the ill form or lightness of the money be an encouragement to counterfeiters.
6. That about one seventh part of ye moneys be in farthings. The first 20 or 30 Tunns may be all in farthings because they are wanted.
8. That ye Master & Worker for the time being be charged & discharged by his Note as in the coinage of gold & silver.
9. That all the charge of copper, coinage, coinage tools, wages & incidents be paid out of the profits of the coinage.
10. That the copper be paid for out of ye produce of the coinage within two three or four months after delivery.
11. That if the copper upon the assay be not good the Master have power to refuse it.
7. That one or more persons be appointed to buy & import the copper by weight & receive it back in money by weight & tale & put the same away.(3)
13. That ye Master for the time being be allowed *5d* per pound weight for coinage by casting the blanks, or [*blank*] per pound weight for coinage by casting milling & sizing of barrs & cutting out the blanks, as is done in the coinage of gold & silver.

14. That ye surveyor of the meltings be paid by the number of melting days & the Tellers by tale.

12. That ye copper be coined without any mixture wch may diminish its price in the market if hereafter the moneys should be called in & melted down. For the price of the metal in the market is the intrinsick value of ye money. And that not above three or four ounces of Tynn be added to an hundred weight of copper in fusion to make it run close.

15. That the Importer above mentioned be paid by the Tunn for all trouble & charge in buying & importing the copper & for coinage & porters & barrels & baggs & paper & packthread & for putting away the copper money & making up the account, & that he bring in no particular account for any of these things.

16. Or if the Master be the Cashier & makes up the account, that he be allowed for his trouble & hazzard therein.

18. If her Maty should think fit by Patent to grant away the profit of the coinage above all charges, there will be no need of an Accountant or Teller or other Importer then the Patentee. But it was thought fit in the reigns of K. Charles and K. James that ye coinage should be upon account. And the copper and coinage will be the better if paid for by her Maty. And the profits above ye Account may be given to whom her Maty pleases if it be thought fit.

17. [*Blank*]

NOTES

(1) Mint Papers II, fo. 352. This paper was obviously written before the death of Queen Anne (21 July 1714, o.s.), otherwise we have as yet no means of dating it. We assume that it belonged to an early stage of Oxford's abortive planning for a copper coinage, and accordingly place it here.

(2) A possibly earlier draft of this paper (*ibid.*, at fo. 318) has 'two thirds' amended to 'three quarters'. In this the clauses are differently numbered, and clause 7 does not appear.

(3) That is, to disperse the copper money into circulation.

APPENDIX
BENTLEY'S ACCOUNTS FOR THE PRINTING OF THE *PRINCIPIA*[1]

On 18 June 1713 Bentley paid Cornelius Crownfield £53 3*s* 1½*d* as the cost of printing the *Principia*, and Crownfield gave his receipt for this sum.

	£	s.	d.
'Printed for ye Revd: Dr Bentley Sir Isaac Newton's Principia &c containing 65½ sheets, and two whole sheets Reprinted, in all 67 sheets and a half, at 15*s* „ 9*d* per sheet for 700 copies in 4to:'	53:	3:	1½

In addition, Bentley presumably paid

'for Printing ye Alma Mater in ye Title Page	00:	7:	0
for ye Carriage of ye Paper of ye Title Sheet up and down again	00:	4:	0
for 112 Ream and a half of Paper, at 11*s*: per Ream	61:	17:	6
for Advertisements in ye Gazette, Post-Boy, Daily Courant and Evening-Posts twice in each Paper'	1:	12:	6
	117:	4:	1½

Finally, there was a small outgoing on packing and postage.

Bentley's receipts were as follows·

Fourteen parcels of books were sold to individuals and booksellers, containing 375 copies,[2] at an average price of about 13*s*. each	205:	19:	0[3]
200 copies were sold to Crownfield at 11*s*. each, probably for sale abroad	110:	00:	0
	315:	19:	0

Of the 711 available copies of the *Principia*, 575 were sold (as above), 65 were given or sold privately by Bentley (including ten given to Newton and a dozen to Cotes) leaving 71 copies for Bentley's disposal at 25 November 1715, when he settled the account with Crownfield (apart from the £110 still outstanding upon bond).

Thus Bentley's profit was *at least* £198, a substantial gift to him from Newton largely earned by Cotes' unpaid labour, even though Bentley seems to have been cheated.[4]

NOTES

(1) Trinity College, Cambridge, MS. Add. a197[4].

(2) The parcels actually add up to 350 copies only.

(3) Addition of the individual sums yields a total of £229 14*s* 9*d*. Dividing the two corrected figures, the average cost per book is 13*s* (and £32 10*s* 0*d* was in fact paid for each parcel of 50). The MS. totals give an average price of slightly over 11*s* per copy, clearly wrong.

(4) Bentley should have been paid £220 10*s* 0*d* *in toto*, and have had 96 volumes in hand for disposal.

INDEX

Bold figures refer to Letter numbers; italic figures refer to page numbers of major biographical notes

ABERCORN, sixth Earl of: *240 n. 5*, 239
acceleration due to gravity: 217, 219, 221, 223, 224 n. 6, 226, 229 n. 3, 249 n. 2, 255–6, 260, 266, 268, 279
Acta Eruditorum
LETTER to Editors from Newton: February 1712/13, **976**, 383–4
references to papers by
Bernoulli; February 1713: 349 n. 2, 400, 401 n. 3
Leibniz; February 1689: 390 n. 5; 1694: 134, 137, 142, 145, 152 n. 40; May 1700: 96–7, 98 n. 4; January 1705: xxiii–xxiv, 115, 116 n. 7, 117, 132, 133, 142, 207–8, 212–13; February 1712: 383
Wolf, 1710: 116 nn. 4 & 5
review of Freind's *Prælectiones Chymicæ*, 1710: 116 n. 6, 204
review of Jones' *Analysis*, February 1712: 383
aether, Newton's concept of: 301 n. 7, 369 n. 3
ALLARDES, GEORGE
and Barons of the Exchequer: 1
and coinage in Scotland: 14–15, 206
payment of: 59, 373, 395
and payments to moneyers: 2
Scott his deputy: 57 n. 4, 316–18, 331
succeeded by Montgomerie: 58 n. 2, 59
and Trial of Scottish Pyx: 12–13, 54
alloys: *see under* Mint
AMBROSE, EDWARD: 358
anonymous LETTER to Newton: 28 March 1710, **774**, 17–24
ANSTIS, JOHN: *254 n. 1*
LETTER to Newton: 24 March 1711/12, **907**, 254
and the tin affair: 254, 381, 388
Apscourt Manor: xlv
ARBUTHNOT, JOHN
LETTERS
to Flamsteed: 14 March 1710/11, **823**, 99–100; 26 March 1711, **827**, 105; 6 April 1711, **831**, 118; 16 April 1711, **832**, 119–20; 21 April 1711, **834**, 122–3
from Flamsteed: 23/25 March 1710/11, **824**, 100–1; 28 March 1711, **828**, 106; 19 April 1711, **833**, 120–2
and *Commercium Epistolicum* committee, xxv
and Flamsteed's *Historia Cœlestis*: xlii–xliii, 99–102, 105–6, 118–23, 129–30, 209, 408
and visitation to Greenwich: 80 n. 3
ARISTARCHUS: 299, 301 n. 4
ARISTOTLE: xxxviii
ASKY: 63
assaying: *see under* Mint
ASTON, FRANCIS
on *Commercium Epistolicum* committee: xxv, xlviii n. 4
and Flamsteed's *Historia Cœlestis*: 162
Astronomer Royal: xli, 79–81; *see also* FLAMSTEED
astronomical instruments
micrometer: 209
mural arch: 209
at Royal Observatory: xli, 80–1, 91, 92 nn. 3 & 4, 209
sextant, Flamsteed's: 194, 209
telescope: 379
voluble quadrant: 209
astronomical observations
and Derham: 381 n. 10
Flamsteed's annual observations: 313, 385
and Flamsteed's Catalogue of Fixed Stars: xli, 80, 90–100, 102, 105–6, 118–21, 130, 166, 194
Newton's use of: xxx–xxxvii
at the Royal Observatory: 80–1, 131
astronomical symbols: *see* vol. IV, 546
attractive forces: 112–14, 392
between small particles: xxxiv, 361, 364
close-range: xxxiv, 112–14, 116 n. 6, 364, 367
Keill's interpretation: 116 n. 6

27-2

attractive forces (*cont.*)
measurement of: 364–5, 367–8
Roberval on: 301 n. 4
and surface tension: 363, 367
see also centripetal forces
Ayscough, Hannah *see* Smith, Hannah
Ayscough, William: 346

Baldwin, Ann: 393
Ball, Robert, 93–4
Barrow, Isaac
Lectiones Geometricæ: 150 n. 10
letters from Newton: 95 n. 7
and Lucasian Professorship: 380 n. 6
and Mercator's *Logarithmotechnia*, 379
his method of tangents: xxv–xxvi, 134–5, 143–4, 150 n. 8, 209 n. 4, 213
and Newton's MS. of *De Analysi*: 150 n. 6
Barton, Catherine: *200 n. 1*, xliv–xlvi, li n. 45, 201 n. 2, 251–2, 353
Barton, Daniel: 358, 360 n. 2
Barton, Hannah: *200 n. 1*, xliii, 201
Barton, Katherine: *200 n. 1*, xliii, 201, 345
Barton, Margaret: *200 n. 1*, 201
Barton, Robert (the elder): *200 n. 1*, xliii–xliv, 199–201
Barton, Robert (the younger): *200 n. 1*, 201, 345 n. 1
Bates, Arundell: 125
Beig, Ulug, and his Catalogue of Stars: 121
Benson, Robert: *371 n. 1*, 72 n. 2
Letter from Newton: 16 January 1712/13, **963**, 370–1
Bentley, Richard
Letters
to Cotes: 5 March 1712/13, **979**, 386; 12 March [1712/13], **984**, 390–1
to Newton: 20 October 1709, **767**, 7–8; [30 June 1713], **1002**, 313–14
from Cotes: 10 March 1712/13, **983**, 389–90
Accounts for the *Principia*: Appendix, 417
and Cambridge Press: 30 n. 2
and Cotes: xxviii–xxix, xlix n. 15, 152, 155, 174–5, 279, 280 n. 17, 305, 307, 315, 347, 351, 356
and Lucasian Professorship: 202–3
and Newton's correspondence with him: xvii–xxix, 301 n. 6
and Newton's portrait: 316 n. 5
and the *Principia*
his part in the second edition: xxviii, 4, 263, 269, 248 n. 2, 256, 400, 413–14
and Cotes' Preface: 386–7, 389–91
Berenger, Moses: 188, 192
Beresford, Francis: 308 n. 5, 314
Beresford *or* Berisford, 72, 308 n. 5
Bernoulli, Jakob: 116 n. 2, 208
Bernoulli, Johann [I]
and the calculus dispute: xxi, 116 n. 2, 208
his correspondence with English mathematicians: xxvii, xxxv, 349, 401 n. 3
his correspondence with Leibniz: 349
meeting with Newton and Halley: 349 n. 2
papers by: *see Acta Eruditorum*
and the *Principia*
his objections to Book ii, Prop. 10; xxxiv–xxxv, xxxvii, 348–50, 400–1
his objections to Book ii, Prop. 16: 28 n. 15
Bernoulli, Nikolaus [I]
Letter from Newton: [? *c.* 1 October 1712], ***951a***, 348–50
and error in Book ii, Prop. 10: xxxiv–xxxv, xxxvii, 348–50
Bianchi, Vendramino: *60 n. 2*, 59, 60 n. 3
Bingley, Baron: *see* Benson
Blackwell, Sir Lambert: *163 n. 2*, 162, 188
Bogotà, the Mint at: 163
Boissier, Guillaume & Co.: 125
Bolingbroke: *see* St. John
Bonet: xxv
Borthwick: 65
Boscawen, Hugh: *10 n. 3*, 10
Boscovich, Roger: xxxix
Boswell, Walter: 57 n. 2, 251 n. 2
Bouillaud, Ismael: *see* Bullialdus
Bowles, William: 57 n. 2
Boyle, Henry: 68–9
Boyle, Robert: 116 n. 6
Bradley, James: 405 n. 6
Brahe, Tycho
astronomical observations: 121
and Halley: 210 n. 4
and lunar theory: xxxi, xxxiii, xxxiv, xlix n. 17, 296 n. 3
Brand, Isaac, 63
Brattel, Charles: *374 n. 3*, 374, 375 n. 1, 403–4

BRATTELL, DANIEL: *374 n. 2*, 374, 375 n. 1
BROWNE, EDWARD
his house as accommodation for Royal Society: 62 n. 4, 76, 78
BRUCE, JAMES: *326 n. 1*, 341–4
LETTERS
to Oxford: 7 October 1712, **948**, 342–3
from Newton: [? September 1712], **947**, 341–2
BRYDGES, JAMES: *326 n. 2*, 342
BULL, SAMUEL: *308 n. 2*
BULLIALDUS, and his astronomical observations: 121
BURNET, WILLIAM: xxv, xxvii
Bushey Park: xlv
BUTLER, JAMES: *see* ORMONDE, Duke of

calculus: *see also* fluxions
dispute between Newton and Leibniz: *see under* LEIBNIZ
Cambridge Press: 30 n. 2
printer at: see CROWNFIELD
CANNON, Dr.: 412
capillarity: 364, 367
CARLETON, Lord: *see* BOYLE, HENRY
Cartesians: *see also* DESCARTES
attack Newton's philosophy: 204
CARTLICH: 87
CASSINI
measurement of Earth's shape: 216, 218, 219 n. 3, 222 n. 19, 255–62
papers by: *see Philosophical Transactions*
table of Jupiter's satellites: 169, 172 n. 11
CATENARO, J. B.: *225 n. 6*
CAVE, JOSEPH: 211, 314
cavitation: 44 n. 11
CAWOOD, FRANCIS: *332 n. 2*, 332–3
LETTER from Newton: [after 28 August 1712], **940a**, 333
undated Letter from Newton: 333 n. 3
centripetal forces: 4 n. 2, 5, 6, 25–8, 31
Cotes' discussion of: 392
Keill's paper on: 116 n. 5
CHAMBERLAYNE, JOHN: *59 n. 1*, 59
LETTER to Newton: 21 August 1710, **799**, 59–60
CHAMBERS: 358
CHANDOS, Duke of: *see* BRYDGES, JOHN
Charta Volans: xxii
Chelsea
Newton's removal to: 8 n. 8, 9
Newton's removal from: 70, 71 n. 2
chemistry
antimony: 85
aqua fortis: 85
and assaying: xlvii, 85–6, 88 n. 2, 198
and attraction of particles: 365, 368
and Boyle: 116 n. 7
of gold refining: 85
iron particles in bodies: 112–14
Keill, Freind, and the explanation of chemical phenomena: 116 n. 6
lead: 85
Chester Mint: 131 n. 4
CHEYNE, GEORGE: *118 n. 2*, 117–18
CHILD, Sir FRANCIS: *64 n. 3*, 63–4
chronology, Newton's drafts on: 354 n. 1
CHURCHILL, AWNSHAM: 212, 224, 254
Civil List, payment of Newton from: 179 n. 2
CLARKE, GEORGE: 106 n. 2
CLARKE, SAMUEL: *413 n. 1*, 8 n. 2
LETTER from Cotes: 25 June 1713, **1001**, 412–13
corrects Cotes' Preface: 387 n. 3, 412–13
Leibniz–Clarke correspondence: 301 n. 5
receives copy of *Principia*, 2nd edition: 413
CLAVELL, WALTER: *23 n. 12*, 18, 19, 78 n. 8
CLAVIUS, CHRISTOPHER: 121
CLERK, JAMES: 211, 314
CLERK, Sir TALBOT: 358
clocks
at Royal Observatory: 209
for Trinity College observatory: 5
CLOTTERBOOKE, JASPER: 63
COCKBURN, WILLIAM: *24 n. 21*, 21
cohesion: 365, 368
coin: *see under* Mint
COLEPRESS, SAMUEL: *244 n. 6*, 243, 264
COLLINS, JOHN
his correspondence: 383
with Leibniz: xxvi, 117 n. 1
with Newton: xxvi, 135–6, 143–4, 150 nn. 11 & 21
with Towneley: 379, 380 n. 3
the *Historiola*: xlix n. 10
and Newton's MSS.: xxvi, 95, 135, 143, 150 nn. 6, 7 & 21, 213
see also Commercium Epistolicum
COLSON, JOHN: 203 n. 5

comets, theory of
  improvements of, in *Principia*, 2nd edition: xxxiv, 112–13, 228, 231, 284, 315, 347, 350–1
  wooden cut for *Principia*: 354, 384, 386, 387, 400
Commercium Epistolicum: 95 n. 7, 389, 393
  committee: xxv–xxvi, xxxix
  distribution of: 380 n. 2, 390 n. 4
  drafts for: l n. 22, 114
  publication of: xxi–xxiii, xxvii, xxxv, xxxviii
compass: 333 n. 2, 340
Comptrollers of the Accounts of the Army
  LETTER from Taylour: 21 August 1712, **936**, 326–7
CONDUITT, JOHN: xvii, xxviii, xliv, xlvi, xlviii, 200 n. 1, 409 n. 4
COOPER, THOMAS: 62
COPERNICUS, NIKOLAUS: 121
COPLEY, GODFREY: 77–8, 78 n. 8
copper: *see under* Mint *and* Edinburgh Mint
COTES, ROGER: *xxviii*
  LETTERS
    to Bentley: 10 March 1712/13, **983**, 389–90
    to Clarke: 25 June 1713, **1001**, 412–13
    to Jones: 15 February 1710/11, **821**, 94–5; 30 September 1711, **873**, 194–5; 11 November 1711, **881**, 204; 3 May 1713, **995**, 405
    to Newton: 18 August 1709, **765**, 3–5; 15 April 1710, **775**, 24–8; 29 April 1710, **777**, 29–30; 7 May 1710, **779**, 33–5; 17 May 1710, **781**, 37; 20 May 1710, **783**, 39–42; 1 June 1710, **786**, 45–7; 11 June 1710, **788**, 48–50; 30 June 1710, **793**, 54–5; 4 September 1710, **800**, 60; 21 September 1710, **805**, 65–8; 5 October, 1710, **809**, 73–4; 26 October 1710, **810**, 74–5; 31 March 1711, **829**, 107–12; 4 June 1711, **844**, 152–3; 9 June 1711, **847**, 156–60; 23 June 1711, **854**, 166–72; 19 July 1711, **857**, 174–5; 30 June 1711, **863**, 183–5; 4 September 1711, **871**, 193–4; 25 October 1711, **879**, 202–3; 7 February 1711/12, **890**, 220–2; 16 February 1711/12, **893**, 225–9; 23 February 1711/12, **896**, 232–7; 28 February 1711/12, **899**, 242–4; 13 March 1711/12, **901**, 246–7; 14 April 1712, **910**, 269–70; 15 April 1712, **911**, 271–3; 24 April 1712, **913**, 275–8, [26 April 1712], **914**, 278–81; 1 May 1712, **916**, 282–4; 3 May 1712, **917**, 284–7; 13 May 1712, **920**, 303–5; 25 May 1712, **921**, 305–6; 20 July 1712, **929**, 315–16; 10 August 1712, **931**, 318–19; 17 August 1712, **935**, 324–6; 28 August 1712, **938**, 329–31; 6 September 1712, **942**, 335–7; 15 September 1712, **944**, 338–40; 23 October 1712, **953**, 351–2; 1 November 1712, **956**, 354; 23 November 1712, **958**, 356; 13 January 1712/13, **962**, 369–70; 8 March 1712/13, **980**, 387; 18 [March] 1712/13, **985**, 391–4
  from Bentley: 12 March [1712/13], **984**, 390–1; 5 March 1712/13, **979**, 386
  from Jones: 25 October 1711, **880**, 203; 15 November 1711, **882**, 204–5; 29 April 1713, **992**, 403–4
  from Newton: 11 October 1709, **766**, 5–6; 1 & 2 May 1710, **778**, 31–2; 13 May 1710, **780**, 35–6; 30 May 1710, **784**, 42–4; 8 June 1710, **787**, 47; 15 June 1710, **789**, 50–1; 1 July 1710, **794**, 55–6; 13 September 1710, **803**, 64; 30 September 1710, **807**, 70–1; 27 October 1710, **811**, 75; 24 March 1710/11, **826**, 103–4; 7 June 1711, **846**, 155–6; 18 June 1711, **851**, 164; 28 July 1711, **860**, 179–80; 2 February 1711/12, **889**, 215–16; 12 February 1711/12, **891**, 222–4; 19 February 1711/12, **894**, 230–1; 26 February 1711/12, **898**, 240–2; 18 March 1711/12, **903**, 248–9; 3 April 1712, **908**, 255–63; 8 April 1712, **909**, 263–4; 22 April 1712, **912**, 273–4; [29 April 1712], **915**, 281–2; 10 May 1712, **919**, 302–3; 27 May 1712, **922**, 307; 12 August 1712, **932**, 320–3; [14 August 1712], **933**, 323; 26 August 1712, **937**, 327–8; 2 September 1712, **941**, 333–5; 13 September 1712, **943**, 337–8; 23 September 1712, **946**, 341; 14 October 1712, **951**, 347–8; 21 October 1712, **952**, 350–1; 2 March

COTES, ROGER: LETTERS (*cont.*)
1712/13, **977**, 384–5; 5 March 1712/13, **979**, 386; 28 March 1713, **988**, 396–9; 31 March 1713, **989**, 400–1
and Bentley: xxviii–xxix, 386, 413
and calculus: 3–4, 114, 203–4, 393
and centripetal force: 392
and the *Commercium Epistolicum*: 380 n. 3
character: 7
and Crownfield: 173–4
on cubic equations: 218 n. 19
experimental lectures: 48
Harmonia Mensurarum: 277 n. 10
and Leibniz, 387 n. 3
and *Logometria*: xxxvi, 42 n. 8, 305–6, 319 n. 2
Lucasian Professorship: 202
lunar theory: xxxi, xxxiv
and Mariotte's experiment: 73–5
and Muys: 166
Newton's
acknowledgement of: xxii, xxix, xxxv–xxxvii, 111, 113–14, 194 n. 2
correspondence with: xv, xvii, xxi, xxiv
De Quadratura Curvarum 3–4
delays in replying to: 193–4, 247, 305
manuscripts: 194–5
portrait: 316
Principia: *see under* NEWTON, *Principia*
relatives of: 55
return from the country: 7, 60
takes orders: 32 n. 9
and the Third Law of motion: 1 n. 26
Cotton House: 62 n. 4
counterfeiting: *see under* Mint
Country Mints: *see under* Mint
COUPLET, CLAUDE-ANTOINE: 258
COX, Sir RICHARD, *240 n. 6*, 239
CRAIGE, JOHN: *118 n. 3*, 117
CROKER, JOHN: *308 n. 2*, 308, 314, 396 n. 2
CROWNFIELD, CORNELIUS: *30 n. 2*, 173–4, 417
LETTER to Newton: 11 July 1711, **856**, 173–4
cubic equations: 281 n. 19, 283
CUMBERLEGE, JOHN: 376
curves of second and higher degree: 116 n. 3
CUTTS BARTON family: 345 n. 1

DE GLOS: 237 n. 12
degree of meridian: *see* Earth, shape of
DE LA HIRE: 235, 237 n. 13, 258
DE L'HOSPITAL: *see* L'HÔPITAL
DE MOIVRE, ABRAHAM: xxv, xxvii, xxxv, 118 n. 2, 348–50, 401 n. 3
Denmark, GEORGE, Prince of: *see* GEORGE, Prince of Denmark
DERHAM, WILLIAM: xlvii, *380 n. 1*, 409 n. 6
LETTER to Newton: 20 February 1712/13, **973**, 379–81
DERING, CHARLES: 239
DES HAYES: 235, 237 n. 12, 240, 258
DESAGULIERS, J. T., xl
DESBOSSES of Cologne: 301 n. 3
DESCARTES, RENÉ: xxxviii–xxxix, 203, 387 n. 3, 391
differential method: *see also* fluxions
Leibniz's use of: xxi, 117, 213–14, 383
and Newton: 136, 145
DIXON, THOMAS: 72
DOPPING, SAMUEL: 239
DRAKE: 253
DRUMMOND, JOHN & Co.: *193 n. 2*, 191–2
DRUMMOND, WILLIAM: 3 n. 5, 54 n. 3, 57 n. 2, 251 n. 2
DU GUERNIER, LOUIS: *225 n. 3*, 224
Dunkirk, payment of British forces at: 326, 341–4
duties and taxes: *see under* Mint

Earth, shape of: xxxiii, 216–23, 224 n. 7, 226, 229 n. 3, 230–6, 240–4, 249, 255–63
Earth's rotation and its effect on *g*: 217, 219, 221, 223, 224 n. 6, 226, 229 n. 3, 256, 260, 279
East India Company: 92–3
Edinburgh Mint
assaymasters: *see* BORTHWICK *and* PENMAN
and Barons of the Exchequer of North Britain: 1
closure of: 3, 212 n. 2
Collectors of Bullion: 1
copper added to silver: 372
engravers: *see* CAVE *and* CLERK
salaries of: 211
financing of: 1
General of: *see* LAUDERDALE
indenture for: 14–16, 58, 211, 212 n. 2
Master of: *see* ALLARDES *and* MONTGOMERIE
melters: 394
moneyers: 2–3, 250–1

Edinburgh Mint (*cont.*)
Pyx, Trial of: 12–13, 51–2, 56–7
bringing the Pyx to London: 53–4, 58–9
recoinage of 1707–8: xlvii, 2–3, 61 n. 1, 65, 372, 394
accounts for: FINAL STATEMENTS OF ACCOUNT, 12 December 1711, **883**, 205–6; 65, 250–1, 316–17, 331, 373
salaries at: 1, 2–3, 211
silver coinage at: 14–16, 372
sweep at: 372
value of Scottish coin: 14–16
Warden of: *see* DRUMMOND, WILLIAM
waste at: 372, 373
EDLESTON, JOSEPH and Newton's MSS.: xvii
elastic medium: 168–9
electric force: 112–14, 363, 366
electric spirit: 362–9
ELLIS, JOHN: 189
LETTERS
to Goldolphin: 10 August 1709, **764**, 2–3; 16 November 1709, **770**, 11–12; 31 January 1709/10, **772**, 13–14; 17 May 1710, **782**, 37–8
to Oxford: 6 June 1711, **845**, 154
to St. John: [late September 1710], **806**, 68–70
to the Treasury: 18 September 1710, **804**, 65; 4 October 1710, **808**, 71–2; 2 February 1710/11, **820**, 93–4; 25 April 1711, **835**, 123–4; 14 May 1711, **838**, 127–8
and counterfeiters: 68–9
and the Edinburgh Mint: 2–3, 16, 65
and the Pyx: 13–14, 53–4
and the tin affair: 11–12, 37–8, 71–2, 154
and the values of plate: 123–4, 127–8, 183, 186
ELLIS, WELBORE, *240 n. 6*, 239
Ely, Bishop of: *see* MOORE, JOHN
ERCKER, LAZARUS: 88 nn. 2 & 3
experimental philosophy: xxxviii
experiments
Cotes' course of: 48
on the efflux of water: 73–5, 103–4, 109–12
on the electric spirit: 363, 366, 368 n. 1
Huygens' experiments: 43 n. 6
Mariotte's experiment: 66–8, 70, 73–5
on motion in a resisting medium: 31, 32 n. 6, 42–4, 159, 164
on the nature of matter: 242
oil-of-oranges experiment: 364–5, 367–8
on pendulums: 32 n. 6, 42–4, 240, 241 n. 2
Picard's experiments on heated wire: 235, 237 n. 13
Sauveur's experiments on sound: 185 n. 5
on surface tension: 364, 367, 369
on thin plates: 195 n. 2
eye: 363, 366
EYRE: 63

fathom, definition of: 222 n. 12
FATIO DE DUILLIER, NICHOLAS: xxviii, 96–8, 208 n. 4
FAUQUIER, JOHN FRANCIS: *72 n. 4*, 72, 331, 332 n. 2, 373
LETTER from H. Scott: 29 January 1712/13, **965**, 373–4
FERMAT, PIERRE DE: 213–14
fermentation: 365, 368
FERNE: 382
FEUILLÉE: 237 n. 20, 240, 258
FIELDING, CHARLES: 239
fits of easy reflection and transmission: 362, 366, 368 n. 2
FLAMSTEED, JOHN: xli–xliii
LETTERS
to Arbuthnot: 23/25 March 1710/11, **824**, 100–1; 28 March 1711, **828**, 106; 19 April 1711, **833**, 120–2
to Sharp: 25 October 1709, **768**, 8–9; 23 January 1710–11, **818**, 91–2; 15 May 1711, **841**, 129–31; 20 September 1711, **872**, 194; 22 December 1711, **885**, 209–10
from Arbuthnot: 14 March 1710/11, **823**, 99–100; 26 March 1711, **827**, 105; 6 April 1711, **831**, 118; 16 April 1711, **832**, 119–20; 21 April 1711, **834**, 122–3
from Halley: 23 June 1711, **853**, 165–6
from Newton: [*c.* 24 March 1710/11], **825** 102
from the Royal Society: 3 July 1712, **926**, 313; 5 March 1712/13, **978**, 385
and Cotes: xxviii
the *Historia Cœlestis*: 8, 80 n. 2, 99–101, 119–20, 408

FLAMSTEED: the *Historia Cœlestis* (*cont.*)
account for: 100 n. 4, 106 n. 2, 212, 224
and Catalogue of Fixed Stars: 92 n. 4, 99–100, 102, 105–6, 118–22, 129, 165–6, 194, 209
his instruments: 91–2, 209–10
and Newton: xli, 8–9, 414 n. 7
his observations: xxxii–xxxiii, 131, 313, 385
and the Royal Observatory: 79, 91–2, 209–10, 385
and the Royal Society: xl, 22 n. 2
FLETCHER, RICHARD: 187 n. 2, 352
fluids: *see* resisting medium
fluxions
Cotes' use of: 39–40
Leibniz's review of *De Quadratura*: 115 n. 2
and methods of tangents: 134, 143
Newton as first inventor: xxiv–xxvi, 116 n. 5, 133–52, 207–8, 213–14
in *Principia*: xxiii, xxxiv, xlix n. 21, 27 n. 2, 114
Raphson's *History of Fluxions*: 95
in Wallis's works: xxii, 96, 97
FOLEY, THOMAS: *23 n. 7*, 18
foreign coin: *see under* Mint
FREE, Mr: 191
FREEMAN, GEORGE: 358
FREIND, JOHN: xxii, 115, 116 n. 6, 203, 204–5, 390 n. 3
function of a function: 151 n. 24

GALILEO: 156, 164
GARDINER, 62
GARDNER, Colonel: 345 n. 1
GARDNER, KATHERINE: *see* BARTON, KATHERINE
General Election of 1710: 72
Genoa: 93, 125, 154, 161 n. 4, 162, 163, 188
GEORGE, Prince of Denmark: xli, 80, 92 n. 4, 99–103, 105, 130, 210 n. 5, 212, 225 nn. 3 & 4, 408
GIBBON, EDWARD: *189 n. 4*, 188
God, rôle of in physical world: xxxi, 228, 244 n. 3, 299–301, 398 n. 1
GODOLPHIN, SIDNEY, Earl of
LETTERS
to the Mint: 4 July 1710, **795**, 56–7
from the Mint: 10 August 1709, **764**, 2–3; 16 November 1709, **770**, 11–12; 31 January 1709/10, **772**, 13–14; 17 May 1710, **782**, 37–8; 29 June 1710, **792**, 53–4
from Newton: 16 February 1709/10, **773**, 14–15
dismissal of: 72 n. 2, 127 n. 4
and the Edinburgh Mint: 1, 12, 14–15
Newton's visit to: 52 n. 2
and Oxford: 161 n. 2
and the tin affair: 9–10, 37–8, 94 n. 4
and the Trial of the Pyx: 13–14, 51–4, 56
gold: *see under* Mint
Goldsmiths' Company
and assaying of Spanish money: 370
and marking of plate: 126, 175
and the Trial of the Pyx: 13 n. 3, 83 n. 2, 86–7, 90 n. 17, 412
and the waste: 372, 395
Gorée: 237 n. 12, 241
graphical methods: 27 n. 5
gravity: xxxviii–xxxix
Cotes on: 412–13
nature of: 231 n. 6, 233, 236 n. 4, 240, 241 n. 2, 248, 387 n. 3, 391–2, 393 nn. 4, 5 & 6, 397
Newton and Leibniz on: 299–300
and surface tension: 364, 367
Great Marlow copper works: 376, 378 n. 2
GREEN, ROBERT: *170 n. 3*, 166, 204, 247
Greenwich Observatory: *see* Royal Observatory
GREENWOOD, GEORGE: 199–201
LETTER from Newton: 9 October 1711, **878**, 199–201
GREENWOOD, KATHERINE: *see* BARTON, KATHERINE
GREGORY, DAVID
Astronomiæ Physicæ et Geometricæ Elementa: xxxii, 298 n. 3
and the calculus dispute: xxvi
and Cheyne: 118 n. 2
and Flamsteed's *Historia Cœlestis*: 102, 408
and Keill, 115 n. 1
and Newton's *Lectiones Opticæ*: 195 n. 2
and Newton's *Principia*: xxviii
and Newton's Theory of the Moon: 222 n. 8, 276, 285, 319, 323 n. 7
GREGORY, JAMES
correspondence with Collins: 379, 380 nn. 7 & 8

GREGORY, JAMES (*cont.*)
Geometriæ pars universalis: 150 n. 9
and his method of tangents: xxv, xlviii n. 1, 134, 143, 150 n. 8, 213
Gresham College
administration of: 409 n. 2
Meeting of Joint Committee for Gresham Affairs: 13 December 1704, **802a**, 62–4, xli
Professor of astronomy at: 408–9
Royal Society at: 1 n. 29, 20, 23 n. 17, 61 n. 4, 76–8
and Woodward: xli, 19, 22 n. 2
GRESHAM, Sir THOMAS, his will: 63, 409 n. 2
GROVNOR, Mr.: 353

Hackney Coaches, Bill for regulating: 175, 177 n. 6, 182; *see also* Two Million Act
HALIFAX, Baron: *see* MONTAGUE, CHARLES
HALLEY, EDMUND
LETTERS
to Flamsteed: 23 June 1711, **853**, 165–6; 3 July 1712, **926**, 313; 5 March 1712/13, **978**, 385
to Oxford: 18 September 1712, **945**, 340–1
from Oxford: 28 August 1712, **940**, 332–3
and Johann [I] Bernoulli: xxvii, 349 n. 2
and Cawood: 332–3, 340
and the Chester Mint: 131 n. 4
and comets, theory of: 351
on *Commercium Epistolicum* committee: xxv, xxvii
Conduitt's opinion of: 409 n. 5
experiments on water jets: 67 n. 4
and Flamsteed's *Historia Cœlestis*: xlii–xliii, l n. 32, 100 n. 6, 120–2, 130, 165–6, 194, 209, 212, 224
and Flamsteed's observations: 313
and Keill, 116 n. 5
and lunar theory: xxxiv, xlix n. 20, 230, 289, 296 nn. 2 & 3, 320
and Machin's recommendation: 408
and navigation: 332–3, 340
and Newton's *Principia*: 413
called Raymer: 210 n. 4
and the Royal Society: 24 n. 20
and the visitation of the Royal Observatory: 80 n. 3, 385
and Woodward: 17–18

Hamburgh: 188, 191–2
HAMILTON, JAMES: *see* ABERCORN, sixth Earl of
Hampton Court: xlvi
HARLEY, ROBERT: *see* OXFORD, Earl of
HARLEY, THOMAS: *196 n. 1*
LETTERS
to the Mint: 5 December 1712, **959**, 357; 2 April 1713, **991**, 403; 2 May 1713, **994**, 404–5
to Newton: 3 October 1711, **874**, 195–6; 6 October 1711, **876**, 196–7; 15 January 1711/12, **887**, 212; 26 February 1711/12, **897**, 238; 9 March 1712/13, **982**, 388–9
to Phelipps: 9 March 1712/13, **982**, 388–9
and appointment of assaymaster: 403
and copper coinage: 357, 405–7
and counterfeiting: 238, 245–6
and Flamsteed's *Historia Cœlestis*: 212, 224
and Mint salaries: 388
and moneyers: 251 n. 2
and the value of Spanish coin: 195–7, 198 n. 1
HARRIS, HENRY: 308 n. 2
HARRIS, JOHN: *22 n. 3*, 17, 18–20
HARRIS, RICHARD: 78 n. 8
HARRISON, EDMOND: 62
HART, JOHN: 62
HARTSOEKER, NICOLAUS: 298, 301 nn. 3 & 4, 393
HAUKSBEE, FRANCIS: xl, 32 n. 6, 369
HEARNE, Sir JOSEPH: 79, 81, 359
heat
expansion of pendulum: 241
Picard's experiments on heated wire: 235, 237 n. 13, 258
vibrations of: 362, 365
HEATHCOT, Sir GILBERT: *64 n. 4*, 63–4
HEVELIUS, JOHANNES: 121
HILL, ABRAHAM: *23 n. 8*, xxv, xlviii n. 4, 18, 80 n. 3, 313, 385
LETTERS to Flamsteed: 3 July 1712, **926**, 313; 5 March 1712/13, **978**, 385
HILL, JOHN: 199–200
HIPPARCHOS' theory of the Moon: xxxi
HODGSON, JAMES: 106 n. 2, 118, 120, 122, 130
HODGSON, Mrs: 166
Holland, and the tin affair: 188, 191
HOOKE, ROBERT: 61, 62 n. 5
HORE, CHARLES: 378, 406–7

HORROX, JEREMIAH: xxxi–xxxiv, xlix n. 20, 289, 293, 296 nn. 2 & 3
HOULTON, NATHANIEL: 63
HOUSDEN, JANE: 68–9
HOW, EMMANUEL SCROOP: 201 n. 3
HOWKINS, EDWARD, and the Cotes–Newton correspondence: xvii, 6 n. 7
HOWSE, RICHARD: 408
HUDDE, JAN, and his method of tangents: 135, 144
HUNT, HENRY: *62 n. 5*, 61, 100, 210
HUSSEY, CHRISTOPHER: *202 n. 4*
HUYGENS, CHRISTIAAN:
  and acceleration due to gravity: 43 n. 6, 219, 229 n. 3, 255, 260
  and the calculus dispute: 207–9
  Cosmotheros: 169
  and Fatio de Duillier: 98
  Horologium oscillatorium: 8 n. 5
  and Saturn's satellites: 169
  and the telescope: 379
HYDE, LAWRENCE, Earl of Rochester: 92 n. 4
hypothesis, Newton's views on: 387 n. 3, 394 n. 7, 396–401

impenetrability of matter: 240, 241 n. 2
infinitesimals: 383
inflection of light rays: 362, 365, 368 n. 2
INGLE, HENRY: *347 n. 2*
  LETTER from Newton: 13 October 1712, **950**, 346–7
INGOLDSBY, RICHARD: *240 n. 4*, 239
integrals, tables of: 3–4, 150 n. 18
integration: *see* quadrature
interpolation methods: 27 n. 5
inverse problem of central forces: 6 n. 5
Ireland
  counterfeiting of Irish gold coin: 238, 245–6
  currency of foreign coin in: 309
  Newton tests a coin from: 323–4
  standards for Irish coin: 323–4
  see also LORDS JUSTICES AND PRIVY COUNCIL OF IRELAND

JANSSEN, THEODORE: *125 n. 1*, 93, 125, 154, 188
JONES, WILLIAM: *95 n. 1*
  LETTERS
    to Cotes: 25 October 1711, **880**, 203; 15 November 1711, **882**, 204–5; 29 April 1713, **993**, 404
    from Cotes: 15 February 1710/11, **821**, 94–5; 30 September 1711, **873**, 194–5; 11 November 1711, **881**, 204; 3 May 1713, **995**, 405
    from Saunderson: 16 March [1711/12], **902**, 247
  Analysis per Quantitatum Series, Fluxiones ac Differentias: xxiii, xlix n. 21, 94–5, 136, 144, 152, 279
    review of: 383
  on *Commercium Epistolicum* committee: xxv, xxvii
  sends *Commercium Epistolicum* to Cotes: 380 n. 2
  Newton's champion: xxii, 203–5
  and Newton's MSS.: 194–5
  and Newton's *Principia*: 386 n. 5, 304–5
Journal des Sçavans: 301 n. 3
Jupiter: 224 n. 2, 226–7, 257, 261
  perturbation of Saturn: 248
Jupiter's satellites: 169, 172 n. 11, 221–2, 224 n. 2, 226, 409 n. 6

KEILL, JOHN: *115 n. 1*
  LETTERS
    to Newton: 3 April 1711, **830**, 115–17
    to Sloane for Leibniz: [May 1711], **843a**, 133–52
  and the calculus dispute: xxii–xxvii, 117, 132–52, 203, 205, 207–9, 212–14
  on centripetal force: 96, 97, 116 n. 5
  and Machin's recommendation: 408
  and Newton: xvii
  sends Newton Leibniz's review of *De Quadratura*: 115
KEMP, JOHN: 376
KEPLER, JOHANNES: xxxi, 121, 297 n. 3
KERRIDGE, WILLIAM
  LETTER to Newton: 16 January 1710/11, **817**, 90–1
KILDARE, Bishop of: *see* ELLIS, WELBORE
KING, Colonel, his *Journal*: 200 n. 1
KING, WILLIAM: 9 n. 3, 23 n. 10
KNAPS: 8–9
KNELLER, Sir GODFREY: 225 n. 4, 316 n. 5
KNIGHT, JOHN: *23 n. 13*, 18

LAGRANGE, JOSEPH LOUIS DE: 350 n. 2
LA HIRE: *see* DE LA HIRE
LALANDE: *see* LEFRANÇAIS DE LALANDE

LAMBERT, JOHN: *see* BLACKWELL, Sir LAMBERT
LAUDERDALE, Earl of: *3 n. 3*, 1–2, 51–4, 57 n. 5
LETTERS from Newton: [? July/August 1709], **763**, 1–2; 22 June 1710, **790**, 51–2
LE CLERK, GABRIEL: *308 n. 4*, 314
LEFRANÇAIS DE LALANDE, J. J.: xxxiii, 298 n. 3
Leghorn: 93–4, 154, 163
LEIBNIZ, G. G.:
LETTERS
to Sloane: 21 Feburary 1710/11, **822**, 96–8; 18 December 1711, **884**, 207–9
from Sloane: [May 1711], **843**, 132
from Keill to Sloane for Leibniz [May 1711], ***843a***, 133–52
and Johann [I] Bernoulli: xxxv, 349 nn. 1 & 2
and the calculus dispute: xxi–xxvii, xxxix, xlviii n. 1, 95 nn. 5 & 7, 96–8, 115–18, 132–52, 207–9, 212–14, 383, 387 n. 3
correspondence with Collins: 117 n. 1
correspondence with Hartsoeker: 298–300, 393
correspondence with Newton: xxi, xlix n. 11
correspondence with Oldenburg: xxvi, xlviii n. 6, 98 n. 8, 117 n. 1
correspondence with Wolf: 116 n. 4
and Cotes' Preface to the *Principia*: 389–93
and Fatio de Duiller, 96, 97, 98 n. 4
and Newton's *De Quadratura*: xxiii, 115 n. 2
papers by: *see Acta Eruditorum*
Tentamen de Motuum Cœlestium Causis: 389
Levant Company: 125 n. 2, 193 n. 7
L'HÔPITAL, Marquis de: 116 n. 2, 208 n. 4, 349 n 2
light: 362–3, 366, 368–9
LILLY: 382
LISTER, MARTIN: 9 n. 3
LIVEBODY: 5, 7, 305, 307
logarithms: 306 n. 5
LORDS JUSTICES AND PRIVY COUNCIL OF IRELAND
LETTER to Ormonde: 8 January 1711/12, ***897a***, 238–40
LOUTTON: 132
LOVEDAY: 63
LOWNDES, WILLIAM
LETTERS
to the Mint: 9 November 1709, **769**, 9–11; 28 April 1710, **776**, 28–9; 28 June 1710, **791**, 53; 25 July 1710, **797**, 58; 5 September 1710, **801**, 60–1; 8 May 1711, **836**, 125; 14 May 1711, **839**, 128–9; 16 August 1711, **866**, 187–8; 7 January 1711/12, **885A**, 211; 18 March 1711/12, **904**, 250; 31 January 1712/13, **967**, 375; 3 February 1712/13, **968**, 375; 4 February 1712/13, **969**, 376; 12 February 1712/13, **971**, 378; 26 February 1712/13, **974**, 381
to Newton: 14 June 1711, **849**, 162
and Robert Ball: 93
and Brattel's petition: 375
and Clerk and Cave's petition: 211
and copper coinage: 378
and the Edinburgh Mint: 58, 60–1
and engraving seals: 396
and foreign coin: 162, 187, 375
and Janssen: 125
and William Palmes: 79
and Penman's petition: 60–1
and silver plate: 127–8
and the tin affair: 9–11, 250, 381
and the Trial of the Pyx: 53–4, 57 n. 5
LOWTHORP, JOHN: *23 n. 16*, 19
Lucasian Professorship: 202, 380 n. 6
Lunar theory: *see under* Moon
LYONS, Sir HENRY; *History of the Royal Society*: xxxix–xl

MACHIN, JOHN: 409 n. 1, xxv, xxvii, 408–9
Newton's recommendation: 12 May 1713, **997**, 408–9
MACKENZIE, GEORGE: 377 n. 2
magnetic force: 112–14, 248
magnetic needle: 333 n. 3, 340
MAITLAND, ALEXANDER: 1 n. 2
MAITLAND, JOHN: *see* LAUDERDALE
MANNS, JOHN A.: 346
MANSELL, Sir THOMAS: 72 n. 2, 126
ROYAL WARRANT to the Mint: 10 May 1711, **837**, 126–7
manuscripts, sources for: xvii
MARCELLUS: 59 n. 1
MARIOTTE, EDMÉ
Mariotte's experiment: 66–7, 68 n. 5, 71 n. 3, 73–4
Traité du Mouvement des Eaux: 68 n. 5, 70
MARLBOROUGH: 161 n. 2, 238 n. 2, 327 n. 3

Mars, motion of aphelion: 227, 230
MARTIN
  LETTER from Newton: [before 1711], **878a**, 201
matter
  density of: 248
  hardness of: 301 n. 6
  in relation to space: 231 n. 6, 233, 236 n. 4, 242
MEAD, RICHARD: *203 n. 3*
  LETTER to Flamsteed: 3 July 1712, **926**, 313
  and Flamsteed: 80 n. 3, 131, 313
  and Newton: 203
  and the Woodward affair: 18, 22 n. 2, 23 n. 11
medicine
  Bezoar stone: 18, 22 n. 2, 23 n. 9
  gall-stone: 18, 22 n. 2
  colic: 18
MEDOWS, Sir PHILIP: *326 n. 1*, 341–4
  LETTERS
    to Oxford: 7 October 1712, **948**, 342–3
    from Newton: [? September 1712], **947**, 341–2
Mémoires de l'Académie des Sciences: 18, 22 n. 2
  references to papers in, 169, 172 n. 14, 235, 237 n. 12, 349 n. 2
Mémoires de Trevoux: 301 n. 3
Memoirs of Literature: 298–301, 393
  LETTER from Newton: [after 5 May 1712], **918**, 298–302
MERCATOR, NICHOLAS: 138, 146
  Logarithmotechnia: 379
Mercers' Company: 23 n. 17, 61 n. 4, 62–4, 409 n. 2
MEREDITH, THOMAS: 201 n. 3
meteorological instruments: 62 n. 5
method of first and last ratios: 383
MIDDLEMORR, MR.: 251
MILLER, ROBERT: 57 n. 2
Mint, the: xiv–xv, xvii, xxiv, xxxvi, xlvi–xlvii, 174, 193
  LETTERS
    to Godolphin: 10 August 1709, **764**, 2–3; 16 November 1709, **770**, 11–12; 31 January 1709/10, **772**, 13–14; 17 May 1710, **782**, 37–8; 29 June 1710, **792**, 53–4
    to Oxford: 6 June 1711, **845**, 154; 21 August 1711, **868**, 189; 16 June 1712, **923**, 307–8; 16 July 1712, **928**, 314–15; December 1712, **960**, 357–61; 9 March 1712/13, **981**, 388; 10 April 1713, **992**, 403–4
    to the Treasury: 2 February 1710/11, **820**, 93–4; 25 April 1711, **835**, 123–4; 14 May 1711, **838**, 127–8
    from Godolphin: 4 July 1710, **795**, 56–7
    from T. Harley: 2 April 1713, **991**, 403
    from Lowndes: 9 November 1709, **769**, 9–11; 28 April 1710, **776**, 28–9; 28 June 1710, **791**, 53; 25 July 1710, **797**, 58; 5 September 1710, **801**, 60–1; 8 May 1711, **836**, 125; 14 May 1711, **839**, 128–9; 16 August 1711, **866**, 187–8; 7 January 1711/12, **885A**, 211; 18 March 1711/12, **904**, 250; 31 January 1712/13, **967**, 375; 3 February 1712/13, **968**, 375; 12 February 1712/13, **971**, 378; 26 February 1712/13, **974**, 381
    from Taylour: 28 January 1709/10, **771**, 12–13
    from the Treasury: 29 November 1710, **813**, 79; 12 February 1712/13, **972**, 378; 21 March 1712/13, **987**, 396
  ROYAL WARRANTS to the Mint: 10 May 1711, **837**, 126–7; 30 July 1711, **862**, 182–3
  accounts: 205–6
  allowances to Mint employees: 58 n. 2, 71–2, 186–7, 206, 307–8, 313–15, 352, 355, 374, 395, 415
    Newton's salary: 210 n. 7
  alloys: 395, 416
  assaying: xlvii, 82–9, 89 n. 6, 198, 371 n. 2, 415
  assaymasters: *see* BRATTEL, CHARLES *and* BRATTEL, DANIEL
    appointment of: 374, 376, 403–4
  bullion: *see* silver *and* gold
  clipping of coin: xlvii, 69, 382
  Coinage Act: 180
  coinage duty: 87, 178, 251, 332, 355
  Coinage of 1711
    Newton's plans for: 75 n. 4
    Coinage of Plate Account: 12 December 1711, **883 I**, 205–6

Mint (*cont.*)
coiner: *see* counterfeiter
Comptroller of the Mint: *see* Ellis, John
copper
addition to silver: 372, 394
standard for: 359, 406
copper coinage: *see also* Mint: farthings *and* halfpennies
and counterfeiting: 382
and William Palmes: 79, 81–2
plans for coinage: xlvii, 357–60, 376, 378 n. 2, 404–7, 415–16
previous copper coinages: 81–2, 359, 406
Swedish copper: 81
Copper Mines, Company of: 358
copper works at Great Marlow: 376, 378 n. 2
Cornish stannaries: xlvii, 9–10, 190, 254 n. 1; *see also* Mint: tin
counterfeiters: xlvii, 68–9, 381–2, 401–2, 406–7, 415
in Ireland: 238–9, 245–6
country mints: 52 n. 4
cross-dollar: 310–12
culling of imported money: 382
Deputy Master and Worker: *see* Fauquier
dollars: 160–3, 197–8, 310–12
ducatoone: 211–12
Dutch coins: 191, 341–4, 375
Dutch monetary system: 342 n. 2
duties and taxes
acceptance of foreign coin as taxes: 239 n. 2
coinage duty: 87, 178, 232, 251, 355
paid to the Exchequer: 378 n. 2
in Scotland: 1 n. 3
on tin and pewter, 38
Two Million Act: 177 n. 6, 182–3
Edinburgh Mint: *see* Edinburgh
engravers: 307–8, 313–15; *see also* Beresford, Francis; Bull; Croker, John; Le Clerk
chief engraver of seals: *see* Roos
farthings, coining of: 404–5; *see also* Mint: copper
foreign coin: 375
circulating in Ireland: 238–9, 245–6, 309–11, 324, 342–4
value of: xlvii, 160–3, 187–9, 191, 195–8, 326 n. 2, 370–1
French coin: 326, 342–4, 375
Louis D'or: 239, 245, 310–12
pistole: 231, 310–12
gold
assay: xlvii, 84–6
coinage of: 360, 374
collection of: 180
counterfeiting of: 238–9, 245–6
medals: 411
melting of: 371
production of: xlvii
refining of: 85
standards: xlvii, 82–3, 87
test plate: 56–7, 82–3
guineas
Come-again-Guineas: 86–7
value of: 245–6
Indenture of the Mint: 14–15, 58, 87, 180, 378
melters: 371–2, 394
Mexico pieces of eight: 160–3, 196–8, 310–12, 371
moneyers
and copper coinage: 358
and Scottish recoinage: 2–3, 16, 206 nn. 2 & 3, 250–1
payment of British forces abroad: 161 n. 4, 197, 326, 341
payments to Mint employees: *see* Mint: allowances
Peru pieces of eight: 161, 163, 197, 312
pewter: *see under* Mint: tin
pieces of eight: *see* Mint: dollars
pillar pieces: 310, 312
Portuguese coin:
crusado: 310–12
moyder: 239, 245–6, 301–12
puncheons: 315
Pyx, Trial of: *see* Mint: Trial of the Pyx
recoinage: *see* Mint: coinage
refining, waste in: 372
remedy: 86–7, 89 n. 10
rix dollar *or* Reichsthaler: 310, 312 n. 2
salaries of Mint employees: *see* Mint: allowances
seals, engraving of: 396
shillings: 126, 175; *see also* Mint: silver, value in Ireland: 310
silver
assay: 84–5, 198

Mint: silver (*cont.*)
coined at Edinburgh: 206, 316–18, 331
coining of: xlvii, 126–7, 173, 175–8, 182, 205, 360
collection of: xlvii, 126–9, 165, 172–3, 175–8, 180–3, 186, 231–2
deficiency: 178 n. 1, 205, 206 n. 1, 232, 251
melting of: 173, 180
South American silver: 161 n. 4, 160–3
standards for: 82–3, 123–4, 126, 176, 394
value of coin: 186, 245, 246, 341–4
sixpences: 126, 175; *see also* Mint: silver
smiths to the Mint: *see* FLETCHER *and* SYLVESTER
payment of: 186–7
Spanish coin: 311–12, 370–1
pistole of gold: 238–40, 245, 311
Seville pieces of eight: 160–3
two Reale piece: 187–8, 189, 195–8
sterling: *see* Mint: silver
sweep: 172–3, 205, 316, 331, 371–2
taxes: *see* Mint: duties
tin: xlvii, 10 n. 3, 71–2, 188–9, 190, 250, 254
accounts for sale of tin 1705–9: 11
added to copper: 416
Cornish tinners: 9–12, 72, 190, 381, 388
duty on tin and pewter: 38
and East India Company: 92–3
export of pewter: 38
export of tin: 38, 93, 125, 154, 191–2
foreign pewter: 38 n. 2
payment of Newton in: 178
Pewterers' Company of London: 28, 37–8, 191
'post groats': 12 n. 4
salaries of those involved in the tin affair: 72
surplus tin: 11–12, 188–9
Swedish tin: 81
tin accounts: 190–3
tin farthings and halfpennies: 81, 357
Trial of the Pyx, 21 August 1710: xlvii, 409
attendance at: 57
Edinburgh Pyx brought to London: 13–14, 51–3
gold test plates used at: 56–7 n. 2, 88 nn. 2 & 3
Keys of the Pyx, 54
procedure at: 13 n. 3, 56–7, 86–7
result of trial: 57 n. 2
Trial of the Pyx, 7 August 1713: 83 n. 1, 409–11
Orders of Council for: 24 June 1713, **1000 II**, 411–12
trial pieces: xlvii, 82–9
value of coin: xlviii, 160–3, 189, 195–9, 239, 245–6, 309–12, 341–4, 370–1, 375, 378
Vigo plate: 180
Warden of the Mint: *see* PEYTON *and* STANLEY
waste: 316–17, 332 n. 3, 371–3, 394–5
wearing: 382
miracles: 299–300
monochord: 184
MONTAGUE, CHARLES, Baron Halifax
and Catherine Barton: xliv–xlvi, 200 n. 1
will of: li n. 34
MONTAGUE, GEORGE: xlv
MONTGOMERIE, JOHN, of Girvan: 54 n. 3, 57 nn. 4 & 5, 58 n. 2, 212 n. 2
LETTER from Newton: [July 1710], **796**, 57
Moon
effect on the tides: 217–18, 220–3, 227, 229 n. 11, 230, 241, 243–4, 264–9, 272 n. 6, 273–5, 278–80
Newton's theory of, published by Gregory: xxxii, 222 n. 8, 276, 285, 319, 323 n. 7
and the 2nd edition of the *Principia*: xxxi–xxxiv, 112–13, 217–37, 240–4, 248–9, 255–307, 315, 318–31
tables of, in Flamsteed's Catalogue: 194
MOORE, JOHN: *414 n. 6*, 413
MOORE, Sir JONAS: *92 n. 3*, 91, 209
MORGAN, WILLIAM: 359, 360
MORICE, JOHN: 62
MORLAND, JOSEPH: *23 n. 14*, 18
MORLAND, Sir SAMUEL: *91 n. 2*, 23 n. 14, 90
MORTIMER, Earl: *see* OXFORD, Earl of
motion in a resisting medium: *see* resisting medium
MOUTON, GABRIEL: xxvi, xlix n. 8
MOYER, Sir SAMUEL: 62
MUNFORD: 62
MUYS, WYER WILLEM, *Elementa Physices*: 152–3, 164, 166, 204

Naperian logarithms: *see* logarithms
navigation: xli, 44 n. 12, 332–3, 340

NEALE, THOMAS: 72 n. 4
nerves: 363, 366
New Style dates (N.S.): xv
NEWSTEAD, JANE: *see* HOUSDEN, JANE
NEWSTEAD, THOMAS: 69
NEWTON, HANNAH; *see* SMITH, HANNAH
NEWTON, Sir ISAAC
  CERTIFICATE by Newton: 14 August 1712, **934**, 323–4
  for LETTERS to Newton *see*: anonymous, ANSTIS, BENTLEY, CHAMBERLAYNE, COTES, CROWNFIELD, DERHAM, T. HARLEY, KEILL, KERRIDGE, LOWNDES, JAMES NEWTON, OXFORD, MARY PILKINGTON, H. SCOTT, SEAFIELD, WOODWARD
  for LETTERS from Newton *see*: *Acta Eruditorum*, BENSON, NIKOLAUS BERNOULLI, BRUCE, CAWOOD, COTES, FLAMSTEED, GODOLPHIN, GREENWOOD, INGLE, LAUDERDALE, MARTIN, *Memoirs of Literature*, MEDOWS, MONTGOMERIE, OXFORD, ST. JOHN, H. SCOTT, SLOANE, TODD, the Treasury; *see also under* Mint: LETTERS
  MEMORANDA by Newton: [*c.* 31 December 1710], **816*a***, 84–90; [before 24 July 1771], **858**, 175–7; [*c.* 28 July 1711], **861**, 180–1; [early October 1712], **948*a***, 343–4; [1713], **1003**, 415–16
  NEWTON AT THE TREASURY: 27 February 1712/13, **975**, 381–2; 8 May 1713, **996**, 405–6
  ROYAL WARRANTS to Newton: 12 December 1710, **814**, 79–81; 20 November 1712, **957**, 355
  'Analysis per Quantitates fluentes et earum Momenta': xlix n. 21, 114 n. 4, 280 n. 17, 315
  annuity: 201
  Arithmetica universalis: 281 n. 18
  and Bianchi: 59
  bribing of: xlviii
  and the calculus dispute: *see* Leibniz: calculus dispute
  and the Cartesians: 203–5
  character: 254
  and Collins: 95
  and Cotes: xv, xvii, xxi, xxiv, xxix, xxxv, 193 n. 2
  De analysi per æquationes infinitas: 136, 141, 144–5, 149, 150 nn. 6 & 17, 213
  De Quadratura Curvarum
    and the calculus dispute: xxiii, xxv, 115 n. 1, 117, 136, 145, 150 nn. 17 & 22, 213–14
    intended publication: xlix n. 21
    Leibniz's review of: 115 n. 2, 383
    tables of integrals, seventh and eighth forms: 3–4, 5
  Enumeratio Linearum tertii ordinis: 281 n. 19
  Epistola Posterior: 137, 145, 149 n. 4, 150 nn. 15 & 16, 151 nn. 25, 26, 28 & 32
  Epistola Prior: 149 n. 4, 151 n. 24
  experiments by, 31, 32 n. 6, 42–4, 75, 103–4, 159, 164
  and Flamsteed: xli–xliii, 120, 131, 209, 313
  visits Godolphin: 52 n. 2
  and Jones: 95
  Lectiones Opticæ: 195 n. 2
  and the Lucasian Professorship: 202
  his mathematical manuscripts: 94–5, 194
  at the Mint: xxiv, xlvi–xlviii
  moves house: 8 n. 8, 9, 70, 71 n. 2, 75
  Opticks: 95 n. 2, 136, 144, 150 n. 22, 195 n. 2, 242, 244 n. 2, 362, 365, 368–9, 399 n. 4
  his philosophy attacked by the Cartesians: 203–5
  Principia (*references refer, where possible, to the numbering in the second edition*): xxvii–xxxviii
    Auctoris Præfatio in Editionem Secundam: 387 n. 3, 400
    Cotes' Preface: xxxvi–xxxviii, l n. 24, 112–14, 386–7, 389–94, 396–400, 412
    Table of contents: 361
    Definition 5: 5, 6 n. 4
    Laws of motion: l n. 26, 393 nn. 4 & 5, 396–7
    Book I
      Sect. I, Schol.: 7
      Props. 2 & 3: 298
      Prop. 4, Coroll. 5: 30
      Prop. 6, Coroll. 4: 398
      Prop. 13, Coroll. 1: 5, 6 n. 5, 7
      Prop. 45, Coroll. 1: 398
      Prop. 66, Coroll. 16: 227, 303–4
      Prop. 91, Coroll. 2: 3, 4 n. 2, 5
      Props. 94–6: 366

NEWTON, Sir ISAAC: *Principia* (*cont.*)
Book II
Lemma 2: 24, 151 n. 38
Lemma 4 (1st edition): 164
Prop. 10: xxx, xxxiv–xxxvii, 27 n. 3, 114 n. 1, 347–51, 354, 356, 361, 369–70; Schol.: 24
Prop. 13: 25
Prop. 14 and Corolls.: 25
Prop. 15 and Corolls.: 25–6, 29–31, 33–5
Prop. 16: 28 n. 15, 30, 31, 36; Coroll. 1: 31, 34–5; Corolls. 2 & 3: 26, 31, 31 n. 5
Prop. 19: 39
Prop. 20: 39, 41 n. 2
Prop. 23, Schol.: 39, 41 n. 3
Prop. 24, Corolls. 5 & 6: 39
Prop. 27, Coroll. 1: 39, 111 n. 3
Prop. 29: 39, 41 n. 7
Prop. 30: 40, 42–3, 45–8, 50
Prop. 31: xxx, 40, 48, 49 n. 4, Schol.: xxx, 42–4, 47–51
Prop. 32 and Corolls.: 43
Prop. 33: 40, 55
Prop. 34: 40
Prop. 36: xxx, 66, 70, 74–5, 104 nn. 2 & 4, 107–9, 111 n. 6, 153 nn. 1 & 5, 156–7, 160 n. 5
Prop. 37: 104 n. 2, 108–10, 112, 152, 153 nn. 1 & 4, 155, 157–60
Prop. 38, Coroll. 2: 66, 67 nn. 3 & 4, 159
Prop. 40 and Schol.: 32 n. 6, 49–50, 67, 68 n. 8, 104 n. 2, 159, 164
Prop. 41: 170
Prop. 44: 170
Prop. 46: 170, 174
Prop. 47 and Coroll.: xxx, 167–71, 174, 179, 185, 193, 215
Prop. 48: 104 n. 2, 169–70, 179
Prop. 49 and Corolls.: 168–9, 171, 179
Prop. 50 and Corolls.: 185 n. 5
Book III: xxxi
Phænomena: 169, 222
Prop. 3: 169, 172, 215, 220, 248
Prop. 4: 215–16, 257; projected Schol.: 216 n. 4, 216–19, 220–1, 223, 226, 248–9
Prop. 5, Corolls.: xxxvi, 228, 233, 240, 242, 248
Prop. 7: 393
Prop. 8: 222–3, 226
Prop. 10: 248
Prop. 12: 222, 227
Prop. 13: 222, 227
Prop. 14, Schol.: 227, 233
Prop. 19: 219 nn. 1, 4 & 6, 221–3, 227, 233, 240, 248–9, 255–7, 259, 262–3, 269–70
Prop. 20: 227, 232–5, 240–2, 248–9, 255, 257–9, 261–3
Prop. 22: xxxii, xxxiii
Prop. 24: 175
Prop. 25: xxxii, 255, 271–3, 275
Props. 26 & 27: xxxiv
Prop. 28: xxxii
Prop. 29: xxxii, xxxiv, 276–7, 281–3, 302–3, 305, 307
Prop. 31: 276–7
Prop. 32: 276–7, 281, 284, 302, 304, 307
Prop. 33: 276, 278 n. 14, 281, 284, 307
Prop. 34: xxxii, 281
Prop. 35: 284–5, 305; Schol.: xxxii–xxxvii, 285–98, 315, 318–30, 333–9, 341
Prop. 36 and Coroll.: 220, 222, 225, 248, 259, 264, 271–2, 276, 279
Prop. 37: 44, 219 nn. 4 & 6, 220, 222–3, 226, 235–6, 241, 243–4, 246, 248–9, 252 n. 4, 259, 264–9, 271–5, 278, 286, 339–40
Prop. 38: 275, 279
Prop. 39: 220, 222–3, 227, 233, 235–6, 248–9, 259, 264, 272–4, 279
Prop. 40: 35
Scholium Generale: xxxiv, 114 n. 3, 302 n. 8, 356, 361–70, 384, 386–7, 394 n. 7, 397, 399–400
Index: 356, 362 n. 6, 386–7, 389, 404–5
Errata: 354
accounts for printing: 417
Amsterdam reprint: xlix n. 21
and Bentley: xxviii–xxix, 4, 7, 348 n. 2, 387 n. 3, 413, 417
and the calculus controversy: 387 n. 3
the Compositor: *see* LIVEBODY
'Corrigenda and Addenda': 354 n. 2

NEWTON, Sir ISAAC: *Principia* (*cont.*)
and Cotes: xxi, xxviii–xxxi, xxxiv–xxxv, xlix n. 16, 4, 60, 413
cuts for: 5, 7, 64, 65, 152, 370, 384, 386–7, 400
DE VI ELECTRICA: **961*a***, 362–9
drafts for
Author's Preface: [? Autumn 1712], **829*a***, 112–14
BOOK III, Prop. 4, Schol.: **889*a***, 216–19
BOOK III, Prop. 35, Schol., first draft: **917*a***, 287–98; revision: **937*a***, 328–9
BOOK III, Prop. 37: **909*a***, 264–9; **912*a***, 274–5
BOOK III, Prop. 38: **912*a***, 274–5
early manuscripts of: 195
fluxions used in: xv, xxiii, 114 n. 4, 116 n. 2, 141, 149, 213
optical propositions in: 369
preparation of: xiv, xxii, xxvii–xxxviii, 247
printing of: xxix, 7, 24, 27 n. 2, 37, 40, 54, 55 n. 2, 64–6, 95, 174, 247 n. 3, 249 n. 3, 296, 351, 356 n. 2, 361 n. 3, 413
rough notes for, on other documents: 211, 246 n. 1, 252 n. 4
Whiston and: 5
Recensio: 384 n. 1
relatives of: xliv, 199–201, 251–4, 345–7, 353–4
reputation abroad: 390 n. 3
and the Royal Society: xxxix–xl, xliii
salary at the Mint, 210 n. 7
scientific method: 396, 401
visit to Seafield: 57
System of the World: xxxiii
'Tractatus de methodis serierum et fluxionum': 4
unpublished MS. on the calculus dispute: 214 n. 1
NEWTON, JAMES
LETTER to Newton: [n.d.], **999**, 410
NEWTON, JOHN: 347
NEWTON, Sir JOHN: 410
NEWTON, ROBERT: 347 n. 2
NOEL, Mr: 253
NORTHEY, EDWARD: 177
NORWOOD, RICHARD: *262 n. 6*, 255, 259, 261, 270

OADHAM, CATESBY: 374 n. 4, 403
occult qualities: xxxviii, 116 n. 6, 299–300, 392
officers of the Mint: *see under* Mint *and* Edinburgh Mint
oil-of-oranges experiment: 364–5, 367–8
Old Style dates (O.S.): xv
OLDENBURG, HENRY: xxvi, xlvii n. 6, 45 n. 6, 98 n. 8, 117, 133–52, 209 n. 4
optic nerve: 363, 366
optics: *see* light *and* eye
Ordnance, Officers of the: 80, 352–3
ORMONDE, Duke of: *238 n. 2*, 238, 245–6, 327 n. 3
LETTER from Lords Justices and Privy Council of Ireland: 8 January 1711/12, **897*a***, 238–40
OXFORD, Earl of: *xlvii*, *161 n. 2*, 72 n. 2
LETTERS
to Halley: 28 August 1712, **940**, 332–3
to Newton: 23 February 1711/12, **895**, 231–2; 21 March 1711/12, **905**, 250–1; 10 July 1712, **927**, 313–14; 28 August 1712, **940**, 332–3
from Bruce: 7 October 1712, **948**, 342–3
from Halley: 18 September 1712, **945**, 340–1
from Medows: 7 October 1712, **948**, 342–3
from the Mint: 6 June 1711, **845**, 154; 21 August 1711, **868**, 160–1; 16 June 1712, **923**, 307–8; 16 July 1712, **928**, 314–15; December 1712, **960**, 357–61; 9 March 1712/13, **981**, 388; 10 April 1713, **992**, 403–4
from Newton: 10 June 1711, **848**, 160–1; [*c.* 15 June 1711], **850**, 162–3; [*c.* 20 June 1711]; **852**, 165; 6 July 1711, **855**, 172–3; 7 August 1711, **865**, 186–7; 20 August 1711, **867**, 188–9; 28 August 1711, **869**, 190 and **870**, 191–3; [5 October 1711], **875**, 196; 14 February 1711/12, **892**, 224–5; 3 March 1711/12, **900**, 244–6; 23 June 1712, **925**, 309–12; 18 September 1712, **945**, 340–1; 7 October 1712, **948**, 342–3; [? October 1712], **949**, 345; 29 October 1712, **954**, 352–3; 29 January 1712/13, **966**, 374; [31 March 1713], **990**, 401–2; [? February 1712/

OXFORD: LETTERS (*cont.*)
13], **996a**, 407; 22 May 1713, **998**, 409–10
from Peyton: 20 August 1711, **867**, 188–9
and assaymaster: 374–5
and Katherine Barton: 345
and Cawood: 332–3, 340
and copper coinage: 357 n. 2
and engravers: 211, 307–8, 313–15
and Flamsteed's *Historia Cœlestis*: 224
and Thomas Harley: 196 n. 1
and Irish coin: 245–6, 309–12
and the moneyers: 250–1
payment of forces abroad: 342–3
and the Pewterers: 38 n. 3
and the political scene: 327 n. 3
and the *Principia*: 413
and salaries of Mint employees: 186–7
and silver plate: 172–3, 181, 231–2
and the smith: 352, 355
and the tin affair: 154, 188–9, 388
and the Trial of the Pyx: 409–10
and the Two Million Act: 165
and the value of foreign coin: 160–3, 161 n. 2, 189
and Weddell: 401–2

PAGET, HENRY: 72 n. 2, 126
LETTER to the Mint: 10 May 1711, **837**, 126–7
PALMES, WILLIAM: 79, 81, 82 n. 4, 359
PAPIN, DENYS: xl
parabolic figure: 151 n. 32
PARRY, CHARLES: 378 n. 2
PASCAL, BLAISE: xlix
PAULETT, Earl: 72 n. 2
PAUNCEFOOT: 382
payment of British forces abroad: *see under* Mint
Peace Medals:
ORDERS OF COUNCIL: 24 June 1713, **1000 I**, 411–12
PELL, JOHN: xxvi
PEMBERTON, HENRY: xxxvi
pendulum: *see also* seconds pendulum
conical pendulums: 44 n. 11
experiments with: 240, 241 n. 2
motion in resisting medium: 32 n. 6, 39–51, 54
PENMAN, JAMES: 57 n. 2, 60–1, 65
PERCIVAL, JOHN: *240 n. 8*, 239
PERCIVAL, THOMAS: 347
PERY, JOHN: 376 n. 3, 378 n. 2
PETTUS, Sir JOHN: 89 n. 2
pewter: *see under* Mint: tin
Pewterers' Company of London: 28, 37
PEYTON, CRAVEN
LETTERS
to Godolphin: 10 August 1709, **764**, 2–3; 16 November 1709, **770**, 11–12; 31 January 1709/10, **772**, 13–14; 17 May 1710, **782**, 37–8
to Oxford: 20 August 1711, **867**, 188–9; 16 June 1712, **923**, 307–8; 16 July 1712, **928**, 314–15; December 1712, **960**, 357–61; 9 March 1712/13, **981**, 388; 10 April 1713, **992**, 403–4
to the Treasury: 2 February 1710/11, **820**, 93–4; 25 April 1711, **835**, 123–4; 14 May 1711, **838**, 127–8; 6 June 1711, **845**, 154
FINAL STATEMENTS OF ACCOUNT: 12 December 1711, **883**, 205–6
and assaymaster: 404
the copper coinage: 357–60
and the Edinburgh Mint: 2–3, 206
and engravers: 307–8, 314–15
and salaries at the Mint: 388
and silver plate: 123–4, 127–8, 181, 183, 186, 205, 382
and the tin affair: 11–12, 37–8, 93 nn. 1 & 3, 154, 188, 388
and the Trial of the Pyx: 13–14, 53–4
and the value of Spanish coin: 189
PHELLIPS, EDWARD: *361 n. 5*, 205, 206, 307–8, 314–15, 357–60, 388, 402 n. 1, 404
LETTERS
to Oxford: 16 June 1712, **923**, 307–8; 16 July 1712, **928**, 314–15; December 1712, **960**, 357–61; 9 March 1712/13, **981**, 388; 10 April 1713, **992**, 403–4
from T. Harley: 9 March 1712/13, **982**, 388–9
FINAL STATEMENTS OF ACCOUNT: 12 December 1711, **883**, 205–6
Philosophical Transactions: 9 n. 3, 17
reference to papers by
Cassini, in vols. **16** & **18**: 169, 172 nn. 11 & 13

Philosophical Transactions (*cont.*)
Cotes (*Logometria*), in vol. **29**: 306 n. 5
Hauksbee, in vol. **27**: 369 nn. 6 & 8
Keill, in vol. **26**: xxii, 96–7, 116 n. 5, 117, 132
Picard, in vol. **10**: 262 n. 4
Sluse, in vol. **7**: 139, 147, 151 n. 27
Taylor, in vol. **27**: 369 n. 6
review of Freind's *Prælectiones Chymicæ*, in vol. **27**: 203, 205 n. 3
PHIPPS, CONSTANTINE: *240 n. 3*, 239
PHOTINUS: 59 n. 1
physiology, Newton's comments on: 300
PICARD, JEAN:
on the Earth's shape: 216, 218, 219 n. 3, 258–9, 261, 262 nn. 3–5
experiments on a heated wire: 235, 237 n. 13
PILKINGTON, GEORGE: 252 n. 1, 353–4
PILKINGTON, MARY: *252 n. 1*, 253–4
PILKINGTON, MARY (the younger): *252 n. 1*, 251–2, 343
LETTERS to Newton: 22 March 1711/12, **906**, 251–2; 30 October 1712, **955**, 353–4
PILKINGTON, THOMAS: 252 n. 1
PILKINGTON, THOMAS (the younger): 252 n. 1
PITCAIRNE, ARCHIBALD: 118 n. 2
PITMAN, MARY: 68–9
plague of 1665–6: 136, 145
plenists: 153
Port Mahon: 161 n. 4, 162–3, 188, 405
Portsmouth Collection: xvii
Potosi: 163
POULETT, JOHN: *127 n. 4*, 126
ROYAL WARRANT to the Mint: 10 May 1711, **837**, 126–7
POUND, JAMES: *409 n. 6*, 381 n. 10, 408
POVEY, JOHN: 411–12
projectiles: 6 n. 4; *see also* resisting medium
Prussian minister: *see* BONET
PTOLEMY: xxxi, xxxiv, 119–22, 130
putrefaction: 363, 365, 366, 368
Pyx, Trials of: *see* Mint: Trial of the Pyx *and* Edinburgh: Trial of the Pyx

Quadrature: 133, 136, 139–42, 144, 147–9, 150 n. 17, 151 n. 29, 166

RAPHSON, JOSEPH: *95 n. 4*, xxii, 94–5
RAWLINSON, THOMAS: 62, 64 n. 2
RAYMARUS, NICOLAUS: 210 n. 4
RAYMER: *see* HALLEY
Recensio: *see* Newton: *Recensio*
reflection: 362, 365, 368 n. 2
refraction: 362, 365
of the atmosphere: 338, 341
REGNAUD, FRANÇOIS: xlix n. 8
RENDA, THOMAS: 358
resistance, negative: 26, 31
resisting medium: 112–13, 164, 166
efflux of fluids: xxx, 66–8, 70, 73–5, 107–12, 152–3, 155–60
Cotes' experiments on: 73–5
Newton's experiments on: 103–4
Mariotte's experiment: 66–8, 70, 73–4, 103
motion of a body in: 24–36, 158–60, 348–50, 369, 400
pendulum moving in: 39–44, 48–51
REYMER: *see* RAYMARUS, NICHOLAUS
ROBARTES, FRANCIS: *23 n. 6*, xxv, 18, 80 n. 3, 102, 385, 408
LETTER to Flamsteed: 5 March 1712/13, **978**, 385
ROBERVAL: 299, 301 n. 4
ROCHESTER, Earl of: *see* HYDE, LAWRENCE
Rochester Free Mathematical School: 202–3
ROOS, JOHN: *396 n. 2*
ROOS, WILLIAM: *see* ROOS, JOHN
ROWLEY, JOHN: 8 n. 3
Royal Mews, Westminster: l n. 29
Royal Observatory
and Flamsteed: 101, 102, 106, 118 n. 1, 120 n. 2, 209–10, 313, 385
instruments at: 92 n. 4
observations at: 291, 295
visitors to: 78–81, 91, 209
Royal Observatory of Paris: 235
Royal Society: xxxix–xliii
LETTERS:
to Flamsteed: 3 July 1712, **926**, 313; 5 March 1712/13, **798**, 385
from Bolingbroke: 7 February 1712/13, **970**, 376–7
ACCOUNT OF PROCEEDINGS AT: 22 November 1710, **812**, 76–9
accommodation for: xl–xli, l n. 29, 61 n. 4, 62 n. 6, 63–4, 75–9
Bianchi's election to: 59–60

Royal Society (*cont.*)
and the calculus dispute: xxiii–xxviii, 96–7, 115 n. 1, 116 n. 7, 132, 134, 143, 208 n. 1
Commercium Epistolicum: 389
Copley Bequest: 77–8, 78 n. 8
at Crane Court: xl–xli, 22 n. 4, 62 nn. 4–6, 76–8
and William Derham: 379–80
embassy to Moscow: 376–7
experiments on surface tension: 364, 367
Flamsteed
his observations: xli–xlii, 131, 209, 313, 385
his opinion of the Royal Society: 9
at Gresham College: 20, 23 n. 17, 62–4
Halley
experiments on water jets: 67 n. 4
lunar theory: 297 n. 3
and Kerridge's water works: 91 n. 1
Newton
discusses density of gold: 89
as president: xxxix, xli, xliii
procedure at meetings: 21–2, 23 nn. 18 & 20, 24 n. 23, 45 n. 3
Statutes: 44, 44 n. 2
telescope for: 379
and visitors to the Royal Observatory: 79–81, 91–2, 209, 385
and the Woodward affair: xl, 17–24, 44–5, 76
ROYAL WARRANTS: 12 December 1710, **814**, 79–81; 10 May 1711, **837**, 126–7; 30 July 1711, **862**, 182–3; 20 November 1712, **957**, 355
RUPERT, Prince: 91
RYMER, THOMAS; *Fœdera*: 254 n. 1

ST. JOHN, HENRY
LETTERS
to the Royal Society: 7 February 1712/13, **970**, 376–7
from Ellis and Newton: [late September 1710], **806**, 68–70
and counterfeiters: 68–9, 70 n.
and Newton, bribery of: xlviii
and the political scene: 327 n. 3
and the Royal Society: 376–7
and the visitors to the Royal Observatory: 80, 91 nn. 2 & 4
saltpetre: 93 n. 4
Saturn: 224 n. 2, 226, 248
satellites of: 169, 172 n. 13, 409 n. 6
SAUNDERSON, NICHOLAS: 202 n. 4, 247
LETTER to Jones: 16 March [1711/12], **902**, **247**
SAURIN, JOSEPH: 205
SAUVEUR, JOSEPH: *185 n. 5*, 184
SAVAGE, MARY: 252, 353
SAVAGE, P.: 239
SCHEUCHZERUS, J. J.: *45 n. 5*, 44
SCOTT, HERCULES
LETTERS
to Fauquier: 29 January 1712/13, **965**, 373–4
to Newton: 26 July 1712, **930**, 316–18; 28 August 1712, **939**, 331–2; 21 March 1712/13, **986**, 394–6
from Newton: [28 January 1712/13], **964**, 371–3
and Allardes' accounts: 316–18, 331, 371–2, 395
SCOTT, PATRICK: *57 n. 4*
and Allardes: 316–18, 331, 373, 395
Trial of Scottish Pyx: 57 n. 2, 58–9
SCOTT, THOMAS: 373
Scottish Mint: *see* Edinburgh Mint
SCROOP, JOHN: 1 n. 2
SEAFIELD, Earl of
LETTER to Newton: 2 August 1710, **798**, 58–9
and Lauderdale: 52 nn. 1 & 2
Newton's visit to: 57
and Trial of the Pyx: 13 n. 2, 14 n. 1, 57–8
second differences: 151 n. 35, 349 n. 1
seconds pendulum: 217, 219, 221, 223, 226, 229 n. 3, 230, 234, 236 n. 9, 241, 255, 257–61
series
binomial series: 138, 146, 150 n. 22, 151 nn. 23 & 24
converging: 213, 276
SHARP, ABRAHAM: xlii, 1 n. 31, 194
LETTERS from Flamsteed: 25 October 1709, **768**, 8–9; 23 January 1710/11, **818**, 91–2; 15 May 1711, **841**, 129–31; 20 September 1711, **872**, 194; 22 December 1711, **885**, 209–10
SHEPARD, N.: 358
SHEPHERD, WILLIAM: 358

ships, Newton's suggestion for model-testing: 44 n. 12
SHORTGRAVE, RICHARD: 62 n. 5
SILLES, Mr.: 252
silver: *see under* Mint *and* Edinburgh Mint
simple harmonic motion: 170 n. 4: 185 n. 3
SLANEY, ABEL: 357–8, 360 n. 2
SLOANE, HANS
 LETTERS
  to Flamsteed: 30 May 1711, **842**, 131; 3 July 1712, **926**, 313; 5 March 1712/13, **978**, 385
  to Leibniz: [May 1711], **843**, 132
  from Keill: [May 1711], ***843a***, 133–52
  from Kerridge: 16 January 1710/11, **817**, 90–1
  from Leibniz: 21 February 1710/11, **822**, 96–8; 18 December 1711, **884**, 207–9
  from Newton: [13 September 1710], **802**, 61–2; [? April 1711], ***830a***, 117–18; [? February 1711/12], **888**, 212–14
 and the calculus dispute: xxiii–xxv, 96–8, 115–17, 132–52, 207–8, 212–14
 and Flamsteed's observations: 131, 209–10, 313
 and the Joint Committee for Gresham Affairs: 63
 and King's satires: 9 n. 3, 23 n. 10
 and Newton: 132 n. 1
 at the Royal Society: xl, 61, 62 n. 6, 77–8, 91
 and the visitors to the Royal Observatory: 80 n. 3, 385
 and waterworks: 91 n. 1
 and the Woodward affair: xl, 17–21, 22 nn. 1 & 2, 45 n. 3
SLUSE, RENÉ FRANÇOIS DE
 correspondence with Towneley: 379, 380 n. 9
 his method of tangents: 134–6, 139, 140, 143–4, 147–8, 150 n. 8, 151 n. 27
 paper by: 139, 147, 151 n. 27
SMITH, BARNABAS: 200 n. 1
SMITH, HANNAH: *see* BARTON, HANNAH
SMITH, HANNAH (the younger): 200 n. 1, 347 n. 1
SMITH, JOHN: xxix, xlix n. 15, 4 n. 4, 55 n. 3
SMITH, JOHN, Lord Chief Baron of Scotland: 1 n. 2
SMITH, JOHN, portrait of Newton by: 316 n. 5
SMITH, MARY: *see* PILKINGTON, MARY
SMITH, ROBERT: xvii, l n. 23
SOPHIA, Electress of Hanover: 196 n. 1
sound
 analogy with light: 363, 365
 soundwaves: xxx, 167–71, 179, 183–5, 215–16
 speed of sound: 184 n. 5
South Sea Company: 125 n. 1, 189 n. 4
SOUTHWELL, EDWARD: *240 n. 9*, 239, 309
SOUTHWELL, Sir ROBERT: 240 n. 9
spectrum: 363, 366
spheroid, general gravitational attraction of: 4
spiral motion: 28 n. 12, 30, 63
STANLEY, Sir JOHN: 16
stannaries: *see under* Mint: Cornish stannaries
STANSFIELD, ROBERT: 9 n. 2
STRATFORD, FRANCIS: 188, 191–2
STREETE, THOMAS; *Astronomia Carolina*: 296
STUART, DANIEL: 2 n. 4
STURMY, SAMUEL: *244 n. 5*, 243, 264
surface tension: 364, 367
Swedish copper and tin: 81
SWIFT, JONATHAN: xliv, li n. 45, 200 n. 1, 327 n. 3
SYLVESTER, THOMAS: *187 n. 2*, 186–7, 352–3, 355

tallow chandlers: 382
tangents, methods of finding: xxv, xlix n. 11, 134, 139, 141, 143, 147, 148, 150 n. 7, 213–14
taxes and duties: *see* Mint: duties and taxes
TAYLOR, BROOK: xxv, 369
TAYLOUR, JOHN: 13, 211, 309, 326, 342
 LETTERS
  to the Comptrollers of the Accounts of the Army: 21 August 1712, **936**, 326–7
  to the Mint: 28 January 1709/10, **771**, 12–13; 11 January 1711/12, **886**, 211–12
  to Newton: 16 June 1712, **924**, 309
telescope: *see* astronomical instruments
TENISON, THOMAS: *45 n. 6*, 44
theology; notes by Newton on: 59 n. 1
THOMPSON, ISAAC: 91 n. 2
THORNHILL, Sir JAMES: 8 n. 9, 316
THORPE, JOHN: *23 n. 19*, 21, 313 n. 1

tides
Colepress' observations of: 244 n. 6, 264
and the theory of the Moon: xxxiii, 112–13, 217–18, 220–3, 227, 229 n. 11, 230, 241, 243–4, 264–9, 272–5, 278–80
Todd
Letter from Newton: [n.d.] **906a**, 253–4
toise: 222 n. 12
Tompion, Thomas: 92 n. 3
Torriano, Alessandro: *409 n. 3*, 408
Torricelli, Evangelista: 67 n. 4, 68 n. 5
Totton, Samuel: 62
Towneley, Richard: *380 n. 3*, 209
correspondence with Collins: 379, 380 n. 3
correspondence with Sluse: 379, 380 n. 9
the Treasury
Letters
to the Mint: 29 November 1710, **813**, 79; 12 February 1712/13, **972**, 378; 21 March 1712/13, **987**, 396
from Ellis: 18 September 1710, **804**, 65; 4 October 1710, **808**, 71–2
from the Mint: 2 February 1710/11, **820**, 93–4; 25 April 1711, **835**, 123–4; 14 May 1711, **838**, 127–8
from Newton: 18 September 1710, **804**, 65; 4 October 1710, **808**, 71–2; 13 December 1710, **815**, 81–2; [*c.* 31 December 1710], **816**, 82–3; 23 January 1710/11, **819**, 92–3; [15 May 1711], **840**, 129
Chancellor of the Exchequer: *see* Benson *and* Seafield
Lord High Treasurer: *see* Godolphin *and* Oxford
for affairs concerning the Mint see Mint
Treaty of Union: *see* Union, Treaty of
Treaty of Utrecht Peace Medals: *see* Peace Medals
Trevor, Thomas: *414 n. 5*, 413
Trinity College Observatory: xxviii, 5
Truro; and salaries of tinners: 11
Tuck, Mr: 69
Two Million Act: 165, 173 n. 3, 177 n. 6, 181–2
Tycho Brahe: *see* Brahe

Union, Treaty of, 1704: xlvii, 1, 13–14, 58, 83 n. 2, 87, 316, 358

vacuum: 228, 229 n. 12, 248
pendulum swinging in: 230
surface tension experiments in: 364, 367
value of coin: *see under* Mint *and* Edinburgh Mint
Varin, M.: 237 n. 12, 258
Vauxhall House: 91 n. 3
vena contracta: 71 n. 3, 104 n. 4
Vertue, George: *225 n. 4*
Vigo Bay: 127 n. 2
vis inertiæ: 228, 233, 236 n. 4, 240–2
visitors to the Royal Observatory: *see* Royal Observatory
vortices, theory of: xxxix, 299, 393

Walker, Hovenden: 199 n. 1
Wall, ?William: *23 n. 15*, 19
Waller, Richard: 77, 262 n. 4
Wallis, John: 98 n. 5, 379, 116 n. 5
Opera Mathematica: xxii–xxiii, 96–7, 137, 141, 145, 149, 150 n. 21, 208 n. 4
Walpole, Sir Robert: 327 n. 3
Wanley, Humfrey: 23 n. 19
Ward, John: 92, 93 n. 2
Warrants: *see* Royal Warrants
water oscillating in a U-tube: 170 n. 4
water-works: 90–1
Webster, Edward: 72
Weddell, Martha: 388 n. 1
Weddell, Robert, 388, 388 n. 1, 401–2
Whiston, William: *202 n. 2*, xxviii, 5, 202, 413
Whitworth, Charles: 377 n. 2
Williams, John: *250 n. 2*, 250, 254
Williamson, Sir Joseph: 203 n. 5
Willys, William: 63
Withers, William: 62
Witty, John: xxviii
Wolf, Christian: *116 n. 4*
and Keill: 115, 116 n. 5, 204
papers by: *see Acta Eruditorum*
Woodward, John: *22 nn. 2 & 3*, xl–xli, 9, 18–21, 24 n. 22, 44–5, 78 n. 8
Letter to Newton: 30 May 1710, **785**, 44–5
Woolsthorpe: xliii, 346–7
Wren, Christopher (junior): *61 n. 2*, 80 n. 3
Wren, Sir Christopher
and accommodation for the Royal Society: 61–3
and Flamsteed: 1 n. 31, 80 n. 3, 102, 106 n. 2, 408